BOLYAI SOCIETY
MATHEMATICAL STUDIES

19

BOLYAI SOCIETY MATHEMATICAL STUDIES

1. **Combinatorics, Paul Erdős is Eighty, Vol. 1**
   D. Miklós, V.T. Sós, T. Szőnyi (Eds.)
2. **Combinatorics, Paul Erdős is Eighty, Vol. 2**
   D. Miklós, V.T. Sós, T. Szőnyi (Eds.)
3. **Extremal Problems for Finite Sets**
   P. Frankl, Z. Füredi, G. Katona, D. Miklós (Eds.)
4. **Topology with Applications**
   A. Császár (Ed.)
5. **Approximation Theory and Function Series**
   P. Vértesi, L. Leindler, Sz. Révész, J. Szabados, V. Totik (Eds.)
6. **Intuitive Geometry**
   I. Bárány, K. Böröczky (Eds.)
7. **Graph Theory and Combinatorial Biology**
   L. Lovász, A. Gyárfás, G. Katona, A. Recski (Eds.)
8. **Low Dimensional Topology**
   K. Böröczky, Jr., W. Neumann, A. Stipsicz (Eds.)
9. **Random Walks**
   P. Révész, B. Tóth (Eds.)
10. **Contemporary Combinatorics**
    B. Bollobás (Ed.)
11. **Paul Erdős and His Mathematics I+II**
    G. Halász, L. Lovász, M. Simonovits, V. T. Sós (Eds.)
12. **Higher Dimensional Varieties and Rational Points**
    K. Böröczky, Jr., J. Kollár, T. Szamuely (Eds.)
13. **Surgery on Contact 3-Manifolds and Stein Surfaces**
    B. Ozbagci, A. I. Stipsicz
14. **A Panorama of Hungarian Mathematics in the Twentieth Century, Vol. 1**
    J. Horváth (Ed.)
15. **More Sets, Graphs and Numbers**
    E. Győri, G. Katona, L. Lovász (Eds.)
16. **Entropy, Search, Complexity**
    I. Csiszár, G. Katona, G. Tardos (Eds.)
17. **Horizons of Combinatorics**
    E. Győri, G. Katona, L. Lovász (Eds.)
18. **Handbook of Large-Scale Random Networks**
    B. Bollobás, R. Kozma, D. Miklós (Eds.)

Martin Grötschel
Gyula O. H. Katona (Eds.)

Building Bridges

Between Mathematics and Computer Science

Springer
JÁNOS BOLYAI MATHEMATICAL SOCIETY

Martin Grötschel
Konrad-Zuse-Zentrum
für Informationstechnik Berlin (ZIB)
Takustr. 7
14195 Berlin
Germany
e-mail: groetschel@zib.de

Gyula O. H. Katona
Hungarian Academy of Sciences
Alfréd Rényi Institute of Mathematics
Reáltanoda u. 13-15
Budapest 1053
Hungary
e-mail: ohkatona@renyi.hu

Managing Editor:
Gábor Sági
Hungarian Academy of Sciences
Alfréd Rényi Institute of Mathematics
Reáltanoda u. 13-15
Budapest 1053
Hungary
e-mail: sagi@renyi.hu

Mathematics Subject Classification (2000): 05-XX, 68-XX

Library of Congress Control Number: 2008932353

ISSN 1217-4696
ISBN 3-540-85218-6 Springer Berlin Heidelberg New York
ISBN 978-963-9453-11-1 János Bolyai Mathematical Society, Budapest

Springer is a part of Springer Science+Business Media
springer.com

Printed in Hungary

Cover design: WMXDesign GmbH, Heidelberg

Printed on acid-free paper 44/3142/db – 5 4 3 2 1 0

CONTENTS

PREFACE

László Lovász, briefly called Laci by his friends, turned sixty on March 9, 2008. To celebrate this special birthday two conferences have been held in Hungary, one in Budapest (August 5–9, 2008) with invited speakers only and one in Keszthely (August 11–15, 2008). Several top mathematicians and computer scientists have not only lectured at these meetings but also dedicated (together with some coauthors) research papers to this occasion. This volume is the collection of their articles. The contributions to the conferences and this book honor a person who has not only made an almost uncountable number of fundamental contributions to mathematics and computer science, but who also broke down many borders between mathematical disciplines and built sustainable bridges between mathematics and computer science.

Laci has been a role model for many young researchers, he inspired lots of colleagues, and guided quite a few scientific careers. In addition, he is an extremely nice person and very pleasant colleague, and that is why so many researchers have come to the "Lovász meetings" in Hungary to present their best recent work and celebrate with him.

In the Fazekas Mihály Gimnázium in Budapest, a breeding place of world class mathematicians, Laci's outstanding talent became visible at very young age. He did not only win various mathematics competitions in Hungary, Lovász also won three gold medals and one silver medal in the International Mathematical Olympiad. The solution of an open problem in lattice theory gained him his first international visibility and soon after, in 1972, his proof of the perfect graph theorem earned him lasting fame in graph theory. An unparalleled sequence of scientific achievements followed and is continuing till today. It is impossible to mention even a small fraction of Laci's results here. The list of his publications (up to summer 2008) is contained in this volume to indicate the breadth and depth of his contributions.

Being a combinatorialist at heart, like so many Hungarian mathematicians, it has been natural for Laci to employ combinatorial techniques in other areas of mathematics; but he also brought topology, algebra, analysis, stochastics and other mathematical fields to combinatorics, often in quite unusual ways. In this way he opened up quite a number of new flourishing fields of research. Algorithmic issues such as polynomial time solvability and general complexity theory opened his eyes for computer science where he particularly contributed to the interface between computer science and discrete mathematics.

After a distinguished academic career with employments and visiting positions in Szeged, Budapest, Waterloo, Bonn, Cornell, Princeton, Yale (and guest professorships in many other places) Lovász had worked for Microsoft Research in Redmond from 1998 to 2006. He returned thereafter to Budapest to become director of the Mathematical Institute of Eötvös Loránd University.

His international reputation is stellar. One proof of this claim is a remarkable series of prestigious honors; among them are the Grünwald, Pólya, Fulkerson, Wolf, Knuth, Gödel, von Neumann, Bolyai, and Széchenyi Prizes and various other distinctions such as honorary degrees and professorships. In 2006 László Lovász has been elected president of the International Mathematical Union for the years 2007 to 2010. A few more details can be found in his (very brief) Curriculum Vitae on pages 11 to 13 in this volume.

Laci has always been a family man, a loving husband, father, and meanwhile also grandfather. With Kati Vesztergombi, his wife for almost 40 years, he has not only shared family and friends; Kati and Laci are also closely linked by their common love for mathematics. Their relationship has not only resulted in three wonderful daughters and a son but also in two joint books and quite a number of papers. Kati has been a mainstay in Laci's life since their common time in high school. That is why we have chosen to display a husband and wife portrait of the couple on page 9 of this volume.

Budapest
August 2008

Martin Grötschel
Gyula Katona

Kati Vesztergombi and László Lovász

BOLYAI SOCIETY
MATHEMATICAL STUDIES, 19

Building Bridges
pp. 11–13.

Curriculum Vitae of László Lovász

Born: March 9, 1948 in Budapest, Hungary

Family: Married, 4 children

Citizenship: Hungary, United States

Degrees:
Dr. rer. nat., Eötvös Loránd University, Budapest, Hungary, 1971;
Candidate of mathematical sciences, Hungarian Academy of Sciences, Budapest, Hungary, 1970;
Dr. of mathematical sciences, Hungarian Academy of Sciences, Budapest, Hungary, 1977.

Academies:
Hungarian Academy of Sciences, corresp. member, 1979; full member, 1985;
European Academy of Arts, Sciences and Humanities, 1981;
Academia Europaea, 1991;
Nordrhein-Westfälische Akademie der Wissenschaften, corresp. member, 1993;
Deutsche Akademie der Naturforscher Leopoldina, 2002;
Russian Academy of Sciences, 2006;
Royal Netherlands Academy of Arts and Sciences, 2006;
Royal Swedish Academy of Sciences, 2007.

Positions in scientific societies:
Executive Committee of the International Mathematical Union, 1987–1994;
Presidium of the Hungarian Academy of Sciences, 1990–1993 and since 2008;
Abel Prize Committee, 2004–2006;
President of the International Mathematical Union, 2007–2010;

Positions held:
Research Associate, Eötvös Loránd University, Budapest, 1971–75;
Docent, József Attila University, Szeged, 1975–78;
Professor, Chair of Geometry, József Attila University, Szeged, 1978–82;

Professor, Chair of Computer Science, Eötvös Loránd University, Budapest, 1983–93;
Professor, Dept. of Computer Science, Yale University, 1993–2000;
Senior Researcher, Microsoft Research, 1999–2006;
Director, Mathematical Institute of the Eötvös Loránd University, Budapest, 2006–.

Visiting positions:
Vanderbilt University, 1972/73;
University of Waterloo, 1978/79;
Universität Bonn, 1984/85;
University of Chicago, Spring 1985;
Cornell University, Fall 1985;
Mathematical Sciences Research Institute, Berkeley, Spring 1986;
Princeton University, Fall 1987, Spring 1989, 1990/91, 1992/93.

Honorary degrees and positions:
Adjunct Professor, University of Waterloo, Waterloo, Ontario, Canada, 1980–1990;
A. D. White Professor-at-Large, Cornell University, Ithaca, NY, 1982–1987;
Honorary Professor, Universität Bonn, 1984; Academia Sinica, 1988;
John von Neumann Professor, Universität Bonn, 1985;
Doctor Honoris Causa: University of Waterloo, Ontario, Canada, 1992; University of Szeged, Hungary, 1999; Budapest University of Technology, 2002; University of Calgary, 2006.

Awards:
Grünwald Géza Prize, Bolyai Society, 1970;
George Pólya Prize, Soc. Ind. Appl. Math., 1979;
Best Information Theory Paper Award, IEEE, 1981;
Ray D. Fulkerson Prize, Amer. Math. Soc. – Math. Prog. Soc., 1982;
State Prize, Hungary, 1985;
Tibor Szele Medal, Bolyai Society, 1992;
Brouwer Medal, Dutch Matematical Society – Royal Netherl. Acad. Sci., 1993;
National Order of Merit of Hungary, 1998;
Bolzano Medal, Czech Mathematical Society, 1998;
Wolf Prize, Israel, 1999;
Knuth Prize, ACM–SIGACT, 1999;
Corvin Chain, Hungary, 2001;
Goedel Prize, ACM–EATCS, 2001;

John von Neumann Medal, IEEE, 2005;
John von Neumann Theory Prize, INFORMS, 2006;
Bolyai Prize, Hungary, 2007;
Széchenyi Grand Prize, Hungary, 2008.

Editorial boards:
Combinatorica (Editor-in-Chief),
Advances in Mathematics,
J. Combinatorial Theory (B),
Discrete Math.,
Discrete Applied Math.,
J. Graph Theory,
Europ. J. Combinatorics, Discrete and Computational Geometry,
Random Structures and Algorithms,
Acta Mathematica Hungarica,
Acta Cybernetica,
Electronic Journal of Combinatorics,
Geometric and Functional Analysis.

Fields of research: Combinatorial optimization, graph theory, theoretical computer science.

Publications: 9 books, 250 research papers.

BOLYAI SOCIETY
MATHEMATICAL STUDIES, 19

Building Bridges
pp. 15–29.

Publications of László Lovász

Books:

[1] Lovász L., Pelikán J., Vesztergombi K.: *Kombinatorika,* Tankönyvkiadó, Budapest, 1977 (German translation: Teubner, 1977; Japanese translation: 1985).

[2] Gács P., Lovász L.: *Algoritmusok,* Műszaki Könyvkiadó, Budapest, 1978; Tankönyvkiadó, Budapest, 1987.

[3] L. Lovász: *Combinatorial Problems and Exercises,* Akadémiai Kiadó – North Holland, Budapest, 1979 (Japanese translation: Tokai Univ.Press, 1988; Hungarian translation: Typotech, 1999; Second edition: North-Holland Publishing Co., Amsterdam, 1993.)

[4] L. Lovász, M.D. Plummer: *Matching Theory,* Akadémiai Kiadó – North Holland, Budapest, 1986 (Russian translation: Mir, 1998).

[5] L. Lovász: *An Algorithmic Theory of Numbers, Graphs, and Convexity,* CBMS-NSF Regional Conference Series in Applied Mathematics **50,** SIAM, Philadelphia, Pennsylvania 1986.

[6] M. Grötschel, L. Lovász, A. Schrijver: *Geometric Algorithms and Combinatorial Optimization,* Springer, 1988; Chinese edition: World Publishing Corp., Beijing, 1990.

[7] B. Korte, L. Lovász, R. Schrader: *Greedoids,* Springer, 1991.

[8] R. Graham, M. Grötschel, L. Lovász (eds.): *Handbook of Combinatorics* Elsevier Science B.V. (1995), 1740–1748.

[9] L. Lovász, J. Pelikán, K. Vesztergombi.: *Discrete Mathematics: Elementary and Beyond,* Springer, New York (2003); Portuguese translation: Sociedade Brazileira de Matemática, Rio de Janiero (2005); German Translation: Springer, Heidelberg (2005); Hungarian Translation: Typotex, Budapest (2006).

Research papers:

[1] Lovász L.: Független köröket nem tartalmazó gráfokról (On graphs containing no independent circuits), *Mat. Lapok,* **16** (1965), 289–299.

[2] L. Lovász: On decomposition of graphs, *Studia Math. Hung.*, **1** (1966), 237–238.

[3] L. Lovász: On connected sets of points, *Annales Univ. R. Eötvös,* **10** (1967), 203–204.

[4] L. Lovász: Über die starke Multiplikation von geordneten Graphen, *Acta Math. Hung.*, **18** (1967), 235–241.

[5] L. Lovász: Operations with structures, *Acta Math. Hung.*, **18** (1967), 321–328.

[6] L. Lovász: Graphs and set-systems, in: *Beiträge zur Graphentheorie,* Teubner, Leipzig (1968), 99–106.

[7] L. Lovász: On chromatic number of graphs and set-systems, *Acta Math. Hung.*, **19** (1968), 59–67.

[8] L. Lovász: On covering of graphs, in: *Theory of Graphs* (ed. P. Erdős, G. Katona), Akad. Kiadó, Budapest (1968), 231–236.

[9] Lovász L.: Kapcsolatok polinomoknak és helyettesítési értékeiknek számelméleti tulajdonságai között, *Mat. Lapok,* **20** (1969), 129–132.

[10] L. Lovász: Generalized factors of graphs, in: *Combinatorial Theory and its Applications,* Coll. Math. Soc. J. Bolyai, **4** (1970), 773–781.

[11] L. Lovász: Subgraphs with prescribed valencies, *J. Comb. Theory,* **8** (1970), 391–416.

[12] L. Lovász: A generalization of König's theorem, *Acta Math. Hung.,* **21** (1970), 443–446.

[13] L. Lovász: A remark on Menger's theorem, *Acta Math. Hung.,* **21** (1970), 365–368.

[14] L. Lovász: The factorization of graphs, in: *Combinatorial Struc. Appl.,* Gordon and Breach (1970), 243–246.

[15] L. Lovász: Representation of integers by norm-forms II, (K. Győry), *Publ. Math. Debrecen,* **17** (1970), 173–181.

[16] L. Lovász: On the cancellation law among finite relational structures, *Periodica Math. Hung.,* **1** (1971), 145–156.

[17] L. Lovász: On finite Dirichlet series, *Acta Math. Hung.,* **22** (1971), 227–231.

[18] L. Lovász: On the number of halving lines, *Annales Univ. Eötvös,* **14** (1971), 107–108.

[19] L. Lovász: Normal hypergraphs and the perfect graph conjecture, *Discrete Math.,* **2** (1972), 253–267; reprinted *Annals of Discrete Math.,* **21** (1984), 29–42.

[20] L. Lovász: On the structure of factorizable graphs, *Acta Math. Hung.,* **23** (1972), 179–195.

[21] L. Lovász: The factorization of graphs II, *Acta Math. Hung.,* **23** (1972), 223–246.

[22] L. Lovász: On the structure of factorizable graphs II, *Acta Math. Hung.,* **23** (1972), 465–478.

[23] L. Lovász: Direct product in locally finite categories, *Acta Sci. Math. Szeged,* **23** (1972), 319–322.

[24] L. Lovász: A characterization of perfect graphs, *J. Comb. Theory,* **13** (1972), 95–98; reprinted in: *Classic Papers in Combinatorics* (ed. I. Gessel and G. C. Rota), Birkhäuser, 1987, 447–450.

[25] L. Lovász: A note on the line reconstruction problem, *J. Comb. Theory,* **13** (1972), 309–310; reprinted in: *Classic Papers in Combinatorics* (ed. I. Gessel and G. C. Rota), Birkhäuser, 1987, 451–452.

[26] L. Lovász: A note on factor-critical graphs, *Studia Sci. Math.,* **7** (1972), 279–280.

[27] L. Lovász, J. Pelikán: On the eigenvalues of trees, *Periodica Math. Hung.,* **3** (1973), 175–182.

[28] L. Lovász: Antifactors of graphs, *Periodica Math. Hung.,* **4** (1973), 121–123.

[29] P. Erdős, L. Lovász, G. J. Simmons, E. G. Strauss: Dissection graphs of planar point sets, in: *A Survey of Comb. Theory* (ed. S. Srivastava), Springer (1973), 139–149.

[30] L. Lovász, P. Major: A note to a paper of Dudley, *Studia Sci. Math.,* **8** (1973), 151–152.

[31] L. Babai, L. Lovász: Permutation groups and almost regular graphs, *Studia Sci. Math.,* **8** (1973), 141–150.

[32] L. Babai, W. Imrich, L. Lovász: Finite homeomorphism groups of the 2-sphere, in: *Topics in Topology,* Coll. Math. Soc. J. Bolyai, **9** (1973), 61–75.

[33] L. Lovász: Connectivity in digraphs, *J. Comb. Theory,* **15** (1973), 174–177.

[34] L. Lovász: Independent sets in critical chromatic graphs, *Studia Sci. Math.,* **8** (1973), 165–168.

[35] L. Lovász, A. Recski: On the sum of matroids, *Acta Math. Hung.,* **24** (1973), 329–333.

[36] L. Lovász: Coverings and colorings of hypergraphs, in: *Proc. 4th Southeastern Conf. on Comb.,* Utilitas Math. (1973), 3–12.

[37] L. Lovász: Factors of graphs, in: *Proc. 4th Southeastern Conf. on Comb.,* Utilitas Math. (1973), 13–22.

[38] L. Lovász: Valencies of graphs with 1-factors, *Periodica Math. Hung.,* **5** (1974), 149–151.

[39] L. Lovász: Minimax theorems for hypergraphs, in: *Hypergraph Seminar* (ed. C. Berge and D. K. Ray-Chaudhuri), Lecture Notes in Math., **411** (1974), Springer, 111–126.

[40] V. Chvátal, L. Lovász: Every directed graph has a semi-kernel, in: *Hypergraph Seminar* (ed. C. Berge and D. K. Ray-Chaudhuri), Lecture Notes in Math., **411** (1974), Springer, 175.

[41] D. Greenwell, L. Lovász: Applications of product coloring, *Acta Math. Hung.,* **25** (1974), 335–340.

[42] L. Lovász, M. D. Plummer: A family of planar bicritical graphs (M. D. Plummer), in: *Combinatorics,* London Math. Soc. Lecture Notes **13** (1974), 103–108; journal version: *Proc. London Math. Soc.,* **30** (1975), 160–176.

[43] P. Erdős, L. Lovász: Problems and results on 3-chromatic hypergraphs and some related questions, in: *Infinite and Finite Sets,* Coll. Math. Soc. J. Bolyai, **11** (1975), 609–627.

[44] L. Lovász, M. D. Plummer: On bicritical graphs, in: *Infinite and Finite Sets,* Coll. Math. Soc. J. Bolyai, **11** (1975), 1051–1079.

[45] R. Appleson, L. Lovász: A characterization of cancellable k-ary structures, *Periodica Math. Hung.,* **6** (1975), 17–19.

[46] L. Lovász: Three short proofs in graph theory, *J. Comb. Theory,* **19** (1975), 269–271.

[47] L. Lovász: Spectra of graphs with transitive groups, *Periodica Math. Hung.,* **6** (1975), 191–195.

[48] L. Lovász: 2-matchings and 2-covers of hypergraphs, *Acta Math. Hung.,* **26** (1975), 433–444.

[49] L. Lovász: On the ratio of optimal fractional and integral covers, *Discrete Math.,* **13** (1975), 383–390.

[50] L. Lovász: A kombinatorika minimax tételeiről (On the minimax theorems of combinatorics), *Mat. Lapok,* **26** (1975), 209–264.

[51] D. E. Daykin, L. Lovász: The number of values of Boolean functions, *J. London Math. Soc.,* **30** (1976), 160–176.

[52] L. Lovász: On two minimax theorems in graph theory, *J. Comb. Theory B,* **21** (1976), 93–103.

[53] S. A. Burr, P. Erdős, L. Lovász: On graphs of Ramsey type, *Ars Combinatoria,* **1** (1976), 167–190.

[54] L. Lovász: On some connectivity properties of Eulerian graphs, *Acta Math. Hung.,* **28** (1976), 129–138.

[55] L. Lovász: Covers, packings and some heuristic algorithms, in: *Combinatorics,* Proc. 5th British Comb. Conf. (ed. C. St. J. A. Nash-Williams, J. Sheehan), Utilitas Math. (1976), 417–429.

[56] L. Lovász, M. Simonovits: On the number of complete subgraphs of a graph, in: *Combinatorics,* Proc. 5th British Comb. Conf. (ed. C. St. J. A. Nash-Williams, J. Sheehan), Utilitas Math. (1976), 439–441.

[57] L. Lovász, M. Marx: A forbidden substructure characterization of Gauss codes, *Bull. Amer. Math. Soc.,* **82** (1976), 121–122; full version *Acta. Sci. Math. Szeged,* **38** (1976), 115–119.

[58] L. Lovász: Chromatic number of hypergraphs and linear algebra, *Studia Sci. Math.,* **11** (1976), 113–114.

[59] P. Gács, L. Lovász: Some remarks on generalized spectra, *Zeitschr. f. math. Logik u. Grundlagen d. Math.,* **23** (1977), 547–554.

[60] L. Lovász: Certain duality principles in integer programming, *Annals of Discrete Math.,* **1** (1977), 363–374.

[61] L. Lovász: A homology theory for spanning trees of a graph, *Acta Math. Hung.,* **30** (1977), 241–251.

[62] L. Lovász, M. D. Plummer: On minimal elementary bipartite graphs, *J. Comb. Theory B,* **23** (1977), 127–138.

[63] L. Lovász: Flats in matroids and geometric graphs, in: *Combinatorial Surveys,* Proc. 6th British Comb. Conf., Academic Press (1977), 45–86.

[64] R. L. Graham, L. Lovász: Polynomes de la matrice des distences d'un arbre, in: *Problemes Combinatoires et Theorie de Graphes,* CNRS (1977), 189–190.

[65] R. L. Graham, L. Lovász: Distance matrices of trees, in: *Theory and Appl. of Graphs,* Lecture Notes in Math., **642** (1978), Springer, 186–190; journal version: Distance matrix polynomials of trees, *Adv. in Math.,* **29** (1978), 60–88.

[66] L. Lovász, V. Neumann-Lara, M. D. Plummer: Mengerian theorems for paths with bounded length, *Periodica Math. Hung.,* **9** (1978), 269–276.

[67] L. Lovász: Some finite basis theorems in graph theory, in: *Combinatorics,* Coll. Math. Soc. J. Bolyai, **18** (1978), 717–729.

[68] L. Lovász, K. Vesztergombi: Restricted permutations and the distribution of Stirling numbers, in: *Combinatorics,* Coll. Math. Soc. J. Bolyai, **18** (1978), 731–738.

[69] L. Lovász: Kneser's conjecture, chromatic number, and homotopy, *J. Comb. Theory A,* **25** (1978), 319–324.

[70] L. Lovász: Topological and algebraic methods in graph theory, in: *Graph Theory and Related Topics,* Academic Press (1979), 1–14.

[71] P. Erdős, L. Lovász, J. Spencer: Strong independence of graphcopy functions, in: *Graph Theory and Related Topics,* Academic Press (1979), 165–172.

[72] L. Lovász: Graph theory and integer programming, *Annals of Discrete Math.,* **4** (1979), 141–158.

[73] L. Lovász: On the Shannon capacity of graphs, *IEEE Trans. Inform. Theory,* **25** (1979), 1–7.

[74] A. Hajnal, L. Lovász: An algorithm to prevent the propagation of certain diseases at minimum cost, in: *Interfaces between Computer Science and Operations Research,* Amsterdam Math. Centr. Tract, **99** (1979), 105–108.

[75] L. Lovász: Gráfelmelet és diszkrét programozás (Graph theory and discrete programming), *Mat. Lapok,* **27** (1979), 69–86.

[76] L. Lovász: Determinants, matchings, and random algorithms, in: *Fundamentals of Computation Theory, FCT'79* (ed. L. Budach), Akademie-Verlag Berlin (1979), 565–574.

[77] R. Aleliunas, R. M. Karp, R. J. Lipton, L. Lovász, C. W. Rackoff: Random walks, universal travelling sequences, and the complexity of maze problems, *Proc. 20th IEEE Ann. Symp. on Found. of Comp. Sci.* (1979), 218–223.

[78] L. Lovász, J. Nesetril, A. Pultr: On a product dimension of graphs, *J. Comb. Theory B,* **29** (1980), 47–67.

[79] L. Lovász: Selecting independent lines from a family of lines in a space, *Acta Sci. Math. Szeged,* **42** (1980), 121–131.

[80] L. Lovász: Matroid matching and some applications, *J. Comb. Theory B,* **28** (1980), 208–236.

[81] L. Lovász: The matroid matching problem, in: *Algebraic Methods in Graph Theory,* Coll. Math. Soc. J. Bolyai, **25** (1980), 495–517.

[82] L. Lovász: Matroids and Sperner's Lemma, *Europ. J. Combin.,* **1** (1980), 65–66.

[83] L. Lovász: Efficient algorithms: an approach by formal logic, in: *Studies on Math. Programming* (ed. A. Prékopa), Akadémiai Kiadó (1980), 119–126.

[84] B. Korte, L. Lovász: Mathematical structures underlying greedy algorithms, in: *Fundamentals of Computation Theory* (F. Gécseg, ed.), Lecture Notes in Comp. Sci., **117** (1981), Springer, 205–209.

[85] L. Lovász, A. Sárközi, M. Simonovits: On additive arithmetic functions satisfying a linear recursion, *Annales Univ. Eòtòs,* **24** (1981), 205–215.

[86] M. Grötschel, L. Lovász, A. Schrijver: The ellipsoid method and its consequences in combinatorial optimization, *Combinatorica,* **1** (1981), 169–197; Corrigendum *Combinatorica,* **4** (1984), 291–295.

[87] L. Lovász, A. Schrijver: Remarks on a theorem of Rédei, *Studia Sci. Math. Hung.*, **16** (1981), 449–454.

[88] A.J. Bondy, L. Lovász: Cycles through given vertices of a graph, *Combinatorica,* **1** (1981), 117–140.

[89] P. Gács, L. Lovász: Khachiyan's algorithm for linear programming, *Math. Prog. Study,* **14** (1981), 61–68.

[90] A. K. Lenstra, H. W. Lenstra, L. Lovász: Factoring polynomials with rational coefficients, *Math. Annalen,* **261** (1982), 515–534.

[91] J. Edmonds, L. Lovász, W. R. Pulleyblank: Brick decompositions and the matching rank of graphs, *Combinatorica,* **2** (1982), 247–274.

[92] L. Lovász, Y. Yemini: On generic rigidity in the plane, *SIAM J. Alg. Discr. Methods,* **1** (1982), 91–98.

[93] L. Lovász: Some combinatorial applications of the new linear programming algorithms, in: *Combinatorics and Graph Theory* (ed. S. B. Rao), Lecture Notes in Math., **885** (1982), Springer, 33–41.

[94] L. Lovász: Bounding the independence number of a graph, *Ann. of Discr. Math.*, **16** (1982), 213–223.

[95] L. Lovász, A. Recski: Selected topics of matroid theory and its applications, *Suppl. Rendiconti del Circ. Mat. Palermo,* **2** (1982), 171–185.

[96] I. Bárány, L. Lovász: Borsuk's Theorem and the number of facets of centrally symmetric polytopes, *Acta Math. Hung.*, **40** (1982), 323–329.

[97] L. Lovász: Perfect graphs, in: *More Selected Topics in Graph Theory* (ed. L. W. Beineke, R. L. Wilson), Academic Press (1983), 55–67.

[98] L. Lovász: Ear-decompositions of matching-covered graphs, *Combinatorica,* **3** (1983), 105–117.

[99] B. Korte, L. Lovász: Structural properties of greedoids, *Combinatorica,* **3** (1983), 359–374.

[100] L. Lovász: Submodular functions and convexity, in: *Mathematical Programming: the State of the Art* (ed. A. Bachem, M. Grötschel, B. Korte), Springer (1983), 235–257.

[101] L. Lovász: Self-dual polytopes and the chromatic number of distance graphs on the sphere, *Acta Sci. Math. Szeged,* **45** (1983), 317–323.

[102] L. Lovász, M. Simonovits: On the number of complete subgraphs of a graph II, in: *Studies in Pure Math.,* To the memory of P. Turán (ed. P. Erdős), Akadémiai Kiadó (1983), 459–495.

[103] L. Lovász: Algorithmic aspects of combinatorics, geometry and number theory, in: *Proc. Int. Congress Warsaw 1982,* Polish Sci. Publishers – North-Holland (1984), 1591–1595.

[104] M. Grötschel, L. Lovász, A. Schrijver: Geometric methods in combinatorial optimization, in: *Progress in Combinatorial Optimization* (ed.W.R.Pulleyblank), Academic Press (1984), 167–183.

[105] B. Korte, L. Lovász: Greedoids – a structural framework for the greedy algorithm, in: *Progress in Combinatorial Optimization* (ed. W. R. Pulleyblank), Academic Press (1984), 221–243.

[106] M. Grötschel, L. Lovász, A. Schrijver: Polynomial algorithms for perfect graphs, *Annals of Discrete Math.,* **21** (1984), 325–256.

[107] W. Cook, L. Lovász, A. Schrijver: A polynomial-time test for total dual integrality in fixed dimension, *Math. Programming Study,* **22** (1984), 64–69.

[108] B. Korte, L. Lovász: Greedoids and linear objective functions, *SIAM J. on Algebraic and Discrete Methods,* **5** (1984), 229–238.

[109] R. Kannan, A. K. Lenstra, L. Lovász: Polynomial factorization and the nonrandomness of bits of algebraic and some transcendental numbers, in: *Proc. 16th ACM Symp. on Theory of Computing* (1984), 191–200.

[110] B. Korte, L. Lovász: Shelling structures, convexity, and a happy end, in: *Graph Theory and Combinatorics* (ed. B. Bollobas), Acad. Press (1984), 219–232.

[111] B. Korte, L. Lovász: A note on selectors and greedoids, *Eur. J. Combinatorics,* **6** (1985), 59–67.

[112] B. Korte, L. Lovász: Posets, matroids, and greedoids, in: *Matroid Theory,* Coll. Math. Soc. J. Bolyai, **40** (ed. L. Lovász, A. Recski), North-Holland (1985), 239–265.

[113] B. Korte, L. Lovász: Polymatroid greedoids, *J. Comb. Theory B,* **38** (1985), 41–72.

[114] B. Korte, L. Lovász: Basis graphs of greedoids and 2-connectivity, *Math. Programming Study,* **24** (1985), 158–165.

[115] L. Lovász: Computing ears and branchings in parallel, *26th IEEE Annual Symp. on Found. of Comp. Sci.* (1985), 464–467.

[116] L. Lovász: Some algorithmic problems on lattices, in: *Theory of Algorithms,* (eds. L. Lovász and E. Szemerédi), Coll. Math. Soc. J. Bolyai 44, North-Holland (1985), 323–337.

[117] L. Lovász: Vertex packing algorithms, *Proc. ICALP Conf.*, Springer (1985).

[118] A. Björner, B. Korte, L. Lovász: Homotopy properties of greedoids, *Advances in Appl. Math.,* **6** (1985), 447–494.

[119] B. Korte, L. Lovász: Relations between subclasses of greedoids, *Zeitschr. f. Oper. Res. A: Theorie,* **29** (1985), 249–267.

[120] L. Lovász: Algorithmic aspects of some notions in classical mathematics, in: *Mathematics and Computer Science* (ed. J. W. de Bakker, M. Hazewinkel, J. K. Lenstra), CWI Monographs **1,** North-Holland, Amsterdam (1986), 51–63.

[121] M. Grötschel, L. Lovász, A. Schrijver: Relaxations of vertex packing, *J. Combin. Theory B,* **40** (1986), 330–343.

[122] L. Lovász, J. Spencer, K. Vesztergombi: Discrepancy of set-systems and matrices, *Europ. J. Combin.,* **7** (1986), 151–160.

[123] B. Korte, L. Lovász: Non-interval greedoids and the transposition property, *Discrete Math.,* **59** (1986), 297–314.

[124] K. Cameron, J. Edmonds, L. Lovász: A note on perfect graphs, *Periodica Math. Hung.,* **17** (1986), 173–175.

[125] N. Linial, L. Lovász, A. Wigderson: A physical interpretation of graph connectivity, *Proc. 27th Annual IEEE Symp. on Found. of Comp. Sci.,* (1986), 39–48.

[126] R. Kannan, L. Lovász: Covering minima and lattice point free convex bodies, *Proc. Conf. on Foundations of Software Technology and Theoretical Comp. Sci.,* Lecture Notes in Comp. Science **241,** Springer (1986) 193–201.

[127] L. Lovász: Connectivity algorithms using rubber bands, *Proc. Conf. on Foundations of Software Technology and Theoretical Comp. Sci.,* Lecture Notes in Comp. Science **241,** Springer (1986), 394–411.

[128] U. Faigle, L. Lovász, R. Schrader, G. Turán: Searching in trees, series-parallel and interval orders, *SIAM J. Computing,* **15** (1986), 1075–1084.

[129] B. Korte, L. Lovász: Homomorphisms and Ramsey properties of antimatroids, *Discrete Appl. Math.,* **15** (1986), 283–290.

[130] N. Alon, P. Frankl, L. Lovász: The chromatic number of Kneser hypergraphs, *Trans. Amer. Math. Soc.,* **298** (1987), 359–370.

[131] A. Dress, L. Lovász: On some combinatorial properties of algebraic matroids, *Combinatorica,* **7** (1987), 39–48.

[132] L. Lovász: Matching structure and the matching lattice, *J. Comb. Theory B,* **43** (1987), 187–222.

[133] A. Björner, L. Lovász: Pseudomodular lattices and continuous matroids, *Acta Sci. Math. Szeged,* **51** (1987), 295–308.

[134] N. Linial, L. Lovász, A. Wigderson: Rubber bands, convex embeddings, and graph connectivity, *Combinatorica,* **8** (1988), 91–102.

[135] L. Lovász: Geometry of numbers: an algorithmic view, in: *ICIAM '87: Proc. 1st Internatl. Conf. on Industr. Appl. Math.* (ed. J. McKenna, R. Teman), SIAM, Philadelphia (1988), 144–152.

[136] L. Lovász, M. Saks: Lattices, Möbius functions and communication complexity, *29th IEEE Annual Symp. on Found. of Comp. Sci.* (1988), 81–90.

[137] P. Erdős, L. Lovász, K. Vesztergombi: The chromatic number of the graph of large distances, in: *Combinatorics,* Proc. Coll. Eger 1987, Coll. Math. Soc. J. Bolyai, **52,** North-Holland (1988), 547–551.

[138] C.A.J.Hurkens, L. Lovász, A. Schrijver, É. Tardos: How to tidy up your set-system? in: *Combinatorics,* Proc. Coll. Eger 1987, Coll. Math. Soc. J. Bolyai, **52,** North-Holland (1988), 309–314.

[139] B. Korte, L. Lovász: The intersection of matroids and antimatroids, *Discrete Math.,* **73** (1988), 143–157.

[140] L. Lovász: Covering minima and lattice point free convex bodies (R. Kannan), *Annals of Math.,* **128** (1988), 577–602.

[141] L. Lovász, K. Vesztergombi: Extremal problems for discrepancy, in: *Irregularities of Partitions,* (ed. G. Halász, V. T. Sós), Algorithms and Combinatorics **8** (1989), Springer, 107–113.

[142] L. Lovász, M. Saks and A. Schrijver: : Orthogonal representations and connectivity of graphs, *Linear Alg. Appl.,* **114/115** (1989), 439–454. A correction: *Linear Algebra Appl.,* **313** (2000), 101–105.

[143] B. Korte, O. Goecke, L. Lovász: Examples and algorithmic properties of greedoids, in: Combinatorial Optimization (ed. B. Simeone), *Lecture Notes in Math.,* **1403,** Springer (1989), 113–161.

[144] R. Anderson, L. Lovász, P. Shor, J. Spencer, É. Tardos, S. Winograd: Disks, balls and walls: The analysis of a combinatorial game, *Amer. Math. Monthly,* **96** (1989), 481–493.

[145] L. Lovász, M. Saks, W. T. Trotter: An on-line graph coloring algorithm with sublinear performance ratio, *Discrete Math.,* **75** (1989), 319–325.

[146] L. Lovász: Faster algorithms for hard problems, *Information Processing '89* (ed. G. X. Ritter), Elsevier (1989), 135–141.

[147] L. Lovász, B. Korte: Polyhedral results for antimatroids, in: *Combinatorial Mathematics,* Proc. 3rd Intern. Conf., (ed. G. S. Bloom, R. L. Graham, J. Malkevitch), Annals of the NY Academy of Sciences **555** (1989), 283–295.

[148] P. Erdős, L. Lovász, K. Vesztergombi: On the graph of large distances, *Discr. Comput. Geometry,* **4** (1989), 541–549.

[149] L. Lovász: Geometry of numbers and integer programming, in: *Mathematical Programming, Recent Developments and Applications,* Kluwer Academic Publishers (1989), 177–201.

[150] L. Lovász: Singular spaces of matrices and their application in combinatorics, *Bol. Soc. Braz. Mat.,* **20** (1989), 87–99.

[151] L. Lovász, M.D. Plummer: Some recent results on graph matching, in: *Graph Theory and its Applications: East and West,* (ed. M. F. Capobianco, M. Guan, D. F. Hsu, F. Tian), Ann. NY Acad. Sci., **576** (1989), 389–398.

[152] I. Bárány, Z. Füredi, L. Lovász: On the number of halving planes, *Proc. 5th Symp. Comp. Geom.,* (1989), 140–144; journal version: *Combinatorica,* **10** (1990), 175–183.

[153] L. Lovász, A. Schrijver: Matrix cones, projection representations, and stable set polyhedra, in: *Polyhedral Combinatorics,* DIMACS Series in Discrete Mathematics and Theoretical Computer Science I (1990), 1–17.

[154] R. Kannan, L. Lovász, H. E. Scarf: The shapes of polyhedra, *Math. of Oper Res.,* **15** (1990), 364–380.

[155] L. Lovász, M. Simonovits: The mixing rate of Markov chains, an isoperimetric inequality, and computing the volume, *Proc. 31st IEEE Annual Symp. on Found. of Comp. Sci.* (1990), 346–354.

[156] L. Lovász: Communication complexity: a survey, in: *Paths, flows, and VLSI-Layout,* (ed. B. Korte, L. Lovász, H. J. Prömel, A. Schrijver), Springer (1990), 235–265.

[157] I. Csiszár, J. Körner, L. Lovász, K. Marton, G. Simonyi: Entropy splitting for antiblocking pairs and perfect graphs *Combinatorica,* **10** (1990), 27–40.

[158] L. Babai, A.J. Goodman, L. Lovász: Graphs with given automorphism group and few edge orbits, *Europ. J. Combin.,* **12** (1991), 185–203.

[159] A. Björner, L. Lovász, P. Shor: Chip-firing games on graphs, *Europ. J. Comb.,* **12** (1991), 283–291.

[160] L. Lovász, A. Schrijver: Cones of matrices and set-functions, and 0-1 optimization, *SIAM J. Optim.,* **1** (1991), 166–190.

[161] L. Lovász: Geometric algorithms and algorithmic geometry, *Proc. of Int. Congress of Math, Kyoto, 1990,* Springer-Verlag (1991), 139–154.

[162] L. Lovász, M. Naor, I. Newman, A. Wigderson: Search problems in the decision tree model, *Proc. 32nd IEEE Annual Symp. on Found. of Comp. Sci.* (1991), 576–585; journal version: *SIAM J. Disc. Math.,* **8** (1995), 119–132.

[163] J. Csima, L. Lovász: A matching algorithm for regular bipartite graphs, *Discrete Appl. Math.,* **35** (1992), 197–203.

[164] L. Lovász: How to compute the volume? *Jber. d. Dt. Math.-Verein, Jubiläumstagung 1990,* B. G. Teubner, Stuttgart (1992), 138–151.

[165] A. Björner, L. Lovász, A. Yao: Linear decision trees, hyperplane arrangements, and Möbius functions, in: *Proc. 24th ACM Symp. on Theory of Computing* (1992), 170–177.

[166] U. Feige, L. Lovász: Two-prover one-round proof systems: their power and their problems, in: *Proc. 24th ACM Symp. on Theory of Computing* (1992), 733–744.

[167] I. Bárány, R. Howe, L. Lovász: On integer points in polyhedra: a lower bound, *Combinatorica,* **12** (1992), 135–142.

[168] L. Lovász, M. Simonovits: On the randomized complexity of volume and diameter, *Proc. 33rd IEEE Annual Symp. on Found. of Comp. Sci.* (1992), 482–491.

[169] L. Lovász, H. Scarf: The generalized basis reduction algorithm, *Math. of OR,* **17** (1992), 751–764.

[170] A. Björner, L. Lovász: Chip-firing games on directed graphs, *J. Algebraic Combinatorics,* **1** (1992), 305–328.

[171] J. Csima, L. Lovász: Dating to marriage, *Discrete Appl. Math.,* **41** (1993), 269–270.

[172] L. Lovász, Á. Seress: The cocycle lattice of binary matroids, *Europ. J. Comb.,* **14** (1993), 241–250.

[173] L. Lovász, M. Simonovits: Random walks in a convex body and an improved volume algorithm, *Random Structures and Alg.,* **4** (1993), 359–412.

[174] L. Lovász, M. Saks: Communication complexity and combinatorial lattice theory, *J. Comp. Sys. Sci.,* **47** (1993), 322–349.

[175] L. Lovász, P. Winkler: Note: on the last new vertex visited by a random walk, *Journal of Graph Theory,* **17** (1993), 593–596.

[176] N. Karmarkar, R. M. Karp, R. Lipton, L. Lovász, M. Luby: A Monte-Carlo algorithm for estimating the permanent, *SIAM J. Comp.,* **22** (1993), 284–293.

[177] L. Lovász: Stable sets and polynomials, *Discrete Math.,* **124** (1994), 137–153.

[178] A. Björner, L. Lovász: Linear decision trees, subspace arrangements, and Möbius functions, *Journal of the Amer. Math. Soc.,* **7** (1994), 677–706.

[179] A. Björner, L. Lovász, S. Vrecica, R. Zivaljevic: Chessboard complexes and matching complexes, *Journal of the London Math. Soc.,* **49** (1994), 25–39.

[180] R. Kannan, L. Lovász, M. Simonovits: Isoperimetric problems for convex bodies and a localization lemma, *Disc. Comput. Geometry,* **13** (1995), 541–559.

[181] L. Lovász: Randomized algorithms in combinatorial optimization, in: Combinatorial Optimization, Papers from the DIMACS Special Year, (ed. W. Cook, L. Lovasz, P. Seymour), DIMACS Series in Discrete Mathematics and Combinatorial Optimization **20,** Amer. Math. Soc., Providence (1995), 153–179.

[182] L. Lovász, P. Winkler: Exact mixing in an unknown Markov chain, *Electronic Journal of Combinatorics,* **2** (1995), paper R15, 1–14.

[183] L. Lovász, P. Winkler: Mixing of random walks and other diffusions on a graph, in: *Surveys in Combinatorics,* 1995 (ed. P. Rowlinson), London Math. Soc. Lecture Notes Series **218,** Cambridge Univ. Press (1995), 119–154.

[184] L. Lovász, P. Winkler: Efficient stopping Rules for Markov Chains, *Proceedings of the 1995 ACM Symposium on the Theory of Computing* (1995), 76–82.

[185] L. Lovász, Á. Seress: The cocycle lattice of binary matroids II, *Linear Algebra and its Applications,* **226–228** (1995), 553–566.

[186] H. v.d.Holst, L. Lovász, A. Schrijver: On the invariance of Colin de Verdière's graph parameter under clique sums, *Linear Algebra and its Applications,* **226–228** (1995), 509–518.

[187] A. Kotlov, L. Lovász: The rank and size of graphs, *J. Graph Theory,* **23** (1996), 185–189.

[188] L. Lovász: Random walks on graphs: a survey, in: *Combinatorics, Paul Erdős is Eighty,* Vol. 2 (ed. D. Miklós, V. T. Sós, T. Szőnyi), János Bolyai Mathematical Society, Budapest, (1996), 353–398.

[189] U. Feige, S. Goldwasser, L. Lovász, S. Safra and M. Szegedy: Approximating clique is almost NP-complete, *Proc. 32nd IEEE Annual Symp. on Found. of Comp. Sci.* (1991), 2–12; journal version: *Journal of the ACM,* **43** (1996), 268–292.

[190] L. Lovász: The membership problem in jump systems, *J. Comb. Theory* (B), **70** (1997), 45–66.

[191] L. Lovász, J. Pach, M. Szegedy: On Conway's thrackle conjecture, *Discrete and Computational Geometry,* **18** (1997), 369–376.

[192] R. Kannan, L. Lovász, M. Simonovits: Random walks and an $O^*(n^5)$ volume algorithm for convex bodies, *Random Structures and Algorithms,* **11** (1997), 1–50.

[193] A. Kotlov, L. Lovász, S. Vempala: The Colin de Verdière number and sphere representations of a graph, *Combinatorica,* **17** (1997), 483–521.

[194] D. Aldous, L. Lovász, P. Winkler: Mixing times for uniformly ergodic Markov chains, *Stochastic Processes and their Applications,* **71** (1997), 165–185.

[195] L. Lovász, A. Schrijver: A Borsuk theorem for antipodal links and a spectral characterization of linklessly embeddable graphs, *Proceedings of the Amer. Math. Soc.,* **126** (1998), 1275–1285.

[196] L. Lovász, P. Winkler: Reversal of Markov chains and the forget time, *Combinatorics, Probability and Computing,* **7** (1998), 189–204.

[197] L. Lovász, P. Winkler: Mixing times, in: *Microsurveys in Discrete Probability* (ed. D. Aldous and J. Propp), DIMACS Series in Discrete Math. and Theor. Comp. Sci., Amer. Math. Soc. (1998), 85–133.

[198] A. Brieden, P. Gritzmann, R. Kannan, V. Klee, L. Lovász, M. Simonovits: Approximation of diameters: randomization doesn't help *Proceedings 39th Ann. Symp. on Found. of Comp. Sci.* (1998), 244–251.

[199] A. Beveridge, L. Lovász: Random walks and the regeneration time, *J. Graph Theory,* **29** (1998), 57–62.

[200] L. Lovász, A. Schrijver: The Colin de Verdière graph parameter, Proc. Conf. Combinatorics, in: *Graph Theory and Combinatorial Biology,* Bolyai Soc. Math. Stud., **7**, János Bolyai Math. Soc., Budapest (1999), 29–85.

[201] F. Chen, L. Lovász, I. Pak: Lifting Markov chains to speed up mixing, in: *Proc. 31st Annual ACM Symp. on Theory of Computing* (1999), 275–281.

[202] R. Kannan, L. Lovász: Faster mixing via average conductance, in: *Proc. 31st Annual ACM Symp. on Theory of Computing* (1999), 282–287.

[203] L. Lovász, A. Schrijver: On the null space of a Colin de Verdière matrix, *Annales de l'Institute Fourier,* **49** (1999), 1017–1026.

[204] L. Lovász: Hit-and-run mixes fast, *Math. Programming, series A,* **86** (1999), 443–461.

[205] L. Lovász: Integer sequences and semidefinite programming *Publ. Math. Debrecen,* **56** (2000), 475–479.

[206] L. Lipták, L. Lovász: Facets With Fixed Defect of the Stable Set Polytope, *Math. Programming,* Series A **88** (2000), 33–44.

[207] J. Kahn, J. H. Kim, L. Lovász, V. H. Vu: The cover time, the blanket time, and the Matthews bound, *Proc. 41st IEEE Ann. Symp. on Found. of Comp. Sci.* (2000),

[208] L. Lipták, L. Lovász: Critical Facets of the Stable Set Polytope, *Combinatorica,* **21** (2001), 61–88.

[209] L. Lovász: Steinitz representations of polyhedra and the Colin de Verdière number, *J. Comb. Theory B,* **82** (2001), 223–236.

[210] L. Lovász: Energy of convex sets, shortest paths and resistance *J. Comb. Theory A,* **94** (2001), 363–382.

[211] N. Alon, L. Lovász: Unextendible product bases, *J. Combin. Theory A,* **95** (2001), 169–179.

[212] A. Brieden, P. Gritzmann, R. Kannan, V. Klee, L. Lovász, M. Simonovits: Deterministic and randomized approximation of radii, *Mathematika,* **48** (2001), 63–105.

[213] I. Benjamini, L. Lovász: Global Information from Local Observation, *Proc. 43rd Ann. Symp. on Found. of Comp. Sci.* (2002), 701–710.

[214] S. Arora, B. Bollobás, L. Lovász: Proving integrality gaps without knowing the linear program *Proc. 43rd Ann. Symp. on Found. of Comp. Sci.* (2002), 313–322; journal version: S. Arora, B. Bollobás, L. Lovász, I. Tourlakis, *Theory of Computing,* **2** (2006), 19–51.

[215] L. Lovász, K. Vesztergombi: Geometric representations of graphs, in: *Paul Erdos and his Mathematics,* (ed. G. Halász, L. Lovász, M. Simonovits, V. T. Sós), Bolyai Soc. Math. Stud., 11, Jnos Bolyai Math. Soc., Budapest, (2002), 471–498.

[216] I. Benjamini, L. Lovász: Harmonic and analytic functions on graphs, *J. of Geometry,* **76** (2003), 3–15.

[217] L. Lovász, S. Vempala: Logconcave Functions: Geometry and Efficient Sampling Algorithms *Proc. 43rd Ann. Symp. on Found. of Comp. Sci.* (2003), 640–649.

[218] L. Lovász, S. Vempala: Simulated Annealing in Convex Bodies and an $O*(n^4)$ Volume Algorithm, *Proc. 43rd Ann. Symp. on Found. of Comp. Sci.* (2003), 650–659; journal version *J. Comput. System Sci.,* **72** (2006), 392–417.

[219] N. Harvey, E. Ladner, L. Lovász, T. Tamir: Semi-Matchings for Bipartite Graphs and Load Balancing, *Algorithms and Data Structures,* Lecture Notes in Comput. Sci., *2748,* Springer, Berlin (2003), 294–306; journal version *Journal of Algorithms,* **59** (2006), 53–78.

[220] L. Lovász: Semidefinite programs and combinatorial optimization, in: *Recent Advances in Algorithms and Combinatorics,* CMS Books Math./Ouvrages Math. SMC *11,* Springer, New York (2003), 137–194.

[221] L. Lovász, K. Vesztergombi, U. Wagner, E. Welzl: Convex quadrilaterals and k-sets, in: *Towards a Theory of Geometric Graphs,* (J. Pach, Ed.), AMS Contemporary Mathematics **342** (2004), 139–148.

[222] J. Chen, R.D. Kleinberg, L. Lovász, R. Rajaraman, R. Sundaram, A. Vetta: (Almost) tight bounds and existence theorems for confluent flows, *Proc. 36th Annual ACM Symposium on Theory of Computing,* ACM, New York (2004), 529–538.

[223] L. Lovász, S. Vempala: Hit-and-run from a corner, *Proc. 36th Annual ACM Symposium on Theory of Computing,* 310–314 (electronic), ACM, New York (2004).

[224] U. Feige, L. Lovász, P. Tetali: Approximating min sum set cover, *Proc. Conf. APPROX* (2002), 94–107; journal version: *Algorithmica,* **40** (2004), 219–234.

[225] L. Lovász: Discrete Analytic Functions: An Exposition, in: *Surveys in Differential Geometry IX, Eigenvalues of Laplacians and other geometric operators* (Ed. Grigoryan A., Yau S.-T.), Int. Press, Somerville, MA (2004), 241–273.

[226] M. Bordewich, M. Freedman, L. Lovász, D. Welsh: Approximate counting and quantum computation, *Combinatorics, Probability and Computing,* **14** (2005), 737–754.

[227] C. Borgs, J. T. Chayes, L. Lovász, V. T. Sós, B. Szegedy and K. Vesztergombi: *Graph Limits and Parameter Testing,* Proc. 38th Annual ACM Symp. on Theory of Computing 2006, 261–270

[228] K. Jain, L. Lovász, P. A. Chou: Building scalable and robust peer-to-peer overlay networks for broadcasting using network coding, *Ann. ACM Symp. Principles of Dist. Comp.* (2005), 51–59.

[229] R. Kannan, L. Lovász, R. Montenegro: Blocking conductance and mixing in random walks, *Combinatorics, Probability and Computing,* **15** (2006), 541–570.

[230] C. Borgs, J. Chayes, L. Lovász, V. T. Sós, K. Vesztergombi: Counting graph homomorphisms, in: *Topics in Discrete Mathematics* (ed. M. Klazar, J. Kratochvil, M. Loebl, J. Matoušek, R. Thomas, P. Valtr), Springer (2006), 315–371.

[231] L. Lovász, M. Saks: A localization inequality for set functions, *J. Comb. Theory* A, **113** (2006), 726–735.

[232] L. Lovász: The rank of connection matrices and the dimension of graph algebras, *Eur. J. Comb.*, **27** (2006), 962–970.

[233] L. Lovász: Graph minor theory, *Bull. Amer. Math. Soc.*, **43** (2006), 75–86.

[234] I. Benjamini, G. Kozma, L. Lovász, D. Romik, G. Tardos: Waiting for a bat to fly by (in polynomial time), *Combinatorics, Probability and Computing,* **15** (2006), 673–683.

[235] L. Lovász, B. Szegedy: Limits of dense graph sequences, *J. Comb. Theory B,* **96** (2006), 933–957.

[236] L. Lovász, S. Arora, I. Newman, Y. Rabani, Y. Rabinovich, S. Vempala: Local versus Global Properties of Metric Spaces, *Proc. 17th Ann. ACM-SIAM Symp. Disc. Alg.* (2006), 41–50.

[237] L. Lovász: Connection matrices, in: Combinatorics, Complexity and Chance, A Tribute to Dominic Welsh, Oxford Univ. Press (2007), 179–190.

[238] L. Lovász, M. Freedman, A. Schrijver: Reflection positivity, rank connectivity, and homomorphisms of graphs *J. Amer. Math. Soc.,* **20** (2007), 37–51.

[239] L. Lovász, S. Vempala: The Geometry of Logconcave Functions and Sampling Algorithms, *Random Struct. Alg.,* **30** (2007), 307–358.

[240] L. Lovász, B. Szegedy: Szemerédi's Lemma for the analyst, *Geom. Func. Anal.,* **17** (2007), 252–270.

[241] A. Blokhuis, L. Lovász, L. Storme, T. Szőnyi: On multiple blocking sets in Galois planes, *Adv. Geom.,* **7** (2007), 39–53.

[242] L. Lovász, V.T. Sós: Generalized quasirandom graphs, *J. Comb. Th. B,* **98** (2008), 146–163.

[243] L. Lovász, A. Schrijver: Graph parameters and semigroup functions, *Europ. J. Comb.,* **29** (2008), 987–1002.

Expository papers:

[1] L. Lovász: A matroidelmélet rövid áttekintése (A short survey of matroid theory), *Mat. Lapok 22* (1971), 249–267.

[2] L. Lovász: A szitaformuláról (On the sieve formula), *Mat. Lapok,* **23** (1972), 53–69.

[3] L. Lovász: Kombinatorikus optimalizáció (Combinatorial optimization), *Magyar Tudomány,* **25** (1980), 736–742.

[4] L. Lovász: A new linear programming algorithm: better or worse than Simplex Method? *Math. Intelligencer,* **2** (1980), 141–146.

[5] L. Lovász: Mit ad a matematikának és mit kap a matematikától a számítógéptudomány? (What does computer science get from mathematics and what does it give to it? *Magyar Tudomány,* **35** (1990), 1041–1047.

[6] L. Lovász: The mathematical notion of complexity, *Proc. IFAC Symposium,* Budapest (1984).

[7] L. Lovász: Algorithmic mathematics: an old aspect with a new emphasis, in: *Proc. 6th ICME, Budapest, J. Bolyai Math. Soc.* (1988), 67–78.

[8] L. Lovász: The work of A. A. Razborov, *Proc. of Int. Congress of Math, Kyoto,* Springer-Verlag (1989), 37–40.

[9] L. Lovász: Features of computer language: communication of computers and its complexity, *Acta Neurochirurgica,* **56** [Suppl.] (1994), 91–95.

[10] L. Lovász: Random walks, eigenvalues, and resistance, Appendix to Chapter 31: Tools from linear algebra, in: Handbook of Combinatorics (ed. R. Graham, M. Grötschel, L. Lovász), Elsevier Science B.V. (1995), 1740–1748.

[11] M. Grötschel, L. Lovász: Combinatorial optimization, Chapter 28 in: *Handbook of Combinatorics* (ed. R. Graham, M. Grötschel, L. Lovász), Elsevier Science B.V. (1995), 1541–1597.

[12] L. Lovász, D.B. Shmoys, É. Tardos: Combinatorics in computer science, Chapter 40 in: *Handbook of Combinatorics* (ed. R. Graham, M. Grötschel, L. Lovász), Elsevier Science B.V. (1995), 2003–2038.

[13] L. Lovász, L. Pyber, D. J. A. Welsh, G. M. Ziegler: Combinatorics in pure mathematics, Chapter 41 in: *Handbook of Combinatorics* (ed. R. Graham, M. Grötschel, L. Lovász), Elsevier Science B.V. (1995), 2039–2082.

[14] L. Lovász: Information and complexity (how to measure them?) in: *The Emergence of Complexity in Mathematics, Physics, Chemistry and Biology* (ed. B. Pullman), Pontifical Academy of Sciences, Vatican City, Princeton University Press (1996), 65–80.

[15] L. Lovász: One mathematics, The Berlin Intelligencer, Mitteilungen der Deutschen Math.-Verein, Berlin (1998), 10–15.

[16] L. Lovász: Egységes tudomány-e a matematika? (Is mathematics a single science?), *Természet Világa, Special issue on Mathematics,* **1998**

[17] L. Lovász: Discrete and Continuous: Two sides of the same? *GAFA, Geom. Funct. Anal., Special volume - GAFA2000,* Birkheuser, Basel (2000), 359–382.

[18] L. Lovász: Véletlen és álvéletlen (Randomness and pseudo-randomness), *Természet Világa,* Special issue in Informatics, (2000), 5–7.

[19] L. Lovász: Nagyon nagy gráfok (Very large graphs), *Természet Világa,* **138** (2007), 98–103.

BOLYAI SOCIETY
MATHEMATICAL STUDIES, 19

Building Bridges
pp. 31–45.

On the Power of Linear Dependencies

IMRE BÁRÁNY

Simple as they may be, linear dependencies have proved very useful in many ways. In this survey several geometric applications of linear dependencies are discussed, focusing on rearrangements of sums and on sums with ± 1 signs.

1. Introduction

Linear algebra is a basic and powerful tool in many areas of mathematics. In combinatorics, for instance, there are several cases when the size of a set can be bounded by a number n because the elements of the set are associated with vectors in R^n, and these vectors turn out to be linearly independent. The excellent book [1] by Babai and Frankl (which is unfortunately, still unpublished) contains thousands of beautiful applications of the so-called linear algebra method.

This article describes another kind of use of linear algebra, this time in geometry. The method uses linear dependencies and is often referred to as the *method of floating variables*. The same method is used at other places as well, for instance in discrepancy theory, in the Beck–Fiala theorem [7] or [8], and in probability theory, [9]. Here we focus on rearrangement of sums and on sums with ± 1 signs.

In what follows the setting is the d-dimensional Euclidean space $\mathbb{R}^d$, together with a (Minkowski) norm, $\|\cdot\|$ whose unit ball is denoted by B or B^d. We write $\mathbb{N}$ for the set of natural numbers and $[n]$ for the set $\{1, 2, \dots, n\}$ where $n \in \mathbb{N}$. We assume that $V \subset B$ is a finite set.

2. The Steinitz lemma

Assume $V \subset B$ is finite and $\sum_{v \in V} v = 0$. The question, due to Riemann and Lévy, is whether there is an ordering, $v_1, v_2, \ldots, v_n$, of the elements of V such that all partial sums along this order are bounded by a number that only depends on B. The answer is yes. An incomplete proof came from Lévy [17] in 1905. The first complete proof, from 1913, is due to Steinitz [20], and that's why it is usually called the Steinitz Lemma.

Theorem 2.1. *Given a finite set $V \subset B$ with $\sum_{v \in V} v = 0$, where B is the unit ball of a norm in $\mathbb{R}^d$, there is an ordering $v_1, v_2, \ldots, v_n$ of the elements of V such that for all $k \in [n]$*

$$\sum_1^k v_i \in dB.$$

So all partial sums are contained in a blown-up copy of the unit ball, with blow-up factor d. We will return to the value of the blow-up factor later. Let's see first the proof of Theorem 2.1, which is our first application of linear dependencies.

Proof. The key step is the construction of sets $V_{d+1} \subset \cdots \subset V_{n-1} \subset V_n = V$ where $|V_k| = k$, together with functions $\alpha_k \colon V_k \to [0,1]$ satisfying

$$\sum_{v \in V_k} \alpha_k(v) v = 0$$

$$\sum_{v \in V_k} \alpha_k(v) = k - d.$$

So the functions $\alpha_k(.)$ are *linear dependencies* on V_k, with coefficients in $[0,1]$ that sum to $k-d$.

The construction goes by backward induction. The starting case $k = n$ is easy: $V_n = V$ and $\alpha_n = (n-d)/n$ satisfy the requirements. Assume now that V_k and α_k have been constructed, and consider the auxiliary system

$$\sum_{v \in V_k} \beta(v) v = 0,$$

$$\sum_{v \in V_k} \beta(v) = k - 1 - d,$$

$$0 \leq \beta(v) \leq 1 \quad \text{for all} \quad v \in V_k.$$

Write P for the set of functions $\beta\colon V_k \to [0,1]$ satisfying this auxiliary system. The elements of P can and will be regarded as vectors in $\mathbb{R}^k$ whose components are indexed by the elements of V_k.

Note that P is non-empty since $\beta(v) = \frac{k-1-d}{k-d}\alpha_k(v)$ belongs to P. Thus P is a convex polytope, lying in the unit cube of $\mathbb{R}^k$. Let $\beta^*(.)$ be an extreme point of P.

We claim now that $\beta^*(v) = 0$ for some $v \in V_k$. Indeed, assume $\beta^*(v) > 0$ for all $v \in V_k$. The auxiliary system has $d+1$ equations and k variables, so at least $k - (d+1)$ of the inequalities $\beta^*(v) \leq 1$ are satisfied as equalities (the inequalities $\beta^*(v) \geq 0$ are all strict). Then $\sum_{v \in V_k} \beta^*(v) > k - d - 1$ (we use $k > d+1$ here), which contradicts one of the conditions defining P.

Let $v^* \in V_k$ be an element with $\beta^*(v^*) = 0$, and define $V_{k-1} = V_k \setminus \{v^*\}$ and $\alpha_{k-1}(v) = \beta^*(v)$ for all $v \in V_{k-1}$. All conditions are satisfied for V_{k-1} and α_{k-1}. The construction is finished.

Now we ready to order the elements of V. For $k = n, n-1, \ldots, d+2$ we set, quite naturally,

$$v_k = V_k \setminus V_{k-1}.$$

The remaining $d+1$ vectors are ordered arbitrarily.

We check, finally, that all partial sums are contained in dB. This is trivial for the first d partial sums. Assume now that $k \geq d+1$.

$$\sum_1^k v_i = \sum_{v \in V_k} v = \sum_{v \in V_k} v - \sum_{v \in V_k} \alpha_k(v)v = \sum_{v \in V_k} \left(1 - \alpha_k(v)\right) v.$$

Taking norms and using that $1 - \alpha_k(v) \geq 0$ and $\|v\| \leq 1$ gives

$$\left\| \sum_1^k v_i \right\| \leq \sum_{v \in V_k} \left(1 - \alpha_k(v)\right) = k - (k-d) = d. \qquad \blacksquare$$

This splendid proof, due to Grinberg and Sevastyanov [13], is a streamlined version of an earlier one by Sevastyanov [21]. Steinitz's original proof also used linear dependencies, and gave constant $2d$ instead of d. Yet another proof, from Bárány [3], using linear dependencies again, gave blow-up factor $1.5d$.

It comes as a surprise that the norm plays a rather marginal role. I describe now another proof, due to Kadets [15], in which the norm is important. The proof works for the Euclidean norm B_2 and gives the weaker blow-up factor $C_d = \sqrt{(4^d - 1)/3}$. It goes by induction on dimension and the case $d = 1$ is very simple: one takes positive elements as long as the sum stays below 1, then one takes negative elements as long as the sum is above -1, and so on. For the induction step $d - 1 \to d$, let $V \subset B_2$ be a finite set with $\sum_{v \in V} v = 0$. We choose a subset $W \subset V$ for which $\|\sum_{v \in W} v\|$ is maximal among all subsets of V, and set $a = \sum_{v \in W} v$. For $v \in V$ let $\overline{v}$ denote its orthogonal projection onto the subspace A orthogonal to a, and set $v_a = v - \overline{v}$. It follows from the maximality of W that, for all $v \in W$, v_a points the same direction as a, and, for all $v \notin W$, v_a points the opposite direction. Further, each $\overline{v} \in A$ and has Euclidean norm at most one in A which is the $d-1$-dimensional Euclidean space. Also, $\sum_{v \in W} \overline{v} = 0$. By the induction hypothesis, there is an ordering $v_1, v_2 \dots$ of the vectors in W with all partial sums of the $\overline{v_i}$ having Euclidean length at most C_{d-1}. The same applies to the set $V \backslash W$, so its elements can be ordered as $w_1, w_2, \dots$ with all partial sums of the $\overline{w_j}$ shorter than or equal to C_{d-1}. The sequences $v_1, \dots$ and $w_1, \dots$ are then interlaced making sure (using the method given for the case $d = 1$) that the absolute value of the a-component of each partial sum is at most 1. Then the square of each partial sum of the interlaced sequence is at most $4C_{d-1}^2 + 1 = C_d^2$ as one can easily see.

This is quite a neat proof, yet the other one is superior: it works for all (even non-symmetric) norms, gives a far better bound, and is much more elegant, as far as I can judge.

3. The story of the Steinitz lemma

The story actually began with Riemann who showed that a conditionally convergent series (of real numbers) can be made to converge, by a suitable rearrangement of the series, to any real number. What happens with a conditionally convergent series of d-dimensional vectors? Let $U = \{u_1, u_2, \dots\}$ be the vectors in the series, and let σU be the set of points that it can be made to converge by rearrangements. It turns out that σU is always an affine subspace of $\mathbb{R}^d$. For $d = 1$ this is equivalent to Riemann's result. In higher dimensions the problem quickly reduces to the statement of what is called now Steinitz Lemma, with arbitrary norm and arbitrary constant depending only on dimension. This is what Steinitz proved in [20].

The smallest constant the Steinitz lemma holds with is a number, to be denoted by $S(B)$, that depends only on the unit ball B. It is called the Steinitz constant of B. Theorem 2.1 says that $S(B) \le d$ for all B in $\mathbb{R}^d$. This norm need not be symmetric: a quick look at the last step of the proof shows that $\|\lambda v\| \le \lambda\|v\|$ was only used with $\lambda \ge 0$. For non-symmetric norms, the estimate $S(B) \le d$ is best possible. In the example proving this, B is a simplex in $\mathbb{R}^d$ with its center of gravity at the origin, and V consists of the vectors pointing to the vertices of this simplex.

Yet, of course, $S(B)$ might be much smaller than d for any particular norm. Write B_p for the unit ball of the ℓ_p norm in $\mathbb{R}^d$, or B_p^d if we wish to stress the dimension of the underlying space. It is easy to see that $S(B_1) \ge d/2$, so the order of magnitude for symmetric norms cannot be improved. In 1931 Bergström [9] proved that $S(B_2) = \sqrt{5}/2$ for $d = 2$, a surprisingly precise result. The lower bound comes from a construction consisting of $n/2$ copies of the vector $\left(\sqrt{1-t^2}, -t\right)$, $n/2$ copies of the vector $\left(-\sqrt{1-t^2}, -t\right)$ where $t = 1/n$, and the vector $(0,1)$. This is essentially $n/2$ copies of a slightly modified e_1 and $-e_1$ compensated by e_2. In higher dimensions, the analogous example shows that $S\left(B_2^d\right) \ge \sqrt{d+3}/2$. It has been conjectured that

$$S\left(B_2^d\right) = O\left(d^{1/2}\right).$$

But even the much weaker $S\left(B_2^d\right) = o(d)$ estimate seems to be out of reach though quite a few mathematicians have tried.

The case of the maximum norm, B_∞, is also open. An example can be built from a $d+1$ by $d+1$ Hadamard matrix: its first row is the all 1 vector, and the vectors in V are the $d+1$ columns of this matrix with the first coordinate deleted. It is not hard to see that the squared Euclidean norm of the sum of k vectors from V is $k(d+1-k)$. This shows, when $k = (d+1)/2$, that one coordinate of the sum is at least $(d+1)/\sqrt{4d}$ in absolute value, implying that the conjecture,

$$S\left(B_\infty^d\right) = O\left(d^{1/2}\right),$$

if true, is best possible. But again, there is no proof in sight even for the much weaker $S\left(B_\infty^d\right) = o(d)$ estimate.

The Steinitz Lemma has many applications. It is used, or can be used to prove Lyapunov's theorem stating that the image of an atomless vector valued measure is always convex. In Operations Research, the Lemma has been applied to scheduling problems. In particular, it was used to find optimal flow shop and job shop schedules in polynomial time under some mild conditions, although these scheduling problems are NP-hard in general. See the excellent survey by Sevastyanov [22], or some of the original works [3], [21]. Halperin and Ando [14] cite 290 papers related to the Steinitz Lemma up to 1989, by now the number must be much higher.

4. Signed sum

In this section $V \subset B$ is a finite set, again, and we want to find signs $\varepsilon(v) = 1$ or -1 for all $v \in V$ such that $\sum_{v\in V} \varepsilon(v)v$ is not too large. In the following theorem, which is from Bárány, Grinberg [4], we work with the Euclidean ball B_2.

Theorem 4.1. *Under the above conditions there are signs $\varepsilon(v)$ such that*

$$\sum_{v\in V} \varepsilon(v)v \in \sqrt{d}B.$$

The example when V consists of d pairwise orthogonal unit vectors shows that the above estimate is best possible.

Proof. The proof is in two steps. For the first one, consider the set of linear dependencies $\alpha\colon V \to [-1,1]$, that is, functions satisfying

$$\sum_{v\in V} \alpha(v)v = 0 \quad \text{and} \quad -1 \le \alpha(v) \le 1 \quad \text{for all} \quad v \in V.$$

These functions form a convex polytope in the ± 1 cube of $\mathbb{R}^V$, of dimension at least $n-d$ (where $|V| = n$). This polytope is non-empty since it contains the origin. At an extreme point, α^* say, of this polytope many $\alpha^*(v) \in \{-1,1\}$. Precisely, the set of vectors $v \in V$ with $-1 < \alpha^*(v) < 1$ are linearly independent since otherwise α^* is not an extreme point. For simpler notation we assume that this happens for the vectors $v_1, \dots, v_k$, where, obviously, $k \le d$. For the rest of the vectors v_i (with $i \in \{k+1, \dots, n\}$), we have $\alpha^*(v_i) = \alpha_i \in \{-1,1\}$.

Define now $u = \sum_{k+1}^{n} \alpha_i v_i$ and set

$$Q = \Big\{ \sum_{1}^{k} \beta_i v_i \colon \beta_i \in [-1, 1] \Big\}. \tag{4.1}$$

Clearly, Q is a parallelotope whose sides have Euclidean length at most 2. What we have shown so far is that $0 \in u + Q$.

The second step in the proof is geometric. We claim that if a point a lies in a parallelotope Q defined by k linearly independent vectors $v_1, \dots, v_k$ as in (4.1) with $\|v_i\| \le 1$, then Q has a vertex at distance at most $\sqrt{k}$ from a. The theorem follows from this since $k \le d$.

We prove the claim by induction on k. The case $k = 1$ is trivial. In the induction step $k - 1 \to k$, we assume that a is in the interior of Q as otherwise a is on a facet of Q which is itself a parallelotope of dimension $k - 1$ and the induction works. Now put a Euclidean ball, $B(a)$, centered at a, into Q and increase its radius as long as you can with $B(a)$ still remaining in Q. The maximal $B(a)$ contained in Q has a point, say b, on the boundary of Q. Then b is contained in a face F of Q which is a $k-1$-dimensional parallelotope, whose defining vectors are of unit length at most. By induction, F has a vertex, w say, at distance at most $\sqrt{k-1}$ from b. Of course, w is a vertex of Q as well. As $B(a)$ touches F at b, $a - b$ is orthogonal to F. Further, $\|a - b\| \le 1$ as otherwise a ball of radius larger than 1 would be contained in Q which is impossible. Now

$$\|a - w\|^2 = (a - w)^2 = (a - b)^2 + (b - w)^2 \le 1 + (k - 1) = k,$$

finishing the proof. ■

The first step of the proof works for every norm but the second, more geometric step, does not. In general, in the second step one can only guarantee distance d from a vertex. A point a in a parallelotope Q of side length 2 in norm B, may be far away from all vertices of Q. The straightforward example of the ℓ_1 norm shows that every vertex of the ± 1 cube in $\mathbb{R}^d$ is at distance d from the origin.

The situation is better for the B_∞ norm, because then every point of the parallelotope Q is closer than $6\sqrt{d}$ from some of its vertices. This is a result due to Spencer [19] and, independently to Gluskin[12], who was relying on earlier work of Kashin [16]. It is an interesting fact that Spencer finds the signs by a combination of the pigeon hole principle and random methods, while Gluskin and Kashin use volume estimates and Minkowski's theorem on lattice points in 0-symmetric convex bodies.

In connection with this we mention a striking question of J. Komlós (cf. [2] or [19]). He asks whether there is a universal constant C such that for every $d \geq 1$ and for every finite $V \subset B_2^d$, there are signs $\varepsilon(v)$ for each $v \in V$ such that $\sum_{v \in V} \varepsilon(v)v \in CB_\infty^d$. The best result so far in this direction is that of Banaszczyk [2]. He showed the existence of signs such that the signed sum lies in $C(d)B_\infty^d$ where the constant $C(d)$ is of order $\sqrt{\log d}$.

5. Signing vector sequences

In this section U will denote a sequence, $u_1, u_2, \dots$ from the unit ball $B \subset \mathbb{R}^d$. This time B is symmetric, and the sequence may be finite or infinite. We wish to find signs $\varepsilon_i \in \{-1, +1\}$ such that all partial sums $\sum_1^n \varepsilon_i u_i$ are bounded by a constant depending only on B. The following result is from Bárány, Grinberg [4].

Theorem 5.1. *Under the above conditions there are signs ε_i such that for all $n \in \mathbb{N}$*

$$\sum_1^n \varepsilon_i u_i \in (2d-1)B.$$

Proof. We will only prove that all partial sums are in $2dB$. The improvement to $2d-1$ is explained in the remark after this proof.

We start again with a construction, which is the prime example of the method of "floating variables".

Define $U_k = \{u_1, u_2, \dots, u_{k+d}\}$, $k = 0, 1, 2, \dots$. We are going to construct mappings $\beta_k \colon U_k \to [-1, 1]$ and subsets $W_k \subset U_k$ with the following properties (for all k):

(i) $\sum_{U_k} \beta_k(u)u = 0$,

(ii) $\beta_k(u) \in \{-1, 1\}$ whenever $u \in W_k$,

(iii) $|W_k| = k$ and $W_k \subset W_{k+1}$ and $\beta_{k+1}(u) = \beta_k(u)$ if $u \in W_k$.

The construction is by induction on k. For $k = 0$, $W_0 = \emptyset$ and $\beta_0(\cdot) = 0$ clearly suffice. Now assume that β_k and W_k have been constructed and satisfy (i), (ii), and $|W_k| = k$ from (iii). The $d+1$ vectors in $U_{k+1} \setminus W_k$ are linearly dependent, so there are $\gamma(u) \in \mathbb{R}$ not all zero such that

$$\sum_{U_{k+1} \setminus W_k} \gamma(u)u = 0.$$

Putting $\beta_k(u_{k+d+1}) = 0$, we have

$$\sum_{W_k} \beta_k(u)u + \sum_{U_{k+1} \setminus W_k} \left(\beta_k(u) + t\gamma(u)\right) u = 0$$

for all $t \in \mathbb{R}$. For $t = 0$ all coefficients lie in $[-1, 1]$. Hence for a suitable $t = t^*$, all coefficients still belong to $[-1, 1]$, and $\beta_k(u) + t\gamma(u) \in \{-1, 1\}$ for some $u^* \in U_{k+1} \setminus W_k$. Set now $W_{k+1} = W_k \cup \{u^*\}$ and $\beta_{k+1}(u) = \beta_k(u)$, if $u \in W_k$, and $\beta_{k+1}(u) = \beta_k(u) + t^*\gamma(u)$, if $u \in U_{k+1} \setminus W_k$. Then W_{k+1} and β_{k+1} satisfy (i) and (ii) and $|W_{k+1}| = k+1$ from (iii). Moreover, $W_k \subset W_{k+1}$ and $\beta_{k+1}(u) = \beta_k(u)$ for all $u \in W_k$.

We now define the signs ε_i. Set $\varepsilon_i = 1$ if $u_i \in W_k$ and $\beta_k(u_i) = 1$ for some k, and set $\varepsilon_i = -1$ if $u_i \in W_k$ and $\beta_k(u_i) = -1$ for some k. As $\beta_k(u)$ stabilizes, that is, $\beta_k(u) = \beta_{k+1}(u) = \beta_{k+2}(u) = \dots$ once $u \in W_k$, this definition is correct for all vectors that appear in some W_k. For the remaining (at most d) vectors one can set $\varepsilon_i = \pm 1$ arbitrarily.

Again, we have to check the partial sums. The first d (actually, the first $2d$) partial sums lie automatically in $2dB$. For $n > d$ define $k = n - d$. Denoting ε_i by $\varepsilon(u_i)$ or simply by $\varepsilon(u)$ we have

$$\begin{aligned}
\sum_1^n \varepsilon_i u_i &= \sum_{u \in U_k} \varepsilon(u)u = \sum_{u \in U_k} \varepsilon(u)u - \sum_{u \in U_k} \beta_k(u)u \\
&= \sum_{u \in U_k} \left(\varepsilon(u) - \beta_k(u)\right) u = \sum_{u \in U_k \setminus W_k} \left(\varepsilon(u) - \beta_k(u)\right) u.
\end{aligned}$$

Note that the last sum has only d terms, because $U_k \setminus W_k$ has exactly d elements. We take the norm:

$$\left\| \sum_1^n \varepsilon_i u_i \right\| \le \sum_{u \in U_k \setminus W_k} \left|\varepsilon(u) - \beta_n(u)\right| \le 2d, \tag{5.1}$$

since every term in the last sum is at most 2. ∎

Remark. Where can one get $2d-1$ instead of $2d$ in this proof? Well, when choosing the suitable t^* which gives $u^* \in U_{k+1}$ the coefficient 1 or -1, we can move from $t=0$ to both positive or negative values of t, and this degree of freedom helps. Here is a sketch of how this can be done. For each $k \geq 1$ one has a *special element* $v \in U_k \setminus W_k$ with the property that $\beta_{k+1}(v) \geq \beta_k(v)$ if $\beta_k(v) > 0$ and $\beta_{k+1}(v) \leq \beta_k(v)$ if $\beta_k(v) < 0$. The special element remains the same as long as $v \notin W_k$. What can be reached this way is that $\beta_k(v)$ has the same sign as long as v is special. Then $\left|\varepsilon(v) - \beta_k(v)\right| \leq 1$ for the special element in the sum over $U_k \setminus W_k$ in equation (5.1) and this is where we get $2d-1$ instead of $2d$. When the special v enters W_k we let $v = u_{k+d}$ be the new special element, and the sign of $\beta_l(v)$ for $l > k$ is going to be the same as that of $\beta_k(v)$. There is no $\beta_l(v)$ for $l < k$ so they can't influence the validity of (5.1) for the previous indices. The choice of the first special element and the case when $\beta_k(v)$ never reaches ± 1 needs extra care which we leave to the interested reader.

The above proof gives, in fact, a good algorithm for finding a suitable sign sequence. It is an almost on-line algorithm: it does not have to foresee the whole sequence. At each moment, it only keeps a buffer of d vectors with undecided signs. In fact, the previous remark shows that a buffer of size $d-1$ suffices. But smaller buffer wouldn't do. This was proved by Peng and Yan [18].

The sign sequence constant, $E(B)$, of the unit ball B is the smallest blow-up factor for which Theorem 5.1 holds. By the same theorem, $E(B) \leq 2d-1$ always holds for every symmetric norm in $\mathbb{R}^d$. For individual norms, of course, much better estimates are possible. The lower bounds for $S(B_p)$ with $p = 1, 2, \infty$ apply also to $E(B)$. We will return to this question in connection with Chobanyan's remarkable transference theorem in Section 7.

One can set up the problem leading to Theorem 5.1 more generally. Namely, assume we are given a sequence of sets $V_1, V_2, \dots,$ with $V_i \subset B$ and $0 \in \operatorname{conv} V_i$ for each $i \in \mathbb{N}$. Can one choose vectors, $u_i \in V_i$ for each i such that each partial sum $\sum_1^n u_i$ lies in cB with a suitable constant c that depends only on the norm. The answer is yes, with $c = 2d$, and the proof is similar to the one above, see [4]. The case of Theorem 5.1 is when $V_i = \{u_i, -u_i\}$. Several other questions treated in this paper have similar generalizations.

6. Partitioning a sequence

We now formulate Theorem 5.1 in a different form, suitable for generalization. We need one more piece of notation: if $U' = \{u_{i_1}, u_{i_2} \dots\}$ is a subsequence of U with $i_1 < i_2 < \dots$, then let $\sum_n U'$ denote the sum of all u_{i_j} with $i_j \le n$. This unusual notation will be very convenient.

With this notation the statement of Theorem 5.1 is that U can be partitioned into two subsequences U^+ and U^- such that for every $n \in \mathbb{N}$

$$\sum_n U^+ - \sum_n U^- \in 2dB, \quad \text{and also} \quad \sum_n U^- - \sum_n U^+ \in 2dB.$$

The two statements here are equivalent since the norm is symmetric. Further, of course, U^+ (U^-) consists of elements of U for which $\varepsilon = +1$ ($\varepsilon = -1$).

Adding $\sum_n U$ to both sides and dividing by 2 gives

$$\sum_n U^+ \in dB + \frac{1}{2}\sum_n U \quad \text{and} \quad \sum_n U^- \in dB + \frac{1}{2}\sum_n U.$$

The new formulation of the Theorem 5.1 is this. Under the same conditions U can be partitioned into two subsequences U^1 and U^2 such that for all $j = 1, 2$ and all $n \in \mathbb{N}$

$$\sum_n U^j \in dB + \frac{1}{2}\sum_n U.$$

Can one partition U into r subsequences with similar properties? The answer is yes. The following theorem is from [5], and is an improved version of a similar result by Doerr and Srivastav [11].

Theorem 6.1. *Assume $U \subset B$ is a sequence, and $r \ge 2$ is an integer. Then U can be partitioned into r subsequences $U^1, \dots, U^r$ such that for all $j \in [r]$ and $n \in \mathbb{N}$*

$$\sum_n U^j \in CdB + \frac{1}{r}\sum_n U,$$

where C is a universal constant which is smaller than 2.0005.

Proof. We only give a sketch. We set $r = r_0 + r_1$ with $r_0, r_1 \in \mathbb{N}$ whose values will be chosen later. Then partition U into two subsequences U_0 and U_1 so that, for all $n \in \mathbb{N}$,

$$r_1 \sum_n U_0 - r_0 \sum_n U_1 \in (r_0 + r_1)dB. \tag{6.1}$$

This is accomplished by the same construction as in the proof of Theorem 5.1, only the bounding interval $[-1, 1]$ is to be changed to $[-r_0, r_1]$. Then we add $r_0 \sum_n U$ to both sides of (6.1) and divide by $r = (r_0 + r_1)$ to obtain

$$\sum_n U_0 \in dB + \frac{r_0}{r} \sum_n U.$$

The same way we have

$$\sum_n U_1 \in dB + \frac{r_1}{r} \sum_n U.$$

The proof proceeds from here recursively, by choosing $r_0 = r_{00} + r_{01}$ and splitting U_0 into subsequences U_{00} and U_{01}, just as U was split into U_0 and U_1, and then splitting U_{00} and U_{01} further.

This recursion gives rise to a recursion tree. It is a binary tree whose root is marked by r, its two children by r_0 and r_1, etc. It has to have r leaves, each marked by 1. For the jth leaf, let b_j denote the sum of the reciprocals of the numbers on the nodes from this leaf to the root (including the leaf but excluding the root). The recursion gives then a partition of U into subsequences $U^1, \ldots, U^r$ such that for all $j \in [r]$ and $n \in \mathbb{N}$ we have

$$\sum_n U^j \in b_j dB + \frac{1}{r} \sum_n U.$$

Thus the recursion tree is to be built in such a way that all b_j be small. This can be achieved, giving $b_j \leq 2.0005$ for all $j \in [r]$, see [11] and [5] for the details. ■

7. A transference theorem

The methods for the Steinitz constant $S(B)$ and the sign-sequence constant $E(B)$ are similar, and so are the bounds. Is there some deeper connection between them? This is answered by the following beautiful result of Chobanyan [10].

Theorem 7.1. *Assume B is the unit ball of symmetric norm in $\mathbb{R}^d$. Then $S(B) \le E(B)$.*

The result shows that the sign-sequence problem is "easier" than the rearrangement problem. One may wonder whether the opposite inequality, that is, $E(B) \le CS(B)$, holds with dimension independent constant C. It does hold with $C = 2d-1$ since $S(B) \ge 1$ trivially and $E(B) \le 2d-1$ by Theorem 5.1, but this is not interesting.

We mention that Theorem 7.1 holds in any normed space, not necessary finite dimensional. The proof below will show this.

Proof. Clearly both $S(B)$ and $E(B)$ are at least one. By the definition of $S(B)$, for every small $\eta > 0$ there is a is a finite $V \subset B$ with $\sum_{v \in V} v = 0$, such that V has an ordering $v_1, \ldots, v_n$ so that every partial sum along this ordering lies in $S(B)B$, but for every ordering of V there is a partial sum (along that ordering) which is outside of $\big(S(B) - \eta\big)B$.

Choose a small $\eta > 0$ together with the finite V with the above properties. Consider now the sign-sequence $\varepsilon_1, \ldots, \varepsilon_n$ such that for all $k \in [n]$,

$$\sum_1^k \varepsilon_i v_i \in E(B)B.$$

By the definition of $E(B)$ such a sign sequence exists. The vectors v_i with $\varepsilon_i = +1$, in the same order, form a sequence $u_1, u_2, \ldots, u_m$, while the vectors v_i with $\varepsilon_i = -1$, in the opposite order, form a sequence $u_{m+1}, u_{m+2}, \ldots, u_n$. The sequence $u_1, \ldots, u_n$ is a rearrangement of V. Then one partial sum, the kth say, has norm greater than $S(B) - \eta$.

Assume $k \le m$. Then, clearly,

$$\sum_1^k u_i = \frac{1}{2}\Big(\sum_1^k v_i + \sum_1^k \varepsilon_i v_i\Big) \in \frac{1}{2}\big(S(B) + E(B)\big)B.$$

This shows that $S(B)-\eta<\frac{1}{2}\big(S(B)+E(B)\big)$ implying

$$S(B) < E(B) + 2\eta. \tag{7.1}$$

Assume now that $k>m$. Then $\sum_1^k u_i = -\sum_{k+1}^n u_i$ is outside $\big(S(B)-\eta\big)B$. Consequently, $\sum_{k+1}^n u_i$ is outside $\big(S(B)-\eta\big)B$ as well. But the last sum is just the sum of the first $n-k$ elements of the sequence $v_1,\dots,v_n$ that go with $\varepsilon_i=-1$. This sum is equal to

$$\frac{1}{2}\Big(\sum_1^{n-k} v_i - \sum_1^{n-k}\varepsilon_i v_i\Big) \in \frac{1}{2}\big(S(B)+E(B)\big)B,$$

again. This proves inequality (7.1) in all cases.

Finally, since (7.1) holds for all $\eta>0$, we have $S(B)\le E(B)$. ■

References

[1] L. Babai, P. Frankl, *Linear Algebra Methods in Combinatorics,* Preliminary version 2. Department of Computer Science, University of Chicago, 1992 (book, 216 pp).

[2] W. Banaszczyk, Balancing vectors and Gaussian measures of n-dimensional convex bodies, *Random Structures and Alg.,* **12** (1998), 315–360.

[3] I. Bárány, A vector-sum theorem and its application to improving flow shop guarantees, *Math. Oper. Res.,* **6** (1981), 445–452.

[4] I. Bárány, V. S. Grinberg, On some combinatorial questions in finite dimensional spaces, *Linear Alg. Appl.,* **41** (1981), 1–9.

[5] I. Bárány, B. Doerr, Balanced partitions of vector sequences, *Linear Alg. Appl.,* **414** (2006), 464–469.

[6] J. Beck, T. Fiala, Roth's estimate of the discrepancy of integer sequences is nearly sharp, *Combinatorica,* **1** (1981), 319–325.

[7] J. Beck, T. Fiala, "Integer-making" theorems, *Discrete Appl. Math.,* **3** (1981), 1–6.

[8] J. Beck, V. T. Sós, Discrepancy theory, in: *Handbook of combinatorics* (ed. R. Graham, M. Grötschel, L. Lovász), Elsevier, Amsterdam, 1995, 1405–1446.

[9] V. Bergström, Zwei Sätze über ebene Vectorpolygone, *Abh. Math. Sem. Univ. Hamburg,* **8** (1931), 148–152.

[10] S. Chobanyan, Convergence a.s. of rearranged random series in Banach space and associated inequalities, in: *Probability in Banach spaces, 9: Proceedings from the 9th international conference on probability in Banach spaces* (ed.: J. Hoffmann-Jorgensen et al.), Prog. Probab. (Birkhäuser, Boston), **35**, 3–29 (1994).

[11] B. Doerr, A. Srivastav, Multicolour discrepancies, *Combinatorics, Prob. Comp.*, **12** (2003), 365–399.

[12] Gluskin, E. D., Extremal properties of orthogonal parallelepipeds and their applications to the geometry of Banach spaces, *Mat. Sbornik,* **136** (1988), 85–96. (in Russian)

[13] V. S. Grinberg, S. V. Sevastyanov, The value of the Steinitz constant, *Funk. Anal. Prilozh.,* **14**(1980), 56–57. (in Russian)

[14] Halperin, I., Ando, T., Bibliography: Series of vectors and Riemann sums, Sapporo, Japan, 1989.

[15] M. I. Kadets, On a property of broken lines in n-dimensional space, *Uspekhi Mat. Nauk,* **8**(1953), 139–143. (in Russian)

[16] B. S. Kashin, On some isometric operators in $L^2(0,1)$, *Comptes Rendus Acad. Bulg. Sciences,* **38**(1985), 1613–1615.

[17] P. Lévy, Sur les séries semi-convergentes, *Nouv. Ann. de Math.,* **64** (1905), 506–511.

[18] H. Peng, C. H. Yan, Balancing game with a buffer, *Adv. in Appl. Math.,* **21** (1998), 193–204.

[19] J. Spencer, Six standard deviations suffice, *Transactions AMS,* **289** (1985), 679–706.

[20] E. Steinitz, Bedingt konvergente Reihen und konvexe Systeme, *J. Reine Ang. Mathematik,* **143** (1913), 128–175, ibid, **144** (1914), 1–40, ibid, **146** (1916), 1–52.

[21] S. V. Sevastyanov, On approximate solutions of a scheduling problem, *Metody Diskr. Analiza,* **32**(1978), 66–75. (in Russian)

[22] S. V. Sevastyanov, On some geometric methods in scheduling theory: a survey, *Discrete Appl. Math.,* **55**(1994), 59–82.

Imre Bárány

Rényi Institute of Mathematics
Hungarian Academy of Sciences
PO Box 127, 1364 Budapest
Hungary

e-mail: barany@renyi.hu

and

Department of Mathematics
University College London
Gower Street, London WC1E 6BT
England

BOLYAI SOCIETY
MATHEMATICAL STUDIES, 19

Building Bridges
pp. 47–102.

Surplus of Graphs and the Lovász Local Lemma

JÓZSEF BECK

1. What is the Surplus?

1. Row-Column Games. The *Surplus* is a game-theoretic graph parameter. To illustrate it in a special case, suppose that two players, called Maker and Breaker, are playing on an $n \times n$ chessboard, and alternately mark previously unmarked little squares. Maker uses (say) mark X and Breaker uses (say) O, exactly like in Tic-Tac-Toe; Maker's goal is to achieve a large *lead* in some line, where a "line" means either a row or a column. Let $\frac{n}{2} + \Delta$ denote the maximum number of Xs ("Maker's mark") in some line at the end of a play; then the difference $\left(\frac{n}{2} + \Delta\right) - \left(\frac{n}{2} - \Delta\right) = 2\Delta$ is Maker's *lead*; Maker wants to maximize $\Delta = \Delta(n)$. Since $\Delta = \Delta(n)$ can be a half-integer (it happens when n is odd), and it is customary to work with integral graph parameters (like chromatic number), I prefer to call $2\Delta = 2\Delta(n)$ the **Surplus** of the $n \times n$ board (and refer to $\Delta = \Delta(n)$ as the half-surplus). That is, the Surplus is the maximum terminal lead that Maker can always achieve against a perfect opponent. (In other words, Surplus is a game-theoretic one-sided discrepancy concept.)

A closely related concept is the Maximum Temporary Lead, which is the largest lead that Maker can always achieve at some instant of the whole course of the play against a perfect opponent. We have the following inequality:

$$\text{Maximum Terminal Lead} \le \text{Maximum Temporary Lead} \le$$
$$\le \text{Maximum Terminal Lead} + 1.$$

The left-hand side inequality is trivial; the right-hand side inequality is almost trivial, since Maker can save his temporary lead by filling up the large-lead line; there is a possible loss of 1 at the end (by parity reason).

Of course, one can replace the $n \times n$ square board with any other $n \times m$ rectangle board (say, $n \leq m$). One can further generalize by allowing an arbitrary number of "holes" in the board: assume that some of the nm little squares of an $n \times m$ rectangle board are "holes", meaning that the players are forbidden to mark "holes"; "holes" are not legitimate moves.

The Row-Column Game played on a rectangle board with holes is equivalent to the following *Degree Game on Bipartite Graphs.* Let G be an arbitrary finite bipartite graph; it is natural to assume that G is simple (no multiple edges and no loops). The Degree Game on G is played by two players, Maker and Breaker, who alternately take previously untaken edges of G, and color them: Maker uses red and Breaker uses blue. At the end of a play Maker owns half of the edges (the red edges), so Maker's graph (the red subgraph) must have degree $m_i \geq d_i/2$ in some vertex i (where d_i is the G-degree of i). Let $\mathrm{Sur}\,(G)$ be the largest integer S such that Maker can always force a red degree $m_i \geq (d_i + S)/2$, where d_i is the G-degree. Formally,

$$\mathrm{Sur}\,(G) = \max_{\mathrm{Str}_M} \min_{\mathrm{Str}_B} \max_i (m_i - b_i), \tag{1.1}$$

meaning that, $\mathrm{Sur}\,(G)$ is the largest integer S such that, playing the Degree Game on G, Maker can always force a terminal lead $\geq S$. That is, Maker has a strategy Str_M with the property that, whatever strategy Str_B is used by Breaker, at the end of the $(\mathrm{Str}_M, \mathrm{Str}_B)$-play there is always a vertex i where Maker's degree m_i is $\geq b_i + S$ (here b_i is Breaker's degree in i).

The last step of generalization is to drop the assumption that G is bipartite: definition (1.1) does make perfect sense for any finite graph G. It is natural to assume that G is simple, that is, no multiple edges and no loops. Of course, the surplus of the empty graph is zero.

If G is regular (i.e., every degree is the same d) then Maker's goal is simply to build a large degree, substantially larger than $d/2$.

Let's return to the special case of the Row-Column Game on an $n \times n$ board, which is equivalent to the Degree Game on the complete symmetric bipartite graph $K_{n,n}$. The family of $2n$ lines (n rows and n columns) forms a particularly simple hypergraph: it is

(1) Almost Disjoint (any two lines intersect in at most one cell); and

(2) has uniform Height 2, i.e., every cell belongs to exactly 2 lines.

Note that I reserve the term Degree for graphs; the analog concept in hypergraphs is called Height throughout this paper. Since Height 1 means a disjoint hypergraph, uniform Height 2 is the first nontrivial case – a Height 2 hypergraph is extremely sparse.

Switching from the $n \times n$ board (i.e. $K_{n,n}$) to an arbitrary finite and simple graph G, the "lines" become the "stars". The star-hypergraph of any graph G also has properties (1)–(2): it is Almost Disjoint (because any two stars have at most one common edge) and has uniform Height 2 (because every edge belongs to two stars).

There is an easy converse: if a finite hypergraph $\mathcal{H}$ is

(1) Almost Disjoint, and

(2) has uniform Height 2,

then $\mathcal{H}$ is the star-hypergraph of a simple graph G. Indeed, let $A_1, A_2, \ldots, A_n$ be the hyperedges of $\mathcal{H}$. To construct the desired graph G, we associate with each A_i a vertex v_i, and two different vertices v_i and v_j are joined by an edge if and only if A_i intersects A_j. Now let S_i be the star of v_i in graph G; then

$$v_j \in S_i \Leftrightarrow v_i v_j \in G \Leftrightarrow A_i \cap A_j \neq \emptyset,$$

and the G-degree of v_i is

$$\big|\{j\colon v_j \in S_i\}\big| = \big|\{j\colon A_i \cap A_j \neq \emptyset\}\big| = |A_i|,$$

where in the last step we used both properties (1)–(2). Since the G-degree of v_i is exactly $|A_i|$, we obtain that the star-hypergraph of G is isomorphic to $\mathcal{H}$.

Any kind of "achieve majority" type game played on a disjoint hypergraph is trivial. The star-hypergraph of a graph represents the simplest nontrivial hypergraph, and as the reader will find out below, this case is already very challenging and interesting.

The Surplus is a natural "majority" type game-theoretic graph parameter. At first glance it may seem as an easy parameter, but a closer look at (1.1) explains the main difficulty: there are a huge number of strategies even for a small graph, which means we cannot carry out exhaustive computer experimentation! The surplus is far from easy, we don't know the exact value of the surplus even for the complete graph!

Our main result is that every dense graph has a (relatively) large Surplus, see Theorem 1 below. This sounds like a simple result, but I don't have a simple proof. The Almost Disjointness of the star-hypergraph (of a graph) plays a key role in the proof. To emphasize this point I show a large class of hypergraphs such that (1) the hypergraphs are very far from being Almost Disjoint and (2) the Surplus is trivially zero. This is the class of *Strictly-Even Hypergraphs.* The board V of a Strictly-Even Hypergraph is an even-size set, say a $2M$-element set, representing the inhabitants of a little town: M married couples, M husbands and M wives. The citizens of this little town have a habit of forming clubs, small and large. The same citizen may have membership in many different clubs at the same time, but there is a rule which is strictly enforced: if a citizen is the member of a club, then his/her spouse is automatically a member, too. Each club represents a hyperedge of a Strictly-Even Hypergraph (and vice versa). In technical terms, a Strictly-Even Hypergraph has an underlying "pairing", and if a hyperedge intersects a "pair", then the hyperedge must contain the whole "pair".

Note that in a Strictly-Even Hypergraph every hyperedge ("winning set") has even size, and in general, the intersection of an arbitrary family of hyperedges has even size, too (explaining the name "strictly-even").

The Maker–Breaker majority achieving game played on an arbitrary Strictly-Even Hypergraph is trivial: Maker cannot even achieve a surplus of one! Indeed, by using a pairing strategy, from each hyperedge Breaker can take the exact half.

The main point here is that Almost Disjoint Hypergraphs are very different from Strictly-Even Hypergraphs; in some sense they are opposite classes. This is the intuitive reason why we can prove large surplus for graphs.

In the rest of the paper "graph" always means a finite and simple graph; also it is always assumed that Maker is the first player (unless I specifically say otherwise).

2. Exact results. It is easy to determine the Surplus of some sparse graphs such as

(1) a cycle C_n of length n has Surplus 2;

(2) a path P_n of length n has Surplus 2 if $n \geq 3$, and has Surplus 1 if $n = 1$ or 2.

There are a few more classes of graphs for which we know the exact value of the Surplus:

(3) all trees;

(4) all 4-regular graphs;

(5) very asymmetric complete bipartite graphs: $K_{d,n}$ with $n > (d+2)2^{d-1}$.

These are less trivial, but still rather easy results. I put them in

Proposition 1.1. (a) *The Surplus of a (nonempty) graph is always* ≥ 1.

(b) *Every tree has Surplus* 1 *or* 2. *The Surplus is* 1 *if and only if the number of even degrees of the tree is* 0 *or* 1.

(c) *Every 4-regular graph has Surplus* 2.

(d) $\mathrm{Sur}\,(K_{d,n}) = d$ *holds for all asymmetric complete bipartite graphs* $K_{d,n}$ *with* $n > (d+2) \cdot 2^{d-1}$.

Proof. By definition (see (1.1)) the surplus is an integer ≥ 0. To prove (a) note that if at least one degree is odd then in that vertex Maker can force a surplus ≥ 1, so trivially $\mathrm{Sur}\,(G) \geq 1$.

Next let G be a connected graph where every degree is even. Note that if the degrees of a graph G have the same parity then the surplus $\mathrm{Sur}\,(G)$ also has the same parity; in particular, if every degree is even then the surplus is also even. Thus after making an arbitrary opening move, in one of the two endpoints of his opening edge Maker can always force a surplus ≥ 2.

(b) First I show that the surplus $\mathrm{Sur}\,(T)$ of a tree T is always ≤ 2. Fix an arbitrary vertex v of T; starting from v (the root) there is a unique orientation $T(\rightarrow)$ of the edges of the tree such that every vertex has in-degree 1 (except the root itself). Here is Breaker's strategy to force $\mathrm{Sur}\,(T) \leq 2$: suppose Maker just took edge e; this edge is an arrow in $T(\rightarrow)$, and let w be its tail, then Breaker's next move is to take an arbitrary unselected edge from vertex w. This way Breaker perfectly balances the out-degree, so the largest surplus is $\leq 1+1=2$, where the first "1" comes from the out-degree (possible parity loss of 1) and the second "1" is the common in-degree in $T(\rightarrow)$.

If a tree T has exactly one even degree vertex, say, v, then by choosing v to be the root, the argument above gives $\mathrm{Sur}\,(T) = 1$.

If every degree of a tree T is odd then the surplus is also odd, and $\mathrm{Sur}\,(T) \leq 2$ implies $\mathrm{Sur}\,(T) = 1$.

If a tree T has ≥ 2 even degree vertices, then I prove $\mathrm{Sur}\,(T) = 2$. Let u and w be two even degree vertices with the extra property that their T-distance is the minimum (on every tree there is a unique well-defined distance). If u and w are neighbors (i.e., the uw-edge is in T) then Maker's first move is the uw-edge, and either in u or in w he can always force a surplus ≥ 2.

If u and w are not neighbors, then let $u, v_1, \ldots, v_k, w$ be the (unique) T-path joining u and w. By the minimum property $v_1, \ldots, v_k$ all have odd degrees; clearly each one is ≥ 3. Maker's first move is the uv_1-edge, then Breaker has to respond in vertex u (otherwise Maker can force a surplus ≥ 2 in u). Maker's second move is the v_1v_2-edge, then Breaker has to respond in vertex v_1 (otherwise Maker can force a surplus ≥ 3 in v_1); $\ldots$; Maker's kth move is the $v_{k-1}v_k$-edge, then Breaker has to respond in vertex v_{k-1} (otherwise Maker can force a surplus ≥ 3 in v_{k-1}); finally, Maker takes the v_kw-edge, which is a trap: Breaker has to respond in both v_k (to prevent a surplus ≥ 3 in v_k) and w (to prevent a surplus ≥ 2 in w), which is impossible. This proves part (b).

Next I prove (c). If G is a 4-regular connected graph then every degree is even, and G contains an Euler trail. Any fixed Euler trail defines an orientation $G(\rightarrow)$ of the edges of G such that every in-degree is 2, and similarly, every out-degree is 2. Breaker's strategy to force $\text{Sur}\,(G) = 2$ is in fact a Pairing Strategy using orientation $G(\rightarrow)$: when Maker just took edge e then Breaker finds the orientation of e in $G(\rightarrow)$, and takes the other out-edge of $G(\rightarrow)$ from the tail of e. This way Breaker has at least one edge in each star, forcing $\text{Sur}\,(G) \leq 3 - 1 = 2$. By parity reason $\text{Sur}\,(G) \geq 1$ has to be even, so we have only one option: $\text{Sur}\,(G) = 2$. The result remains true for non-connected G; I leave the trivial modification in the argument to the reader.

Finally, I prove (d). I go back to the more geometric Row-Column Game representation: the board is an extremely long rectangle with d rows and n columns where $n > (d+2)2^{d-1}$. First I note that the upper bound $\text{Sur}\,(K_{d,n}) \leq d$ is trivial: Breaker just plays "row-wise", that is, Breaker always responds in the same row where Maker's last move was. The lower bound $\text{Sur}\,(K_{d,n}) \geq d$ is less trivial: it is based on a simple Halving Argument.

Initially Maker keeps playing in the first row until it is all filled. At this point, if Maker's surplus in the (long) first row is $\geq d$, then of course Maker is already done. Otherwise, there are at least $(n-d)/2$ columns that are Breaker-free and contain exactly one mark of Maker at the top of each; these $\geq (n-d)/2$ columns are the only ones that Maker cares about in the rest of the play.

For the second row, Maker keeps on playing only in these $\geq (n-d)/2$ columns, and he continues on afterward until this "relevant" part of the second row is all filled. At this point – which is the end of the secound round – if Maker did not get the surplus d in the second row, then out of the $\geq (n-d)/2$ columns of the first round, there are

$$\geq \frac{\frac{n-d}{2}-\frac{d}{2}}{2}=\frac{n-2d}{4}$$

columns that are Breaker-free and contain exactly two marks of Maker at the top of each; these $\geq (n-2d)/4$ columns are the only ones that Maker cares about in the rest of the play.

Proceeding by induction, at the end of the ith round, if Maker did not get the required surplus d in some row yet, there are

$$\geq \frac{\frac{n-2^{i-2}d}{2^{i-1}}-\frac{d}{2}}{2}=\frac{n-2^{i-1}d}{2^i}$$

columns that are Breaker-free and contain exactly i marks of Maker at the top of each. Hence Maker can guarantee a surplus d if

$$\frac{n-2^{d-1}d}{2^d}>0,$$

which is equivalent to $n>(d+2)\cdot 2^{d-1}$. This completes the proof of Proposition 1.1. ■

Notice that in Propostion 1.1 the surplus is either bounded (≤ 2), or it is logarithmically small in terms of the number of vertices, see part (d). Are there graphs with relatively large surplus, larger than logarithmic? What happens for an arbitrary graph? I will prove that every dense graph has a relatively large surplus, much larger than logarithmic. Roughly speaking, the surplus is larger than the square root of the maximum local density of the graph.

3. The Core-Density and the Surplus. The Core-Density is simply the maximum local density of a nonempty graph G. Let C_G be a *densest* subgraph of G, that is, $C_G \subseteq G$ is a subgraph for which the Edge/Vertex ratio attains its maximum. We call $C_G(\subseteq G)$ a *Core* of graph G. Note that the Core itself is not necessarily uniquely defined (a graph may have several Cores), but the maximum Edge/Vertex ratio is uniquely defined. The density of a Core is exactly what I call the **Core-Density;** I denote it with $\mathrm{cd}(G)$. (If G is the empty graph then $\mathrm{cd}(G) = 0$.)

I call the Average Degree of a Core C_G the Core-Degree; of course it is $2\,\mathrm{cd}(G)$, that is, the Core-Degree is twice as large as the Core-Density. Here are a couple of easy facts.

If G is a connected d-regular graph then its Core is itself, the Core-Degree is d, and so the Core-Density is $\mathrm{cd}(G) = d/2$.

If T is a tree then again its Core is itself, and $\mathrm{cd}(T) < 1$.

Here is another easy fact about the Core that we are going to use repeatedly.

Proposition 1.2. *If $C_G \subseteq G$ is any Core of graph G, then the minimum C_G-degree is $\geq \mathrm{cd}(G)$.*

Proof. Let v be an arbitrary vertex of C_G with C_G-degree d. Removing v from C_G gives the new Edge/Vertex ratio $(E - d)/(V - 1)$, where E is the number of edges and V is the number of vertices of C_G. By the maximum density property of the Core C_G:

$$\frac{E-d}{V-1} \leq \frac{E}{V} = \mathrm{cd}(G),$$

which implies $d \geq \mathrm{cd}(G)$. (Here I assumed that $\mathrm{cd}(G) > 1/2$; otherwise the statement is trivial.) ■

It is easy to give an upper bound on the Surplus in terms of the Core-Degree$= 2\,\mathrm{cd}(G)$; we simply generalize the argument of the proof of Proposition 1.1(b). This argument gives that for every graph G the surplus

$$\mathrm{Sur}(G) \leq 1 + 2\,\mathrm{cd}(G) \tag{1.2}$$

To show (1.2) we basically repeat the well-known proof of the fact that every planar graph has chromatic number ≤ 6. Let v_1 be a minimum degree vertex of G; the G-degree of v_1 is clearly $\leq 2\,\mathrm{cd}(G)$ (the average degree of the densest subgraph). By removing v_1 from G (of course we remove all edges from v_1) we obtain a subgraph G_1. Let v_2 be a minimum degree vertex of G_1; the G_1-degree of v_2 is clearly $\leq 2\,\mathrm{cd}(G)$ (the average degree of the densest subgraph). By removing v_2 from G_2 (of course we remove all edges from v_2 in G_1) we obtain a new subgraph G_2, and so on. At the end of this process we obtain a permutation $v_1, v_2, v_3, \ldots, v_n$ of the vertices of G such that, with $G_i = G \setminus \{v_1, \ldots, v_i\}$, v_{i+1} has G_i-degree $\le 2\,\mathrm{cd}\,(G)$ (the average degree of the densest subgraph). Now Breaker applies the following strategy: if Maker just took the edge $v_i v_j$ with $1 \le i < j \le n$, then Breaker takes an arbitrary unoccupied edge of the type $v_k v_j$ where $1 \le k < j \le n$ (v_j is a common vertex!); if there is no such available edge then Breaker makes an arbitrary move. To show why this strategy guarantees that $\mathrm{Sur}\,(G) \le 1 + 2\,\mathrm{cd}\,(G)$, consider an arbitrary vertex v_j; there are 2 types of edges from v_j: (1) $v_i v_j$-edges with $1 \le i < j \le n$, and (2) $v_j v_l$-edges with $1 \le j < l \le n$. There are at most $2\,\mathrm{cd}\,(G)$ second type edges, and Breaker has no control over these edges; in the first type Breaker's strategy guarantees a perfect balance, or possible a deficit ≤ 1 (due to parity reasons); altogether the surplus is $\le 1 + 2\,\mathrm{cd}\,(G)$.

What Breaker uses to enforce (1.2) is an Orientation Strategy. Next I show a nice trick that improves (1.2) by about a factor of two – I learned this trick from Benjamin Sudakov. The new idea is to combine the Orientation Strategy with the following "Orientation Lemma". (Is it folklore? Sudakov doesn't remember the source of the lemma.)

Proposition 1.3 (Orientation Lemma). *Let G be an arbitrary graph; then there is an orientation of the edges of G such that the in-degree of every vertex is at most $\lceil \mathrm{cd}\,(G) \rceil$, the upper integral part of the Core-Density.*

Proof. We apply Hall's theorem ("Marriage Lemma") to a particular bipartite graph B that I am going to construct as follows. As usual we call the two vertex classes "girls" and "boys". The "girls" are the edges of G; the "boys" are the vertices of G in several copies: each one in $k = \lceil \mathrm{cd}\,(G) \rceil$ copies. A "boy" and a "girl" are joined by an edge in our bipartite graph B if and only if the vertex, representing the "boy", is an endpoint of the edge representing the "girl". The choice of multiplicity $k = \lceil \mathrm{cd}\,(G) \rceil$ guarantees that there is no "boy-shortage": each subset of girls have enough boys to choose from. Formally: if $H \subseteq G$ is an induced subgraph, the $E(H)$ edges, representing a subset of girls, altogether have $k \cdot V(H)$ endpoints, representing $k \cdot V(H)$ boys (to choose from). Since

$$k = \lceil \mathrm{cd}\,(G) \rceil \ge \mathrm{cd}\,(G) = \max_{H \subseteq G} \frac{E(H)}{V(H)},$$

the inequality $E(H) \le k \cdot V(H)$ is trivial, that is, there is no "boy-shortage". Thus Hall's theorem applies, and gives a perfect matching in bipartite graph B. The perfect matching supplies the choice of "heads" in the desired orientation of G (each edge of G becomes an "arrow" with a "head" and a "tail"). ■

If Breaker uses the orientation in Proposition 1.3, the corresponding Orientation Strategy yields the following improvement of (1.2):

$$\text{Sur}\,(G) \leq 1 + \lceil \text{cd}\,(G) \rceil . \tag{1.3}$$

Indeed, we just repeat the end of the proof of (1.2): by using the Orientation Strategy Breaker forces a perfect balance in the out-degree of every vertex – with a possible loss of 1 by parity reason – and has no control over the in-degree, which is $\leq \lceil \text{cd}\,(G) \rceil$, so altogether the surplus is $\leq 1 + \lceil \text{cd}\,(G) \rceil$, proving (1.3).

The first main result about the surplus is

Theorem 1. *For every graph G*

$$c_0 \sqrt{\text{cd}\,(G)} \leq \text{Sur}\,(G) \leq 1 + \lceil \text{cd}\,(G) \rceil \tag{1.4}$$

with some positive absolute constant c_0; for example, $c_0 = 10^{-3}$ is a good choice.

4. Remarks. The upper bound in (1.4) is simply a restating of (1.3). Theorem 1 states, very roughly speaking, that the Surplus is "around" the Core-Density. Since the Core-Density is exactly the maximum local density of a graph, Theorem 1 clarifies the vague statement that "dense graphs have large surplus".

It is worth mentioning that the Core-Density (and the Core-Degree) is almost identical to some other well-known graph-theoretic concepts such as

(1) the Arboricity, and

(2) the Greedy Coloring Number, or Degeneracy.

The Arboricity of graph G (here G can be any multigraph) is the forest-partition number, that is, the minimum number of forests (set of disjoint trees) forming a partition of G. The connection between the Surplus and the Arboricity is clear from the fact that every forest has Surplus ≤ 2 (see Proposition 1.1(b))

If a multigraph G can be partitioned into k forests then, of course, the inequality

$$\frac{|G(U)|}{|U|-1} \leq k \tag{1.5}$$

holds for every (at least 2-element) subset U of the vertex-set of G; here $G(U)$ denotes the induced subgraph (the set of edges of G with both endpoints in U). A well-known theorem of Nash-Williams [7] demonstrates the perfect converse: if inequality (1.5) holds for every (at least 2-element) vertex-set U, then multigraph G can be partitioned into $\le k$ forests. It follows that

$$\text{Arboricity}\,(G) = \max_{H \subseteq G} \left\lceil \frac{E(H)}{V(H)-1} \right\rceil \tag{1.6}$$

(upper integral part) where $E(H)$ and $V(H)$ are, respectively, the edge-number and the vertex-number of the nonempty subgraph $H \subseteq G$.

The other maximum local density type concept, the Greedy Coloring Number, comes from the simplest greedy way of finding a proper vertex-coloring of a graph G. If G is d-regular then the chromatic number of G is trivially $\le 1 + d$. If G is not regular then we have the analog inequality

$$\text{Chromatic Number}\,(G) \le 1 + \max_{H \subseteq G} d_{\min}(H), \tag{1.7}$$

where $d_{\min}(H)$ denotes the minimum degree of subgraph H. We call $\max_{H \subseteq G} d_{\min}(H)$ the Degeneracy of graph G, and the one larger value $1 + \max_{H \subseteq G} d_{\min}(H)$ is usually called the Greedy Coloring Number of graph G. (Note that the left side and the right side in (1.7) can be very far from each other; for example, if $G = K_{n,n}$ then the left side is 2 and the right side is n.)

Trivially $\max_{H \subseteq G} d_{\min}(H) \le 2\,\text{cd}\,(G)$ =Core-Degree; on the other hand, by Proposition 1.2, the minimum degree of a Core C_G of G is $\ge \text{cd}\,(G)$. It follows that

$$\text{cd}\,(G) \le \text{Degeneracy}\,(G) = \max_{H \subseteq G} d_{\min}(H) \le 2\,\text{cd}\,(G). \tag{1.8}$$

Another simple inequality comes from (1.6) and the definition of the Core-Density:

$$\text{cd}\,(G) = \frac{E(C_G)}{V(C_G)} = \max_{H \subseteq G} \frac{E(H)}{V(H)} < \tag{1.9}$$

$$< \max_{H \subseteq G} \left\lceil \frac{E(H)}{V(H)-1} \right\rceil = \text{Arboricity}\,(G) \le$$

$$\leq \max_{H \subseteq G} \frac{2E(H)}{V(H)} = 2\,\mathrm{cd}\,(G).$$

(The last inequality in (1.9) fails for $E = E(H) = 1$ and $V = V(H) = 3$, but we can clearly assume that H is connected, implying $E \geq V - 1 \geq 1$, and then the last inequality holds.) Comparing (1.8)–(1.9) we can justly say that the five concepts: (1) the Arboricity, (2) the Greedy Coloring Number, (3) the Degeneracy, (4) the Core-Degree, and (5) the Core-Density are basically the same (differ by a factor ≤ 2). They describe the very same property – the maximum local density – of a graph in slightly different terms.

It is very interesting to know that the Core-Density also shows up in the theory of Random Graphs as a *critical exponent.* Let $\mathbf{R}(K_n, p)$ denote the Random Graph on n vertices with edge-inclusion probability $0 < p < 1$. Note that a Random Graph $\mathbf{R}(K_n, p)$ – a random variable really – has about $\binom{n}{2}p$ edges. Now let G be an arbitrary fixed "goal graph"; we study the event $\{G \subset \mathbf{R}(K_n, p)\}$, meaning that "$G$ is a subgraph of the Random Graph $\mathbf{R}(K_n, p)$", as $n \to \infty$ and p goes from 0 to 1 ("evolution of the Random Graph"). A classic theorem of Erdős and Rényi [5] states that the threshold probability for the event $\{G \subset \mathbf{R}(K_n, p)\}$ is

$$p \approx n^{-1/\mathrm{cd}\,(G)},$$

as $n \to \infty$, explaining the term "critical exponent".

5. Regular graphs – local randomness. The special case of regular graphs is particularly interesting for a couple of reasons: (1) for regular graphs the Core-Degree equals the common degree, and (2) the most interesting and natural special case $G = K_{d,d}$ – which corresponds to the Row-Column Game on a square board – is regular.

If G is d-regular then Theorem 1 gives the following lower bound for the Surplus: $\mathrm{Sur}\,(G) \geq c_0\sqrt{d}$ where $c_0 = 10^{-3}$. In the special case of regular graphs the bad constant factor c_0 can be substantially improved. Our second main result is

Theorem 2. *If G is a d-regular graph then we have the lower bound* $\mathrm{Sur}\,(G) \geq \frac{2}{15}\sqrt{d}$.

If $d \geq 200$ then we have the upper bound

$$\mathrm{Sur}\,(G) \leq 4\sqrt{d} \cdot (\log d)^2, \tag{1.10}$$

assuming the vertex-number n of G is less than the gigantic upper bound

$$n \le 2^{2^{2^{\cdot^{\cdot^{\cdot^{2^d}}}}}} = \text{Giant}\,(d), \tag{1.11}$$

where the height of the tower in (1.11) *is* $1 + \log d$ *(natural logarithm): the tower consists of* $\log d$ *2s and an extra* d *on the top.*

The main point here is that the lower bound and the upper bound are both around $\sqrt{d}$, assuming $n \le \text{Giant}\,(d)$. This $\sqrt{d}$ – a square root law – suggests "local randomness"; I will return to this vague intuition below.

Notice that the function Giant (d), defined on the right side of (1.11), deserves to be called gigantic: Giant (d) asymptotically beats

$$2^d \quad \text{and} \quad 2^{2^d} \quad \text{and} \quad 2^{2^{2^d}} \quad \text{and} \quad 2^{2^{2^{2^d}}} \quad \text{and}$$

all other towers of fixed height as $d \to \infty$.

But, whatever gigantic the function Giant (d) is, Theorem 2 still has an obvious handicap: the upper bound *does* depend, though extremely weakly, on the global size (i.e., the vertex-number n). What is the best upper bound that does not depend on the global size n and depends only on the local size d (i.e. the degree)? Of course, we have the general upper bound

$$\text{Sur}\,(G) \le 1 + \lceil \text{cd}\,(G) \rceil = 1 + \lceil d/2 \rceil \tag{1.12}$$

from (1.3). Note that (1.12) was discovered by Tibor Szabó a few years ago, in fact in the slightly stronger form $\text{Sur}\,(G) \le d/2$ if d is divisible by 4 (by using the Euler trail argument); he also raised the following question: Can one improve the upper bound $d/2$ for large d? Of course $d/2$ is best possible for $d = 4$, but how about large values of d? Unfortunately there is no progress in this very attractive problem.

Theorem 2 strongly suggests the possibility of a tremendous improvement, at least asymptotically. I conjecture that $d/2$ can be improved to something like $10\sqrt{d} \cdot \log d$ or $10\sqrt{d} \cdot (\log d)^2$. (But I don't know how to get rid of the annoying global condition vertex-number $\le$ Giant (d) in (1.11).)

6. Can we use the Lovász Local Lemma in games? In view of (1.12) the best known general upper bound on the Surplus of d-regular graphs is $d/2$, and it is based on an Euler trail argument. Breaker's strategy is to always reply in the (say) tail of Maker's last move (the Euler trail defines an orientation of the edges, so each edge becomes an "arrow" with a "tail" and a "head"). Perhaps a random choice between the two endpoints of Maker's last move is more efficient than using a fixed Euler trail. This vague idea motivates the following

Probabilistic Intuition. In each turn Breaker tosses a fair coin, and the outcome determines the endpoint (of Maker's last move) where Breaker is going to respond: Breaker's next move is to take a new edge from the chosen endpoint. Let's pick an arbitrary vertex v in the d-regular graph G. If Maker owns x edges from v, then by the law of large numbers in probability theory, roughly half of the time in the x random choices endpoint v should come up, so Breaker must have at least $\big(1-o(1)\big)\,x/2$ edges from v. Since $x+\big(1-o(1)\big)\,x/2 \le d$, we conclude that $x \le \big(1+o(1)\big)\,2d/3$, that is, the surplus in v is at most $\big(1+o(1)\big)\,2d/3 - \big(1-o(1)\big)\,d/3 = \big(1+o(1)\big)\,d/3$. This vague intuition, due to Tibor Szabó, may suggest the improvement $\big(1+o(1)\big)\,d/3$ from $d/2$. It is a natural idea to try to save this argument by involving the well-known Lovász Local Lemma (a sophisticated tool of the Probabilistic Method).

The Lovász Local Lemma (see Erdős–Lovász [4]) is a remarkable probabilistic sieve argument, which is usually applied to prove the *existence* of certain very complicated structures that we are unable to construct directly. To be precise, let $E_1, E_2, \dots, E_s$ denote events in a probability space. In the applications, the E_is are "bad" events, and we want to avoid all of them, that is, we wish to show that $\text{Prob}\,(\cup_{i=1}^{s} E_i) < 1$. A trivial way to guarantee this is to assume $\sum_{i=1}^{s} \text{Prob}\,(E_i) < 1$. A completely different way to guarantee $\text{Prob}\,(\cup_{i=1}^{s} E_i) < 1$ is to assume that $E_1, E_2, \dots, E_s$ are mutually independent and all $\text{Prob}\,(E_i) < 1$. Indeed, we then have $\text{Prob}\,(\cup_{i=1}^{s} E_i) = 1 - \prod_{i=1}^{s}\big(1-\text{Prob}\,(E_i)\big) < 1$. The Lovász Local Lemma applies in the very important case when we don't have mutual independence, but "independence dominates" in the sense that each event is independent of all but a small number of other events.

Lovász Local Lemma. *Let $E_1, E_2, \dots, E_s$ be events in a probability space. If* $\text{Prob}\,(E_i) \le p < 1$ *holds uniformly for all i, and each event is* independent *of all but at most $\frac{1}{4p}$ other events, then* $\text{Prob}\,(\cup_{i=1}^{s} E_i) < 1$.

Every attempt we made along these lines – trying to use the Lovász Local Lemma in games – failed so far. Can the reader turn the Probabilistic Intuition into a precise proof? Can the reader involve the Lovász Local Lemma and improve the upper bound $d/2$ in (1.12) to something better?

7. When the Lovász Local Lemma successfully predicts the truth: examples in Tic-Tac-Toe like games. Perhaps the best example is the (K_n, K_q) Clique Achievement Game and its Avoidance version, that I often call the Reverse Game. In both games the players alternately take new edges of the complete graph K_n; in the Achievement version the players are called Maker and Breaker, Maker's goal is to occupy a large clique K_q; Breaker simply wants to stop Maker (i.e., Breaker does not want to build a clique of his own). In the reverse game (Avoidance version) the two players are called Avoider and Forcer, Forcer's goal is to force the reluctant Avoider to occupy a K_q and Avoider's goal is to avoid occupying a K_q (Forcer does not want to build a clique of his own).

If $q = q(n)$ is "very small" in terms of n, then Maker (or Forcer) can easily win. On the other hand, if $q = q(n)$ is "not so small" in terms of n, then Breaker (or Avoider) can easily win. Where is the game-theoretic breaking point? We call the breaking point the Clique Achievement (Avoidance) Number.

For "small" n's no one knows the answer, but for "large" n's I know the exact value of the breaking point! Indeed, assume that n is sufficiently large like $n \geq 2^{10^{10}}$. If one takes the lower integral part

$$q = \lfloor 2\log_2 n - 2\log_2 \log_2 n + 2\log_2 e - 3 \rfloor$$

($\log_2$ is the base 2 logarithm), then Maker (or Forcer) wins. On the other hand, if one takes the upper integral part

$$q = \lceil 2\log_2 n - 2\log_2 \log_2 n + 2\log_2 e - 3 \rceil,$$

then Breaker (or Avoider) wins.

For example, if $n = 2^{10^{10}}$ then

$$2\log_2 n - 2\log_2 \log_2 n + 2\log_2 e - 3 =$$

$$= 2 \cdot 10^{10} - 66.4385 + 2.8854 - 3 = 19,999,999,933.446,$$

and so the largest clique size that Maker can build (Forcer can force Avoider to build) is $19,999,999,933$.

This level of accuracy is particularly striking, because for smaller values of n I don't know the Clique Achievement Number. For example, if $n = 20$ then all that I know is that it can be either 4 or 5 or 6 (which one?); if $n = 100$ then it can be either 5 or 6 or 7 or 8 or 9 (which one?); if $n = 2^{100}$

then it can be either 99 or 100 or 101 or ... or 188 (which one?), that is, there are 90 possible candidates. Even less is known about the small Avoidance Numbers. I think we will (probably!) never know the exact values of these game numbers for $n = 20$, or for $n = 100$, or for $n = 2^{100}$, but we know the exact value for a monster number like $n = 2^{10^{10}}$. This is truly surprising! This is the complete opposite of the usual induction way of discovering patterns from analysing the small cases.

The explanation for this unusual phenomenon comes from my proof technique, which is a "fake probabilistic method". Probability theory is a collections of Laws of Large Numbers. Converting the probabilistic arguments into a (deterministic!) potential strategy leads to certain "error terms", and these "error terms" become negligible compared to the "main term" if the board is large.

It is also very surprising that the Clique Achievement Game and its reverse have *exactly* the same breaking point: Clique Achievement Number=Clique Avoidance Number. I feel this contradicts common sense. I think one would rather expect that an eager Maker in the "straight" game has a better chance to build a large clique than a reluctant Avoider in the Reverse version, but this "natural" expectation turnes out to be wrong. I cannot give any *a priori* reason why the two breaking points coincide. All that I can say is that the highly technical proof of the "straight" case (around 30 pages long) can be easily adapted (e.g., *maximum* is replaced by *minimum*) to yield the same breaking point for the Reverse game, but this is hardly the answer that we are looking for.

What is the mysterious expression $2\log_2 n - 2\log_2\log_2 n + 2\log_2 e - 3$? An expert of the theory of Random Graphs immediately recognizes that $2\log_2 n - 2\log_2\log_2 n + 2\log_2 e - 3$ is exactly 2 less than the Clique Number of the symmetric Random Graph $\mathbf{R}(K_n, 1/2)$ (1/2 is the edge probability).

A combination of the first and second moment methods (standard Probability Theory) shows that the clique number $\omega\big(\mathbf{R}(K_n, 1/2)\big)$ of the Random Graph has a very strong concentration. Typically it is concentrated on a *single* integer with probability $\to 1$ as $n \to \infty$ (and even in the worst case there are at most two values). Indeed, the expected number of q-cliques in $\mathbf{R}(K_n, 1/2)$ equals

$$f(q) = f_n(q) = \binom{n}{q} 2^{-\binom{q}{2}}.$$

The function $f(q)$ drops under 1 around $q \approx 2\log_2 n$. The real solution of the equation $f(q) = 1$ is

$$q = 2\log_2 n - 2\log_2\log_2 n + 2\log_2 e - 1 + o(1), \tag{1.13}$$

which is exactly 2 more than the game-theoretic breaking point

$$q = 2\log_2 n - 2\log_2\log_2 n + 2\log_2 e - 3 + o(1) \tag{1.14}$$

mentioned above.

Building a clique K_q of size (1.13) by Maker (or Avoider in the Reverse game) on the board K_n is that I call the majority outcome. The size of the majority play outcome differs from the size of the optimal play outcome by a mere additive constant 2, and this additive constant is motivated by the Lovász Local Lemma – it is a local versus global phenomenon.

Notice that (1.14) is the real solution of the equation

$$\binom{n}{q}2^{-\binom{q}{2}} = f(q) = \frac{\binom{n}{2}}{2^{\binom{q}{2}}}. \tag{1.15}$$

The intuitive meaning of (1.15) is that the overwhelming majority of the edges of the random graph are covered by exactly one copy of K_q. In other words, the random graph may have a large number of copies of K_q, but they are well-spread (un-crowded), in fact, there is room enough to be typically pairwise edge-disjoint. This suggests the following *intuition*. Assume that we are at a "last stage" of playing a clique game where Maker (playing the Achievement Game) has a large number of "almost complete" K_q's: "almost complete" in the sense that, (a) in each "almost complete" K_q all but *two edges* are occupied by Maker, (b) all of these edge-pairs are unoccupied yet, and (c) these extremely dangerous K_q's are pairwise edge-disjoint. If (a)–(b)–(c) hold then Breaker can still escape from losing: he can block these disjoint unoccupied edge-pairs by a simple Pairing Strategy.

This intuition emphasizes the "local size" over the "global size", and this is where the Lovász Local Lemma enters the story. A typical application of the Lovász Local Lemma goes as follows.

Erdős–Lovász 2-Coloring Theorem (1975). *Let $\mathcal{F} = \{A_1, A_2, A_3, \dots\}$ be an n-uniform hypergraph. Suppose that each A_i intersects at most 2^{n-3} other $A_j \in \mathcal{F}$ ("local size"). Then there is a 2-coloring of the "board" $V = \bigcup_i A_i$ such that no $A_i \in \mathcal{F}$ is monochromatic.*

The conclusion (almost!) means that, by playing generalized Tic-Tac-Toe (the player who occupies a whole winning set first is the winner) on the hypergraph, there exists a *drawing terminal position.* I wrote "almost", because I cheated a little bit here: in a drawing terminal position the two color-classes must have equal size. The very surprising message of the Erdős–Lovász 2-coloring theorem is that the "global size" of hypergraph $\mathcal{F}$ is irrelevant (it can even be infinite!), only the "local size" matters.

Of course, the existence of a single (or even several) drawing terminal position does *not* guarantee at all the existence of a drawing strategy in the generalized Tic-Tac-Toe game, or a Breaker's (or Avoider's) winning strategy. But perhaps it is true that, under the Erdős–Lovász condition (or under some similar but slightly weaker local condition), Breaker (or Avoider) has a winning strategy, i.e. he can block every winning set in the Achievement (or Avoidance) game on $\mathcal{F}$. I refer to this "blocking strategy" as a Strong Draw strategy.

This "exponentially bounded local size implies Strong Draw" conjecture is a wonderful general problem; I call it the Neighborhood Conjecture. Unfortunately the conjecture is still open in general, in spite of all efforts trying to prove it during the last 25 years.

We know, however, several partial results, see the vaguely stated Local Criterion below. A very important special case, when the conjecture is "nearly proved", is the class of Almost Disjoint hypergraphs: where any two hyperedges have at most one common point – this is certainly the case for geometric lines, the winning sets of the n^d Tic-Tac-Toe.

By the way, what do we know about the multidimensional n^d Tic-Tac-Toe? We know that it is a draw game even if the dimension d is as large as $d = c_1 n^2/\log n$, that is, nearly quadratic in terms of (the winning size) n. What is more, the draw is a Strong Draw: the second player can mark every winning line (if they play till the whole board is occupied).

Note that this bound is nearly best possible: if $d > c_2 n^2$ then the second player *cannot* force a Strong Draw (because the first player can occupy a whole winning line, but not necessarily first – I call it a Weak Win).

How come that for the Clique Game we know the *exact* value of the breaking point, but for the multi-dimensional Tic-Tac-Toe we couldn't even find the asymptotic truth (due to the extra factor of $\log n$ in the denominator)? The answer is somewhat technical. The winning lines in the multi-dimensional n^d Tic-Tac-Toe form an extremely *irregular* hypergraph: the maximum degree is *much* larger than the average degree. This is why one cannot apply the Blocking Criterions directly to the "n^d hypergraph". First one has to employ a Truncation Procedure to bring the maximum degree close to the average degree, and the price that one pays for this degree reduction is the loss of a factor of $\log n$.

However, if one considers the n^d *Torus* Tic-Tac-Toe, then the corresponding hypergraph becomes perfectly uniform (since the torus is a *group*). For example, every point of the n^d *Torus* Tic-Tac-Toe has $(3^d - 1)/2$ winning lines passing through it. This uniformity explains why for the n^d *Torus* Tic-Tac-Toe we can prove asymptotically sharp thresholds.

A "winning line" in the n^d Tic-Tac-Toe is a set of n points on a straight line forming an n-term Arithmetic Progression. This motivates the "Arithmetic Progression Game": the board is the interval $1, 2, \ldots, N$, and the goal is to build an n-term Arithmetic Progression. The corresponding hypergraph is "nearly regular"; this is why we can prove asymptotically sharp results.

Let us return to the n^d Torus Tic-Tac-Toe. If "winning line" is replaced by "winning plane" (or "winning subspace of dimension ≥ 2" in general), then we can go far beyond "asymptotically sharp": we can even determine the *exact* value of the game-theoretic threshold, just like in the Clique Game. For example, a "winning plane" is an $n \times n$ lattice in the n^d Torus. This is another rapidly changing 2-dimensional configuration: if n switches to $n+1$, then $n \times n$ switches to $(n+1) \times (n+1)$, which is again a "square-root size" increase just like in the case of the cliques. This formal similarity to the Clique Game (both have "2-dimensional goals") explains why there is a chance to find the *exact* value of the game-theoretic breaking point (the actual proofs are rather different).

It is very difficult to visualize the d-dimensional torus if d is large; here is an easier version: a game with 2-dimensional goal sets played on the plane.

Two-dimensional Arithmetic Progression Game. A natural way to obtain a 2-dimensional arithmetic progression is to take the Cartesian product. The Cartesian product of two q-term arithmetic progressions with the same gap is a $q \times q$ Aligned Square Lattice. Let ($N \times N, q \times q$ square lattice) denote the game where the board is the $N \times N$ chessboard, and the winning sets are the $q \times q$ Aligned Square Lattices. Again we know the exact value of the game-theoretic breaking point: if

$$q = \left\lfloor \sqrt{\log_2 N} + o(1) \right\rfloor, \tag{1.16}$$

then Maker can always build a $q \times q$ Aligned Square Lattice, and this is the best that Maker can achieve: Breaker can always prevent Maker from building a $(q+1) \times (q+1)$ Aligned Square Lattice. Again the error term $o(1)$ becomes negligible if N is large. For example, $N = 2^{10^{40}+10^{20}}$ is large enough, and then

$$\sqrt{\log_2 N} = \sqrt{10^{40} + 10^{20}} = 10^{20} + \frac{1}{2} + O(10^{-20}),$$

so $\sqrt{\log_2 N}$ is not too close to an integer (in fact, it is almost exactly in the middle), which guarantees that $q = 10^{20}$ is the largest aligned square lattice size that Maker can build.

Similarly, $q = 10^{20}$ is the largest aligned square lattice size that Forcer can force Avoider to build.

Another way to look at this "local versus global" phenomenon is to consider it as some kind of "game-theoretic independence"; I can summarize it in a nutshell as follows. We are studying Tic-Tac-Toe like games for which the local size is much smaller than the global size. Even if the game starts out as a coherent entity, either player can force it to develop into smaller, local size composites. A sort of intuitive explanation behind it is the Erdős–Lovász 2-coloring theorem, which itself is a sophisticated application of statistical independence. Our "game-theoretic independence" is about how to sequentialize statistical independence.

Let's return to (1.14) and (1.16) one more time. These results describe the exact values of infinitely many Achievement and Avoidance Numbers, and they are special cases of the following vaguely stated

Game-theoretic Local Criterion for (at least) two-dimensional goals. In each one of the "exact solution games" (games with two-dimensional goals) the "phase transition" from Weak Win (i.e., Maker' win) to Strong Draw (i.e., Breaker's win) happens when the winning set size equals the binary logarithm of the Set/Point ratio of the hypergraph, formally, $\log_2\left(|\mathcal{F}|/|V|\right)$.

For example, in the (K_n, K_q) Clique Game the *Set/Point* ratio is $\binom{n}{q}\binom{n}{2}^{-1}$, and the equation

$$\binom{q}{2} = \log_2\left(\binom{n}{q}\binom{n}{2}^{-1}\right)$$

has the real solution

$$q = q(n) = 2\log_2 n - 2\log_2 \log_2 n + 2\log_2 e - 3 + o(1),$$

which is exactly (1.14). By contrast I call (1.13) the Majority-Play Number: it gives the largest clique size in the majority of all possible plays (of course, most of the plays are dull).

In the Aligned Square Lattice Game on an $N \times N$ board the Local Criterion gives the equation

$$q^2 = \log_2 \left(\frac{N^3}{3(q-1)} \cdot N^{-2} \right),$$

which has the real solution

$$q = q(N) = \sqrt{\log_2 N} + o(1),$$

that is, exactly (1.16).

We know that the Clique Achievement Number is 2 less than the Majority-Play Clique Number – these two thresholds are strikingly close to each other (see (1.13)–(1.14)). The Aligned Square Lattice Game is very different: instead of a negligible additive constant, for the lattice game we have a substantial constant factor larger than 1 Indeed, if the goal is a $q \times q$ aligned square lattice, the Achievement Number is

$$q_1 = q_1(N) = \left\lfloor \sqrt{\log_2 N} + o(1) \right\rfloor,$$

and the corresponding Majority-Play Number is the solution of the equation

$$\frac{N^3}{3(q-1)} = 2^{q^2}$$

in $q = q(N)$, which gives

$$q = q_2 = q_2(N) = \left\lfloor \sqrt{3\log_2 N} + o(1) \right\rfloor.$$

Notice that they are not that close: the ratio q_2/q_1 is $\sqrt{3} > 1$.

In these examples the Achievement Number is always less than the Majority-Play Number: slightly less in the clique game and substantially less in the lattice game. This is explained by a very general result, due to

Erdős and Selfridge [6], which basically says that the Achievement Number is *always* less than the Majority-Play Number; it holds for every finite hypergraph. The Erdős–Selfridge theorem – a global inequality with a simple proof – hardly ever gives the truth; our Local Criterion, on the other hand, seems to predict the exact value of the game-theoretic breaking point for surprisingly large and natural classes of games. Unfortunately, the Local Criterion (that I didn't even formulate precisely(!), missing many technical conditions) has a very difficult proof (motivated by the Lovász Local Lemma). Nevertheless one cannot overestimate the importance of the Erdős–Selfridge theorem: it was the pioneering application of the potential technique, the starting point of a long line of research.

For more about this (including precise statements, detailed proofs and many more results) I refer the reader to my new book *Tic-Tac-Toe Theory*.

8. How sharp is Theorem 1? Theorem 2 is (nearly) satisfying in the sense that for d-regular graphs the surplus is simply around $\sqrt{d}$, unless the global size is truly gigantic. Theorem 1, on the other hand, has a linear upper bound and a square-root size lower bound; these bounds are obviously far from each other. In spite of this weakness, I have to defend Theorem 1: it is (nearly) sharp for the family of *all* graphs. I begin with the linear upper bound: it is sharp for the family of very asymmetric complete bipartite graphs $K_{d,n}$ with $n > (d+2)2^{d-1}$. Indeed, for such a $K_{d,n}$ the Core is the whole graph, so

$$\operatorname{cd}(K_{d,n}) = \frac{dn}{n+d} \approx \frac{dn}{n} = d;$$

on the other hand, by Proposition 1.1(d) the surplus $\operatorname{Sur}(K_{d,n}) = d$. That is, for the graphs $G = K_{d,n}$ with $n > (d+2)2^{d-1}$ the upper bound $1 + \lceil \operatorname{cd}(G) \rceil$ in Theorem 1 is $d + O(1)$, and the surplus is d – basically best possible result.

Next consider the square-root size lower bound in Theorem 1: it is nearly best possible for the family of d-regular graphs – see Theorem 2 – assuming the global size of the graph is not gigantic.

For the most natural graph – the symmetric complete bipartite graph $K_{d,d}$, i.e., the Row-Column Game on a square board – the following is known.

Proposition 1.4. *We have*

$$\frac{2}{15}\sqrt{d} \le \operatorname{Sur}(K_{d,d}) \le 2\sqrt{d\log d} \tag{1.17}$$

and

$$\frac{2}{15}\sqrt{d} \leq \text{Sur}\,(K_d) \leq 2\sqrt{d \log d}. \tag{1.18}$$

We postpone the proof to Section 2.

It is somewhat embarrassing that we cannot determine the surplus for the complete graphs K_d and $K_{d,d}$ (which usually represent the "simplest case"). It seems hard to get rid of the (small) factor $\sqrt{\log d}$. My excuse is that the surplus is not as easy as it looks; in fact, games are surprisingly difficult!

Here is a simple but important observation that we are constantly using: the Surplus is essentially monotone. Formally,

$$\text{if} \quad H \subset G \quad \text{then} \quad \text{Sur}\,(H) \leq \text{Sur}\,(G) + 1. \tag{1.19}$$

Indeed, let Str denote Maker's strategy such that, playing on subgraph H he can always force a Surplus $\geq$ Sur (H). Playing on G Maker can use strategy Str as follows: (1) he starts in H according to Str, (2) whenever Breaker answers in subgraph H then Maker responds by using Str, (3) if Breaker's edge is disjoint from H then Maker responds arbitrarily, (4) if Breaker's edge is not in H but has one endpoint in H then Maker takes another edge from the same endpoint in H, an edge which is not entirely in H (as long as he can do it; otherwise moves arbitrarily). This way at the end Maker can force a Surplus $\geq$ Sur $(H) - 1$, which is (1.19).

I conclude the section with a few historic notes. As far as I know the question of determining the value of the surplus for the complete graph goes back to Erdős; he frequently asked this question in his graph theory lectures starting from the 1970s. An early publication is Székely [8]; the halving argument in Proposition 1.1(d) is due to him. Székely also raised three problems. I solved these problems starting with Beck [2] (where I showed that Sur $(K_n) > c \cdot \sqrt{n}$), and the rest was published in Section 16 of my book *Tic-Tac-Toe Theory.* By the way, there is a considerable overlapping between this paper and my book (especially Sections 16–17). However, the two main results of this paper – Theorem 1 (the lower bound) and Theorem 2 (the upper bound) – are new, and are published here for the first time. The upper bound proof of Theorem 2 is based on a technique that I originally developed in my book (I call it the "Big-Game&small-game decomposition"); the new ingredient here is "iteration". Finally a technical question to the reader: my proof of Theorem 1 in the general case (see

Section 3) is much harder than in the special case of regular graphs (see Section 2). Do I overlook something simple here?

Most of the rest of the paper is devoted to the proofs of Theorems 1–2; each one is about ten pages long.

2. Potential Technique

1. We use potential functions for proving game-theoretic upper and lower bounds, see (2.2) and (2.3) below. Unfortunately the upper and lower bounds do not coincide, not even for the complete graph (but at least then they are close to each other, differ by a logarithmic factor). I start with the upper bound; in fact, with a general hypergraph upper bound. The following result, that I proved in 1981 (see Lemma 3 in Beck [1]), immediately gives the desired upper bound in Proposition 1.4 (see (1.17)–(1.18)), and also it will be a key ingredient in the long and complicated proof of the upper bound in Theorem 2 (see Section 4).

Proposition 2.1. *Let $\mathcal{F}$ be an arbitrary finite hypergraph, and let ε with $0 < \varepsilon \leq 1$ be an arbitrary real number. There are two players, Balancer and Unbalancer, who play the $(1:1)$ game on $\mathcal{F}$: they alternate, and each player takes one new point per move. Unbalancer's goal is to achieve an ε-majority: he wins if he owns $\geq \frac{1+\varepsilon}{2}$ part of some $A \in \mathcal{F}$; otherwise Balancer wins. Here is a Balancer's win criterion: if*

$$\sum_{A\in\mathcal{F}} \left((1+\varepsilon)^{1+\varepsilon}(1-\varepsilon)^{1-\varepsilon}\right)^{-|A|/2} < 1, \tag{2.1}$$

then Balancer, as the first player, has a winning strategy.

If the upper bound 1 in (2.1) is replaced by $\frac{1}{1+\varepsilon}$ then Balancer can be the second player, and still has a winning strategy.

I challenge the reader to show that the base $(1+\varepsilon)^{1+\varepsilon}(1-\varepsilon)^{1-\varepsilon}$ is greater than 1 for every $0 < \varepsilon \leq 1$ – this fact is critical in the applications.

Notice that in the special case $\varepsilon = 1$, where Unbalancer's goal is to occupy a whole set $A \in \mathcal{F}$, Proposition 2.1 gives back the well-known Erdős–Selfridge theorem [6].

Not surprisingly, the **proof** of Proposition 2.1 is very similar to that of the Erdős–Selfridge theorem. Assume that we are in the middle of a play:

Balancer occupied $b_1, b_2, \ldots, b_t$, and Unbalancer occupied $u_1, u_2, \ldots, u_t$; t is the time parameter. Write $B(t) = \{b_1, b_2, \ldots, b_t\}$, $U(t) = \{u_1, u_2, \ldots, u_t\}$, and consider the function

$$P_t = \sum_{A \in \mathcal{F}} (1+\varepsilon)^{|A \cap U(t)| - \frac{1+\varepsilon}{2}|A|}(1-\varepsilon)^{|A \cap B(t)| - \frac{1-\varepsilon}{2}|A|}, \tag{2.2}$$

which is very sensitive (in fact "exponentially sensitive") to Unbalancer's lead. The core idea, taken from the Erdős–Selfridge proof, is that Balancer can force the monotone *decreasing* property

$$P_0 \geq P_1 \geq P_2 \geq \cdots \geq P_{\text{end}}, \tag{2.3}$$

so $P_{\text{start}} = P_0 \geq P_{\text{end}}$. Indeed, Balancer's next move b_{t+1} decreases sum (2.2) and Unbalancer's next move u_{t+1} increases sum (2.2), where the "decrease sum" is the same kind of term as the "increase sum" with the natural distinction that the "decrease sum" depends on b_{t+1} and the "increase sum" depends on u_{t+1}. However, since Balancer comes first, he can choose the "best" option, and so the damage caused by his move b_{t+1} is larger than the gain caused by the opponent's move u_{t+1} (also note that if a hyperedge contains both b_{t+1} and u_{t+1}, then the common effect $(1-\varepsilon)(1+\varepsilon) < 1$ is again a decrease).

Here come the details. The effect of the $(t+1)$st moves b_{t+1} (by Balancer) and u_{t+1} (by Unbalancer) is the following:

$$\begin{aligned} P_{t+1} = P_t - \varepsilon & \sum_{A \in \mathcal{F}:\ b_{t+1} \in A} (1-\varepsilon)^{|A \cap B(t)| - \frac{1-\varepsilon}{2}|A|}(1+\varepsilon)^{|A \cap U(t)| - \frac{1+\varepsilon}{2}|A|} \\ + \varepsilon & \sum_{A \in \mathcal{F}:\ u_{t+1} \in A} (1-\varepsilon)^{|A \cap B(t)| - \frac{1-\varepsilon}{2}|A|}(1+\varepsilon)^{|A \cap U(t)| - \frac{1+\varepsilon}{2}|A|} \\ - \varepsilon^2 & \sum_{A_0 \in \mathcal{A}_0} (1-\varepsilon)^{|A_0 \cap B(t)| - \frac{1-\varepsilon}{2}|A_0|}(1+\varepsilon)^{|A_0 \cap U(t)| - \frac{1+\varepsilon}{2}|A_0|}, \end{aligned}$$

where $A_0 \in \mathcal{F}_0$ denotes the hyperedges containing both b_{t+1} and u_{t+1}.

Since Balancer's b_{t+1} is selected before Unbalancer's u_{t+1}, Balancer can select the "best" point in the following sense: Unbalancer chooses that $b_{t+1} = z$ for which the sum

$$\sum_{A \in \mathcal{F}:\ z \in A} (1-\varepsilon)^{|A \cap B(t)| - \frac{1-\varepsilon}{2}|A|}(1+\varepsilon)^{|A \cap U(t)| - \frac{1+\varepsilon}{2}|A|}$$

attains its *maximum.* Then

$$P_{t+1} \leq P_t - \varepsilon^2 \sum_{A_0 \in \mathcal{A}_0} (1-\varepsilon)^{|A_0 \cap B(t)| - \frac{1-\varepsilon}{2}|A_0|}(1+\varepsilon)^{|A_0 \cap U(t)| - \frac{1+\varepsilon}{2}|A_0|} \leq P_t,$$

proving (2.3).

Notice that (2.3) easily completes the proof. Assume that Unbalancer wins, then by (2.2) we have $P_{\text{end}} \geq 1$, but this is impossible, since (2.1) and (2.3) give $P_{\text{end}} \leq P_{\text{start}} = P_0 < 1$. This is how (2.3) forces Balancer's win. ■

Let us apply Proposition 2.1 to the star hypergraph of K_n: the hyperedges are the n stars (each star has $n-1$ edges), so $\mathcal{F}$ is an $(n-1)$-uniform hypergraph with $|\mathcal{F}| = n$. By choosing

$$\varepsilon = \sqrt{\frac{c \log n}{n}}$$

with some unspecified (yet) constant $c > 0$, criterion (2.1), applied to the star hypergraph of K_n, gives

$$\sum_{A \in \mathcal{F}} \left((1+\varepsilon)^{1+\varepsilon}(1-\varepsilon)^{1-\varepsilon}\right)^{-|A|/2} = ne^{-((1+\varepsilon)\log(1+\varepsilon)+(1-\varepsilon)\log(1-\varepsilon))\frac{n-1}{2}} =$$

$$= ne^{-\left((1+\varepsilon)\left(\varepsilon - \frac{\varepsilon^2}{2} \pm \cdots\right) + (1-\varepsilon)\left(-\varepsilon - \frac{\varepsilon^2}{2} - \cdots\right)\right)\frac{n-1}{2}} =$$

$$= ne^{-(\varepsilon^2 + O(\varepsilon^3))\frac{n-1}{2}} = ne^{-\frac{c \log n}{2}(1+O(\varepsilon))} = n^{1-\frac{c}{2}(1+O(\varepsilon))},$$

which is less than 1 if $c > 2$ and n is sufficiently large. So Proposition 2.1 applies with

$$\varepsilon = \sqrt{\frac{\big(2+o(1)\big)\log n}{n}},$$

and yields the upper bound in (1.18) for K_n.

The same calculation applies for $K_{n,n}$, and yields the upper bound in (1.17).

In general we have the following corollary of Proposition 2.1.

Proposition 2.2. *Let $\mathcal{F}$ be an n-uniform hypergraph, and consider the Balancer-Unbalancer game played on hypergraph $\mathcal{F}$ where Unbalancer's goal is to own at least $\frac{n+\Delta}{2}$ points from some $A \in \mathcal{F}$. If*

$$\Delta = \left(1 + O\left(\sqrt{\frac{\log|\mathcal{F}|}{n}}\right)\right)\sqrt{2n\log|\mathcal{F}|},$$

then Balancer has a winning strategy.

Proof. Let $\varepsilon = \Delta/n$; in view of Proposition 2.1 we have to check the inequality

$$(1+\varepsilon)^{(n+\Delta)/2} \cdot (1-\varepsilon)^{(n-\Delta)/2} \geq |\mathcal{F}|.$$

Note that

$$(1+\varepsilon)^{(n+\Delta)/2} \cdot (1-\varepsilon)^{(n-\Delta)/2} = (1-\varepsilon^2)^{n/2} \cdot \left(\frac{1+\varepsilon}{1-\varepsilon}\right)^{\Delta/2} \approx$$

$$\approx e^{-\varepsilon^2 n/2 + \varepsilon\Delta} = e^{\Delta^2/2n}.$$

More precisely, we have

$$e^{(1+O(\Delta/n))\frac{\Delta^2}{2n}} = (1+\varepsilon)^{(n+\Delta)/2} \cdot (1-\varepsilon)^{(n-\Delta)/2} \geq |\mathcal{F}| = e^{\log|\mathcal{F}|},$$

which implies

$$\left(1 + O(\Delta/n)\right)\frac{\Delta^2}{2n} \geq \log|\mathcal{F}|,$$

or equivalently,

$$\Delta \geq \left(1 + O\left(\sqrt{\frac{\log|\mathcal{F}|}{n}}\right)\right)\sqrt{2n\log|\mathcal{F}|},$$

which proves Proposition 2.2. ■

2. Next I switch to the lower bound in Theorem 2. Working with the star hypergraph of a graph, the lower bound problem is equivalent to the following.

Proposition 2.3. *Let $\mathcal{F}$ be a hypergraph which is (1) n-uniform, (2) Almost Disjoint: $|A_1 \cap A_2| \leq 1$ for any two different elements of hypergraph $\mathcal{F}$, and (3) the common height of $\mathcal{F}$ is 2: every point of the hypergraph is contained in exactly two hyperedges. Maker and Breaker play the usual $(1:1)$ game on $\mathcal{F}$. Then, at the end of the play, Maker can occupy at least $\frac{n}{2} + c\sqrt{n}$ points from some $A \in \mathcal{F}$, where $c = 1/15$.*

Remark. Proposition 2.3 is "Theorem 16.2" in my book *Tic-Tac-Toe Theory.* The proof below is my "second proof" there. The "first proof" there is the motivation for the more complicated argument in the next section to prove the general case of Theorem 1. The lower bound in Theorem 2 is "Theorem 16.3" in my book.

Proof. Assume we are in the middle of a play, Maker already occupied $x_1, x_2, \ldots, x_t$ (t is the time) and Breaker occupied $y_1, y_2, \ldots, y_t$. Let $X(t) = \{x_1, x_2, \ldots, x_t\}$ and $Y(t) = \{y_1, y_2, \ldots, y_t\}$. We work with the exponential expression

$$P_t = \sum_{A \in \mathcal{F}} (1+\varepsilon)^{|A \cap X(t)|}(1-\varepsilon)^{|A \cap Y(t)|}, \tag{2.4}$$

which is the perfect analog of (2.2) for uniform hypergraphs. What is the effect of the $(t+1)$st moves x_{t+1} (by Maker) and y_{t+1} (by Breaker)? Well, the answer is easy:

$$\begin{aligned} P_{t+1} = P_t &+ \varepsilon \sum_{A \in \mathcal{F}:\, x_{t+1} \in A} (1+\varepsilon)^{|A \cap X(t)|}(1-\varepsilon)^{|A \cap Y(t)|} \\ &- \varepsilon \sum_{A \in \mathcal{F}:\, y_{t+1} \in A} (1+\varepsilon)^{|A \cap X(t)|}(1-\varepsilon)^{|A \cap Y(t)|} \\ &- \varepsilon^2 \cdot \delta(x_{t+1}, y_{t+1}) \cdot (1+\varepsilon)^{|A_0 \cap X(t)|}(1-\varepsilon)^{|A_0 \cap Y(t)|}, \end{aligned} \tag{2.5}$$

where $\delta(x_{t+1}, y_{t+1}) = 1$ if there is an $A \in \mathcal{F}$ containing both x_{t+1} and y_{t+1}; Almost Disjointness yields that, if there is one, then there is exactly one: let A_0 be this uniquely determined $A \in \mathcal{F}$; finally let $\delta(x_{t+1}, y_{t+1}) = 0$ if there is *no* $A \in \mathcal{F}$ containing both x_{t+1} and y_{t+1}.

Since Maker's x_{t+1} is selected before Breaker's y_{t+1}, Maker can select the "best" point: Maker chooses that $x_{t+1} = z$ for which the sum

$$\sum_{A \in \mathcal{F}:\, z \in A} (1+\varepsilon)^{|A \cap X(t)|}(1-\varepsilon)^{|A \cap Y(t)|}$$

attains its *maximum*. Then by (2.5)

$$P_{t+1} \geq P_t - \varepsilon^2 \cdot \delta(x_{t+1}, y_{t+1}) \cdot (1+\varepsilon)^{|A_0 \cap X(t)|}(1-\varepsilon)^{|A_0 \cap Y(t)|}. \tag{2.6}$$

Let $\Delta = \Delta(n)$ denote the largest positive discrepancy that Maker can achieve; it means $\frac{n}{2} + \Delta$ points from some $A \in \mathcal{F}$. If Δ is the maximum discrepancy then the inequality $|A \cap X(t)| - |A \cap Y(t)| \leq 2\Delta$ must hold during the whole play (meaning every t) and for every $A \in \mathcal{F}$. Indeed, if $|A \cap X(t)| - |A \cap Y(t)| > 2\Delta$ then Maker can keep this lead for the rest of the play till the end, contradicting the maximum property of Δ. Combining this observation with (2.6) we have

$$P_{t+1} \geq P_t - \varepsilon^2 \cdot \delta(x_{t+1}, y_{t+1}) \cdot (1+\varepsilon)^{z_t+\Delta}(1-\varepsilon)^{z_t-\Delta}, \tag{2.7}$$

where

$$z_t = \frac{|A_0 \cap X(t)| + |A_0 \cap Y(t)|}{2}.$$

Since $P_0 = P_{\text{start}} = |\mathcal{F}| = N$ and "total time"$= T = nN/4$, from (2.7) we obtain

$$P_{\text{end}} = P_T \geq P_0 - \varepsilon^2 \sum_{t=1}^{nN/4} (1+\varepsilon)^{z_t+\Delta}(1-\varepsilon)^{z_t-\Delta} = \tag{2.8}$$

$$= N - \varepsilon^2 \left(\frac{1+\varepsilon}{1-\varepsilon}\right)^{\Delta} \left(\sum_{t=1}^{nN/4} (1-\varepsilon^2)^{z_t}\right) \geq$$

$$\geq N - \varepsilon^2 \left(\frac{1+\varepsilon}{1-\varepsilon}\right)^{\Delta} \frac{nN}{4}.$$

On the other hand, by definition

$$P_{\text{end}} = P_T \leq N(1+\varepsilon)^{n/2+\Delta}(1-\varepsilon)^{n/2-\Delta} = \tag{2.9}$$

$$= N \left(\frac{1+\varepsilon}{1-\varepsilon}\right)^{\Delta} (1-\varepsilon^2)^{n/2}.$$

Combining (2.8) and (2.9)

$$N - \varepsilon^2 \left(\frac{1+\varepsilon}{1-\varepsilon}\right)^{\Delta} \frac{nN}{4} \leq N \left(\frac{1+\varepsilon}{1-\varepsilon}\right)^{\Delta} (1-\varepsilon^2)^{n/2},$$

or equivalently,

$$\left(\frac{1+\varepsilon}{1-\varepsilon}\right)^{\Delta} \geq \frac{1}{\frac{\varepsilon^2 n}{4} + (1-\varepsilon^2)^{n/2}}. \tag{2.10}$$

We want to minimize the denominator in the right-hand side of (2.10): we are looking for an optimal ε in the form $\varepsilon = \sqrt{2\beta/n}$ where β is an unspecified constant (yet); then

$$\frac{\varepsilon^2 n}{4} + (1-\varepsilon^2)^{n/2} \approx \frac{\beta}{2} + e^{-\beta} = \frac{1+\log 2}{2}$$

if $\beta = \log 2$. With this choice of β (2.10) becomes

$$e^{2\varepsilon\Delta} \approx \left(\frac{1+\varepsilon}{1-\varepsilon}\right)^{\Delta} \geq \frac{1}{\frac{\beta^2 n}{4} + (1-\beta^2)^{n/2}} \geq \frac{1}{\frac{1+\log 2}{2}},$$

implying

$$2\sqrt{2\log 2}\cdot\frac{\Delta}{\sqrt{n}} \geq \log\left(\frac{2}{1+\log 2}\right),$$

that is,

$$\Delta \geq \frac{\log\left(\frac{2}{1+\log 2}\right)}{2\sqrt{2\log 2}}\sqrt{n} \geq \frac{\sqrt{n}}{15}.$$

This proves Proposition 2.3 with $c = 1/15$. ■

Proposition 2.3 implies the lower bound in Theorem 2, and also the lower bound in (1.17)–(1.18) (as a special case). Since Proposition 2.2 implies the upper bound in (1.17)–(1.18), the proof of Proposition 1.4 is complete.

The missing upper bound in Theorem 2 will be proved in Section 4.

I conclude this section with two remarks: (1) "socialism works perfectly on graphs", and (2) "the discrepancy is much easier than the surplus"; I explain these vague statements as follows.

3. Inevitable surplus is trivial. Perhaps the reader is wondering: Is Surplus the right concept to study? Well, here is an alternative concept that the reader may find equally natural and interesting; I call it the "inevitable surplus". Assume that each one of the two players (Maker and Breaker) is an idealist (or socialist) who strongly believes in Equality and Fairness. They are willing to cooperate to avoid large Surplus (that they consider unfair). Two such idealist players, playing the Row-Column Game on an $n \times n$ board with even n, can easily achieve perfect equality as follows. They cooperate: Maker takes only the white cells and Breaker takes only the black cells in the usual chessboard 2-coloring. This settles the special case $K_{n,n}$. The result can be easily generalized to large classes of graphs. For example, let G be an arbitrary d-regular bipartite graph with d even. Repeated application of Hall's theorem gives that G falls apart into d perfect matchings. If the players cooperate, and each player takes $d/2$ perfect matchings, the Surplus is zero.

If G is an odd cycle, then Maker (the first player) must have Surplus 2 even if the two players are willing to cooperate to avoid it. This motivates the concept of **Inevitable Surplus**; every odd cycle has Inevitable Surplus 2. What is the largest possible Inevitable Surplus? Well, I have noticed that the Inevitable Surplus is a dull/bounded concept, and my proof was based on a very general hypergraph theorem that we proved with T. Fiala a long time ago (see Beck–Fiala [1]).

Hypergraph Balancing Theorem. *Let $\mathcal{H}$ be an arbitrary finite hypergraph with Height $\leq h$ (i.e., every point is contained in at most h hyperedges). Then there is a 2-coloring of the points of $\mathcal{H}$ red and blue such that*

$$\big|\mathrm{Red}\,(A) - \mathrm{Blue}\,(A)\big| < 2h$$

holds for every $A \in \mathcal{H}$.

A simple application of this general theorem gives that every finite graph G has Inevitable Surplus ≤ 9. Indeed, let $\{1, 2, 3, \ldots, n\}$ denote the vertex set of G, and let $S_1, S_2, S_3, \ldots, S_n$ be the stars of G. I define a hypergraph $\mathcal{H}$ as follows: the "points" of $\mathcal{H}$ are the edges of G, and the hyperedges are the stars $S_1, S_2, S_3, \ldots, S_n$ plus the whole graph G itself. $\mathcal{H}$ has Height 3, so by the Balancing Theorem above, we can decompose G into two parts G_1 and G_2 such that

$$\big||G_1| - |G_2|\big| \leq 5 \quad \text{and} \quad \big||G_1 \cap S_i| - |G_2 \cap S_i|\big| \leq 5 \text{ for all } 1 \leq i \leq n.$$

We modify G_1 and G_2 by relocating at most two edges from the larger one to the smaller one; thus we obtain a slightly modified new decomposition

G_1' and G_2' of G such that

$$\big||G_1'| - |G_2'|\big| \le 5 - 2 - 2 = 1 \quad \text{and}$$

$$\big||G_1 \cap S_i| - |G_2 \cap S_i|\big| \le 5 + 2 + 2 = 9 \quad \text{for all} \quad 1 \le i \le n.$$

Later Paul Seymour pointed out to me that my (clumsy) upper bound ≤ 9 can be pushed down to the best possible

$$\text{(2.11)} \qquad \text{Inevitable Surplus} \le 2$$

by using a surprisingly simple graph-specific argument (instead of the very general hypergraph balancing theorem above).

The proof of (2.11) is an Euler trail argument with an extra technical trick. The following edge-two-coloring result, which is clearly equivalent to (2.11), was proved by Seymour in the 1970s (Seymour remarks that the result might have been done earlier by others).

Proposition 2.4. *Let G be a connected graph, and let u be any vertex of G. There is a partition (A, B) of the edge set $E(G)$ such that*

(1) *for every vertex v different from u, the number of A-edges at v differs by at most one from the number of B-edges at v;*

(2) *the number of A-edges at u differs by at most two from the number of B-edges at vertex u;*

(3) *$|A|$ and $|B|$ differ by at most one.*

Remark. If the two players, playing on G, cooperate: one takes A and the other one takes B, the Inevitable Surplus is ≤ 2. Socialism does work on graphs! The worst case scenario is when G is an odd cycle.

Proof. If G has a closed Euler trail, i.e., every degree is even, let $G = H$. If G has odd degrees (always an even number) then we apply the following trick: let H be an extension of G by adding a new vertex x to G, x is adjacent to every vertex of G with odd degree. Thus in either case H contains a closed Euler trail. Take an Euler trail starting and ending at u, and color its edges alternately C and D. (Thus the first and last edges of the trail have the same/different color if $\big|E(H)\big|$ is odd/even.) Let $A = C \cap E(G)$ and $B = D \cap E(G)$, i.e., we throw away the edges from the extra vertex x. We claim (A, B) satisfies requirements (1)–(2)–(3).

Certainly $\big||C|-|D|\big| \le 1$, and $|C| = |D|$ if $\big|E(H)\big|$ is even. At the extra vertex x (if it exists) the number of Cs and Ds are the same, and so $\big||A|-|B|\big| \le 1$, proving (3).

Moreover, at every vertex v of H different from u the number of Cs and Ds are the same, and so the number of As and Bs incident with v in G differ by at most one, proving (1).

Finally, at u, the number of Cs and Ds in H differ by at most two, and by two only if u has even degree in H. So in G the number of As and Bs incident with u differ by at most two, proving (2). ■

What makes the concept of Surplus particularly interesting is that large Surplus is not inevitable, it is not automatic. Proposition 2.4 tells us that, if both players want to avoid surplus, then, by cooperation, they can always achieve the inequality Surplus ≤ 2. Large surplus does not come for free: one has to make a serious effort to achieve it!

4. Discrepancy and variance. The concept of Discrepancy means either Surplus or Deficit, that is, Discrepancy is simply a deviation from the expected value, but it does not distinguish between positive and negative.

Another way to put it is that Surplus is the one-sided version of Discrepancy where Maker has the majority.

The following result demonstrates that Maker can rather easily force a large Discrepancy, i.e., large Surplus or large Deficit. What is surprisingly hard is how to force a large Surplus, i.e., large one-sided discrepancy (of course we always assume that the underlying graph is dense enough).

Proposition 2.5. *Let G be an arbitrary graph. Playing the Maker-Breaker game on G, Maker can always achieve a Discrepancy $\ge \sqrt{E/V}$, where E/V is the Density, that is, the Edge/Vertex ratio in graph G.*

Remark. If Maker restricts the play to a Core of graph G, the Density becomes the Maximum Local Density, that is, the Core-Density cd (G) of G.

Proof. Let $1, 2, 3, \ldots, n$ denote the vertices of G. Assume that we are in the middle of a play: each player owns (say) r edges from G; $m_i = m_i(r)$ and $b_i = b_i(r)$ denote, respectively, Maker's degree and Breaker's degree in vertex i at this stage of the play – I call it stage r. Of course

$$\sum_{i=1}^{n} m_i = \sum_{i=1}^{n} b_i = 2r.$$

Maker (the first player) simply wants to maximize the Discrepancy $\max_i |m_i - b_i|$, he does not care that it is a large Surplus or a large Deficit. It is natural, therefore, to introduce the quadratic sum

$$Q(r) = \sum_{i=1}^{n} (m_i - b_i)^2, \tag{2.12}$$

which is a game-theoretic analog of the Variance in probability theory. What happens to this quadratic sum after each player makes his $(r+1)$st move? Let e and f be the $(r+1)$st move of Maker and Breaker, respectively; let j, k be the endpoints of Maker's edge e and let p, q be the endpoints of Breaker's edge f.

Case 1: Assume that j, k, p, q are four different vertices.

Then we have

$$\begin{aligned} Q(r+1) &= Q(r) + \left((m_j + 1 - b_j)^2 - (m_j - b_j)^2\right) + \\ &+ \left((m_k + 1 - b_k)^2 - (m_k - b_k)^2\right) + \\ &+ \left((m_p - b_p - 1)^2 - (m_p - b_p)^2\right) + \left((m_q - b_q - 1)^2 - (m_q - b_q)^2\right) = \\ &= Q(r) + 4 + 2(m_j + m_k - b_j - b_k) - 2(m_p + m_q - b_p - b_q). \end{aligned} \tag{2.13}$$

Here the sum $(m_j + m_k - b_j - b_k)$ is the "total signed lead of Maker" at the two endpoints of his edge e, and the other sum $(m_p + m_q - b_p - b_q)$ is the "total signed lead of Maker" at the two endpoints of the opponent's edge f – both at stage r. Equality (2.13) suggests Maker's strategy: since he comes first, in his $(r+1)$st move he can choose the *best* unoccupied edge. Here *best* means the untaken edge with *maximum* "total signed lead". Then

$$\text{best} = m_j + m_k - b_j - b_k \geq m_p + m_q - b_p - b_q,$$

and so by (2.13),

$$Q(r+1) \geq Q(r) + 4. \tag{2.14}$$

Next comes

Case 2: The four endpoints j, k, p, q represent only three different vertices.

Then, by repeating the argument of Case 1, Maker can enforce (at least) the following weaker version of (2.14):

$$Q(r+1) \ge Q(r) + 2. \tag{2.15}$$

Let $|G|$ denote the number of edges of graph G; then each player has $|G|/2$ turns, and by repeated application of (2.14)–(2.15) we obtain

$$Q(\text{end}) = Q\big(|G|/2\big) \ge Q(\text{start}) + 2 \cdot \frac{|G|}{2} = |G|,$$

or equivalently, the inequality below

$$Q(\text{end}) = \sum_{i=1}^{n} (m_i - b_i)^2 \ge |G| \tag{2.16}$$

holds at the end of the play. By (2.16)

$$\max_i (m_i - b_i)^2 \ge \frac{|G|}{n} = \text{Density},$$

completing the proof. ∎

Unfortunately, the squaring in the quadratic sum (2.16) kills the sign. The argument above doesn't give any information whether the large Discrepancy comes from a large Surplus or from a large Deficit. Comparing Theorem 1 with Proposition 2.5, we can roughly say that "the large discrepancy can be converted into a large one-sided discrepancy (i.e., surplus) by a mere constant factor loss".

Note that there are many examples where the proof of a large one-sided discrepancy (i.e., surplus) is much harder than the proof of a large discrepancy. For example, Szemerédi's famous density theorem can be interpreted as a one-sided version of van der Waerden's theorem on arithmetic progressions. There is a general agreement that Szemerédi's theorem is much deeper than van der Waerden's theorem. For example, van der Waerden's theorem can be proved on 2 pages; on the other hand, Szemerédi's theorem has several proofs, but each one is at least 20 pages long.

A second example is a famous old problem in classical Fourier Analysis: estimating cosine sums. Let $1 \leq k_1 < k_2 < \cdots < k_n$ be an arbitrary set of n integers, and consider the cosine series

$$f(x) = \sum_{j=1}^{n} \cos(2\pi k_j x).$$

By Parseval's formula

$$\int_0^1 f^2(x)\, dx = \frac{n}{2},$$

implying

$$\max_x |f(x)| \geq \sqrt{n/2}.$$

What happens if we drop the absolute value? What can we say about the minimum? Is it true that the inequality

$$|\min_x f(x)| \geq c \cdot \sqrt{n} \tag{2.17}$$

holds with some (small) positive absolute constant factor c? Perhaps (2.17) is too much to expect; how about the much weaker lower bound

$$|\min_x f(x)| \geq n^c \tag{2.18}$$

with some (small) positive absolute constant power $0 < c < 1/2$? Unfortunately both (2.17) and (2.18) remain unsolved. The best known result has the following sub-polynomial order of magnitude

$$|\min_x f(x)| \geq e^{(\log n)^c}, \tag{2.19}$$

It was Bourgain who first proved (2.19); he had a small constant $c > 0$ (very hard proof!). The current record is due to I. Ruzsa who proved it with $c = 1/2$. It is shocking to see that for cosine sums a one-sided problem, namely estimating the minimum, is so much more difficult than the discrepancy problem.

There are many more examples of this strange phenomenon. The basic challenge is that there is no general recipe to convert large discrepancy to large one-sided discrepancy. In each particular case one has to invent a new technique.

3. Proof of Theorem 1

1. What is the difficulty in general? Unfortunately the proof technique of Proposition 2.3 heavily relies on uniformity, and it is not clear how to adapt it for non-uniform hypergraphs (which corresponds to non-regular graphs, i.e., the general case in Theorem 1).

Indeed, assume that G is an arbitrary (not necessarily regular) finite simple graph, and repeat the proof of Proposition 2.3 to the star-hypergraph $\mathcal{F} = \mathcal{F}(G)$ of G. Then we obtain the following perfect analog of (2.8):

$$P_{\text{end}} \geq N - \varepsilon^2 \left(\frac{1+\varepsilon}{1-\varepsilon}\right)^{\Delta} \frac{dN}{4}, \tag{3.1}$$

where $N = |\mathcal{F}|$ is the vertex-number of graph G, $d = (d_1 + \cdots + d_N)/N$ is the average degree of G, and $dN/4$ is the total length of the play.

On the other hand, if Δ denotes the largest positive discrepancy that Maker can achieve at the end of the play (it means $\frac{d_i}{2} + \Delta$ edges from some star of degree d_i), then we obtain the following version of (2.9):

(3.2)

$$P_{\text{end}} \leq \sum_{j=1}^{N} (1+\varepsilon)^{d_j/2+\Delta} \cdot (1-\varepsilon)^{d_j/2-\Delta} = \left(\frac{1+\varepsilon}{1-\varepsilon}\right)^{\Delta} \cdot \left(\sum_{j=1}^{N} (1-\varepsilon^2)^{d_j/2}\right),$$

where d_j, $j = 1, 2, \ldots, N$ are the degrees of graph G. Combining (3.1) and (3.2)

$$N - \varepsilon^2 \left(\frac{1+\varepsilon}{1-\varepsilon}\right)^{\Delta} \frac{dN}{4} \leq \left(\frac{1+\varepsilon}{1-\varepsilon}\right)^{\Delta} \cdot \left(\sum_{j=1}^{N} (1-\varepsilon^2)^{d_j/2}\right). \tag{3.3}$$

Simplifying (3.3) by N, and using the approximation $1 \pm \delta \approx e^{\pm\delta}$, we obtain

$$e^{2\varepsilon\Delta} \geq \frac{1}{\frac{\varepsilon^2 \cdot d}{4} + \frac{1}{N}\sum_{j=1}^{N} e^{-\varepsilon^2 d_j/2}}. \tag{3.4}$$

Now we are ready to understand the technical difficulty in the general (non-regular) case. If the overwhelming majority of degrees d_j are very close to

$d/2$, i.e., half of the average degree, then the right hand side of (3.4) is close to

$$\frac{1}{\frac{\varepsilon^2 \cdot d}{4} + \frac{1}{N}\sum_{j=1}^{N} e^{-\varepsilon^2 (d/2)/2}} = \frac{1}{y + e^{-y}} \quad \text{with} \quad y = \frac{\varepsilon^2 d}{4}. \tag{3.5}$$

Unfortunately the function $f(y) = y + e^{-y}$ is monotone increasing for $y \geq 0$ (since the derivative $f'(y) = 1 - e^{-y} \geq 0$ for $y \geq 0$), implying $f(y) \geq 1$ for all $y \geq 0$, which means that inequality (3.5) is useless. Whatever way we choose ε, we just cannot derive from (3.5) a nontrivial lower bound for Δ.

Note that there exist graphs for which the overwhelming majority of degrees d_j come very close to $d/2$, i.e., to the half of the average degree. Consider, for example, the very asymmetric complete bipartite graph $K_{m,r}$ where m is much larger than r. Then the average degree is $2mr/(m+r) \approx 2r$, and the overwhelming majority of the degrees are r, coming very close to the half of the average degree.

2. An alternative approach. To find a way around the technical difficulty outlined above, we are going to define a complicated potential function, which combines the variance with an exponential sum. The purpose of the extra exponential sum (see (3.8) below) is to compensate for the bad effect of squaring (which cannot distinguish "deficit" from "surplus"). Let G be an arbitrary simple finite graph, and let $H = C_G$ be a core of G. Let d denote the average degree of H, and let N denote the vertex-number of graph H. By Proposition 1.2 the minimum degree of H is $\geq d/2$. This fact will be used in the definition (3.6) below.

We work with the star-hypergraph $\mathcal{H} = \mathcal{H}(H)$ of graph H. Notice that $|\mathcal{H}| = N$. First we divide the hyperedges of $\mathcal{H}$ into classes according to their sizes. We have a power-of-two classification: let

$$\mathcal{H}_j = \left\{ A \in \mathcal{H} \colon d \cdot 2^{j-1} \leq |A| < d \cdot 2^j \right\} \quad \text{where} \quad j = 0, 1, 2, \ldots, r. \tag{3.6}$$

Since every hyperedge $A \in \mathcal{H}$ represents a star, and has the size of the degree, so trivially $|A| < |\mathcal{H}| = N$ (where N is the vertex-number); it follows that parameter r in (3.6) satisfies the inequality $r \leq \log_2(N/d)$ (base 2 logarithm).

Let

$$V_j = \bigcup_{A \in \mathcal{H}_j} A$$

denote the union set of $\mathcal{H}_j$, and let W_j denote the set of those elements in V_j which are contained by two sets $A \in \mathcal{H}_j$ (the height of $\mathcal{H}$ is two, so two is the maximum here). Thus we have the multi-set equality

$$\bigcup_{A\in\mathcal{H}_j}^{*} A = V_j \cup^* W_j, \tag{3.7}$$

where the mark * indicates that the elements of V_j are counted with multiplicity (i.e., with 1 or 2).

Assume we are in the middle of a play, Maker already occupied $x_1, x_2, \ldots, x_t$ and Breaker occupied $y_1, y_2, \ldots, y_t$; I refer to the elements as "points" and call t the time. Let $X(t) = \{x_1, x_2, \ldots, x_t\}$ and $Y(t) = \{y_1, y_2, \ldots, y_t\}$.

We define a modification of (2.12): the new potential function is

$$\begin{aligned} P(t) = \sum_{A_i\in\mathcal{H}} \Big(\big(|A_i \cap X(t)| - |A_i \cap Y(t)|\big)^2 - \\ - \alpha \cdot |A_i| \cdot (1-\lambda_i)^{|A_i\cap X(t)|}(1+\lambda_i)^{|A_i\cap Y(t)|}\Big) - \\ - \sum_{j=0}^{r} \alpha \cdot |V_j| \cdot (1-\mu_j)^{|V_j\cap X(t)|}(1+\mu_j)^{|V_j\cap Y(t)|} \\ - \sum_{j=0}^{r} \alpha \cdot |W_j| \cdot (1-\omega_j)^{|W_j\cap X(t)|}(1+\omega_j)^{|W_j\cap Y(t)|}, \end{aligned} \tag{3.8}$$

where $0 < \alpha < 1$, $0 < \lambda_i < 1$ for $1 \le i \le N = |\mathcal{H}|$, $0 < \mu_j < 1$ and $0 < \omega_j < 1$ hold for $0 \le j \le r$ (note that $r \le \log_2(N/d)$); these parameters will be specified later.

The basic idea, which will become very clear later, is that *exponential beats quadratic*, and this way we will be able to turn a large deficit into our benefit. Maker is going to force a *global balance* in every large set V_j and W_j, and this way a large amount of deficit automatically implies large surplus. We refer to this idea as Global Balancing.

By using potential function (3.9) we prove

Lemma 1. *Maker can always achieve that at the end of the play*

$$P(\text{end}) \ge Nd\left(\frac{1}{2} - 2\alpha\right),$$

where $N = |\mathcal{H}|$ and d is the average size of the hyperedges of $\mathcal{H}$.

Proof. We study the effect of the $(t+1)$st moves x_{t+1} (by Maker) and y_{t+1} (by Breaker) in potential (3.8). Since (3.8) is very complicated, for simplicity, first consider only the quadratic part in (3.8):

$$Q(t) = \sum_{A \in \mathcal{H}} \left(|A \cap X(t)| - |A \cap Y(t)| \right)^2, \tag{3.9}$$

and study the effect of the $(t+1)$st moves x_{t+1} (by Maker) and y_{t+1} (by Breaker) in (3.9). By using the trivial fact $(m \pm 1)^2 = m^2 \pm 2m + 1$, we have

(3.10)

$$\begin{aligned} Q(t+1) = Q(t) + &\sum_{A \in \mathcal{H}:\ x_{t+1} \in A \not\ni y_{t+1}} \Big(2\big(|A \cap X(t)| - |A \cap Y(t)|\big) + 1 \Big) + \\ &+ \sum_{A \in \mathcal{H}:\ y_{t+1} \in A \not\ni x_{t+1}} \Big(-2\big(|A \cap X(t)| - |A \cap Y(t)|\big) + 1 \Big) = \\ &= Q(t) + 2\Bigg(\sum_{A \in \mathcal{H}:\ x_{t+1} \in A} \big(|A \cap X(t)| - |A \cap Y(t)|\big) - \\ &- \sum_{A \in \mathcal{H}:\ y_{t+1} \in A} \big(|A \cap X(t)| - |A \cap Y(t)|\big) \Bigg) + \\ &+ \sum_{A \in \mathcal{H}:\ x_{t+1} \in A \not\ni y_{t+1}} 1 + \sum_{A \in \mathcal{H}:\ y_{t+1} \in A \not\ni x_{t+1}} 1. \end{aligned}$$

Since the height of hypergraph $\mathcal{H}$ is uniformly two, there are two hyperedges $A \in \mathcal{H}$ with $x_{t+1} \in A$, and by the almost disjointness of $\mathcal{H}$ at least one of them does not contain y_{t+1}. So we have

$$\sum_{A \in \mathcal{H}:\ x_{t+1} \in A \not\ni y_{t+1}} 1 \geq 1, \tag{3.11}$$

and similarly,

$$\sum_{A \in \mathcal{H}:\ y_{t+1} \in A \not\ni x_{t+1}} 1 \geq 1. \tag{3.12}$$

Now we are ready to handle the long potential (3.8). We have the long equality (a variant of (3.10)):

$$P(t+1) = P(t) + \tag{3.13}$$

$$+2\Bigg(\sum_{A_i\in\mathcal{H}:\ x_{t+1}\in A_i}\big(|A_i\cap X(t)|-|A_i\cap Y(t)|\big)-$$

$$-\sum_{A_i\in\mathcal{H}:\ y_{t+1}\in A_i}\big(|A_i\cap X(t)|-|A_i\cap Y(t)|\big)\Bigg)+$$

$$+\sum_{A_i\in\mathcal{H}:\ x_{t+1}\in A_i\not\ni y_{t+1}}1+\sum_{A_i\in\mathcal{H}:\ y_{t+1}\in A_i\not\ni x_{t+1}}1+$$

$$+\alpha\sum_{A_i\in\mathcal{H}:\ x_{t+1}\in A_i}|A_i|\cdot\lambda_i\cdot(1-\lambda_i)^{|A_i\cap X(t)|}(1+\lambda_i)^{|A_i\cap Y(t)|}-$$

$$-\alpha\sum_{A_i\in\mathcal{H}:\ y_{t+1}\in A_i}|A_i|\cdot\lambda_i\cdot(1-\lambda_i)^{|A_i\cap X(t)|}(1+\lambda_i)^{|A_i\cap Y(t)|}+$$

$$+\alpha\sum_{j:\ x_{t+1}\in V_j}|V_j|\cdot\mu_j\cdot(1-\mu_j)^{|V_j\cap X(t)|}(1+\mu_j)^{|V_j\cap Y(t)|}-$$

$$-\alpha\sum_{j:\ y_{t+1}\in V_j}|V_j|\cdot\mu_j\cdot(1-\mu_j)^{|V_j\cap X(t)|}(1+\mu_j)^{|V_j\cap Y(t)|}+$$

$$+\alpha\sum_{j:\ x_{t+1}\in W_j}|W_j|\cdot\omega_j\cdot(1-\omega_j)^{|W_j\cap X(t)|}(1+\omega_j)^{|W_j\cap Y(t)|}-$$

$$-\alpha\sum_{j:\ y_{t+1}\in W_j}|W_j|\cdot\omega_j\cdot(1-\omega_j)^{|W_j\cap X(t)|}(1+\omega_j)^{|W_j\cap Y(t)|}+$$

$$+\text{ some positive terms.}$$

Here the "some positive terms" comes from the fact that $1-(1-\lambda)(1+\lambda) = \lambda^2 \ge 0$. Ignoring the "some positive terms", we can rewrite the long equality (3.13) in the form of a much shorter inequality:

$$P(t+1) \ge P(t) + \sum_{A_i\in\mathcal{H}:\ x_{t+1}\in A_i\not\ni y_{t+1}}1+\sum_{A_i\in\mathcal{H}:\ y_{t+1}\in A_i\not\ni x_{t+1}}1+ \tag{3.14}$$

$$+F_t(x_{t+1})-F_t(y_{t+1}),$$

where $F_t(z)$ is some positive function defined for all unoccupied points $z \notin X(t) \cup Y(t)$ before the $(t+1)$st turn.

Since Maker's x_{t+1} is selected before Breaker's y_{t+1}, Maker can select the "best" point: Maker chooses that $x_{t+1} = z \notin X(t) \cup Y(t)$ for which the function $F_t(z)$ attains its *maximum*. Then $F_t(x_{t+1}) \geq F_t(y_{t+1})$, and by (3.14),

$$(3.15) \quad P(t+1) \geq P(t) + \sum_{A_i \in \mathcal{H}:\ x_{t+1} \in A_i \not\ni y_{t+1}} 1 + \sum_{A_i \in \mathcal{H}:\ y_{t+1} \in A_i \not\ni x_{t+1}} 1 \geq$$

$$\geq P(t) + 2.$$

In the last step we used (3.11)–(3.12).

Since each player takes one point per turn, the total time of the play is $T = Nd/4$ where $d = (d_1 + \cdots + d_N)/N$ is the average hyperedge size (note that $|A_i| = d_i$ and 2 is the common height). By repeated application of (3.15) we obtain

$$(3.16) \qquad P(\text{end}) = P(T) \geq P(0) + 2T = P(0) + \frac{Nd}{2}.$$

By definition (see (3.8))

$$P(0) = -\alpha \sum_{i=1}^{N} |A_i| - \alpha \sum_{j=1}^{r} \left(|V_j| + |W_j| \right) = -2\alpha \sum_{i=1}^{N} |A_i| = -2\alpha N d.$$

Using this in (3.16) we conclude

$$P(\text{end}) \geq Nd \left(\frac{1}{2} - 2\alpha \right),$$

which completes the proof of Lemma 1. ■

3. Global balancing. First we apply a standard average argument in Lemma 1 as follows. By using the power-of-two decomposition of hypergraph $\mathcal{H}$ in (3.6), and applying the two convergent (telescoping) series

$$\frac{1}{1 \cdot 2} + \frac{1}{2 \cdot 3} + \frac{1}{3 \cdot 4} + \cdots = \left(1 - \frac{1}{2}\right) + \left(\frac{1}{2} - \frac{1}{3}\right) + \left(\frac{1}{3} - \frac{1}{4}\right) + \cdots = 1$$

and

$$\frac{1}{3\cdot 4}+\frac{1}{4\cdot 5}+\frac{1}{5\cdot 6}+\cdots=\left(\frac{1}{3}-\frac{1}{4}\right)+\left(\frac{1}{4}-\frac{1}{5}\right)+\left(\frac{1}{5}-\frac{1}{6}\right)+\cdots=\frac{1}{3}$$

in Lemma 1, we obtain that there is an integer j_0 in the interval $0 \le j_0 \le r$ (where $r \le \log_2(N/d)$) such that

$$\sum_{A_i\in\mathcal{H}_{j_0}} \left((M_i-B_i)^2-\alpha|A_i|\cdot(1-\lambda_i)^{M_i}(1+\lambda_i)^{B_i}\right)- \tag{3.17}$$

$$-\alpha\cdot|V_{j_0}|\cdot(1-\mu_{j_0})^{M(V_{j_0})}(1+\mu_{j_0})^{B(V_{j_0})}$$

$$-\alpha\cdot|W_{j_0}|\cdot(1-\omega_{j_0})^{M(W_{j_0})}(1+\omega_{j_0})^{B(W_{j_0})}\ge$$

$$\ge\frac{\frac{3}{4}Nd\left(\frac{1}{2}-2\alpha\right)}{\min\left\{(r-j_0+1)(r-j_0+2),(j_0+3)(j_0+4)\right\}},$$

where $M_i=\big|A_i\cap X(t)\big|$ is Maker's part in A_i and $B_i=\big|A_i\cap Y(t)\big|$ is Breaker's part in A_i at this stage of the play; similarly, $M(V_j)=\big|V_j\cap X(t)\big|$, $B(V_j)=\big|V_j\cap Y(t)\big|$, $M(W_j)=\big|W_j\cap X(t)\big|$, and $B(W_j)=\big|W_j\cap Y(t)\big|$.

It is time now to specify at least the first two of the parameters $0<\alpha<1$, $0<\lambda_i<1$ (for $1\le i\le N=|\mathcal{H}|$), $0<\mu_j<1$ and $0<\omega_j<1$ (for $0\le j\le r$ where $r\le\log_2(N/d)$) in (3.8).

Let $\lambda_i=3/\sqrt{d_i}$ where $|A_i|=d_i$. Then

$$(1-\lambda_i)^{M_i}(1+\lambda_i)^{B_i}=(1-\lambda_i^2)^{d_i/2}\left(\frac{1+\lambda_i}{1-\lambda_i}\right)^{(B_i-M_i)/2}\approx \tag{3.18}$$

$$\approx e^{-9/2+3(B_i-M_i)/\sqrt{d_i}}.$$

Next let $\alpha=1/10$. Using the easy fact (I challenge the reader to verify it)

$$10u^2\le e^{-9/2+3u}\quad\text{for all}\quad u\ge 3 \tag{3.19}$$

with $u=(B_i-M_i)/\sqrt{d_i}$, by (3.17)–(3.19) we easily have

$$\sum_{A_i\in\mathcal{H}_{j_0}:\,(M_i-B_i)\ge-3\sqrt{d_i}} (M_i-B_i)^2- \tag{3.20}$$

$$-\frac{1}{10}|V_{j_0}|\cdot(1-\mu_{j_0})^{M(V_{j_0})}(1+\mu_{j_0})^{B(V_{j_0})}$$

$$- \frac{1}{10}|W_{j_0}| \cdot (1-\omega_{j_0})^{M(W_{j_0})}(1+\omega_{j_0})^{B(W_{j_0})} \geq$$

$$\geq \frac{\frac{3}{4}Nd\left(\frac{1}{2}-\frac{2}{10}\right)}{\min\left\{(r-j_0+1)(r-j_0+2),(j_0+3)(j_0+4)\right\}} =$$

$$= \frac{\frac{9}{40}Nd}{\min\left\{(r-j_0+1)(r-j_0+2),(j_0+3)(j_0+4)\right\}}.$$

Now we can see the first benefit of the complicated potential function (3.8). The negative exponential terms involving λ_i neutralize the effect of the very large deficits in the quadratic part; only the moderately large deficits (namely those with $M_i - B_i \geq -3\sqrt{d_i}$) survive in (3.20).

We still face the annoying possibility that the main contribution in the quadratic part of (3.20) may well come exclusively from the moderately large deficits; in that case it is not clear how to guarantee large surplus. To overcome this second type of difficulty we use the two remaining exponential terms involving the large sets V_{j_0} and W_{j_0} in (3.20). I call this technique "global balancing", and it goes as follows.

I begin with a trivial corollary of (3.20):

(3.21)
$$\frac{1}{10}|V_{j_0}| \cdot (1-\mu_{j_0})^{M(V_{j_0})}(1+\mu_{j_0})^{B(V_{j_0})} \leq \sum_{A_i \in \mathcal{H}_{j_0}:\ (M_i-B_i)\geq -3\sqrt{d_i}} (M_i-B_i)^2.$$

If in the right side of (3.21) there is a term $(M_i - B_i) \geq 3\sqrt{d_i}$ for some i, then of course Maker can keep this temporary lead till the end of the play, and achieves a surplus $\geq 3\sqrt{d_i} \geq 3\sqrt{d/2}$, which is a much better constant factor than what we claimed in Theorem 1. We can thus assume that

(3.22)
$$\sum_{A_i \in \mathcal{H}_{j_0}:\ (M_i-B_i)\geq -3\sqrt{d_i}} (M_i-B_i)^2 \leq \sum_{A_i \in \mathcal{H}_{j_0}} 9d_i =$$

$$= 9 \sum_{A_i \in \mathcal{H}_{j_0}} |A_i| \leq 9|\mathcal{H}_{j_0}|d2^{j_0},$$

since $|A_i| = d_i \leq d2^j$ holds for all $A_i \in \mathcal{H}_j$ (see (3.6)).

On the other hand, since the common height of $\mathcal{H}$ is two, we trivially have

$$|V_j| = \left|\bigcup_{A\in\mathcal{H}_j} A\right| \geq \frac{1}{2}\sum_{A\in\mathcal{H}_j} |A| \geq \frac{1}{2}|\mathcal{H}_j|d2^{j-1}. \tag{3.23}$$

Using (3.23) with $j = j_0$, and comparing it to (3.22), we obtain the following consequence of (3.21):

$$(1-\mu_{j_0})^{M(V_{j_0})}(1+\mu_{j_0})^{B(V_{j_0})} \leq 360,$$

or equivalently,

$$(1-\mu_{j_0}^2)^{|V_{j_0}|/2}\left(\frac{1+\mu_{j_0}}{1-\mu_{j_0}}\right)^{(B-M)/2} \leq 360, \tag{3.24}$$

where $B = B(V_{j_0})$ and $M = M(V_{j_0})$.

By choosing $\mu_j = 3/\sqrt{|V_j|}$ in (3.8), we can approximate the left side of (3.24) as follows:

$$360 \geq (1-\mu_{j_0}^2)^{|V_{j_0}|/2}\left(\frac{1+\mu_{j_0}}{1-\mu_{j_0}}\right)^{(B-M)/2} \approx e^{-9/2+3(B-M)/\sqrt{|V_{j_0}|}}. \tag{3.25}$$

Comparing the two ends of (3.25) we obtain

$$B - M = B(V_{j_0}) - M(V_{j_0}) \leq 4\sqrt{|V_{j_0}|}, \tag{3.26}$$

which means that Breaker can never achieve a substantial lead in the large set V_{j_0}. This is the "global balancing" that I was talking about.

By choosing $\omega_j = 3/\sqrt{|W_j|}$ we can repeat the same argument for W_{j_0}, and obtain

$$B(W_{j_0}) - M(W_{j_0}) \leq 4\sqrt{|W_{j_0}|} \leq 4\sqrt{|V_{j_0}|}, \tag{3.27}$$

By (3.7) and (3.26)–(3.27):

$$\sum_{A_i\in\mathcal{H}_{j_0}} (M_i - B_i) = \big(M(V_{j_0}) - B(V_{j_0})\big) + \big(M(W_{j_0}) - B(W_{j_0})\big) \geq -8\sqrt{|V_{j_0}|}. \tag{3.28}$$

4. An average argument. Combining (3.28) with (3.20) we can easily finish the proof of Theorem 1 by using a standard average argument. First we rewrite (3.28) as follows:

$$\sum_{A_i\in\mathcal{H}_{j_0}:\,M_i>B_i}(M_i-B_i)+8\sqrt{|V_{j_0}|}\geq \tag{3.29}$$

$$\geq\sum_{A_i\in\mathcal{H}_{j_0}:\,B_i>M_i}(B_i-M_i)\geq\sum_{A_i\in\mathcal{H}_{j_0}:\,0\leq B_i-M_i\leq 3\sqrt{d_i}}(B_i-M_i).$$

Since $|A_i|=d_i\leq d2^{j_0}$ holds for every $A_i\in\mathcal{H}_{j_0}$, (3.29) trivially implies

$$\sum_{A_i\in\mathcal{H}_{j_0}:\,0\leq B_i-M_i\leq 3\sqrt{d_i}}(B_i-M_i)^2\leq$$

$$\leq 3\sqrt{d2^{j_0}}\left(\sum_{A_i\in\mathcal{H}_{j_0}:\,M_i>B_i}(M_i-B_i)+8\sqrt{|V_{j_0}|}\right).$$

Combining the last inequality with (3.20)

$$\sum_{A_i\in\mathcal{H}_{j_0}:\,M_i>B_i}(M_i-B_i)^2\geq \tag{3.30}$$

$$\geq\frac{\frac{9}{40}Nd}{\min\big\{(r-j_0+1)(r-j_0+2),(j_0+3)(j_0+4)\big\}}-$$

$$-3\sqrt{d2^{j_0}}\sum_{A_i\in\mathcal{H}_{j_0}:\,M_i>B_i}(M_i-B_i)-24\sqrt{d2^{j_0}|V_{j_0}|}.$$

It is more convenient to rewrite (3.30) in the form

$$\sum_{A_i\in\mathcal{H}_{j_0}:\,M_i>B_i}(M_i-B_i)^2+3\sqrt{d2^{j_0}}\sum_{A_i\in\mathcal{H}_{j_0}:\,M_i>B_i}(M_i-B_i)\geq \tag{3.31}$$

$$\geq\frac{\frac{9}{40}Nd}{\min\big\{(r-j_0+1)(r-j_0+2),(j_0+3)(j_0+4)\big\}}-24\sqrt{d2^{j_0}|V_{j_0}|}.$$

The obvious benefit of (3.31) is that the left side does not contain any deficit. Of course we have to make sure that the right side of (3.31) is large positive. To check this, first notice that hypergraph $\mathcal{H}$ has exactly $Nd/2$ points, so trivially $|V_{j_0}| \le Nd/2$. Another trivial inequality is $d2^r \le N$, and so

$$2^{j_0} \le 2^{j_0-r} \cdot 2^r \le 2^{j_0-r} \cdot \frac{N}{d}.$$

Combining these trivial facts we have

$$d2^{j_0}|V_{j_0}| \le 2^{j_0-r} d \frac{N}{d} \frac{Nd}{2} = 2^{j_0-r-1} N^2 d,$$

and so

$$\sqrt{d2^{j_0}|V_{j_0}|} \le 2^{(j_0-r-1)/2} N\sqrt{d}. \tag{3.32}$$

By (3.31) and (3.32)

$$\sum_{A_i\in\mathcal{H}_{j_0}:\ M_i>B_i} (M_i - B_i)^2 + 3\sqrt{d2^{j_0}} \sum_{A_i\in\mathcal{H}_{j_0}:\ M_i>B_i} (M_i - B_i) \ge \tag{3.33}$$

$$\ge \frac{\frac{9}{40}Nd}{\min\left\{(r-j_0+1)(r-j_0+2), (j_0+3)(j_0+4)\right\}} - 24\cdot 2^{(j_0-r-1)/2} N\sqrt{d} \ge$$

$$\ge \frac{\frac{1}{6}Nd}{\min\left\{(r-j_0+1)(r-j_0+2), (j_0+3)(j_0+4)\right\}},$$

assuming $d \ge 10^6$; the worst case is $j_0 = r$.

The last trivial step is to distinguish two cases.

The first case is when

$$\sum_{A_i\in\mathcal{H}_{j_0}:\ M_i>B_i} (M_i - B_i)^2 \ge \tag{3.34}$$

$$\ge \frac{\frac{1}{42}Nd}{\min\left\{(r-j_0+1)(r-j_0+2), (j_0+3)(j_0+4)\right\}}.$$

We need the trivial fact

$$|\mathcal{H}_{j_0}|d2^{j_0-1} \le \sum_{i=1}^{N} |A_i| = Nd,$$

which implies

$$|\mathcal{H}_{j_0}| \le 2^{-j_0+1}N. \tag{3.35}$$

Combining (3.34) and (3.35) we have

$$\max_{A_i\in\mathcal{H}_{j_0}:\, M_i>B_i} (M_i-B_i)^2 \ge \frac{1}{|\mathcal{H}_{j_0}|} \sum_{A_i\in\mathcal{H}_{j_0}:\, M_i>B_i} (M_i-B_i)^2 \ge \tag{3.36}$$

$$\ge \frac{2^{j_0-1}}{N} \frac{\frac{1}{42}Nd}{\min\{\,(r-j_0+1)(r-j_0+2),(j_0+3)(j_0+4)\}} =$$

$$= \frac{\frac{1}{42}2^{j_0-1}}{\min\{\,(r-j_0+1)(r-j_0+2),(j_0+3)(j_0+4)\}} d.$$

It follows that

$$\max_{A_i\in\mathcal{H}_{j_0}:\, M_i>B_i} (M_i-B_i) \ge \frac{\sqrt{d}}{200},$$

which is a better constant factor than what we claimed in Theorem 1.

The second case is when (3.34) fails; then we clearly have

$$3\sqrt{d2^{j_0}} \sum_{A_i\in\mathcal{H}_{j_0}:\, M_i>B_i} (M_i-B_i) \ge \tag{3.37}$$

$$\ge \frac{\frac{1}{7}Nd}{\min\{\,(r-j_0+1)(r-j_0+2),(j_0+3)(j_0+4)\}}.$$

Then

$$\max_{A_i\in\mathcal{H}_{j_0}:\, M_i>B_i} (M_i-B_i) \ge \frac{1}{3\sqrt{d2^{j_0}}|\mathcal{H}_{j_0}|} \sum_{A_i\in\mathcal{H}_{j_0}:\, M_i>B_i} (M_i-B_i) \ge \tag{3.38}$$

$$\ge \frac{1}{3\sqrt{d2^{j_0}}|\mathcal{H}_{j_0}|} \frac{\frac{1}{7}Nd}{\min\{\,(r-j_0+1)(r-j_0+2),(j_0+3)(j_0+4)\}}.$$

Again using (3.35) we obtain the following lower bound for the last term in (3.38):

(3.39)

$$\frac{1}{3\sqrt{d2^{j_0}}2^{-j_0+1}N} \frac{\frac{1}{7}Nd}{\min\{\,(r-j_0+1)(r-j_0+2),(j_0+3)(j_0+4)\}} \ge \frac{\sqrt{d}}{1000},$$

since $j_0 \geq 0$. Combining (3.38)–(3.39), Theorem 1 follows. ■

4. Proof of Theorem 2: the Upper Bound

Again we work with a "danger function", which has the unfortunate side-effect of (weakly) depending on the global size (i.e., the vertex number) of the d-regular graph. We develop the potential technique of Propositions 2.1-2 in a new direction, that I call the

Big-Game&small-game decomposition. The first basic idea is a decomposition of the Degree Game – played on a d-regular graph – into two non-interacting games: the Big Game and the small game. The second basic idea is iteration; in other words, the proof has a nested structure. Let n be the vertex number of graph G, let $\{S_i : i = 1, 2, \ldots, n\}$ be the n stars of G, each containing d edges. In the star-hypergraph $\mathcal{H} = \mathcal{H}_G$ of G the "points" are the edges of G, and the hyperedges are the stars S_i, $1 \leq i \leq n$. So $\mathcal{H}_G$ is a d-uniform hypergraph, $|\mathcal{H}_G| = n$, $\mathcal{H}_G$ is Almost Disjoint, and has uniform Height 2.

A play in the Degree Game on G is clearly equivalent to a play on the star-hypergraph $\mathcal{H}_G$; I prefer to work with the hypergraph setup.

Consider now a concrete play on $\mathcal{H}_G$; let $x_1, x_2, x_3, \ldots$ denote the points (i.e., edges of G) taken by Maker and let $y_1, y_2, y_3, \ldots$ denote the points taken by Breaker. A hyperedge $S \in \mathcal{H}_G$ (i.e., a star of G) becomes "dangerous" when Maker's lead equals $4\sqrt{d}\log d$ for the first time; formally,

$$\big| |X(t) \cap S| - |Y(t) \cap S| \big| = 4\sqrt{d}\log d, \tag{4.1}$$

where $X(t) = \{x_1, x_2, \ldots, x_{t+1}\}$ and where $Y(t) = \{y_1, y_2, \ldots, t_t\}$ (t is the time parameter). The unoccupied part $S \setminus \big(X(t) \cup Y(t)\big)$ of S is called an "emergency set". Whenever an "emergency set" arises, its points are removed from the Big Board (i.e., the board of the Big Game), and they are added to the small board (i.e., the board of the small game). This is why the Big Game is "shrinking" and the small game is "growing".

The "growing" small board is exactly the union of all emergency sets. If a hyperedge $S \in \mathcal{H}_G$ intersects the small board, but S is not "dangerous" (yet), then I call S an "innocent bystander", or simply an "innocent".

Every emergency set $E = S \setminus \big(X(t) \cup Y(t)\big)$ has a unique superset $S \in \mathcal{H}_G$; uniqueness follows from the almost disjointness of the hypergraph. I denote this unique superset S (containing E) as $\widetilde{E}$.

Let

$$\mathcal{H}_{\text{small}} = \{\widetilde{E} \in \mathcal{H}_G \colon E \text{ is an emergency set } \} \cup \\ \cup \{I \colon I \in \mathcal{H}_G \colon I \text{ is an innocent}\}.$$

I call $\mathcal{H}_{\text{small}}$ the *small hypergraph;* it is a sub-hypergraph of $\mathcal{H}_G$, formally, $\mathcal{H}_{\text{small}} \subset \mathcal{H}_G$.

The small hypergraph $\mathcal{H}_{\text{small}}$ may fall apart into several components $\mathcal{C}_1$, $\mathcal{C}_2$, $\mathcal{C}_3, \ldots$. Here comes the **core idea** of the proof: Breaker wants to force that each component $\mathcal{C}_i$ of $\mathcal{H}_{\text{small}}$ consist of relatively few sets.

In order to control the components of $\mathcal{H}_{\text{small}}$, Breaker defines an auxiliary *Big Hypergraph* $\mathcal{B}$. To motivate the (nontrivial!) definition of $\mathcal{B}$, we assume that at some stage of the play some component of $\mathcal{H}_{\text{small}}$, say, $\mathcal{C}_1$, has "too many" sets. Assume that $\mathcal{C}_1$ has M sets where $M = E^* + I^*$: E^* emergency sets and I^* innocents, and M is "large".

Step 1. *We prove the inequality $E^* \geq I^*/d$, where d is the degree of the regular graph G*

Notice that the proof of this inequality is a routine double-counting argument. Let P denote the number of intersecting pairs $(\widetilde{E}, I)$ where $\widetilde{E} \in \mathcal{C}_1$ and $I \in \mathcal{C}_1$ $(\widetilde{E} \cap I \neq \emptyset)$. Clearly $E^*d \geq P \geq I^*$. Indeed, every innocent $I \in \mathcal{C}_1$ intersects an emergency set; this proves the first half $P \geq I^*$. On the other hand, every emergency superset $\widetilde{E}$ intersects at most d innocents (because of the Almost Disjointness and the Height 2 property of hypergraph $\mathcal{H}_G$). This proves the other half $E^*d \geq P$, completing $E^*d \geq I^*$.

Next comes

Step 2 *We select many disjoint emergency supersets by a greedy algorithm*

Let $\widetilde{E_1} \in \mathcal{C}_1$ be an arbitrary emergency superset; if $|\mathcal{C}_1| = M \geq d+1$ then by Step 1 we have such an $\widetilde{E_1}$. Assume that the 3rd $\mathcal{C}_1$-neighborhood of $\widetilde{E_1}$ is nonempty. If the 3rd neighborhood contains an emergency superset then let $\widetilde{E_2}$ be one of them. If the 3rd neighborhood entirely consists of innocents, then select one, say, I. Since every innocent intersects an emergency set, there is an emergency superset $\widetilde{E_2}$ that is either the 2nd or 3rd or 4th neighbor of $\widetilde{E_1}$. In any case $\widetilde{E_1}$ and $\widetilde{E_2}$ are disjoint.

Next assume that the 3rd $\mathcal{C}_1$-neighborhood of $\{\widetilde{E_1}, \widetilde{E_2}\}$ is nonempty. If the 3rd neighborhood contains an emergency superset then let $\widetilde{E_3}$ be one of them. If the 3rd neighborhood entirely consists of innocents, then select one, say, I. Since every innocent intersects an emergency set, there is an emergency superset $\widetilde{E_3}$ that is either the 2nd or 3rd or 4th neighbor of $\{\widetilde{E_1}, \widetilde{E_2}\}$. In any case $\widetilde{E_1}$, $\widetilde{E_2}$, and $\widetilde{E_3}$ are disjoint.

Repeating this argument we obtain a sequence of disjoint emergency supersets $\widetilde{E_1}, \widetilde{E_2}, \ldots, \widetilde{E_k}, \ldots$ from component $\mathcal{C}_1$ such that the $\mathcal{C}_1$-distance of $\widetilde{E_k}$ from $\{\widetilde{E_1}, \ldots, \widetilde{E_{k-1}}\}$ is either 2 or 3 or 4. This way we can sequentially build a *tree*, where the edges are labeled by 2 or 3 or 4. The numbers represent the $\mathcal{C}_1$-distance from the nearest neighbor among the predecessors.

Step 3 *Analysis of the greedy algorithm*

Suppose that the greedy algorithm described in Step 2 already produced a sequence of disjoint emergency supersets $\widetilde{E_1}, \widetilde{E_2}, \ldots, \widetilde{E_k}$ from component $\mathcal{C}_1$; what can prevent us from obtaining a new member $\widetilde{E_{k+1}}$? The union

$$\widetilde{E_1} \cup \widetilde{E_2} \cup \ldots \cup \widetilde{E_k}$$

has kd points, so the first $\mathcal{C}_1$-neighborhood of $\{\widetilde{E_1}, \widetilde{E_2}, \ldots, \widetilde{E_k}\}$ has at most kd^2 sets, and the first and second $\mathcal{C}_1$-neighborhoods of $\{\widetilde{E_1}, \widetilde{E_2}, \ldots, \widetilde{E_k}\}$ together have at most kd^3 sets. Therefore, if $|\mathcal{C}_1| > kd^3$ then the 3rd $\mathcal{C}_1$-neighborhood of $\{\widetilde{E_1}, \widetilde{E_2}, \ldots, \widetilde{E_k}\}$ is nonempty, and by the greedy algorithm we can find a new member $\widetilde{E_{k+1}}$. It follows that the greedy algorithm produces at least r disjoint emergency supersets $\widetilde{E_1}, \widetilde{E_2}, \ldots, \widetilde{E_r}$ from component $\mathcal{C}_1$ with $r = |\mathcal{C}_1|d^{-3}$.

Next comes

Step 4 *A counting argument*

We say that a sub-hypergraph $\{S_1, S_2, \ldots, S_r\} \subset \mathcal{H}_G$ has the $(2,4)$-*property* if for every $j = 2, 3, \ldots, r$, the $\mathcal{H}_G$-distance of S_j from $\{S_1, S_2, \ldots, S_{j-1}\}$ is 2 or 3 or 4. (It is a byproduct of ≥ 2 that the r sets are pairwise disjoint.)

If a sub-hypergraph $\{S_1, S_2, \ldots, S_r\} \subset \mathcal{H}_G$ has the $(2,4)$-property then we can represent it with a (2-or-3-or-4 labeled) tree, where every distance is 2 or 3 or 4. I call this tree the *underlying tree* of the sub-hypergraph.

Given any r, we can easily estimate from above the number of r-element sub-hypergraphs $\{S_1, S_2, \ldots, S_r\} \subset \mathcal{H}_G$ satisfying the $(2,4)$-property:

$$\text{total number} \leq 4^{r-1} \cdot n \cdot \left(d^4\right)^{r-1}. \tag{4.2}$$

To prove (4.2) first note that there are less than 4^{r-1} unlabeled rooted trees on r vertices. This well-known fact is a byproduct of the "depth-first search" algorithm, which visits every edge of the rooted tree twice (starting from the root). It associates with every rooted tree a $(+1,-1)$ sequence of length $2(r-1)$ ("$+1$" for forward visit and "-1" for backward visit), and the total number of such sequences is $2^{2(r-1)} = 4^{r-1}$.

Thus there are less than 4^{r-1} ways to fix the underlying tree of $\{S_1, S_2, \ldots, S_r\}$. There are $|\mathcal{H}_G| = n$ ways to choose the first set S_1. Since the $\mathcal{H}_G$-distance of S_2 from S_1 is ≤ 4, there are $\leq d^4$ ways to choose S_2. Working with a fixed underlying tree, there are $\leq d^4$ ways to choose S_3, there are $\leq d^4$ ways to choose S_4, and so on. This proves (4.2).

Step 5 *Defining the Big Sets $B \in \mathcal{B}$*

Now we are ready to define the Big Hypergraph $\mathcal{B}$. Every Big Set B arises as the union set

$$B = \bigcup_{i=1}^{r} S_i \tag{4.3}$$

of an arbitrary r-element sub-hypergraph $\{S_1, S_2, \ldots, S_r\} \subset \mathcal{H}_G$ satisfying the $(2,4)$-property. We will specify the value of parameter r later.

The Big Hypergraph $\mathcal{B}$ is simply the family of all possible Big Sets B (defined by (4.3)).

By (4.2)

$$|\mathcal{B}| \leq 4^{r-1} \cdot n \cdot \left(d^4\right)^{r-1}. \tag{4.4}$$

Step 6 *Balancing Lemma*

The following is just a restatement of Proposition 2.2.

Balancing Lemma. *Let $\mathcal{F}$ be a k-uniform hypergraph, and consider the Balancer-Unbalancer game played on hypergraph $\mathcal{F}$ where Unbalancer's* *goal is to own at least* $\frac{k+\Delta}{2}$ *points from some* $A \in \mathcal{F}$. *If* $\log|\mathcal{F}| < k/8$ *and*

$$\Delta = 2\sqrt{2k \log|\mathcal{F}|},$$

then Balancer has a winning strategy, that is, Balancer can prevent Unbalancer from achieving his goal.

The proof is a straightforward application of Proposition 2.1; I leave the details to the reader.

Step 7 *Applying the Balancing Lemma to the Big Hypergraph*

To prevent the appearance of a "too large" component $\mathcal{C}_1$ of $\mathcal{H}_{\text{small}}$, Breaker plays an auxiliary Big Game. The board of the Big Game, called the Big Board, is the board of $\mathcal{H}_G$ minus the emergency sets. In the Big Game, Breaker restricts himself to the Big Board, and tries to prevent any Big Set from becoming overwhelmingly owned by Maker.

Assume that, at some stage of the play, there is a "too large" component, say, $\mathcal{C}_1$ of $\mathcal{H}_{\text{small}}$. Here "too large" means $|\mathcal{C}_1| \geq rd^3$. Then by Step 3 the greedy algorithm applied to $\mathcal{C}_1$ produces r disjoint emergency supersets $\widetilde{E_1}, \widetilde{E_2}, \ldots, \widetilde{E_r}$, which satisfies the $(2,4)$-property. By Step 5 the union set

$$B = \bigcup_{i=1}^{r} \widetilde{E_i}$$

is a Big Set, and the intersection of this B with the Big Board (meaning: we throw away the emergency sets) has a Maker's lead $r \cdot 4\sqrt{d}\log d$ (see the definition of "dangerous" in (4.1)). On the other hand, if Breaker applies the potential strategy of the Balancing Lemma (see Step 6) for the Big Hypergraph $\mathcal{B}$ in the Big Game, then Breaker (as Balancer) can force that

$$\text{Maker's maximum lead} \leq 2\sqrt{2d \cdot r \cdot \log|\mathcal{B}|},$$

assuming $\log|\mathcal{B}| < d \cdot r/8$.

Therefore, if

$$4r\sqrt{d}\log d > 2\sqrt{2d \cdot r \cdot \log|\mathcal{B}|} \tag{4.5}$$

and

$$\log|\mathcal{B}| < \frac{dr}{8}, \tag{4.6}$$

then we obtain a contradiction, so the assumption $|\mathcal{C}_1| \geq rd^3$ is impossible. In other words, if Breaker plays rationally in the Big Game, and (4.5)–(4.6) hold, then every component $\mathcal{C}_i$ of $\mathcal{H}_{\text{small}}$ has less than rd^3 sets. This is how, by playing rationally in the Big Game, Breaker can control the components of $\mathcal{H}_{\text{small}}$, making sure that no component grows too large.

Step 8 *Specifying parameter r*

Let $r = \log n = \log|\mathcal{H}_G|$. We show that if $d \geq 200$ then inequality (4.6) holds. Indeed, we recall (4.4):

$$|\mathcal{B}| \leq 4^{r-1} \cdot n \cdot \left(d^4\right)^{r-1}, \tag{4.7}$$

and combining the logarithm of (4.7) with $r = \log n$ we have

$$\begin{aligned} \log|\mathcal{B}| &\leq \log n + (r-1)(4\log d + \log 4) = \\ &= r + (r-1)(4\log d + \log 4) < \frac{rd}{8} \end{aligned} \tag{4.8}$$

if $d \geq 200$.

Next we check (4.5): if $r = \log n$ then using the calculation in (4.8) we have

$$\begin{aligned} 2\sqrt{2d \cdot r \cdot \log|\mathcal{B}|} &\leq 2\sqrt{2d \cdot r \cdot \log\left(r + (r-1)(4\log d + \log 4)\right)} < \\ &< \sqrt{d} \cdot r \cdot 4\sqrt{2\log d + 1} < 4r\sqrt{d}\log d \end{aligned} \tag{4.9}$$

if $d \geq e^3$. This shows that (4.5) holds if $d \geq e^3$.

Combining (4.8)–(4.9) we see that, if Breaker plays rationally in the Big Game, then he can force that, at every stage of the play, every component of $\mathcal{H}_{\text{small}}$ has less than $rd^3 = d^3 \log n$ sets ("set" means emergency supersets and innocents) if $d \geq 200$.

Step 9 *Iteration*

Suppose that we are in the middle of a play, hypergraph $\mathcal{H}_{\text{small}}$ falls apart into components $\mathcal{C}_1, \mathcal{C}_2, \ldots, \mathcal{C}_j, \ldots$. Let V_j denote the union $\bigcup E$ of the emergency sets E for all $\widetilde{E} \in \mathcal{C}_j$. I refer to V_j as the jth Emergency Room.

Let V denote the board of $\mathcal{H}_G$, that is, the set of all edges of graph G; we could even write "$V = G$". The term "iteration" means that we are going to repeat the whole argument above for every component $(V_j, \mathcal{C}_j)$ in $\mathcal{H}_{\text{small}}$ (which is the "small game sub-hypergraph" of $\mathcal{H}_G$), where V_j plays the role of V and $\mathcal{C}_j$ plays the role of $\mathcal{H}_G$. And we are going to repeat the whole argument for every such component of every component $(V_j, \mathcal{C}_j)$, and we are going to repeat the whole argument for every such component of every such component of every component $(V_j, \mathcal{C}_j)$, and so on.

Let's go back to the beginning of the iteration: consider an arbitrary component $(V_j, \mathcal{C}_j)$. Playing in a component Breaker follows the Same Board Rule. What it means is that, if Maker's actual last move was in the Big Board then Breaker replies – by using the potential strategy of the Balancing Lemma, see Step 6 – in the Big Board. And similarly, if Maker's actual last move was in the small board, in particular in the jth Emergency Room V_j, then Breaker replies in the same V_j.

Assume that we are in the middle of an actual play in a component $(V_j, \mathcal{C}_j)$. Of course we can repeat the key definitions: a hyperedge $A \in \mathcal{C}_j$ is said to become *dangerous* when Maker's lead in the intersection $A \cap V_j$ equals $4\sqrt{d}\log d$ (see (4.1)); similarly, the unoccupied part in $A \cap V_j$ is called an *emergency set*, and so on.

The main point is the **enormous** reduction in the maximum size of the "components" in the next level:

$$(V, \mathcal{H}_G) \longrightarrow (V_j, \mathcal{C}_j),\ j = 1, 2, 3, \ldots \longrightarrow \cdots \longrightarrow \cdots$$

which leads to the rapidly decreasing sequence (see Step 7)

$$\begin{aligned}(4.10)\qquad n \to d^3 \log n \to d^3 \log(d^3 \log n) &= d^3 \log(\log\log n + 3\log d) \to \\ \to d^3 \log\left(d^3 \log(\log\log n + 3\log d)\right) &= \\ = \left(1 + o(1)\right) d^3(\log\log\log n + 4\log d) &\to \cdots .\end{aligned}$$

The key point in (4.10) is the sequence $\log n$, $\log\log n$, $\log\log\log n, \ldots$. It means, roughly speaking, that in every step of the iteration the size of the new component is the *logarithm* of the previous size. We choose the number of iterations to be $\log n$; this explains the definition of the tower function Giant (d) in (1.11).

Under the conditions of Theorem 2, if Breaker plays rationally in every subgame, the surplus is clearly estimated from above by the product

$$4\sqrt{d}\log d \cdot \log d = 4\sqrt{d}(\log d)^2,$$

where the first factor $4\sqrt{d}\log d$ comes from the definition of "dangerous" in (4.1), and the second factor $\log d$ comes from the number of iterations. Thus the proof of the upper bound in Theorem 2 is complete. ■

References

[1] J. Beck, Van der Waerden and Ramsey type games, *Combinatorica,* **2** (1981), 103–116.

[2] J. Beck, Deterministic graph games and a probabilistic intuition, *Combinat. Probab. Comput.,* **3** (1993), 13–26.

[3] J. Beck, *Combinatorial Games: Tic-Tac-Toe Theory,* Encyclopedia of Mathematics and its Applications 114, Cambridge University Press (2008).

[4] P. Erdős and L. Lovász, Problems and results on 3-chromatic hypergraphs and some related questions, in: *Infinite and Finite Sets* (eds.: A. Hajnal et al.), *Colloq. Math. Soc. J. Bolyai,* **11**, North-Holland, (Amsterdam, 1975), 609–627.

[5] P. Erdős and A. Rényi, On the evolution of random graphs, *Mat. Kutató Int. Közl.,* **5** (1960), 17–60.

[6] P. Erdős and J. Selfridge, On a combinatorial game, *Journal of Combinatorial Theory, Series A,* **14** (1973), 298–301.

[7] C. St. J. A. Nash-Williams, Edge-disjoint spanning trees of finite graphs, *Journ. London Math. Soc.,* **36** (1961), 445–450; see also *Journ. London Math. Soc.,* **39** (1964), 12.

[8] L. A. Székely, On two concepts of discrepancy in a class of combinatorial games, in: *Colloq. Math. Soc. János Bolyai 37 "Finite and Infinite Sets",* North-Holland (Eger, Hungary, 1981), 679–683.

József Beck

Department of Mathematics
Rutgers University
110 Frelinghuysen Road
Piscataway
New Jersey 08854-8019
U.S.A.

e-mail: jbeck@math.rutgers.edu

BOLYAI SOCIETY
MATHEMATICAL STUDIES, 19

Building Bridges
pp. 103–135.

Deformable Polygon Representation and Near-Mincuts

ANDRÁS A. BENCZÚR* and MICHEL X. GOEMANS†

We derive a necessary and sufficient condition for a symmetric family of sets to have a geometric representation involving a convex polygon and some of its diagonals. We show that cuts of value less than 6/5 times the edge-connectivity of a graph admit such a representation, thereby extending the cactus representation of all mincuts.

1. Introduction

Given an undirected graph $G = (V, E)$ possibly with multiple edges (or nonnegative weights on the edges), let $d(S)$ represent the size (or the weight) of the cut $(S\colon \overline{S}) = \big\{(i,j) \in E\colon \big|\{i,j\} \cap S\big| = 1\big\}$ (where $\overline{S} = V \setminus S$). The edge-connectivity λ of G is equal to $\min_{\emptyset \neq S \neq V} d(S)$, and any cut $(S, \overline{S})$ achieving the minimum is called a mincut. A cut $(S, \overline{S})$ for which $d(S) < \alpha\lambda$ for some $\alpha > 1$ is called an α-near-mincut. In 1976, Dinitz, Karzanov and Lomonosov [4] have given a compact representation of all mincuts; this is known as the cactus representation. Informally, the cactus representation is a multigraph H in which every edge is in exactly one cycle[1] and every vertex of G is mapped to a node[2] of H (see Figure 1). This mapping does not need to be bijective, surjective or injective. A node of H can correspond

*Research supported under grant OTKA NK 72845.

†Research supported under NSF grant CCF-0515221 and ONR grant N00014-05-1-0148.

[1]In descriptions of the cactus representation, the cycles of length 2 are sometimes replaced by a single edge (bridge) of weight 2.

[2]To easily distinguish them, we use vertices for the graph G and nodes for the cactus H.

to one, several or even no vertex of G; in the latter case, the node is said to be empty. The set of all mincuts of size 2 in H, i.e. those obtained by removing any two edges of the same cycle, correspond to the set of all mincuts in G. Because of the presence of empty nodes, observe that several mincuts in H can correspond to the same cut in G.

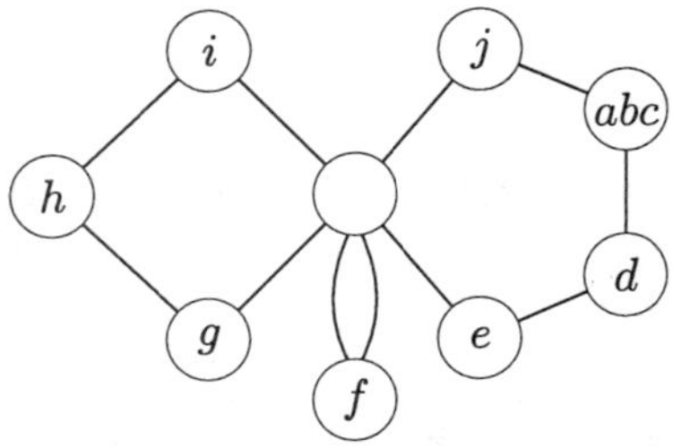

Fig. 1. A cactus of a graph with vertex set $V = \{a, b, c, d, e, f, g, h, i, j\}$.

The cactus is not necessarily unique, and Nagamochi and Kameda [15] describe two canonical cactus representations, one with no cycles of length 3 and the other without precisely 3 cycles meeting at the same empty node. Nagamochi and Kameda also show that these canonical representations (and many others) have at most $n = |V|$ empty nodes. A cactus representation can be constructed efficiently, see [11, 8, 20, 16, 3, 7, 18, 17]. A 2-level cactus representing all cuts of value λ and $\lambda + 1$ in an unweighted graph has been derived by Dinitz and Nutov [5].

In this paper, we consider extensions of the cactus representation to arbitrary symmetric[3] families $\mathcal{F} \subseteq 2^V$, in which every cycle is replaced by a convex polygon P with some of its diagonals drawn and the elements of V are mapped to the cells defined by the diagonals within the polygon, see Figure 2. A cell can have 0, 1 or many elements mapped to it. To every diagonal, one can associate a pair $(S, \overline{S})$ of complementary sets corresponding to those elements mapped to either side of the diagonal. Furthermore, we will focus on the situation when the existence of the mapping of V to the cells of the polygon does not depend on the exact (convex) location of the polygon vertices; we call such polygons *deformable,* see Section 2. Our representation links deformable polygons in a tree fashion, in an almost identical way as the cactus links cycles. To derive this tree structure, we consider the *cross graph* associated with a symmetric family of sets: Its vertex set has a representative for each pair of complementary sets in $\mathcal{F}$ and two such pairs $(S, \overline{S})$ and $(T, \overline{T})$ are joined by an edge if S and T cross.

[3] $A \in \mathcal{F}$ iff $\overline{A} \in \mathcal{F}$.

Recall that two sets S and T are said to cross if $S \cap T$, $\overline{S \cup T}$, $S \setminus T$ and $T \setminus S$ are all non-empty. We show that representations for each connected component of the cross graph can be linked in a tree structure of linear size; this is explained in Section 5, where we provide a geometric proof of a (slightly modified) cactus structure of mincuts.

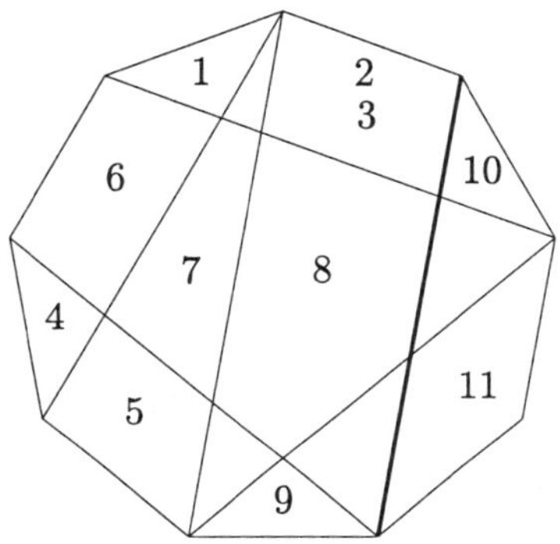

Fig. 2. A polygon representation. The bold diagonal corresponds to the sets $\{1, 2, \ldots, 9\}$ and $\{10, 11\}$. This polygon is deformable; for example, this will follow from our characterization.

The main result of this paper is to give a necessary and sufficient condition for a symmetric family to admit a representation as a tree of deformable polygons. This characterization is in terms of excluded configurations. We show that there are 3 families consisting of 4 pairs of complementary sets each ($\binom{4}{[2]}$, $\mathcal{C}_1$ and $\mathcal{C}_2$, see the forthcoming Figure 5), and the family can be represented by a tree of deformable polygons if and only if none of the three families appear as an induced subfamily. This is stated in Section 2 and proved in Sections 3 and 5. We also show in Section 4 that for any (weighted) undirected graph, the family of 6/5-near-mincuts — those cuts of value less than 6/5 times edge-connectivity λ — satisfies the condition of our main result, and therefore can be represented by a tree of deformable polygons. This, for example, implies that there are at most $\binom{n}{2}$ 6/5-near-mincuts, see Section 6; this is already known even for 4/3-near-mincuts [19, 9].

This paper focuses on a characterization of those families with a tree of deformable polygons, and does not consider efficient algorithms for its construction or implications for connectivity problems, such as speeding up algorithms for graph augmentation problems or the existence of splitting-off which maintains near-mincuts. This is covered in the Ph.D. dissertation [3] of the first author, see also [1, 2].

Many proofs of the existence of the cactus have appeared. The approach that is most useful in the context of this paper is due to Lehel, Maffray and Preissmann [12] who show using characterizations of interval hypergraphs that, for any undirected graph $G = (V, E)$, there exists a cyclic ordering of V such that every mincut corresponds to a partition of this cyclic ordering into two (cyclic) intervals. This will be the basis of our construction of deformable polygons, as we will first place a carefully selected subset of the elements along the sides of our polygon in a circular way and then add the remaining elements in cells farther inside the polygon (using Helly's theorem).

2. Deformable Polygon Representation

Given an arrangement of lines in $\mathbb{R}^2$ and a set V of points in $\mathbb{R}^2$ none of them being on any of the lines, each line partitions $\mathbb{R}^2$ into two halfplanes and hence the set V into two sets S and $\bar{S}$. We associate to this arrangement of lines the symmetric family $\mathcal{F}$ of all sets defined by these lines. We say that this family is *representable as an arrangement of lines.* For example, in Figure 3:(a), a representation of $\binom{4}{[2]} = \{\{1,2\},\{1,3\},\{1,4\},\{2,3\},\{2,4\},\{3,4\}\}$, the family of all subsets of $\{1,2,3,4\}$ of cardinality 2, is given.

Fig. 3. (a): An arrangement of lines representing the family $\mathcal{F} = \binom{4}{[2]} = \{\{1,2\},\{1,3\},\{1,4\},\{2,3\},\{2,4\},\{3,4\}\}$. (b): An equivalent polygon representation.

A classical result of Schläfli [22] says that the maximum number of partitions of an n-element set in $\mathbb{R}^2$ by lines is $\binom{n}{2}+1$, and thus at most $\binom{n}{2}$ once we don't count the trivial partition for which $S = \emptyset$ or $S = V$. This provides a bound on the size of any family representable as an arrangement of lines.

Instead of considering representations as arrangements of lines, we consider a bounded variant of it. In this case, we have a convex polygon P with vertices $a_1, a_2, \ldots, a_k$ (for some suitable k), a subset of diagonals $D \subseteq \{ [a_i, a_j] : 1 \le i < j \le k \}$, and the elements of V are placed in the cells of $P \setminus D$ defined by the diagonals within the polygon. The sets represented correspond to the sets of elements on either side of a diagonal. We refer to such a representation as a *polygon representation.* Clearly, a polygon representation can be transformed into an arrangement of lines representation, and vice versa, see Figure 3 for a simple example.

For certain polygon representations, the polygon can be arbitrarily deformed in a convex manner without the cells containing elements of V vanishing. In this case, the actual positions of the polygon vertices $a_1, \cdots, a_k$ are irrelevant, provided they are in convex position. We refer to this as a *deformable polygon representation.* This is not the case for the polygon representation of $\binom{4}{[2]}$ given in Figure 3:(b), as shown in Figure 4. In fact, it is easy to show that the family $\binom{4}{[2]}$ does not admit a deformable polygon representation.

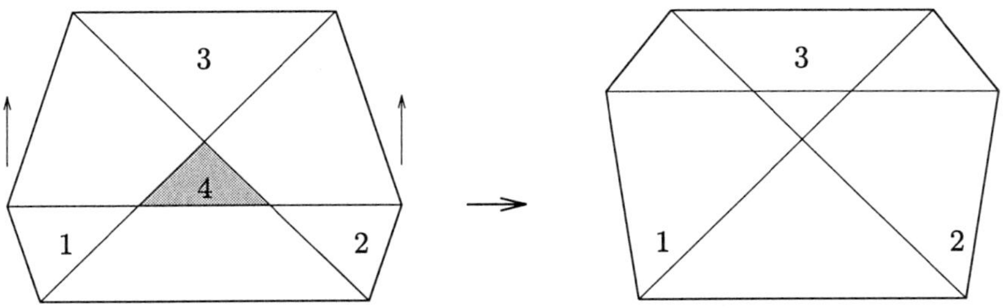

Fig. 4. If the convex polygon is deformed, the cell containing the element 4 might disappear.

In this paper, we provide a characterization in terms of excluded configurations of those symmetric families that admit a deformable polygon representation. To state our result, we first need to introduce two families that do not have a polygon representation (or a representation by an arrangement of lines). In fact, they cannot be represented by convex sets, as any such representation for a symmetric family can be transformed into a representation as an arrangement of lines by simply considering the line separating the convex sets assigned to S and $\overline{S}$.

Lemma 1. *The following two families of subsets of* $\{1,2,3,4,5,6\}$ *(see Figure 5) do not have a polygon representation.*

1. $\mathcal{C}_1 = \{\{1,2,3\}, \overline{\{1,2,3\}}, \{1,4\}, \overline{\{1,4\}}, \{2,5\}, \overline{\{2,5\}}, \{3,6\}, \overline{\{3,6\}}\}$,
2. $\mathcal{C}_2 = \{\{1,2,3\}, \overline{\{1,2,3\}}, \{1,5,6\}, \overline{\{1,5,6\}}, \{2,4,6\}, \overline{\{2,4,6\}}, \{3,4,5\}, \overline{\{3,4,5\}}\}$
$= \{\{1,2,3\}, \{2,3,4\}, \{3,4,5\}, \{4,5,6\}, \{5,6,1\}, \{6,1,2\}, \{1,3,5\}, \{2,4,6\}\}$.

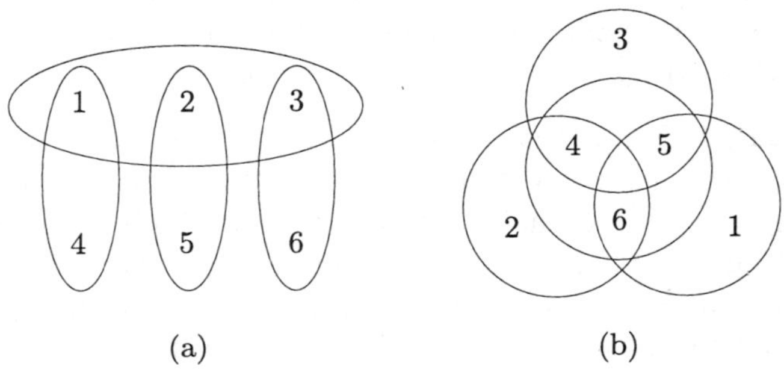

Fig. 5. Half of the sets (a) in $\mathcal{C}_1$ and (b) in $\mathcal{C}_2$. Combs, to be defined in Definition 2, have either an induced $\mathcal{C}_1$ or an induced $\mathcal{C}_2$.

Proof. Assume that $\mathcal{C}_k$ for either $k = 1$ or $k = 2$ has a representation by an arrangement of lines. We start with some notation for both cases. We simply let $1, 2, \ldots, 6$ denote the points in $\mathbb{R}^2$ in the cells of the arrangement of lines corresponding to the elements of the ground set. For a pair of complementary sets $(U, \overline{U})$ in $\mathcal{C}_k$, let $\ell(U) = \ell(\overline{U})$ be the line in the arrangement for $\mathcal{C}_k$ separating U from $\overline{U}$; furthermore let $\ell^+(U)$ and $\ell^-(U)$ denote the open halfplanes containing U and $\overline{U}$, respectively. For $i \in \{1,2,3\}$ and $j \in \{4,5,6\}$, let $t_{ij} \in \ell(\{1,2,3\})$ be the intersection point between $\ell(\{1,2,3\})$ and the line segment extending between points i and j; t_{ij} is well-defined as i and j are separated by $\{1,2,3\} \in \mathcal{C}_k$.

Consider first the arrangement for $\mathcal{C}_1$. Consider the points t_{14}, t_{25} and t_{36}. As $t_{14} \in \ell^+(\{1,4\})$ while $t_{25}, t_{36} \in \ell^-(\{1,4\})$ and similarly for 14 replaced by 25 or 36, the points t_{14}, t_{25} and t_{36} are distinct. Without loss of generality, we can assume that t_{25} is between t_{14} and t_{36}. But, by convexity, $t_{14}, t_{36} \in \ell^-(\{2,5\})$ implies that $t_{25} \in \ell^-(\{2,5\})$ contradicting $t_{25} \in \ell^+(\{2,5\})$.

Assume now we have a representation for $\mathcal{C}_2$. Consider two complementary sets in our family, say $U = \{1,3,5\}$ and $\overline{U} = \{2,4,6\}$. The fact that

$U \subset \ell^+(U)$ and $\overline{U} \subset \ell^-(U)$ implies that the line segments $[t_{15}, t_{35}]$ and $[t_{24}, t_{26}]$ do not intersect, and in particular that $t_{15} \neq t_{26}$. We can similarly deduce that $t_{15} \neq t_{34} \neq t_{26}$. Without loss of generality, let us assume that t_{26} is between t_{15} and t_{34}. But the disjointness of $[t_{15}, t_{35}]$ and $[t_{24}, t_{26}]$ implies that $t_{26} \notin [t_{15}, t_{35}]$, while the same argument with the complementary sets $\{1, 2, 6\}$ and $\{3, 4, 5\}$ implies that $t_{26} \notin [t_{34}, t_{35}]$, which means that $t_{26} \notin [t_{15}, t_{34}]$, a contradiction. ∎

Given a family $\mathcal{F}$ of sets on V and a subset $S \subseteq V$, we define the family $\mathcal{F}|_S$ to be the family $\{F \cap S \colon F \in \mathcal{F}\}$. We say that a family $\mathcal{F}$ of V *contains* a family $\mathcal{G}$ as an *induced family* or simply contains the *induced family* $\mathcal{G}$ if there exists S such that $\mathcal{F}|_S$ contains $\mathcal{G}$, i.e. $\mathcal{F}|_S \supseteq \mathcal{G}$. In this case, if $\mathcal{G}$ does not have a polygon representation then neither does $\mathcal{F}$, and similarly for deformable polygon representations. Thus, any family $\mathcal{F}$ which contains $\binom{4}{[2]}$, $\mathcal{C}_1$ or $\mathcal{C}_2$ as an induced family does not have a deformable polygon representation. We will show that the converse to that statement is also true for families with connected cross graph.

Theorem 2. *Let $\mathcal{F}$ be a symmetric family of sets with connected cross graph. Then $\mathcal{F}$ admits a deformable polygon representation if and only if $\mathcal{F}$ does not contain $\binom{4}{[2]}$, $\mathcal{C}_1$ or $\mathcal{C}_2$ as an induced family.*

The "only if" part follows from Lemma 1 and the fact that $\binom{4}{[2]}$ does not admit a deformable polygon representation. The proof of the "if" part is constructive, and will be the focus of the next section. Here is a brief sketch of the construction proving the existence of the polygon representation. First we identify a set $O \subseteq V$ of elements that we would like to place along the sides of the polygon; we refer to the elements of O as *outside* elements. One non-trivial property of this set O is that every set $S \in \mathcal{F}$ contains at least one outside element but not all of them; this is indeed a property to expect if we place these outside elements along the sides of the polygon and represent sets by diagonals. We then use Tucker's characterization [23] of interval hypergraphs to show that there exists a circular ordering of the outside elements such that any set in $\mathcal{F}|_O$ corresponds to an interval in that circular ordering. As a first trial, one could take a k-gon P where $k = |O|$, place the outside elements in their circular order along the sides of P, and for each set $S \in \mathcal{F}$ add the diagonal that separates $S \cap O$ from $O \setminus S$. This already provides a polygon representation of $\mathcal{F}|_O$. However, placing the remaining elements of $\overline{O}$ appropriately inside the polygon in order to represent $\mathcal{F}$ is not always possible for the following reason. Several sets of $\mathcal{F}$ could have the same intersection with O and thus would be mapped to the same diagonal. We can, however, show that sets having the same intersection with O form a chain, and we can then add vertices to the polygon P so as to make the diagonals distinct (and non-crossing), and still represent $\mathcal{F}_O$. We then need to consider the placement of the remaining *inside* elements, those of $\overline{O}$. For each inside element v, we know on which side of each diagonal we would like to place it. We show that the intersection of all these halfplanes is non-empty by proving that any two or three of them have a non-empty intersection and then using Helly's theorem. This non-empty intersection gives a non-empty cell where to place v, and this completes the construction and the proof.

In Section 5, we show that the connected components of any cross graph can be arranged in a tree structure, and when each component can be represented by a deformable polygon, we obtain a representation that we call a *tree of deformable polygons*. If a family contains $\binom{4}{[2]}$, $\mathcal{C}_1$ or $\mathcal{C}_2$ as an induced family then one of the connected components of its cross graph must contain one of these subfamilies as an induced family. Therefore, after proving Theorem 2 and deriving our tree structure, we will have shown the following theorem.

Theorem 3. *Let $\mathcal{F}$ be a symmetric family of sets. Then $\mathcal{F}$ admits a representation as a tree of deformable polygons if and only if $\mathcal{F}$ does not contain $\binom{4}{[2]}$, $\mathcal{C}_1$ or $\mathcal{C}_2$ as an induced family.*

We would like to point out that we do not know of any necessary and sufficient condition for the existence of a (not necessarily deformable) polygon representation (or a representation by an arrangement of lines).

3. Construction and its Proof

In this Section, we focus on the case where the symmetric family has a connected cross graph, or if this is not the case, we redefine $\mathcal{F}$ to be the family corresponding to a single connected component of the cross graph.

Before we start, observe that we can group together any pair or set of elements which are not separated by any set in our family. Indeed, this does not affect the existence of a polygon representation as all these equivalent elements would fall in the same cell defined by the diagonals of the polygon. More formally, define two elements u and v to be equivalent if, for every set S_i in our family, $u \in S_i$ iff $v \in S_i$. Let the equivalence classes be called *atoms*. Our representation is on the atoms of our family. For simplicity, for the rest of this section, we simply refer to them as the elements; we'll use atoms when we consider several connected components of the cross graph in Section 5.

We will first restate Theorem 2 without using any induced families. For this, we need a few definitions.

Definition 1. 3 subsets C_1, C_2 and C_3 form a 3-cycle (see figure 6) if (i) $\overline{C_1 \cup C_2 \cup C_3} \neq \emptyset$ and (ii) $(C_i \cap C_{i+1}) \setminus C_{i-1} \neq \emptyset$, for all $i \in \{1, 2, 3\}$.

Everywhere in the paper, indices should always be considered cyclic; for example, C_4 represents C_1 in the above definition.

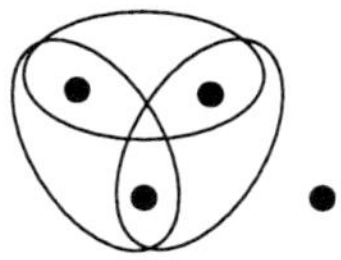

Fig. 6. A 3-cycle. Solid dots in a Venn diagram denote non-empty intersections; other intersections could be empty or not.

Observe that a symmetric family $\mathcal{F}$ contains 3 sets that form a 3-cycle if and only if it contains an *induced* $\binom{4}{[2]}$. We also need to define another configuration of sets.

Definition 2. Four subsets H, T_1, T_2 and T_3 form a comb with handle H and teeth T_1, T_2 and T_3 if (i) either $H \cap \big(T_i \setminus (T_{i-1} \cup T_{i+1})\big) \neq \emptyset$ for all $i = 1, 2, 3$ or $H \cap (T_{i-1} \cap T_{i+1} \setminus T_i) \neq \emptyset$ for all $i = 1, 2, 3$ and (ii) the same holds for H replaced by $\bar{H}$.

The above definition is such that any family of sets which contains (sets forming) a comb must contain either an induced $\mathcal{C}_1$ or an induced $\mathcal{C}_2$, and vice versa, see Figure 5; in both cases, the handle H gets restricted to $\{1, 2, 3\}$ or $\{4, 5, 6\}$. Thus, alternatively, we could reformulate Theorem 2 as follows.

Theorem 4. *Let $\mathcal{F}$ be a family of sets with connected cross graph. Then $\mathcal{F}$ admits a deformable polygon representation if and only if $\mathcal{F}$ does not contain any 3-cycle or comb.*

More generally, we define a k-cycle for $k > 3$ as follows.

Definition 3. k subsets $C_1, \ldots, C_k$ for $k > 3$ form a k-cycle (or simply cycle) if (i) $\overline{\cup_{i=1}^k C_i} \neq \emptyset$, (ii) $P_i = (C_i \cap C_{i+1}) \neq \emptyset$ for $i = 1, \ldots, k$, and (iii) C_i and C_j are disjoint for $j \notin \{i-1, i, i+1\}$. See Figure 7.

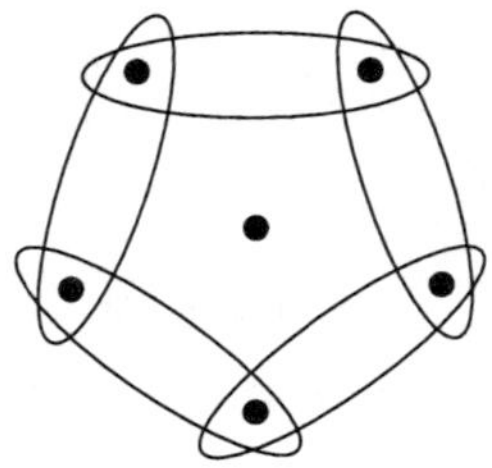

Fig. 7. A 5-cycle.

Because of condition (iii) in the definition of a k-cycle for $k > 3$ (which we didn't have for a 3-cycle), a family could contain an induced k-cycle but no k-cycle itself. However, the next proposition shows that when there are no 3-cycles and combs, we do not need to differentiate between cycles and induced cycles.

Proposition 5. *Consider a collection $\mathcal{F}$ of sets that does not contain any 3-cycle or comb. Then any induced cycle $C_1, \ldots, C_k$ contains a subcollection which forms a cycle.*

The proof is technical and can be skipped at first reading.

Proof. We assume that the collection is minimal, i.e. no subcollection of the C_i's forms an induced cycle. Let W be such that the C_i's induce a cycle in W. Define

$$P_i = W \cap \left[(C_i \cap C_{i+1})\right] = W \cap \left[(C_i \cap C_{i+1}) \setminus \bigcup_{j \notin \{i,i+1\}} C_j\right] \neq \emptyset$$

for $i = 1, \ldots, k$, the last equality following from the fact that $C_i \cap W$ and $C_j \cap W$ are disjoint if i and j are neither consecutive nor equal.

Assume there exist two indices i and $j \notin \{i-1, i, i+1\}$ such that $\big(C_i \setminus (C_{i+1} \cup C_{i-1})\big) \cap C_j \neq \emptyset$, i.e. $(C_i \cap C_j) \setminus (C_{i+1} \cup C_{i-1}) \neq \emptyset$. If j is $i-2$, we have a 3-cycle consisting of C_j, C_{i-1} and C_i, a contradiction. Similarly, if j is $i+2$, we have the 3-cycle (C_j, C_i, C_{i+1}). Thus we can assume $j \notin \{i-2, i-1, i, i+1, i+2\}$. We claim we have a comb with handle C_i and teeth C_{i-1}, C_{i+1} and C_j, a contradiction. This is because the following sets are all non-empty:

- $(C_j \setminus (C_{i+1} \cup C_{i-1})) \cap C_i = (C_i \cap C_j) \setminus (C_{i+1} \cup C_{i-1}) \neq \emptyset$,
- $(C_{i+1} \setminus (C_j \cup C_{i-1})) \cap C_i \supseteq P_i$ as $j \notin \{i, i+1\}$,
- $(C_{i-1} \setminus (C_j \cup C_{i+1})) \cap C_i \supseteq P_{i-1}$ as $j \notin \{i-1, i\}$,
- $(C_j \setminus (C_{i+1} \cup C_{i-1})) \cap \overline{C_i} \supseteq P_j$ as $j \notin \{i-2, i-1, i, i+1\}$,
- $(C_{i+1} \setminus (C_j \cup C_{i-1})) \cap \overline{C_i} \supseteq P_{i+1}$ as $j \notin \{i+1, i+2\}$,
- $(C_{i-1} \setminus (C_j \cup C_{i+1})) \cap \overline{C_i} \supseteq P_{i-2}$ as $j \notin \{i-2, i-1\}$.

We can thus assume that for all i and all $j \notin \{i-1, i, i+1\}$, we have

$$C_i \cap C_j \subseteq (C_{i-1} \cup C_{i+1}). \tag{1}$$

Observe that the C_i's form a cycle unless condition (iii) of Definition 3 is violated, i.e. there exists i and $j \notin \{i-1, i, i+1\}$ with $C_i \cap C_j \neq \emptyset$. Combining this with (1), we can assume that there exist i and j with $j \notin \{i-1, i, i+1\}$ such that $\emptyset \neq C_i \cap C_j \subseteq (C_{i-1} \cup C_{i+1})$. This means that either $C_i \cap C_{i+1} \cap C_j \neq \emptyset$ or $C_{i-1} \cap C_i \cap C_j \neq \emptyset$. Depending on the value of j and after possibly changing i, (i) we either have i and $j \notin \{i-2, i-1, i, i+1\}$ with $C_{i-1} \cap C_i \cap C_j \neq \emptyset$, or (ii) we have i with $C_{i-2} \cap C_{i-1} \cap C_i \neq \emptyset$. We consider both cases separately.

1. Assume that $C_{i-1} \cap C_i \cap C_j \neq \emptyset$ for some i and j with $j \notin \{i-2, i-1, i, i+1\}$. We claim that C_{i-1}, C_i and $\overline{C_j}$ form a 3-cycle, again a contradiction. Indeed $C_{i-1} \cap C_i \cap C_j \neq \emptyset$, $C_{i-1} \setminus (C_i \cup C_j) \supseteq P_{i-2} \neq \emptyset$ as $j \notin \{i-2, i-1\}$, $C_i \setminus (C_{i-1} \cup C_j) \supseteq P_i \neq \emptyset$ as $j \notin \{i, i+1\}$, and $C_j \setminus (C_{i-1} \cup C_i) \supseteq P_{j-1} \neq \emptyset$ as $j \notin \{i-1, i, i+1\}$.
2. Assume that $C_{i-2} \cap C_{i-1} \cap C_i \neq \emptyset$ for some i. If $C_{i-1} \setminus (C_{i-2} \cup C_i) \neq \emptyset$, we have a 3-cycle $(C_{i-2}, \overline{C_{i-1}}, C_i)$. Thus assume that $C_{i-1} \subseteq C_{i-2} \cup C_i$. Furthermore, we can assume that $C_{i-2} \cap C_i = C_{i-2} \cap C_{i-1} \cap C_i$ since, otherwise, C_{i-2}, C_{i-1} and C_i would form a 3-cycle. By our minimality assumption, we cannot remove C_{i-1} from our collection and still have an induced cycle; this implies that $C_{i-2} \cap C_{i-1} \cap C_i = C_{i-2} \cap C_i \subseteq \left[\bigcup_{j \notin \{i-2, i-1, i\}} C_j\right]$. Thus, let $l \notin \{i-2, i-1, i\}$ be such that $C_l \cap C_{i-2} \cap C_{i-1} \cap C_i \neq \emptyset$. If $k > 4$ then C_{i-2}, C_i and $\overline{C_l}$ can be seen to form a 3-cycle (since we have $(C_{i-2} \cap C_i) \setminus \overline{C_l} = C_{i-2} \cap C_i \cap C_l \neq \emptyset$, $(C_{i-2} \cap \overline{C_l}) \setminus C_i \supseteq P_{i-2}$, $(C_i \cap \overline{C_l}) \setminus C_{i-2} \supseteq P_{i-1}$, and $\overline{C_i \cup \overline{C_l} \cup C_{i-2}}$ contains either P_l or P_{l-1}). On the other hand, if $k = 4$ and thus $l = i+1$ then $\overline{C_{i-2}}$, C_{i-1}, $\overline{C_i}$ and C_{i+1} form a comb with C_{i-1} as handle. Indeed, we have

- $\left(\overline{C_{i-2}} \setminus (\overline{C_i} \cup C_{i+1})\right) \cap C_{i-1} \supseteq P_{i-1}$,
- $\left(\overline{C_i} \setminus (\overline{C_{i-2}} \cup C_{i+1})\right) \cap C_{i-1} \supseteq P_{i-2}$,
- $\left(C_{i+1} \setminus (\overline{C_{i-2}} \cup \overline{C_i})\right) \cap C_{i-1} = C_{i-2} \cap C_{i-1} \cap C_i \cap C_{i+1} \neq \emptyset$,
- $\left((\overline{C_{i-2}} \cap \overline{C_i}) \setminus C_{i+1}\right) \cap \overline{C_{i-1}} = \overline{C_{i-2} \cup C_{i-1} \cup C_i \cup C_{i+1}} \neq \emptyset$,
- $\left((\overline{C_i} \cap C_{i+1}) \setminus \overline{C_{i-2}}\right) \cap \overline{C_{i-1}} \supseteq P_{i+1}$,
- $\left((\overline{C_{i-2}} \cap C_{i+1}) \setminus \overline{C_i}\right) \cap \overline{C_{i-1}} \supseteq P_i$. ■

Definition 4. An element $v \in V$ of the family $\mathcal{F} \subset 2^V$ is said to be *inside* if there exists a cycle $C_1, \ldots, C_k$ of $\mathcal{F}$ such that $v \notin \cup_i C_i$. Otherwise, v is said to be outside. The set of outside elements is denoted by O.

A few remarks are in order. Given Proposition 5, we can replace "cycle" by "induced cycle" in the above definition, provided that $\mathcal{F}$ does not contain any comb or 3-cycle. For the rest of this section, we assume throughout that $\mathcal{F}$ has no 3-cycle or comb, even if it is not explicitly stated. Also, at this point, it is not obvious that O is non-empty; this will follow from Proposition 10. This would not be true if our family could have 3-cycles; $\binom{4}{[2]}$ for example has no outside elements. In fact, we will deduce from Corollary 14 that either $|O| \geq 4$ or our family with its connected cross graph consists of only one pair of complementary sets $(S, \overline{S})$ (thus separating two outside elements/atoms from each other). The latter case trivially gives rise to a deformable polygon; just take a 4-gon with one diagonal. Therefore, we will often implicitly assume in this Section that our family consists of more than one complementary pair. Observe also that if our family does not contain any cycles (as is the case for the family of mincuts in a graph, see Section 4) then all elements are outside, i.e. $O = V$.

As a first step towards the contruction of the polygon representation, we show now that the family restricted to the outside elements, $\mathcal{F}|_O$, is a *circular representable hypergraph* [21] or a *circular arc hypergraph*, i.e. there exists a circular ordering of O such that all sets in $\mathcal{F}|_O$ correspond to arcs of the circle. This is similar to the proof of the existence of the cactus representation of all minimum cuts due to Lehel et al. [12].

Proposition 6. *Consider a symmetric family $\mathcal{F}$ of sets with no cycles or combs. Then $\mathcal{F}$ is a circular representable hypergraph.*

By definition of outside elements, we can then derive the following Corollary.

Corollary 7. *Consider a symmetric family $\mathcal{F}$ of sets that does not contain any 3-cycle or comb, and let O be its set of outside elements. Then $\mathcal{F}|_O$ is a circular representable hypergraph.*

Proposition 6 and Corollary 7 follow easily from Tucker's characterization [23] of *interval hypergraphs*, i.e. hypergraphs (or families of sets) for which there exists a total ordering of the elements of the ground set such that every hyperedge (set) corresponds to an interval in the ordering. Tucker gives a necessary and sufficient condition for a 0-1 matrix to have the consecutive 1's property, see Duchet [6] for a short proof of Tucker's result in terms of hypergraphs.

Theorem 8 (Tucker [23]). *A family of sets define an interval hypergraph if and only if it does not contain any of the families listed in Figure 8 as an induced subfamily.*

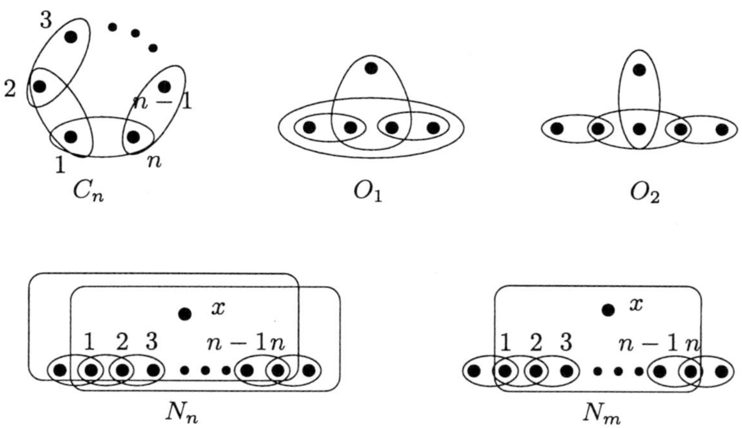

Fig. 8. List of excluded subhypergraphs for interval hypergraphs: C_n for $n \geq 3$, O_1, O_2, N_n for $n \geq 1$ and M_n for $n \geq 1$. Solid dots represent non-empty intersections.

Proof of Proposition 6. Select an element v_0 arbitrarily and consider the family $\mathcal{C} = \{S \in \mathcal{F} : v_0 \notin S\}$. We need to show that $\mathcal{C}$ is an interval hypergraph. By Tucker's Theorem, we only need to show that $\mathcal{C}$ does not contain any of the subfamilies of Figure 8. Observe that C_n is an induced cycle not containing v_0, and therefore is not present by assumption. Similarly, an induced cycle can be obtained from M_n and N_n by complementing the large sets; these cycles do not contain the element marked X in Figure 8. O_2 is a comb, and a comb can be obtained from O_1 by complementing the 4-element set which would then contain the special element v_0. As we have no comb in our family, O_1 and O_2 cannot arise. Therefore, $\mathcal{C}$ is an interval hypergraph, and $\mathcal{F}$ is a circular representable hypergraph. ■

To construct the polygon representation, we start with a convex k-gon with vertices $a_1, a_2, \ldots, a_k$ (clockwise) where $k = |O|$, and place the outside elements along the k sides in the circular order given by Proposition 7 so that any diagonal of the k-gon partitions the outside elements into 2 intervals in the circular ordering. From Proposition 10 stated and proved below, we can derive that any set $S \in \mathcal{F}$ partitions O non-trivially, i.e. $S \cap O \neq \emptyset$ and $O \setminus S \neq \emptyset$. This means that we can associate with S one of the diagonals of our polygon, corresponding to the way it partitions the outside elements. Since we were assuming that any two elements (including consecutive elements of O in the circular ordering) were separated by a set $S \in \mathcal{F}$ (by our definition of atoms at the beginning of this Section), we have at least one diagonal incident to every vertex of our k-gon.

The trouble though is that several pairs of complementary sets might be assigned to the same diagonal. As an example, consider $\mathcal{F}_4$ consisting of a 4-cycle together with the complementary sets:

$$\mathcal{F} = \{\{1,2\},\{2,3\},\{3,4\},\{1,4\},\{3,4,5\},\{1,4,5\},\{1,2,5\},\{2,3,5\}\}.$$

Element 5 is inside, while 1,2,3 and 4 are outside; however $\{1,2\}$, $\{3,4,5\}$ and $\{1,2,5\}$, $\{3,4\}$ separate $O = \{1,2,3,4\}$ in the same way, and there is no space between the corresponding diagonals to place element 5. We will prove in Proposition 15 that all sets S having the same intersection with O, say $S \cap O = A$, form a chain $S_1 \supseteq S_2 \supseteq \cdots \supseteq S_p$. So far, all these sets correspond to the same diagonal, say from a_u to a_v. We modify the polygon by replacing a_u by p vertices, say $a_u^1, \ldots, a_u^p$ clockwise, by replacing a_v by p vertices, say $a_v^1, \ldots, a_v^p$ anticlockwise and by assigning S_i to the diagonal $[a_u^i, a_v^i]$ for $i = 1, \ldots, p$, see Figure 9. The sets $S_1, \ldots, S_p$ are now assigned to p non-crossing diagonals (i.e. which do not intersect in the interior of the polygon). Other diagonals incident to a_u, like $[a_u, a_w]$, are moved to be incident to either a_u^1 if w is between v and u (clockwise) or to a_u^p if w is between u and v (clockwise), see Figure 9. The same process is repeated for every diagonal which corresponds to more than one set. Overall, this creates a polygon with $|O| + |\mathcal{F}| - \left|\mathcal{F}|_O\right|$ vertices. To reduce this number of vertices, we could have replaced only a_u (or a_v) by p vertices instead of replacing both; the proofs also carry through in that case.

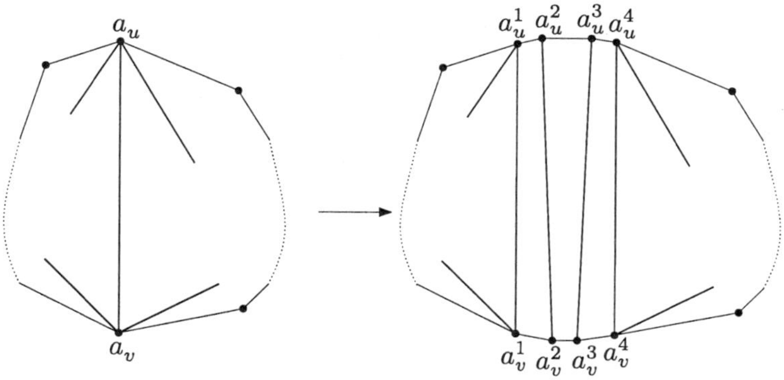

Fig. 9. For sets S having the same intersection with O, we expand the polygon and introduce non-crossing diagonals.

In the example of $\mathcal{F}_4$ (4-cycle plus the complementary sets), we first take a (convex) quadrilateral as there are 4 outside elements and then add 4 additional vertices to duplicate the two diagonals, see Figure 10.

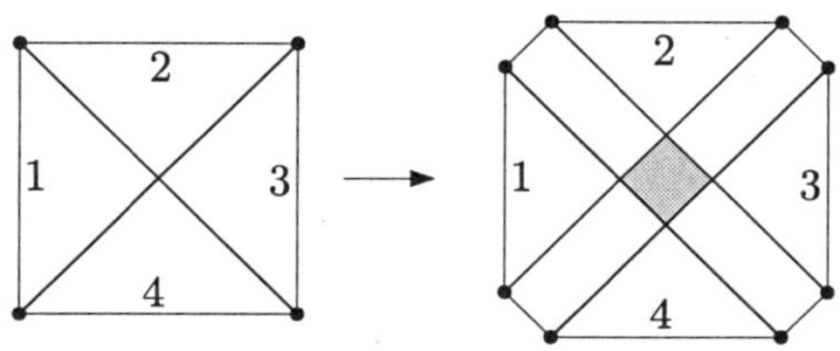

Fig. 10. Construction of the polygon and placement of the outside elements for the symmetrized 4-cycle; observe that there is a shaded cell to correctly place element 5.

At this point, we have constructed a convex q-gon $a_1, \ldots, a_q$, placed all the outside elements, and assigned every set $S \in \mathcal{F}$ to one of its diagonal, say $\ell(S) = \left[a_{i(S)}, a_{i(\overline{S})}\right]$ for some indices $i(S), i(\overline{S})$. Observe that this polygon has the following important property (in addition to representing $\mathcal{F}|_O$):

Proposition 9. *Let $S_1, S_2 \in \mathcal{F}$. Then the corresponding diagonals $\ell(S_1)$ and $\ell(S_2)$ do not cross (i.e. do not intersect in the interior of P) if and only if the sets $S_1 \cap O$ and $S_2 \cap O$ do not cross in O (i.e. $S_1 \cap O \subseteq S_2 \cap O$, $S_2 \cap O \subseteq S_1 \cap O$, $S_1 \cap S_2 \cap O = \emptyset$, or $S_1 \cup S_2 \supseteq O$).*

Indeed, this is true for our initial k-gon with $k = |O|$, and remains true as we introduce additional vertices and diagonals.

We now need to prove (i) every $S \in \mathcal{F}$ partitions O non-trivially and (ii) that sets that have the same intersection with O form a chain, and then prove that (iii) our (deformable) polygon representation can be correctly completed, i.e. that inside elements can be placed appropriately within cells of the arrangement of diagonals of our polygon and this should be true independently of the position of the vertices $a_1, \ldots, a_q$. We start with a stronger statement than (i) which will be also useful for (ii) and (iii).

Proposition 10. *Consider a family $\mathcal{F}$ of sets with a connected cross graph that does not contain any 3-cycle or comb. Let S be a minimal set in $\mathcal{F}$. Then $S \subseteq O$.*

This is the first time we require that the family has a connected cross graph. This proposition implies that any (not necessarily minimal) set $S \in \mathcal{F}$ must contain outside elements since all elements in any minimal subset of S will be outside. Applying the same proposition to $\bar{S}$, we see that every set $S \in \mathcal{F}$ partitions O non trivially.

As a first (and main) step in the proof, we show the following lemma.

Lemma 11. *Consider a family $\mathcal{F}$ that does not contain any 3-cycle or comb. Let $S \in \mathcal{F}$ contain an inside element v, and let $C_1, \ldots, C_k$ be a cycle for v (i.e. $v \notin (\cup C_i)$). Then either*

1. *there exists i such that $C_i \subset S$, or*
2. *S is disjoint from $\cup_i C_i$.*

Observe that we did not impose that the cross graph was connected. In fact, in Lemma 12, we will show that 2. can only happen if S and the cycle belong to different connected components of the cross graph.

Proof. Let us assume that $C_i \not\subset S$ for all $i = 1, \ldots, k$. We proceed in several steps.

Claim 1. For all i, $(C_i \cap C_{i+1}) \setminus S \neq \emptyset$. If not (see Figure 11, (a)), there exists i such that $C_i \cap C_{i+1} \subset S$ and we claim that C_i, C_{i+1} and $\bar{S}$ form a 3-cycle. Indeed $C_i \cap C_{i+1} \cap S = C_i \cap C_{i+1} \neq \emptyset$, $v \in S \setminus (C_i \cup C_{i+1})$, $C_i \setminus (C_{i+1} \cup S) = C_i \setminus S \neq \emptyset$ by our assumption, and similarly $C_{i+1} \setminus (C_i \cup S) \neq \emptyset$.

Claim 2. For all i, $\left(S \setminus (C_{i-1} \cup C_{i+1})\right) \cap C_i = \emptyset$. If not (see Figure 11, (b)), we have a comb with handle C_i and teeth C_{i-1}, C_{i+1} and S. Indeed, $\left[C_{i\mp1} \setminus (C_{i\pm1} \cup S)\right] \cap C_i = [C_{i\mp1} \setminus S] \cap C_i \neq \emptyset$ by claim 1, $\left[S \setminus (C_{i-1} \cup C_{i+1})\right] \cap C_i \neq \emptyset$ by assumption, $\left[C_{i\mp1} \setminus (C_{i\pm1} \cup S)\right] \cap \overline{C_i} = [C_{i\mp1} \setminus S] \cap \overline{C_i} \supseteq (C_{i\mp1} \cap C_{i\mp2}) \setminus S \neq \emptyset$ by claim 1, and $v \in \left[S \setminus (C_{i-1} \cup C_{i+1})\right] \cap \overline{C_i}$.

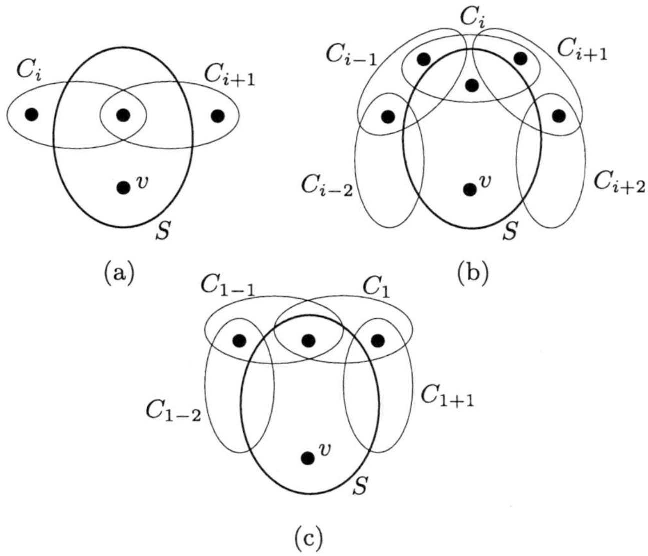

Fig. 11. Cases in the proof of Lemma 11.

Claim 3. For all i, S is disjoint from C_i. If not, there exists i with $S \cap C_i \neq \emptyset$. By Claim 2, $\emptyset \neq S \cap C_i = S \cap C_i \cap (C_{i-1} \cup C_{i+1}) = (S \cap C_i \cap C_{i-1}) \cup (S \cap C_i \cap C_{i+1})$. Thus, $S \cap C_{l-1} \cap C_l \neq \emptyset$ for $l = i$ or $l = i + 1$. This implies that C_{l-1}, C_l and $\bar{S}$ form a 3-cycle (see Figure 11, (c)) since $v \in S \setminus (C_{l-1} \cup C_l)$, $C_l \setminus (C_{l-1} \cup S) \supseteq (C_l \cap C_{l+1}) \setminus S \neq \emptyset$ by claim 1 and $C_{l-1} \setminus (C_l \cup S) \supseteq (C_{l-1} \cap C_{l-2}) \setminus S \neq \emptyset$ also by claim 1.

This completes the proof of the lemma. ∎

Lemma 12. *Consider a family $\mathcal{F}$ of sets with a connected cross graph that does not contain any 3-cycle or comb. Let $S \in \mathcal{F}$ contain an inside element v, and let $C_1, \ldots, C_k$ be a cycle for v. Then there exists i such that $C_i \subset S$.*

Proof. From Lemma 11, assume that S is disjoint from all C_i's. Since the cross graph is connected, there exists a path in the cross graph from S to one of the C_i's. Take a shortest path from S to one of the C_i's and consider the last two sets P and Q on it (P might be S), see Figure 12. We therefore have P disjoint from all C_i's, Q crossing one of them, and P and Q crossing. By Lemma 11 applied to Q and $\bar{Q}$, we derive that $Q \supset C_s$ for some s and that Q is disjoint from C_t for some t. Therefore, we can find two non-consecutive (and hence disjoint) sets C_i and C_j which both cross Q. However, this is a contradiction since we now have three sets, C_i, C_j and P, all disjoint and all crossing Q, which therefore form a comb. ∎

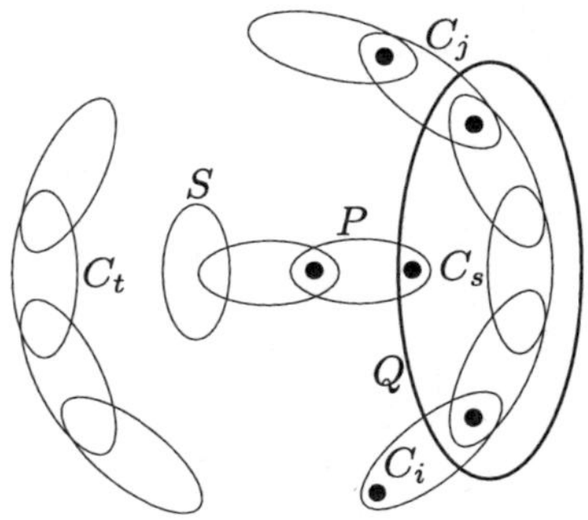

Fig. 12. Setting in the proof of Lemma 12.

Proposition 10 now follows straightforwardly from Lemma 12. We now need to consider the intersection of two sets in our family; this will be useful both for showing (ii) that sets having the same intersection with O form a chain, and (iii) that inside elements can be placed appropriately in the interior of the polygon. Throughout the rest of this section, we assume that $\mathcal{F} \subset 2^V$ is a symmetric family of sets with a connected cross graph that does not contain any 3-cycle or comb. For brevity, the assumption will not be stated in every statement.

Proposition 13. *Let S_1, S_2 be two sets in $\mathcal{F}$ with $S_1 \cap S_2 \neq \emptyset$ and minimal in the following sense: there are no $S_3, S_4 \in \mathcal{F}$ with $S_3 \subsetneq S_1$, $S_4 \subsetneq S_2$ and $\emptyset \neq S_3 \cap S_4 \neq S_1 \cap S_2$. Then either $S_1 \cup S_2 = V$ or $S_1 \cap S_2 \subseteq O$, i.e. $S_1 \cap S_2$ only contains outside elements.*

Proof. Assume that $S_1 \cap S_2$ contains an inside element v, corresponding to a cycle $C_1, \ldots, C_k$. We would like to show that $S_1 \cup S_2 = V$. By Lemma 12 applied to S_1 and S_2, we know the existence of $C_s \subset S_1$ and $C_t \subset S_2$. We claim that $C_s \subseteq S_1 \setminus S_2$; if not, replacing S_1 by C_s would contradict the minimality of S_1, S_2 as $C_s \cap S_2 \subsetneq S_1 \cap S_2$. Similarly, $C_t \subseteq S_2 \setminus S_1$.

Let p and q be such that $C_{p+1}, \ldots, C_s, \ldots, C_{q-1} \subseteq S_1 \setminus S_2$, but $C_p \not\subseteq S_1 \setminus S_2$ and $C_q \not\subseteq S_1 \setminus S_2$. Since p and q cannot be consecutive (because of the existence of t), we have that C_p and C_q are disjoint. Moreover, because of minimality, we also have that $C_p \not\subset S_1$ and $C_q \not\subset S_1$. The fact that C_p crosses $C_{p+1} \subseteq S_1 \setminus S_2$ implies that (a) $C_p \cap (S_1 \setminus S_2) \neq \emptyset$ and (b)

$S_1 \setminus (S_2 \cup C_p) \neq \emptyset$, and the same holds for C_p replaced with C_q (and C_{p+1} replaced with C_{q-1}).

We consider two cases.

1. Assume first that $C_p \setminus (S_1 \cup S_2) \neq \emptyset$. If $C_q \cap (S_2 \setminus S_1) \neq \emptyset$ then C_q, S_1 and S_2 would form a 3-cycle, a contradiction: $(S_1 \cap C_q) \setminus S_2 \neq \emptyset$ by (a) for C_q, $(S_2 \cap C_q) \setminus S_1 \neq \emptyset$ by assumption, $(S_1 \cap S_2) \setminus C_q \neq \emptyset$ as it contains v, and $\overline{S_1 \cup S_2 \cup C_q} \supseteq C_p \setminus (S_1 \cup S_2) \neq \emptyset$ since we assumed it.

   Thus we can assume that $C_q \cap (S_2 \setminus S_1) = \emptyset$, which implies that $C_q \setminus (S_1 \cup S_2) = C_q \setminus S_1 \neq \emptyset$. We now claim that the teeth C_p, C_q and S_2 together with the handle S_1 form a comb, a contradiction. Indeed, the following six sets are non-empty:

   - $\big(C_p \setminus (C_q \cup S_2)\big) \cap S_1 = C_p \cap (S_1 \setminus S_2) \neq \emptyset$ by (a),
   - similarly for $\big(C_q \setminus (C_p \cup S_2)\big) \cap S_1$,
   - $\big(S_2 \setminus (C_p \cup C_q)\big) \cap S_1$ as it contains v,
   - $\big(C_p \setminus (C_q \cup S_2)\big) \cap \overline{S_1} = C_p \setminus (S_1 \cup S_2) \neq \emptyset$ by our assumption,
   - $\big(C_q \setminus (C_p \cup S_2)\big) \cap \overline{S_1} = C_q \setminus (S_1 \cup S_2) \neq \emptyset$ as we have derived,
   - $\big(S_2 \setminus (C_p \cup C_q)\big) \cap \overline{S_1} = S_2 \setminus (S_1 \cup C_p \cup C_q) = S_2 \setminus (S_1 \cup C_p) \supseteq C_t \setminus C_p \neq \emptyset$ (the second equality following from $C_q \cap (S_2 \setminus S_1) = \emptyset$).

2. We can therefore assume that $C_p \subseteq (S_1 \cup S_2)$ and $C_q \subseteq (S_1 \cup S_2)$. Now, S_1, S_2 and C_p (or C_q) form a 3-cycle ($v \in S_1 \cap S_2 \setminus C_p$, $C_p \cap S_1 \setminus S_2 \neq \emptyset$ by (a), $C_p \cap S_2 \setminus S_1 = C_p \setminus S_1 \neq \emptyset$) unless $S_1 \cup S_2 = V$, proving the result. ■

Corollary 14. *Let S_1, S_2 be two sets in $\mathcal{F}$ with $S_1 \cap S_2 \neq \emptyset$. Then either $S_1 \cup S_2 = V$ or $S_1 \cap S_2 \cap O \neq \emptyset$.*

The corollary follows from Proposition 13 by considering a minimal pair (S_3, S_4) within (S_1, S_2).

As a side remark, Corollary 14 implies that if there exists an inside element v then we must have at least 4 outside elements; a cycle $C_1, \ldots, C_k$ for v (where $k \geq 4$) shows the existence of outside elements in each $C_i \cap C_{i+1}$. On the other hand, if we have no inside element and fewer than 4 outside elements then we must have $|V| = |O| = 2$ as a family on 3 elements could not have a connected cross graph.

We can also deduce from Corollary 14 that sets with the same intersection with O form a chain.

Proposition 15. *Let $A \subseteq O$, and let $\mathcal{C} = \{S \in \mathcal{F} : S \cap O = A\}$. Then $\mathcal{C}$ is a chain, i.e. the members of $\mathcal{C}$ can be ordered such that $S_1 \subset S_2 \subset \cdots \subset S_p$.*

Proof. Consider any two members S and T in $\mathcal{C}$, and let $S \setminus T \neq \emptyset$ (if $S \subseteq T$ then simply exchange S and T). By applying Corollary 14 to S and $\overline{T}$, we obtain that $S \cup \overline{T} = V$, i.e. $T \subseteq S$. As this is true for any two sets in $\mathcal{C}$, we have that $\mathcal{C}$ is a chain. ∎

What remains now is to show that any inside element v can be placed in one of the cells of $P \setminus \{\ell(S) : S \in \mathcal{F}\}$. Fix an inside element v, and let $\mathcal{F}_v = \{S \in \mathcal{F} : v \in S\}$. For any $S \in \mathcal{F}$, let $R(S)$ be the intersection of the interior of the polygon with the open halfplane on the left of the line through a_S and $a_{\overline{S}}$ (i.e. the halfplane already containing the elements $S \cap O$). To prove that v can be placed adequately, we need to prove that $\cap_{S \in \mathcal{F}_v} R(S) \neq \emptyset$. Interestingly, this is an implication of Helly's theorem in 2 dimensions (see e.g. [14]):

Theorem 16 (Helly). *Let a collection of convex subsets of $\mathbb{R}^d$ have the property that any collection of up to $d+1$ of them have a non-empty intersection. Then all of them have a common intersection.*

To apply Helly's theorem, we first consider the intersection of two such regions, and then show that the intersection of three regions essentially reduces to the intersection of two regions. In the proofs below, $cl(\cdot)$ denotes the closure operator.

Proposition 17. *Let $S_1, S_2 \in \mathcal{F}_v$. Then $R(S_1) \cap R(S_2) \neq \emptyset$.*

Proof. If $S_1 \cup S_2 = V$ then, by Proposition 9, the diagonals $\ell(S_1)$ and $\ell(S_2)$ do not cross. If $R(S_1)$ and $R(S_2)$ were disjoint then the fact that $S_1 \cup S_2$ contains all outside elements would imply that S_1 and $\overline{S_2}$ have identical intersections with O. But this case was taken care of when we introduced additional polygon vertices, as we made sure that if $S_1 \cap O = \overline{S_2} \cap O$ and $S_1 \cap S_2 \neq \emptyset$ then $R(S_1) \cap R(S_2) \neq \emptyset$. In fact, $R(S_1) \cap R(S_2)$ is a strip extending between $\ell(S_1)$ and $\ell(S_2)$; more formally, $\ell(S_i) \subset cl\big(R(S_1) \cap R(S_2)\big)$ for $i = 1, 2$.

If $S_1 \cup S_2 \neq V$ then Corollary 14 implies that $S_1 \cap S_2$ contains an outside element w, and the result follows trivially as the cell containing w will be in $R(S_1) \cap R(S_2)$. ∎

Finally, we consider the intersection of 3 regions defined by sets in $\mathcal{F}_v$.

Proposition 18. *Let $S_1, S_2, S_3 \in \mathcal{F}_v$. Then $R(S_1) \cap R(S_2) \cap R(S_3) \neq \emptyset$.*

Proof. If $S_1 \cap S_2 \cap S_3$ contains an outside element, we are done. Thus, we assume that $S_1 \cap S_2 \cap S_3 \cap O = \emptyset$. We consider two cases.

Case 1. For every pair (i, j), $S_i \cup S_j \neq V$. In this case, Corollary 14 implies that $T_{ij} = S_i \cap S_j \cap O \neq \emptyset$ for $1 \leq i < j \leq 3$. As the intersection of any two of the T_{ij}'s is empty ($S_1 \cap S_2 \cap S_3 \cap O = \emptyset$), we have that the three sets $S_{i-1} \cap S_{i+1} \setminus S_i$ are non-empty for $i \in \{1, 2, 3\}$. As S_1, S_2, S_3 do not form a 3-cycle, we must have that $S_1 \cup S_2 \cup S_3 = V$. We also know that $\overline{S_1}, \overline{S_2}, \overline{S_3}$ do not form a 3-cycle, and as $v \notin \bigcup_i \overline{S_i}$, we have that $\overline{S_i} \cap \overline{S_j} \setminus \overline{S_k} = \emptyset$ for some permutation i, j, k of $\{1, 2, 3\}$. The fact that $\bigcap_l \overline{S_l} = \emptyset$ then implies that $\overline{S_i} \cap \overline{S_j} = \emptyset$, i.e. that $S_i \cup S_j = V$ contradicting our assumption.

Case 2. $S_i \cup S_j = V$ for some $i, j \in \{1, 2, 3\}$, $i \neq j$. As in the first part of the proof of Proposition 17, we derive that both diagonals $\ell(S_i)$ and $\ell(S_j)$ are in $cl\big(R(S_i) \cap R(S_j)\big)$. Let k be the remaining index. If $\ell(S_k)$ crosses either $\ell(S_i)$ or $\ell(S_j)$ then a segment of $\ell(S_k)$ is in $cl\big(R(S_1) \cap R(S_2) \cap R(S_3)\big)$ showing non-emptyness of the intersection. On the other hand, if $\ell(S_k)$ crosses neither $\ell(S_i)$ nor $\ell(S_j)$ then either $R(S_k) \supseteq R(S_i) \cap R(S_j)$ (and we are done) or $R(S_k) \cap R(S_i) = \emptyset$ or $R(S_k) \cap R(S_j) = \emptyset$. In these latter cases, we obtain a contradiction from Proposition 17. ■

Helly's theorem thus shows that every inside element can be placed in one of the cells. As we did not make any assumptions on the position of the vertices $a_1, \ldots, a_q$ (except that they are in convex position), the polygon representation obtained is deformable, and this completes our proof of Theorem 2.

We now state a few more properties of the polygon representation. First, we can strengthen Proposition 9.

Proposition 19. *Consider a symmetric family with a connected cross graph and no 3-cycles or combs, and consider its deformable polygon representation. Let $S_1, S_2 \in \mathcal{F}$. Then the following are equivalent:*

1. *the diagonals $\ell(S_1)$ and $\ell(S_2)$ do not cross,*
2. *the sets $S_1 \cap O$ and $S_2 \cap O$ do not cross in O,*
3. *the sets S_1 and S_2 do not cross.*

Indeed, Proposition 9 says that 1. and 2. are equivalent, 3. always implies 2., and 1. implies 3. simply by the existence of the polygon representation.

In Proposition 15, we have shown that sets with the same intersection with O form a chain, and this was the basis for introducing new polygon vertices after having placed the outside elements. We now show that this can only happen if we have 4-cycles.

Proposition 20. *If $S_1, S_2 \in \mathcal{F}$ with $S_1 \neq S_2$ and $S_1 \cap O = S_2 \cap O$ then $\mathcal{F}$ contains a 4-cycle.*

Proof. By Proposition 15, we can assume that $S_1 \subset S_2$. Let $v \in S_2 \setminus S_1$; v must be inside as $(S_2 \setminus S_1) \cap O = \emptyset$. Let $C_1, \ldots, C_k$ be a cycle for v. Remember that $O \subseteq \cup_i C_i$ and that $C_i \cap C_{i+1} \cap O \neq \emptyset$ (by Corollary 14). Thus, there is an index i so that C_i contains elements of both A and $O \setminus A$, where $A = S_1 \cap O = S_2 \cap O$. If C_i contains all elements of A then C_i and S_2 do not cross in O (as $S_2 \cap O = A$), and by Proposition 19, they do not cross (in V); thus either $S_2 \subseteq C_i$ or $C_i \subseteq S_2$. This is a contradiction as $v \in S_2 \setminus C_i$ and $S_2 \cap O \subsetneq C_i \cap O$. Therefore, C_i crosses S_1 and S_2. Similarly, this implies that there exists another index j such that C_j crosses S_1 and S_2. If i and j are consecutive then we have a 3-cycle C_i, C_j and S_1, a contradiction. Otherwise, we have a 4-cycle composed of C_i, S_1, C_j and $\overline{S_2}$. ∎

If we have a chain $S_1 \subset \cdots \subset S_k$ with $S_1 \cap O = S_k \cap O$, we say that the inside elements in $S_k \setminus S_1$ are *sandwiched* by this chain. The next proposition shows that any inside element can be sandwiched by at most 2 chains; otherwise, we would have a 3-cycle, a contradiction. This will be useful when establishing a tight bound on the size of families with no 3-cycles or combs in Proposition 26.

Proposition 21. *For $v \in V$, let $\mathcal{A} = \big\{ \{A, O \setminus A\} \colon \exists\, S_1, S_2 \in \mathcal{F}$ s.t. $v \notin S_1$, $v \notin S_2$, $S_1 \cap O = A$, and $S_2 \cap O = O \setminus A \big\}$. Then $|\mathcal{A}| \leq 2$.*

Proof. Assume on the contrary that a given $v \in V$ is sandwiched by 3 chains. Thus, we have $S_{ik} \in \mathcal{F}$ for $i = 1, 2, 3$ and $k = 1, 2$ with $v \notin S_{ik}$ for all i and k, $S_{i1} \cap O = A_i$ for all i, $S_{i2} \cap O = O \setminus A_i$ for all i, and the 6 sets A_i's and $O \setminus A_i$'s are all distinct. Furthermore, possibly exchanging S_{i1} and S_{i2} (replacing A_i by $O \setminus A_i$), we can assume that $w \in A_1 \cap A_2 \cap A_3$ for some $w \in O$.

We claim that, for any $i, j \in \{1,2,3\}$ with $i \neq j$, A_i and A_j cross in O. Otherwise, suppose that $A_i \subset A_j$. Applying Proposition 19 to $\overline{S_{i2}}$ and S_{j1} (recall that $\overline{S_{i2}} \cap O = A_i$ and $S_{j1} \cap O = A_j$, we get that $\overline{S_{i2}} \subset S_{j1}$, a contradiction as $v \in \overline{S_{i2}} \setminus S_{j1}$.

Therefore, the A_i's are pairwise mutually crossing in O and correspond all to circular intervals of O containing w. This means that their start elements (of their circular interval) in O are distinct, so are their end elements, and these elements are ordered in the same way among the 3 sets. Say that A_2 corresponds to the middle interval. Then A_1, $O \setminus A_2$ and A_3 are such that the intersection of any two of them minus the third one is non-empty. This implies that S_{11}, S_{22} and S_{31} form a 3-cycle for v, a contradiction. ■

4. Mincuts and Near-Mincuts

In this section, we show that the main configurations discussed in the previous sections do not exist for families of cuts of sufficiently small value compared to the edge-connectivity. In particular, we show that each connected component of the cross graph of 6/5-near-mincuts has a deformable polygon representation. Recall that λ denotes the edge-connectivity and an α-near-mincut is a cut whose value is (strictly) less than $\alpha\lambda$.

In what follows, let $d(X,Y)$ denote the total weight of edges connecting $X, Y \subset V$ in a weighted graph $G = (V, E)$. We first show that sufficiently near-mincuts have no short cycles as defined in Definitions 1 and 3.

Lemma 22 (Excluded Cycles). *Let k δ-near-mincut sides C_i for $i \leq k$ form a k-cycle. Then $\delta > 1 + 1/k$.*

Proof. For $k = 3$ the result follows from 3-way submodularity of the cut function, see Lovász [13], exercise 6.48 (c):

$$3\delta\lambda > d(C_1) + d(C_2) + d(C_3) \geq d(C_1 \cap C_2 \setminus C_3) + d(C_1 \cap C_3 \setminus C_2)$$
$$+ d(C_2 \cap C_3 \setminus C_1) + d(\overline{C_1 \cup C_2 \cup C_3}) \geq 4\lambda.$$

3-way submodularity can be established by observing that the contribution of any edge is at least as large on the left-hand-side as on the right-hand-side.

For $k \geq 4$, a similar edge counting argument gives:

$$d(\cup_i C_i) + \sum_{i \leq k} d(C_i \cap C_{i+1}) \leq \sum_{i \leq k} d(C_i).$$

Indeed, (i) an edge that contributes to the left-hand-side also contributes to the right-hand-side,(ii) no edge contributes more than twice to the left-hand-side, and (iii) an edge that contributes twice to the left-hand-side must be either between some $C_i \cap C_{i+1}$ and $\overline{\cup_j C_j}$ or between $C_i \cap C_{i+1}$ and $C_j \cap C_{j+1}$ for $i \neq j$; in all cases, it contributes at least twice to the right-hand-side. Since $d(C_i) < \delta\lambda$, $d(C_i \cap C_{i+1}) \geq \lambda$ for $i \leq k$ and $d(\cup_i C_i) \geq \lambda$, the above inequality implies that $\delta > 1 + \frac{1}{k}$ as required. ■

In particular, 6/5-near-mincuts may not contain k-cycles for $k \leq 5$, 4/3-near-mincuts have no 3-cycles, and mincuts contain no cycles at all.

Lemma 23 (Excluded Combs). *There are no four 6/5-near-mincut sides T_1, T_2, T_3 and H which form a comb as defined in Definition 2.*

Proof. Definition 2 gives 3 disjoint and nonempty subsets C_1, C_2 and C_3 of H: Either C_i is $H \cap \big(T_i \setminus (T_{i-1} \cup T_{i+1})\big)$ or C_i is $H \cap (T_{i-1} \cap T_{i+1} \setminus T_i)$. Similarly, it gives 3 disjoint nonempty subsets D_1, D_2 and D_3 of $\overline{H}$. By a counting argument, one observes that

$$\text{(2)} \quad d(T_1) + d(T_2) + d(T_3) + 2d(H) \geq d(C_1) + d(C_2) + d(C_3) + d(D_1) + d(D_2) + d(D_3).$$

Indeed, either an edge does not cross H and the counting argument is similar to the derivation of 3-way submodularity, or the edge crosses H in which case its multiplicity is at most 2 on the right-hand-side and at least 2 on the left-hand-side. Inequality (2) now implies that $5\delta\lambda > 6\lambda$. ■

We have thus derived that the symmetric family of 6/5-near-mincut sides does not contain any comb and any k-cycles for $k \leq 5$. Thus each connected component of its cross graph admits a deformable polygon representation.

5. Tree Hierarchy

In this section, we show that the connected components of the cross graph of any symmetric family $\mathcal{F} \subseteq V$ of sets can be arranged in a tree structure. We first derive it for families with no combs and cycles, and in so doing, rederive the tree structure of a (slightly modified) cactus representation. Then we show that the tree structure depends only on the connected components of the cross graph and thus can be applied to arbitrary symmetric families, including those having deformable polygon representations for each connected component of the cross graph. This tree structure will then be used to derive bounds on the size of the representation and (in the next section) on the cardinality of symmetric families with no 3-cycles or combs.

To fix notation, let $\mathcal{F}_1, \ldots, \mathcal{F}_q$ represent the sets in each of the connected components of the cross graph. Also, for any $1 \leq i \leq q$, let P_i denote the partition of V induced by $\mathcal{F}_i$, i.e. the members of P_i correspond precisely to the atoms of $\mathcal{F}_i$ as defined at the beginning of Section 3. Observe that, for any i, the number of atoms in P_i can either be 2 or greater or equal to 4. Indeed, if $|P_i|$ was 3 the corresponding pairs of sets would not be crossing.

The tree structure on the connected components essentially follows from laminarity. It is similar to the usual cactus representation for the mincuts of a graph, except that, there, the cuts represented by a cycle of the cactus do not quite form a single connected component of the cross graph. Indeed the cut corresponding to a single atom (i.e. two consecutive edges of the cycle) does not cross any of the other cuts represented by the cycle and thus do not belong to the same connected component of the cross graph. Here, however, we redefine the cactus representation and assume that the cuts represented by a cycle of length k with $k \neq 2$ of the cactus are the cuts obtained by removing two *non-consecutive* edges of the cycle. The rest of the definition is unchanged. It is rather easy to see using classical arguments that the mincuts of a graph admit a cactus representation for this slightly modified notion of a cactus representation. For completeness, we provide here a somewhat different proof, of a geometric nature as in the rest of this paper. For generality and to be able to later apply it to our deformable polygons, we state it in terms of symmetric families with no cycles or combs (recall that from Lemmas 22 and 23, mincut sides do not have any cycles or combs).

Proposition 24. *Let $\mathcal{F} \subseteq 2^V$ be a symmetric family of sets with no cycles and no combs. Let $\mathcal{F}_i$ be the connected components of its cross graph and let P_i be the partition of V corresponding to the atoms of $\mathcal{F}_i$. Then there exists a cactus $H = (N, A)$ and a mapping $\phi\colon V \to N$ such that*

1. *H has no cycle of length 3,*
2. *there is a 1-to-1 correspondence between the connected components $\mathcal{F}_i$ of the cross graph and the cycles C_i of H,*
3. *the removal of the edges of $C_i = u_1 - u_2 \cdots - u_k - u_1$ break H into k (depending on i) connected components, $A_1, \dots, A_k \subset N$ where $u_j \in A_j$ such that $P_i = \{\phi^{-1}(A_j)\colon 1 \le j \le k\}$,*
4. *for each set $S \in \mathcal{F}$, there is a unique cycle C_i in H and two edges of C_i which are non-consecutive if the cycle is not of length 2, whose removal partitions N into U and $N \setminus U$ with $S = \phi^{-1}(N)$.*

To complete the proof that mincuts admit a (modified) cactus representation, it remains to show that the removal of *any* 2 edges of a cycle – non-consecutive if the cycle has length different from 2 – gives a mincut of G. This follows from submodularity of the cut function (union or intersection of crossing mincuts is a mincut); this is left to the reader.

Proof. From Proposition 6 and Lemmas 22 and 23, we know that $\mathcal{F}$ is a circular representable hypergraph. Consider one such circular ordering $v_1, \dots, v_n$ where $n = |V|$. If there are several, take one in which the elements in each atom of $\mathcal{F}$ appear consecutively in the ordering (one could for example first shrink the atoms of $\mathcal{F}$).

We provide a geometric construction of the cactus. Take a circle and divide it into n arcs representing the vertices in the circular ordering. For each pair of complementary sets $S, \overline{S}$, draw the corresponding chord that has S on one side and $\overline{S}$ on the other. Observe that two chords corresponding to S_1 and S_2 will cross in the geometric sense (i.e. will have an intersection in their relative interior) if and only if the sets S_1 and S_2 cross.

Consider any of the connected components of the cross graph, say $\mathcal{F}_i$ with partition P_i. The chords defining $\mathcal{F}_i$ connect $k = |P_i|$ points on the circle, say p_1, $p_2, \dots, p_k$, and the arcs between these points correspond to the sets in P_i. See Figure 13. Let R_i be the convex hull of these points. The fact that sets from different components do not cross imply that the chord for a set of a different component $\mathcal{F}_j$ can only intersect R_i on its boundary. This means that the relative interiors of any two such regions, R_i and R_j, will be disjoint. Replace all the chords corresponding to $\mathcal{F}_i$ by a star with root r_i and k spokes connected to $p_1, \ldots, p_k$. To make sure that the stars for different components do not cross, place r_i in the relative interior of R_i. This gives a plane graph D with the outside region delimited by our circle. Now define H to be its dual graph except that we do not create a node of H for the outside face of D. The nodes of H corresponding to the inside faces of D along the circle are labelled with the atoms of $\mathcal{F}$. The node set N of H is thus the set of *bounded* faces of D. Observe that H will be the union of cycles, one for each vertex r_i. Furthermore, every edge is in precisely one cycle; thus, H is a cactus. All the claims in the statement of the Proposition follow easily from the construction itself. ■

This construction of this modified cactus representation has slightly more empty nodes than the usual constructions; the size is still linear as shown below.

Proposition 25. *The modified cactus representation for a symmetric family $\mathcal{F} \subseteq 2^V$ with no cycles and no combs has at most $3n-4$ nodes and at most $5n-8$ edges where $n = |V| \geq 2$.*

Proof. Let $N_p(n)$ and $E_p(n)$ resp. represent the maximum number of nodes and edges resp. of our cactus representation for families with at most p *non-trivial* connected components, where we define a connected component as non-trivial if it contains more than one pair of complementary sets. We proceed by induction on p.

If $\mathcal{F}$ has no crossing sets ($p = 0$) then the number of chords in our construction is at most $2n-3$ (the maximum cardinality of a laminar family with no complementary sets). Each chord will lead to 2 edges of the cactus and the cactus will be a tree of cycles of length 2. Thus, its number of edges will be at most $E_0(n) = 4n-6 \leq 5n-8$ for $n \geq 2$ and the number of nodes will be at most $N_0(n) = 2n-2 \leq 3n-4$.

Suppose $\mathcal{F}$ has p non-trivial connected components for $p > 0$. Let $\mathcal{F}_i$ be a connected component of the cross graph which induces a partition P_i with $k \geq 4$ atoms. Let $n_1, \ldots, n_k$ be the cardinalities of these atoms. Geometrically, we can partition $\mathcal{F} \setminus \mathcal{F}_i$ into k symmetric families $\mathcal{G}_1, \ldots, \mathcal{G}_k$ in a natural way: for the jth family $\mathcal{G}_j$, we only keep those chords that are on the side of $[p_j, p_{j+1}]$ opposite to R_i. Observe that $[p_j, p_{j+1}]$ might be one of the chords represented by this family. Collectively, the representations for the $\mathcal{G}_j$'s account for all nodes and edges of the representation for $\mathcal{F}$, except

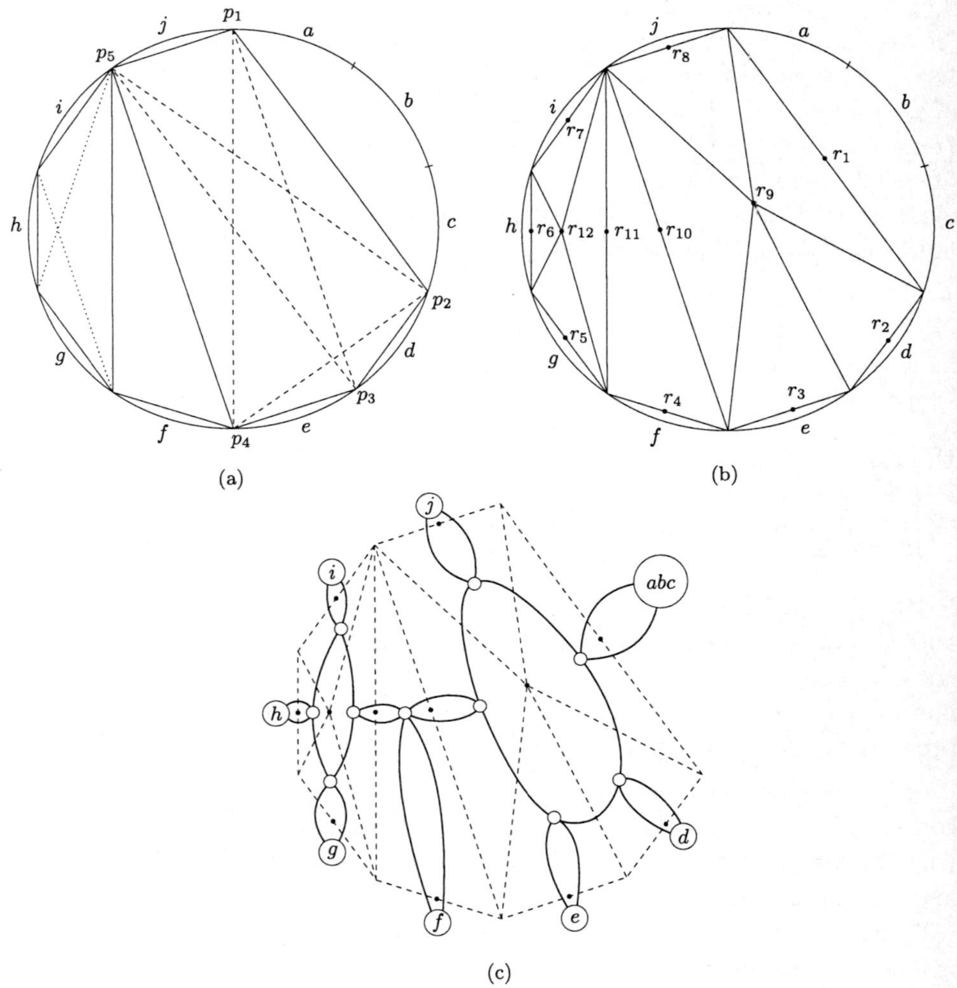

Fig. 13. Obtaining a cactus. (a): The cyclic ordering. The sets are represented by chords. There are 12 connected components, the dashed one, the dotted one, and 10 others with only one chord. The dashed component connects $p_1, \ldots, p_5$. (b): Replacing every component by a star with root r_i. (c): Inside faces along the circle are labelled with the atoms of $\mathcal{F}$, and the dual graph is the cactus.

for the k edges of the cycle corresponding to $\mathcal{F}_i$. $\mathcal{G}_j$ has at most $n_j + 1$ atoms (and fewer non-trivial components), and by induction, we therefore get that:

$$N_p(n) \leq \sum_{j=1}^{k} N_{p-1}(n_j + 1) \leq \sum_j (3n_j - 1) = 3n - k \leq 3n - 4$$

and

$$E_p(n) \leq \sum_{j=1}^{k} E_{p-1}(n_j + 1) + k \leq \sum_j (5n_j - 3) + k = 5n - 2k \leq 5n - 8,$$

since $k \geq 4$. ■

The bounds given in Proposition 25 are tight whenever n is even, and this is achieved when we have $(n-1)/2$ disjoint cycles of length 4 in the cactus linked by cycles of length 2, see Figure 14.

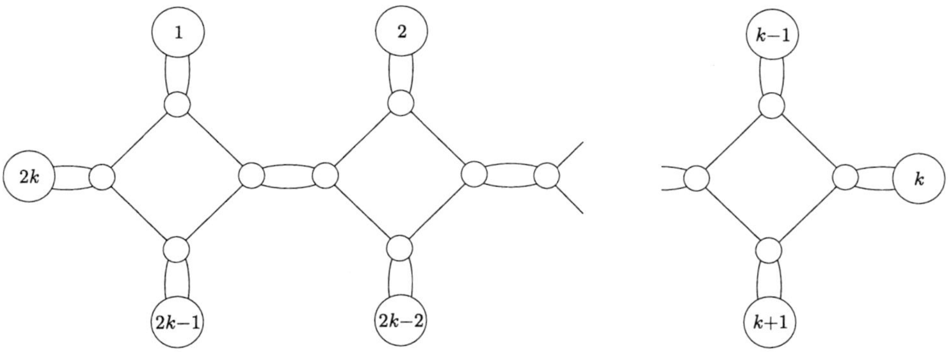

Fig. 14. A cactus with $3n - 4$ nodes and $5n - 8$ edges for a symmetric family with no cycles or combs defined on a ground set of size $n = 2k$.

This cactus representation provides a way to combine representations (i.e. cycles on partitions of V) for each connected component of the cross graph in a tree structure in which certain atoms of different connected components (i.e. cycles) are identified (those corresponding to the same node of the cactus). To highlight the tree T (and get rid of the cycles), we can replace each cycle of the cactus H by a star rooted at a new node representing this connected component of the cross graph. In other words, the nodes of T consist of (i) one node in C for each connected component of the cross graph and also (ii) one node in N for each node of the cactus; the latter ones correspond to atoms that have been identified from one or several connected components of the cross graph. T has an edge between a node c in C and a node u in N if one of the atoms of the connected component corresponding to c is associated with node u of the cactus.

This can be generalized to any symmetric family (independently of whether each connected component of the cross graph can or cannot be represented by polygons). For any symmetric family $\mathcal{F}$, consider the connected components $\mathcal{F}_i$ of the cross graph and let P_i be the atoms of $\mathcal{F}_i$. Now, for each i, arbitrarily choose a cyclic ordering on the atoms in P_i and define $\mathcal{G}_i$ to be the family of sets (with the same atoms as $\mathcal{F}_i$) corresponding to cyclic intervals containing at least 2 and at most $k_i - 2$ of the atoms in P_i, where $k_i = |P_i| \geq 4$ is the number of atoms of $\mathcal{F}_i$; if $k_i = 2$ (i.e. the component is trivial), we simply let $\mathcal{G}_i = \mathcal{F}_i$. Observe that the family $\mathcal{G} = \cup \mathcal{G}_i$ has no cycles or combs since (i) the connected components of the cross graph of $\mathcal{G}$ are still the $\mathcal{G}_i$'s, (ii) any cycle or comb would need to be contained within a connected component of the cross graph and (iii) the $\mathcal{G}_i$'s have no combs or cycles by construction. By Proposition 24, the family $\mathcal{G}$ has a cactus representation, and this means that atoms of different connected components of the cross graph of $\mathcal{G}$, and thus of $\mathcal{F}$, are identified. This gives a tree T for $\mathcal{G}$ and thus also for $\mathcal{F}$. If each connected component of the cross graph has a deformable polygon representation, these polygon representations together with the tree T form our representation as a *tree of deformable polygons*.

6. Size of the Family

In this section, we deduce from the representation as a tree of deformable polygons that any symmetric family of sets $\mathcal{F} \subset 2^V$ with no 3-cycles and no combs has at most $\binom{n}{2}$ complementary pairs where $n = |V|$. For near-mincuts, this shows that there are at most $\binom{n}{2}$ 6/5-near-mincuts, although this is known even for 4/3-near-mincuts [19, 9] and a direct proof is simpler. See also [10] for a proof that there are at most $O(n^2)$ 3/2-near-mincuts.

Proposition 26. *Let $\mathcal{F} \subseteq 2^V \setminus \{\emptyset, V\}$ be a symmetric family of sets with no 3-cycles or combs. Then $|\mathcal{F}| \leq n(n-1)$ where $n = |V|$, i.e. the number of complementary pairs is at most $\binom{n}{2}$.*

Proof. We will first focus on a connected component $\mathcal{F}_i$ of the cross graph with $k \geq 4$ atoms, and show that our construction of a deformable polygon gives rise to at most $k(k-3)/2$ diagonals. Let $k_{\text{out}} \geq 4$ and k_{in} be the number of outside and inside atoms respectively; thus $k = k_{\text{out}} + k_{\text{in}}$. If $\mathcal{F}_i$ has no 4-cycles then our polygon has k_{out} sides (see Proposition 20) and its number of diagonals is at most $k_{\text{out}}(k_{\text{out}} - 3)/2 \leq k(k-3)/2$. If we have 4-cycles then by Proposition 21, the number of diagonals is at most $k_{\text{out}}(k_{\text{out}}-3)/2+2k_{\text{in}} \leq k(k-3)/2$ as $(k_{\text{out}}+1)(k_{\text{out}}-2)/2-k_{\text{out}}(k_{\text{out}}-3)/2 \geq 2$ for $k_{\text{out}} \geq 3$. Thus, a connected component of the cross graph on $k \geq 4$ atoms has at most $k(k-3)/2$ diagonals.

Consider now the various connected components of the cross graph and, as in the proof of Proposition 25, we proceed by induction on p, the number of non-trivial connected components. Let $S_p(n)$ denote the maximum number of complementary pairs for families with at most p non-trivial components. If $p = 0$ then we have $S_0(n) \leq 2n - 3 \leq \binom{n}{2}$ complementary pairs, see Proposition 25. If $p > 0$, let $\mathcal{F}_i$ be one of those non-trivial components on $k \geq 4$ atoms with cardinalities $n_1, \ldots, n_k$ where $n = \sum_{j=1}^{k} n_j$. We use the same notation as in Proposition 25. From the tree structure, we get that:

$$
\begin{aligned}
S_p(n) &\leq \sum_{j=1}^{k} S_{p-1}(n_j+1) + \frac{k(k-3)}{2} \leq \sum_{j=1}^{k} \binom{n_j+1}{2} + \frac{k(k-3)}{2} \\
&\leq (k-1) + \frac{(n-k+2)(n-k+1)}{2} + \frac{k(k-3)}{2} \\
&= \binom{n}{2} - (n-k)(k-2) \leq \binom{n}{2},
\end{aligned}
$$

the third inequality follows from the fact that the maximum of a convex function over $n_j \geq 1$ for $j = 1, \ldots, k$ and $\sum_j n_j = n$ is attained for all but one n_j equal to 1. ∎

We should point out that the existence of a (non-deformable) polygon representation is not enough to prove the bound of $\binom{n}{2}$; for example, all cuts of K_4 admit a representation as a tree of (non-deformable) polygons although there are $7 > \binom{4}{2}$ of them. One can, however, prove a slightly weaker bound by observing that a connected component of the cross graph on k atoms has at most $\binom{k}{2}$ complementary pairs (by Schläfli's result) and the same argument as in the proof above gives an overall bound of $\binom{n}{2} + 2n - 4$.

7. Conclusion

We have derived a representation for symmetric families that do not contain 3-cycles or combs, and this applies to the family of 6/5-near-mincuts in a graph. We refer the reader to the Ph.D. thesis [3] of the first author for algorithmic issues and applications of this representation. It would be interesting to find a representation for families that may contain combs but do not have 3-cycles; this would allow to represent 4/3-near-mincuts.

Acknowledgements. The second author would like to thank the hospitality of the Research Institute for Mathematical Sciences, Kyoto University, where a great deal of the writing of this paper was done.

References

[1] A. A. Benczúr, *On the structure of near-minimum edge cuts,* Technical report MIT/LCS/TR-639 (1994).

[2] A. A. Benczúr, A representation of cuts witihin 6/5 times the edge connectivity with applications, in: *Proc. 34th Annual Symp. on Found. of Comp. Sci.* (1995), pages 92–102.

[3] A. A. Benczúr, *Cut Structures and Randomized Algorithms in Edge-Connectivity Problems,* Ph.D. Thesis, M.I.T. (1997).

[4] Y. Dinitz (E. A. Dinits), A. V. Karzanov and M. L. Lomonosov, On the Structure of a Family of Minimal Weighted Cuts in a Graph, in: *Studies in Discrete Optimization* (in Russian), A. A. Fridman (Ed), Nauka, Moscow (1976), pages 290–306.

[5] Y. Dinitz and Z. Nutov, A 2-level cactus model for the minimum and minimum+1 edge-cuts in a graph and its incremental maintenance, in: *Proc. of the 27th Symposium on Theory of Computing* (1995), pages 509–518.

[6] P. Duchet, Classical Perfect Graphs: An Introduction with Emphasis on Triangulated and Interval Graphs, *Annals of Discrete Mathematics,* **21** (1984), 67–96.

[7] L. Fleischer, Building Chain and Cactus Representations of All Minimum Cuts from Hao–Orlin in the Same Asymptotic Run Time, *Journal of Algorithms,* **33** (1999), 51–72.

[8] H. N. Gabow, *Applications of a Poset Representation to Edge Connectivity and Graph Rigidity,* in: *Proc. of the 32nd Annual Symp. on Found. of Comp. Sci.* (1991), pages 812–821.

[9] M. X. Goemans and V. S. Ramakrishnan, Minimizing Submodular Functions over Families of Sets, *Combinatorica,* **15** (1995), 499–513.

[10] M. Henzinger and D. P. Williamson, On the Number of Small Cuts in a Graph, *Information Processing Letters* (1996), **59**, 41–44.

[11] A. V. Karzanov and E. A. Timofeev, Efficient Algorithms for Finding all Minimal Edge Cuts of a Nonoriented Graph, *Cybernetics* (1986), 156–162. Translated from *Kibernetika,* **2** (1986), 8–12.

[12] J. Lehel, F. Maffray and M. Preissmann, Graphs with Largest Number of Minimum Cuts, *Discrete Applied Mathematics,* **65** (1996), pages 387–407.

[13] L. Lovász, *Combinatorial Problems and Exercises,* North-Holland, Amsterdam (1979).

[14] J. Matoušek, *Lectures on Discrete Geometry,* Springer (2002).

[15] H. Nagamochi and T. Kameda, Canonical Cactus Representation for Minimum Cuts, *J. of Japan Society for Industrial and Applied Mathematics,* **11** (1994), 343–361.

[16] H. Nagamochi and T. Kameda, Constructing Cactus Representation for all Minimum Cuts in an Undirected Network, *Operations Research Society of Japan,* **39** (1996), pages 135–158.

[17] H. Nagamochi, S. Nakamura and T. Ishii, Constructing a Cactus for Minimum Cuts of a Graph in $O(mn+n^2 \log n)$ time and $O(m)$ space, *Inst. Electron. Inform. Comm. Eng. Trans. Information and Systems,* **E86-D** (2003), 179–185.

[18] H. Nagamochi, Y. Nakao and T. Ibaraki, A Fast Algorithm for Cactus Representations of Minimum Cuts, *J. of Japan Society for Industrial and Applied Mathematics,* **17** (2000), 245–264.

[19] H. Nagamochi, K. Nishimura and T. Ibaraki, Computing all small cuts in undirected networks, *SIAM Disc. Math.,* **10** (1997), 469–481.

[20] D. Naor and V. V. Vazirani, Representing and Enumerating Edge Connectivity Cuts in RNC, in: *Proc. Second Workshop on Algorithms and Data Structures,* Lecture Notes in Computer Science **519**, Springer–Verlag (1991), pages 273–285.

[21] A. Quilliot, Circular Representation Problems on Hypergraphs, *Discrete Mathematics,* **51** (1984), pages 251–264.

[22] L. Schläfli, Theorie der vielfachen Kontinuität, 1901. In: *Gesammelte Mathematische Abhandlungen,* volume I (1950), pages 167–387. Verlag Birkhäuser.

[23] A. Tucker, A Structure Theorem for the Consecutive 1's Property, *JCT B,* **12** (1972), pages 153–162.

András A. Benczúr
Computer and Automation Research Institute of the Hungarian Academy of Sciences
e-mail: benczur@sztaki.hu

Michel X. Goemans
M.I.T., Department of Mathematics
e-mail: goemans@math.mit.edu

BOLYAI SOCIETY
MATHEMATICAL STUDIES, 19

Building Bridges
pp. 137–164.

Variations for Lovász' Submodular Ideas

KRISTÓF BÉRCZI and ANDRÁS FRANK*

Dedicated to Lovász Laci on the occasion of his 60'th birthday by his youngest and oldest mathematical descendants

In [18], L. Lovász provided simple and short proofs for two classic min-max theorems of graph theory by inventing basic techniques to handle sub- or supermodular functions. In this paper, we want to demonstrate that these ideas are alive after thirty years of their birth.

1. Introduction

Sub- and supermodular set functions play an important role in proving theorems in graph theory. L. Lovász [18] introduced a submodular technique to derive the disjoint arborescences theorem of J. Edmonds [2] and another one to prove a min-max result of C. Lucchesi and D. Younger [19] on minimum coverings of dicuts of a digraph. It appears that this paper is the first occurance of the so called uncrossing procedure (apart from a Hungarian report reviewing Lovász' solution to a problem of a math student competition, see [16]). Uncrossing became later a particularly efficient proof techniqe in submodular optimization.

In the last fifteen years it turned out that several results and techniques developed for sub- or supermodular set functions can be extended to functions defined on pairs of sets or on bi-sets. Given a ground-set V, we call a pair $X = (X_O, X_I)$ of subsets a **bi-set** if $X_I \subseteq X_O \subseteq V$ where X_O is the

*Research supported by the Hungarian National Foundation for Scientific Research, OTKA K60802. The work was completed while the second author visited the Research Institute for Mathematical Sciences, Kyoto University, 2008.

outer member and X_I is the inner member of X. By a bi-set function we mean a function defined on the set of bi-sets of V. We will tacitly identify a bi-set $X = (X_O, X_I)$ for which $X_O = X_I$ with the set X_I and hence bi-set functions may be considered as straight generalizations of set functions.

While supermodular set functions are typically used for handling only edge-connectivity problems, supermodular bi-set functions can be applied for handling both node- and edge-connectivity problems. For example, the directed edge-connectivity augmentation problem was solved in [9] via crossing supermodular set functions while a solution to its node-connectivity counterpart was derived from a min-max theorem on covering crossing supermodular bi-set functions [11]. Similarly, an answer to the cheapest rooted k-edge-connected subgraph problem followed from a min-max result on covering intersecting supermodular functions by digraphs [5, 6] while the rooted k-node-connected version was derived from an analogous result on supermodular bi-set functions [10].

One goal of this work is to exhibit the evolution of Lovász' proof technique given for proving Edmonds' arborescences theorem. In particular, we extend a theorem of L. Szegő [22] on disjoint coverings of set systems to those of bi-set systems. This will imply a recent theorem of N. Kamiyama, N. Katoh, and A. Takizawa [14] which is a proper extension of Edmonds' disjoint arborescences theorem.

Second, by using the uncrossing technique, a new min-max theorem will be proved on minimal coverings of two fully supemodular bi-set functions by digraphs. This may be considered as a generalization of (the cardinality version of) Edmonds' (poly)matroid intersection theorem [1, 3]. It also provides an answer to a simultaneous connectivity augmentation problem where two given digraphs on the same node set is to be simultaneously augmented by adding a minimum number of new edges so that the resulting digraphs includes k_i g_i-independent paths from s_i to t_i $(i = 1, 2)$ where g_i-independence of paths is a notion including both edge-disjoint and node-disjoint paths.

In the sequel we use the following notions and notation. The set of all bi-sets on ground-set V is denoted by $\mathcal{P}_2(V) = \mathcal{P}_2$. The intersection $\cap$ and the union $\cup$ of bi-sets is defined in a staightforward manner: for $X, Y \in \mathcal{P}_2$ let $X \cap Y := (X_O \cap Y_O, X_I \cap Y_I)$, $X \cup Y := (X_O \cup Y_O, X_I \cup Y_I)$. We write $X \subseteq Y$ if $X_O \subseteq Y_O$, $X_I \subseteq Y_I$ and this relation is a partial order on $\mathcal{P}_2$. Accordingly, when $X \subseteq Y$ or $Y \subseteq X$, we call X and Y **comparable**. A family of pairwise comparable bi-sets is called a **chain**. Two bi-sets X

and Y are **independent** if $X_I \cap Y_I = \emptyset$ or $V = X_O \cup Y_O$. A set of bi-sets is independent if its members are pairwise independent. We call a set of bi-sets a **ring-family** if it is closed under taking union and intersection. Two bi-sets are **intersecting** if $X_I \cap Y_I \neq \emptyset$ and **properly intersecting** if, in addition, they are not comparable. Note that $X_O \cup Y_O = V$ is allowed for two intersecting bi-sets. In particular, two sets X and Y are properly intersecting if none of $X \cap Y, X - Y, Y - X$ is empty. A family of bi-sets is called **laminar** if it has no two properly intersecting members. A family $\mathcal{F}$ of bi-sets is **intersecting** if both the union and the intersection of any two intersecting members of $\mathcal{F}$ belong to $\mathcal{F}$. In particular, a family $\mathcal{L}$ of subsets is intersecting if $X \cap Y,\ X \cup Y \in \mathcal{L}$ whenever $X, Y \in \mathcal{L}$ and $X \cap Y \neq \emptyset$. A laminar family of bi-sets is obviously intersecting. Two bi-sets are **crossing** if $X_I \cap Y_I \neq \emptyset$ and $X_O \cup Y_O \neq V$ and **properly crossing** if they are not comparable. A bi-set (X_O, X_I) is **trivial** if $X_I = \emptyset$ or $X_O = V$. We will assume throughout the paper that the bi-set functions in question are integer-valued and that their value on trivial bi-sets is always zero. In particular, set functions are also integer-valued and zero on the empty set.

A directed edge **enters** or **covers** X if its head is in X_I and its tail is outside X_O. An edge **covers** a family of bi-sets if it covers each member of the family. For a bi-set function p, a digraph $D = (V, A)$ is said to cover p if $\varrho_D(X) \geq p(X)$ for every $X \in \mathcal{P}_2(V)$ where $\varrho_D(X)$ denotes the number of edges of D covering X. For a vector $z\colon A \to \mathbf{R}$, let $\varrho_z(X) := \sum \big[z(a)\colon a \in A,\ a \text{ covers } X \big]$. A vector $z\colon A \to \mathbf{R}$ **covers** p if $\varrho_z(X) \geq p(X)$ for every $X \in \mathcal{P}_2(V)$.

A bi-set function p is said to satisfy the **supermodular inequality** on $X, Y \in \mathcal{P}_2$ if

$$p(X) + p(Y) \leq p(X \cap Y) + p(X \cup Y). \tag{1}$$

If the reverse inequality holds, we speak of the **submodular** inequality. p is said to be **fully supermodular** or supermodular if it satisfies the supermodular inequality for every pair of bi-sets X, Y. If (1) holds for intersecting (crossing) pairs, we speak of **intersecting (crossing) supermodular** functions. Analogous notions can be introduced for submodular functions. Sometimes (1) is required only for pairs with $p(X) > 0$ and $p(Y) > 0$ in which case we speak of **positively supermodular** functions. Positively intersecting or crossing supermodular functions are defined analogously. A typical way to construct a positively supermodular function is replacing each negative value of a fully supermodular functions by zero.

Proposition 1.1. *The in-degree function ϱ_D on $\mathcal{P}_2$ is submodular.* ■

2. Packing Arborescences

2.1. Basic cases

An arborescence is defined to be a directed tree in which every node is reachable from a specified root-node r_0. The starting point is a classical result of J. Edmonds [2]. A digraph D is called **rooted** (more specifically, **r_0-rooted**) **k-edge-connected** with respect to a root-node $r_0 \in V$ if the in-degree of every non-empty subset of $V - r_0$ is at least k. By the directed edge-version of Menger's theorem this is equivalent to requiring that there are k edge-disjoint paths from r_0 to every node of D.

Theorem 2.1 (Edmonds' disjoint arborescences: weak form). *Let $D = (V, A)$ be a directed graph with a designated root-node r_0. D has k disjoint spanning arborescences of root r_0 if and only if D is rooted k-edge-connected, that is,*

$$\varrho(X) \geq k \quad \textit{whenever} \quad X \subseteq V - r_0, \quad X \neq \emptyset. \tag{2}$$ ■

Edmonds actually proved his theorem in a stronger form where the goal was packing k edge-disjoint branchings of given root-sets. A branching is a directed forest in which the in-degree of each node is at most one. The set of nodes of in-degree 0 is called the root-set of the branching. Note that a branching with root-set R is the union of $|R|$ node-disjoint arborescences (where an arborescence may consist of a single node and no edge but we always assume that an arborescence has at least one node). For a digraph $D = (V, A)$ and root-set $\emptyset \subset R \subseteq V$ a branching (V, B) is called a **spanning R-branching** of D if its root-set is R. In particular, if R is a singleton consisting of an element s, then a spanning branching is a spanning arborescence of root s.

Theorem 2.2 (Edmonds' disjoint branchings). *In a digraph $D = (V, A)$, let $\mathcal{R} = \{R_1, \ldots, R_k\}$ be a family of k non-empty* (not necessarily disjoint or distinct) *subsets of* V. *There are* k *edge-disjoint spanning branchings of* D *with root-sets* $R_1, \dots, R_k$, *respectively, if and only if*

$$\varrho(X) \geq p(X) \quad \textit{whenever} \quad \emptyset \subset X \subseteq V \tag{3}$$

where $p(X)$ *denotes the number of root-sets* R_i *disjoint from* X. ■

Remark. In the special case of Theorem 2.2 when each root-set R_i is a singleton consisting of the same node r_0, we are back at Theorem 2.1. Conversely, when the R_i's are singletons (which may or may not be distinct), then Theorem 2.2 easily follows from Theorem 2.1. However, for general R_i's no reduction is known.

The original proof of Edmonds is pretty complex and does not seem to transform into a polynomial time algorithm. However, R. E. Tarjan observed [23] that Theorem 2.2 itself gives rise to such an algorithm provided an MFMC subroutine is available. It should be emphasized that this approach does make use of the theorem and does not provide an alternative proof of it. On the other hand, L. Lovász [18] gave a simple proof of Edmonds' theorem and this proof is algorithmic. Although Lovász derived only the weak form of Edmonds' theorem, his proof carries over to the strong one almost word for word.

It is interesting to formulate Edmonds' Theorem 2.2 in the following equivalent form. Let A_0 denote the set of edges of $D = (V, A)$ leaving the root-node r_0 and let $A^* := A - A_0$.

Theorem 2.3 (Edmonds' disjoint arborescences: strong form). *Let* $D = (V, A)$ *be a directed graph with a designated root-node* r_0. *Let* A_0 *denote the set of edges leaving* r_0 *and* $A^* := A - A_0$. *Let* $\mathcal{A} = \{A_1, \dots, A_k\}$ *be a partition of* A_0 *into* k *sets. Then* D *has* k *disjoint spanning arborescences* $F_1, \dots, F_k$ *of root* r_0 *so that* $F_i \cap A_0 \subseteq A_i$ *for* $i = 1, \dots, k$ *if and only if* $\varrho_{A^*}(X) \geq p(X)$ *for every non-empty subset* X *of* $V - r_0$ *where* $p(X)$ *denotes the number of those members of* $\mathcal{A}$ *which contain no edges entering* X. ■

Note that if the requested arborescences exist, then they can be chosen in such a way that $F_i \cap A_0 = A_i$. Yet another equivalent formulation of the strong theorem is as follows.

Theorem 2.4 (Edmonds' disjoint arborescences: equivalent strong form). *Let* $D = (V, A)$ *be a digraph whose node set is partitioned into a root-set*

$R = \{r_1, \dots, r_k\}$ and a terminal set T. Suppose that no edge of D enters any node of R. There are k disjoint arborescences $F_1, \dots, F_k$ in D so that F_i is rooted at r_i and spans $T + r_i$ for each $i = 1, \dots, k$ if and only if $\varrho_D(X) \geq |R - X|$ for every subset $X \subseteq V$ for which $X \cap T \neq \emptyset$. ■

This follows easily by applying Theorem 2.2 to the subgraph D' of D induced by T with the choice $R_i = \{v\colon$ there is an edge $r_i v \in A\}$ $(i = 1, \dots, k)$. The same construction shows the reverse implication, too.

It has been tempting to find further extensions of the strong version of Edmonds' theorem but straightforward attempts failed. One natural conjecture, for example, was already disproved by Lovász in his original paper: even if there are $k(= 2)$ edge-disjoint paths from a root-node r_0 to every element of a specified terminal set $T \subseteq V - r_0$, the digraph not necessary includes k edge-disjoint arborescences of root r_0 so that each of them contains every node of T. Actually this problem can be shown to be NP-complete. In another possible variation, there are k specified subsets V_i of V each containing a root-node r_i. The problem consists of finding disjoint arborescences F_i $(i = 1, \dots, k)$ so that each F_i is rooted at r_i, contains no node outside V_i, and spans V_i. This problem is NP-complete even in the very special case when $k = 2$, $V_1 = V$ and $V_2 = V - t$ for a specified node t. Indeed, it can be shown that a polynomial algorithm to this special case gives rise to a polynomial algorithm for the two edge-disjoint paths problem of a digraph, a well-known NP-complete problem.

However, we point out that the strong form of Edmonds' theorem implies its sharpening when the following result of Frank and Tardos [7] (which, incidentally, had been motivated by another old paper of Lovász [17]) is used.

Theorem 2.5. *Let $G = (V, U; E)$ be a simple bipartite graph, $p\colon 2^V \to \mathbf{Z}_+$ a positively intersecting supermodular function, and $g\colon V \to \mathbf{Z}_+$ an upper bound function. There is a subset $F \subseteq E$ of the edges of G for which*

$$|\Gamma_F(X)| \geq p(X) \quad \textit{for every} \quad X \subseteq V \tag{4}$$

and

$$d_F(v) \leq g(v) \quad \textit{for every node} \quad v \in V \tag{5}$$

if and only if

$$|\Gamma_E(X)| + g(Z) \geq p(X \cup Z) \tag{6}$$

holds for every pair of disjoint subsets X and Z of V where $\Gamma_F(X)$ denotes the set of neighbours of X in the graph induced by $F \subseteq E$. ■

Now the extension of Theorem 2.2 is as follows. Note that none of the equivalent Theorems 2.2, 2.3, 2.4 implies it immediately.

Theorem 2.6. *Let $D = (V, A)$ be a directed graph and $g\colon V \to \mathbf{Z}_+$ an upper bound function. Let $\mathcal{U} = \{U_1, \dots, U_k\}$ be a family of k subsets of V. There is a family $\mathcal{R} = \{R_1, \dots, R_k\}$ of non-empty subsets of V and k disjoint spanning branchings of D with root sets $R_1, \dots, R_k$, respectively, in such a way that $R_i \subseteq U_i$ for $i = 1, \dots, k$ and each node $v \in V$ belongs to at most $g(v)$ members of $\mathcal{R}$ if and only if*

$$u(X) + g(Z) \geq k - \varrho_D(X \cup Z) \tag{7}$$

for every pair X, Z of disjoint subsets of V with non-empty union where $u(X)$ denotes the number of U_i's intersecting X.

Proof. Suppose first that the requested family $\mathcal{R}$ and the k branchings exist. For disjoint subsets X and Z of V, at most $u(X)$ members of $\mathcal{R}$ intersect X due to $R_i \subseteq U_i$, and at most $g(Z)$ members of $\mathcal{R}$ intersect Z since each element z of Z belongs to at most $g(z)$ members of $\mathcal{R}$. Therefore there must be at least $k - u(X) - g(Z)$ members of $\mathcal{R}$ which are disjoint from $X \cup Z$ and hence the in-degree of $X \cup Z$ must be at least this number, that is, (7) is necesseary.

To see the sufficiency, construct a bipartite graph $G = (V, U; E)$ where $U = \{u_1, \dots, u_k\}$ and a node u_i is connected with $v \in V$ precisely if $v \in U_i$. Let a set function p on V be defined by $p(X) = k - \varrho_D(X)$ if $\emptyset \subset X \subseteq V$ and $p(\emptyset) = 0$. Then p is intersecting supermodular. Since $u(X) = |\Gamma_E(X)|$, (7) and (6) are equivalent. Hence Theorem 2.5 implies the existence of a subset F of E meeting (4) and (5). For each u_i, let R_i denote the neighbours of u_i in the subgraph induced by F. By the construction $R_i \subseteq U_i$, (4) is equivalent to (3), while (5) implies that each node $v \in V$ belongs to at most $g(v)$ members of $\mathcal{R}$. By Theorem 2.2 the requested branchings exist. ■

For the special case $g \equiv 1$, we formulate the result in the following equivalent version.

Theorem 2.7. *Let $D = (V, A)$ be a directed graph with a designated root-node r_0. Let A_0 denote the set of edges leaving r_0 and let $\mathcal{A} = \{A_1, \dots, A_k\}$ be a family of k* (not-necessarily disjoint) *subsets of A_0. D has k disjoint* *spanning arborescences $F_1, \dots, F_k$ of root r_0 so that $F_j \cap A_0 \subseteq A_j$ if and only if*

$$(8) \qquad \varrho_{A^*}(Z) + \varrho'(Z) \geq h$$

for every non-empty subset Z of $V^* := V - r_0$ and for every choice of h members $A_{i_1}, \dots, A_{i_h}$ of $\mathcal{A}$ where $\varrho'(X)$ denotes the number of edges in $A_{i_1} \cup \dots \cup A_{i_h}$ entering X.

The following corollary is still a proper extension of Theorem 2.4.

Theorem 2.8. *Let $D = (V, A)$ be a digraph whose node set is partitioned into a root-set $R = \{r_1, \dots, r_q\}$ and a terminal set T. Suppose that no edge of D enters any node of R. Let $m\colon R \to \mathbf{Z}_+$ be a function and let $k = m(R)$. There are k disjoint arborescences in D so that $m(r)$ of them are rooted at r and spanning $T + r$ for each $r \in R$ if and only if*

$$(9) \quad \varrho_D(X) \geq m(R - X) \ \text{ for every subset } \ X \subseteq V \ \text{ for which } \ X \cap T \neq \emptyset.$$

Proof. Contract R into a new node r_0 and define k subsets of the edge set A_0 leaving r_0 as follows. For each $r \in R$, take $m(r)$ copies of the subset of A_0 corresponding to the set of edges of D leaving r. Then (9) is equivalent to (8) and the result follows from Theorem 2.7. ■

2.2. Extensions

One may be wondering whether there is a direct proof of Theorem 2.7 which follows the original lines of Lovász' proof without relying on Theorem 2.5. To understand better its nature, it has been tempting to extend Lovász' technique to more abstract settings. For example, the following 'abstract form' of the weak Edmonds theorem was derived in [5].

Theorem 2.9. *Let $D = (V, A)$ be a digraph and $\mathcal{F}$ an intersecting family of subsets of V. It is possible to partition A into k coverings of $\mathcal{F}$ if and only if the in-degree of every member of $\mathcal{F}$ is at least k.* ■

Obviously, when $\mathcal{F}$ consists of every non-empty subset of $V - r_0$, we obtain the weak form of Edmonds' theorem. A disadvantage of Theorem 2.9 is that it does not imply the strong version of Edmonds' theorem. The following result of L. Szegő [22], however, overcame this difficulty.

Theorem 2.10 (Szegő). *Let $\mathcal{F}_1, \ldots, \mathcal{F}_k$ be intersecting families of subsets of nodes of a digraph $D = (V, A)$ with the following mixed intersection property:*

$$X \in \mathcal{F}_i,\ Y \in \mathcal{F}_j,\ X \cap Y \neq \emptyset \ \Rightarrow\ X \cap Y \in \mathcal{F}_i \cap \mathcal{F}_j.$$

Then A can be partitioned into k subsets $A_1, \ldots, A_k$ such that A_i covers $\mathcal{F}_i$ for each $i = 1, \ldots, k$ if and only if $\varrho_D(X) \geq p_1(X)$ for all non-empty $X \subseteq V$ where $p_1(X)$ denotes the number of $\mathcal{F}_i$'s containing X. ■

When the k families are identical, we are back at Theorem 2.9. When $\mathcal{F}_i = 2^{V-R_i} - \{\emptyset\}$, we obtain Edmonds' Theorem 2.2. The proof of Szegő is based on the observation that the mixed intersection property implies that p_1 is positively intersecting supermodular and this is why Lovász' approach works again. But Szegő's theorem is still not general enough to imply Theorem 2.7.

As a new contribution of the present work, we extend Szegő's theorem to k families of bi-sets and this will immeadiately yield Theorem 2.7. The proof uses again the same technique. We say that the bi-set families $\mathcal{F}_1, \ldots, \mathcal{F}_k$ satisfy the **mixed intersection** property if

$$X \in \mathcal{F}_i,\ Y \in \mathcal{F}_j,\ X_I \cap Y_I \neq \emptyset \ \Rightarrow\ X \cap Y \in \mathcal{F}_i \cap \mathcal{F}_j.$$

For a bi-set X, let $p_2(X)$ denote the number of indices i for which $\mathcal{F}_i$ contains X. For $X \in \mathcal{F}_i$, $Y \in \mathcal{F}_j$, the inclusion $X \subseteq Y$ implies $X = X \cap Y \in \mathcal{F}_j$ and hence p_2 is monotone non-increasing in the sense that $X \subseteq Y$, $p_2(X) > 0$ and $p_2(Y) > 0$ imply $p_2(X) \geq p_2(Y)$. We will need the following preparatory lemma.

Lemma 2.11. *If $p_2(X) > 0$, $p_2(Y) > 0$ and $X_I \cap Y_I \neq \emptyset$, then $p_2(X) + p_2(Y) \leq p_2(X \cap Y) + p_2(X \cup Y)$. Moreover, if there is an $\mathcal{F}_i$ for which $X \cap Y \in \mathcal{F}_i$ and $X, Y \notin \mathcal{F}_i$, then strict inequality holds.*

Proof. Consider the contribution of one family $\mathcal{F}_i$ to the two sides of the claimed inequality. If this contribution to the left hand side is two, that is, if both X and Y are in $\mathcal{F}_i$, then so are $X \cap Y$ and $X \cup Y$ and hence the contribution to the right hand side is also two. Suppose now that X belongs to $\mathcal{F}_i$ but Y does not. Since $p_2(Y) > 0$ is assumed, Y belongs to an $\mathcal{F}_j$. But then $X \cap Y$ belongs to $\mathcal{F}_i$ due to the mixed intersection property, that is, in this case the contribution of $\mathcal{F}_i$ to the right hand side is at least one.

An $\mathcal{F}_i$ with the properties in the second part contributes only to the right hand side ensuring this way the strict inequality. ■

Theorem 2.12. *Let $D = (V, A)$ be a digraph and $\mathcal{F}_1, \ldots, \mathcal{F}_k$ intersecting families of bi-sets on ground set V satisfying the mixed intersection property. The edges of D can be partitioned into k parts $F_1, \ldots, F_k$ in such a way that F_i covers $\mathcal{F}_i$ for each $i = 1, \ldots, k$ if and only if*

(10) $$\varrho_D(X) \geq p_2(X) \quad \text{for every bi-set} \quad X.$$

Proof. The condition is clearly necessary. We prove the sufficiency by induction on $\sum_i |\mathcal{F}_i|$. There is nothing to prove if this sum is zero so we may assume that $\mathcal{F}_1$, say, is non-empty. Let U be a maximal member of $\mathcal{F}_1$. Call a bi-set **tight** if $\varrho(X) = p_2(X) > 0$.

Claim 2.13. *There is an edge e entering U in such a way that each tight bi-set covered by e is in $\mathcal{F}_1$.*

Proof. Suppose indirectly that no such an edge exists. Then each edge e entering U enters some tight bi-set $M \notin \mathcal{F}_1$. By the mixed intersection property, we cannot have $M \subseteq U$. Select a minimal tight bi-set $M \notin \mathcal{F}_1$ which intersects U. Since p_2 is monotone non-increasing, we know that $p_2(U \cap M) \geq p_2(M)$. Here, in fact, strict inequality must hold since $U \cap M \in \mathcal{F}_1$ and $M \notin \mathcal{F}_1$. The inequality $p_2(U \cap M) > p_2(M)$ implies that D has an edge $f = uv$ for which $u \in M - U$, $v \in U \cap M$. By the indirect assumption, f enters some tight bi-set $Z \notin \mathcal{F}_1$. Lemma 2.11 implies that the intersection of M and Z is tight. Since neither of M and Z is in $\mathcal{F}_1$, the second part of the lemma implies that $M \cap Z$ is not in $\mathcal{F}_1$ either, contradicting the minimal choice of M. ■

Let e be an edge ensured by the Claim. Let $\mathcal{F}'_1 := \{X \in \mathcal{F}_1 : e \text{ does not enter } X\}$. Then $\mathcal{F}'_1$ is an intersecting family of bi-sets. We claim that the mixed intersection property holds for the families $\mathcal{F}'_1, \mathcal{F}_2, \ldots, \mathcal{F}_k$. Indeed, let $X \in \mathcal{F}'_1$ and $Y \in \mathcal{F}_i$ be two intersecting bi-sets for some $i = 2, \ldots, k$. Since $\mathcal{F}'_1 \subseteq \mathcal{F}_1$, one has $X \cap Y \in \mathcal{F}_i$. If indirectly $X \cap Y$ is not in $\mathcal{F}'_1$, then e enters $X \cap Y$. Since e enters U and U was selected to be maximal in $\mathcal{F}_1$, it follows that $X \subseteq U$. But then e must enter X as well, contradicting the assumption $X \in \mathcal{F}'_1$.

Let $p'_2(X)$ denote the number of these families containing X (that is, $p'_2(X) = p_2(X) - 1$ if $X \in \mathcal{F}_1$ and e enters X and $p'_2(X) = p_2(X)$ otherwise).

Let ϱ' denote the in-degree function on bi-sets with respect to $D' := D - e$. The choice of e implies $\varrho' \geq p'_2$. By induction, the edge set of D' can be partitioned into k parts $F'_1, \dots, F_k$ in such a way that F'_1 covers $\mathcal{F}_1$ and F_i covers $\mathcal{F}_i$ for $i = 2, \dots, k$. By letting $F_1 := F'_1 + e$, we obtain a partition of A requested by the theorem. ■ ■

Though not needed in the sequel, we point out that Theorem 2.12 can be reformulated in terms of set families. For a subset $T \subseteq V$, we say that a family $\mathcal{F}$ of subsets of V is **T-intersecting** if $X, Y \in \mathcal{F}$ and $X \cap Y \cap T \neq \emptyset$ imply $X \cap Y, X \cup Y \in \mathcal{F}$.

Theorem 2.14. *Let $D = (V, A)$ be a digraph with a specified subset T of nodes containing the head of every edge of D. Let $\mathcal{F}_1, \dots, \mathcal{F}_k$ be T-intersecting families of subsets of nodes of a digraph D with the following mixed intersection property: $X \in \mathcal{F}_i$, $Y \in \mathcal{F}_j$, $X \cap Y \cap T \neq \emptyset \Rightarrow X \cap Y \in \mathcal{F}_i \cap \mathcal{F}_j$. Then A can be partitioned into k subsets $A_1, \dots, A_k$ such that A_i covers $\mathcal{F}_i$ for each $i = 1, \dots, k$ if and only if $\varrho_D(X) \geq p_1(X)$ for all non-empty $X \subseteq V$ where $p_1(X)$ denotes the number of $\mathcal{F}_i$'s containing X.*

Proof. The necessity is evident again. For the sufficiency, define a family $\mathcal{F}'_i$ of bi-sets as follows. For each set $X \in \mathcal{F}_i$ let the bi-set $(X, X \cap T)$ be a member of $\mathcal{F}'_i$. Then each $\mathcal{F}'_i$ is intersecting and they meet the mixed intersection property. Since the head of every edge is in T, an edge enters a subset X precisely when it enters the bi-set $(X, X \cap T)$. Hence the partition of A into k sets ensured by Theorem 2.12 meets the requirement of the theorem. ■

The reverse implication is equally simple and is left to the reader.

Alternative proof of the sufficiency in Theorem 2.7. Let D' be a digraph arising from D by subdividing first each edge $e \in A_0$ by a node v_e and deleting then r_0. Let V_0 denote the set of the subdividing nodes and let V_i denote the subset of V_0 corresponding to the set A_i $(i = 1, \dots, k)$.

For each $j = 1, \dots, k$, let $\mathcal{F}_j$ be a family of bi-sets (X_O, X_I) for which $\emptyset \neq X_I \subseteq V^*$, $X_I = X_O \cap V^*$, $X_O \subseteq V_0 \cup V^*$ and $X_O \cap V_j = \emptyset$. Then $\mathcal{F}_j$ is an intersecting family of bi-sets and it follows from the definition that these k families meet the mixed intersecting property. It is also straightforward that (8) is equivalent to requiring that the number of edges entering a bi-set X is at least $p_2(X)$, the number of $\mathcal{F}_j$'s containing X. By Theorem 2.12, there are disjoint subsets $F'_1, \dots, F'_k$ of the edge set of D' so that F'_j covers $\mathcal{F}_j$. We may assume that each $\mathcal{F}_j$ is a minimal covering of $\mathcal{F}_j$ (with respect to inclusion) which implies that an edge uv with $u \in V_0, v \in V^*$ can belong to F'_j only if $u \in V_j$. By the construction, the edges set F_j of D corresponding to F'_j is a spanning arborescence of D rooted at r_0 so that $F_j \cap A_0 \subseteq A_j$. ■

Recently, N. Kamiyama, N. Katoh, and A. Takizawa [14] were able to find a surprising new proper extension of Theorem 2.7 (and hence the strong Edmonds theorem). We are going to show that their result can also be derived from Theorem 2.12. This is, however, a bit trickier due to the fact that the corresponding set function p_1 in their theorem is no more supermodular (and for the same reason their original proof is rather complicated). Similarly to Edmonds' theorem, this new result has also several equivalent formulations. One of them is as follows.

Theorem 2.15 (Kamiyama, Katoh, and Takizawa [14]). *Let $D = (V, A)$ be a directed graph and let $R = \{r_1, r_2, \dots, r_k\} \subseteq V$ be a list of k possibly not distinct root-nodes. Let S_i denote the set of nodes reachable from r_i. There are edge-disjoint r_i-arborescences A_i spanning S_i for $i = 1, \dots, k$ if and only if*

(11) $$\varrho_D(Z) \geq p_1(Z) \quad \textit{for every subset} \quad Z \subseteq V$$

where $p_1(Z)$ denotes the number of sets S_i's for which $S_i \cap Z \neq \emptyset$ and $r_i \notin Z$.

Proof. The necessity of the condition is evident.

For brevity, we call a strongly connected component of D an **atom**. It is known that the atoms form a partition of the node set of D and that there is a so-called topological ordering of the atoms so that there is no edge from a later atom to an earlier one. By a subatom we mean a subset of an atom. Clearly, a subset $X \subseteq V$ is a subatom if and only if any two elements of X are reachable in D from each other. Note that any atom is disjoint from or included in S_i for each $i = 1, \dots, k$.

Define k bi-set families $\mathcal{F}_i$ for $i = 1, \dots, k$ as follows. Let $\mathcal{F}_i := \{(X_O, X_I) \colon X_O \subseteq V - r_i,\ X_I = X_O \cap S_i \neq \emptyset$ a non-empty subatom$\}$. For each bi-set X, let $p_2(X)$ denote again the number of $\mathcal{F}_i$'s containing X. It follows immediately that $\mathcal{F}_i$ is an intersecting bi-set family.

Proposition 2.16. *The bi-set families $\mathcal{F}_i$ meet the mixed intersecting property.*

Proof. Let $X = (X_O, X_I)$ and $Y = (Y_O, Y_I)$ be members of $\mathcal{F}_i$ and $\mathcal{F}_j$, respectively, and suppose that X and Y are intersecting, that is, $X_I \cap Y_I \neq \emptyset$. Since a subatom and a subset with no leaving edges are never properly intersecting, we obtain that $X_O \cap S_i \subseteq S_i \cap S_j$ and $Y_O \cap S_j \subseteq S_i \cap S_j$. This implies for the sets $Z_O := X_O \cap Y_O$ and $Z_I := X_I \cap Y_I$ that $Z_O \cap S_i = Z_I = Z_O \cap S_j$ and hence $X \cap Y = (Z_O, Z_I) \in \mathcal{F}_i \cap \mathcal{F}_j$, as required. ■

Proposition 2.17. *$\varrho(X) \geq p_2(X)$ for each bi-set X.*

Proof. Let $q := p_2(X)$ and suppose that X belongs to $\mathcal{F}_1, \mathcal{F}_2, \ldots, \mathcal{F}_q$. Let $Z := \big(V - (S_1 \cup S_2 \cup \cdots \cup S_q)\big) \cup X_I$. Since no edge leaves any S_i, every edge entering Z must enter X_I and hence also the bi-set X. Therefore $\varrho(X) \geq \varrho(Z)$. By (11), $\varrho(Z) \geq p_1(Z)$. It follows from the definition of Z that $p_1(Z) \geq q = p_2(X)$, and hence $\varrho(X) \geq p_2(X)$. ■

Therefore Theorem 2.12 applies and hence the edges of D can be partitioned into sets $F_1, \ldots, F_k$ so that F_i covers $\mathcal{F}_i$ for $i = 1, \ldots, k$.

Proposition 2.18. *Each F_i includes an r_i-arborescence A_i which spans S_i.*

Proof. If the requested arborescence does not exist for some i, then there is a non-empty subset Z of $S_i - r_i$ so that F_i contains no edge from $S_i - Z$ to Z. Consider a topological ordering of the atoms and let Q be the earliest one intersecting Z. Since no edge leaving a later atom can enter Q, no edge with tail in Z enters Q.

Let $X_O := (V - S_i) \cup (Z \cap Q)$ and $X_I := X_O \cap S_i$. Then $X_I = Z \cap Q$ is a subatom and $X = (X_O, X_I)$ belongs to $\mathcal{F}_i$. Therefore there is an edge $e = uv$ in F_i which enters X. It follows that $v \in X_I \subseteq Z$ and that $u \in S_i - X_I$. Since u is not in Z and not in $V - S_i$, it must be in $S_i - Z$, that is, e is an edge from $S_i - Z$ to $X_I \subseteq Z$, contradicting the assumption that no such an edge exists. ■ ■

Note that Theorem 2.8 can immediately be obtained from Theorem 2.15. To this end, add $m(i)$ new root-nodes to D and add an edge from each of them to r_i for $i = 1, \ldots, q$. This way we will get k distinct (new nodes) and each node of V is reachable from every new root. In this setting the necesseary conditions in Theorems 2.15 and 2.7 coincide and each of the k maximal arborescences ensured by Theorem 2.15 will span the whole V.

To describe the original form of the theorem of Kamiyama et al., we call a branching B of D maximal if no edge of D leaves the node set of B.

Theorem 2.19 (Kamiyama, Katoh, Takizawa [14]). *In a digraph $D = (V, A)$, let $\mathcal{R} = \{R_1, \dots, R_q\}$ be a family of non-empty* (not necessarily disjoint or distinct) *subsets of V and let S_i denote the set of nodes of D reachable from R_i. Let $m_1, \dots, m_q$ be positive integers whose sum is k. There are k edge-disjoint maximal branchings of D so that R_i is the root-set of m_i of them for $i = 1, \dots, q$ if and only if*

$$(12) \qquad \varrho(X) \geq \sum [m_i\colon R_i \cap X = \emptyset \text{ and } X \text{ is reachable from } R_i] \quad \text{for every } X \subseteq V.$$

Proof. For each root-set R_i, let $r_i^1, \dots, r_i^{m_i}$ be new nodes and extend the digraph by adding k new parallel edges from r_i^j to every element of R_i for $i = 1, \dots, q$. An easy calculation shows that (11) is equivalent to (12) and the k disjoint arborescences ensured by Theorem 2.15 when restricted to V provide the requested maximal branchings of D. ■

In [12], Frank, Király, and Kriesell observed that Edmonds' disjoint arborescences theorem can be extended to dypergraphs. A subset F of a ground-set V with a specified head-node in F is called a directed hyperedge, or briefly a dyperedge. F is said to enter a subset $X \subseteq V$ if its head is in X but $F \not\subseteq X$. A dypergraph $D = (V, \mathcal{D})$ is a hypergraph consisting of dyperedges in which $\varrho_D(X)$ denotes the number of dyperedges entering a subset X. We call D **rooted k-edge-connected** with respect to a root-node r_0 if the in-degree of every non-empty subset of $V - r_0$ is at least k. In the special case $k = 1$, the dypergraph is **root-connected**. In [12], with a rather easy reduction to Edmonds' disjoint arborescences theorem, it was shown that the dyperedges of a rooted k-edge-connected dypergraph can always be decomposed into k root-connected dypergraphs. With a similar approach, we can derive the following result.

Theorem 2.20. *Let $D = (V, \mathcal{D})$ be a dypergraph and $R = \{r_1, \dots r_k\}$ a root-set. Let S_i denote the set of nodes reachable from r_0 in D. Then D includes k disjoint dypergraphs $D_1 = (S_1, \mathcal{D}_1), \dots, D_k = (S_k, \mathcal{D}_k)$ so that each D_i is root-connected at r_i if and only if $\varrho_D(X) \geq p(X)$ for every $X \subseteq V$ where $p(X)$ denotes the number of roots r_i for which $r_i \notin X$ and $S_i \cap X \neq \emptyset$.*

■

3. Covering Supermodular Bi-set Functions by Digraphs

As mentioned in the introduction, the uncrossing technique was invented by Lovász [18] in order to obtain a short proof of the Lucchesi-Younger theorem. Later the method has become an indispensible tool for deriving combinatorial min-max theorems concerning sub- or supermodular set functions.

As a new application of the uncrossing procedure, we derive a result on covering simultaneously two supermodular bi-set functions by a digraph. (Recall that these functions were assumed to have positive values only on non-trivial bi-sets and they are integer-valued.) There have been two earlier results of this kind. Frank and Jordán [11] proved (in an equivalent form) the following result on minimum coverings of crossing supermodular bi-set functions (whose special case for set-functions appeared in [9]).

Theorem 3.1. *Let p be a positively crossing supermodular bi-set function. The minimum number of directed edges covering p is equal to* $\max\Big\{\sum\big[p(X)\colon X\in\mathcal{F}\big] : \mathcal{F}$ *an independent set of bi-sets*$\Big\}$. ■

The other result of similar vein concerns cheapest coverings of intersecting supermodular bi-set functions (generalizing its set-function version from [5, 6]).

Theorem 3.2 [10]. *Let $D=(V,A)$ be a digraph. Let $p\colon \mathcal{P}_2\to \mathbf{Z}$ be a positively intersecting supermodular bi-set function and $g\colon A\to \mathbf{Z}_+\cup\{\infty\}$ a non-negative upper bound on the edges of D that covers p. The linear system*

$$(13)\qquad \varrho_x(Z)\ge p(Z)\quad \text{for every bi-set}\quad Z\in\mathcal{P}_2,\ \ 0\le x\le g$$

is totally dual integral.

3.1. Simultaneous coverings

Both theorems were motivated by and have several applications in network design. Our new contribution is a min-max theorem on smallest simultaneous coverings of two fully supermodular bi-set functions. It is neither a special case nor a generalization of the two previous results and has no special set function version known earlier. In what follows, we work throughout with a ground-set V of cardinality n. Let $D^* = (V, A^*)$ denote the complete digraph on V where $A^* := \{uv\colon u, v \in V\}$ denotes the set of all the $n(n-1)$ directed edges on V. Recall that $\mathcal{P}_2(V) = \mathcal{P}_2$ denoted the set of all bi-sets. A bi-set function p is positively supermodular if the supermodular inequality holds for every pair $\{X, Y\}$ of bi-sets for which $p(X) > 0$, $p(Y) > 0$. For example, if p is supermodular on a ring-family, and its value is zero otherwise, then p is positively supermodular.

Theorem 3.3. *Let p_1 and p_2 be two positively supermodular bi-set functions which may be positive only on non-trivial bi-sets. Let $p := \max\{p_1, p_2\}$ where p is defined by $p(X) := \max\{p_1(X), p_2(X)\}$. Then p can be covered by γ* (possibly parallel) *directed edges if and only if*

$$p_1(X) + p_2(Y) \leq \gamma \tag{14}$$

for every pair of independent bi-sets X, Y.

Note that, due to $p(\emptyset, \emptyset) = 0$, (14) includes the necessary conditions $p_1(X) \leq \gamma$ and $p_2(Y) \leq \gamma$ so they need not be mentioned explicitly and a similar statement holds for later variations of the theorem.

It is more convenient to prove this result in a slightly more general form. We call a bi-set function p positively **2/3-supermodular** if for any choice of three bi-sets with positive p-value there are two of them that satisfy the supermodular inequality. Clearly, the maximum of two supermodular functions is 2/3-supermodular, but it turns out that there are 2/3-supermodular functions not arising this way.

Theorem 3.4. *A positively 2/3-supermodular bi-set function p can be covered by γ* (possibly parallel) *directed edges if and only if $p(X)+p(Y) \leq \gamma$ for every pair of independent bi-sets X, Y. Equivalently, the minimum number $\tau(p)$ of edges covering p is equal to $\nu(p) := \max\{p(X) + p(Y)\colon \{X, Y\}$ independent bi-sets$\}$.*

Proof. The necessity of the condition is obvious since an edge can cover at most one of two independent bi-sets.

Lemma 3.5. *Let $\mathcal{C}_1$ and $\mathcal{C}_2$ be two chains of nontrivial bi-sets and let $\mathcal{F} = \mathcal{C}_1 \cup \mathcal{C}_2$* (in the sense that a bi-set belonging to both chains occurs in two copies in $\mathcal{F}$)*. Suppose that no edge of D^* covers more than h members* *of $\mathcal{F}$. Then the members of $\mathcal{F}$ can be coloured by h colours so that each edge enters at most one member of each colour class. Furthermore, each colour class consists of at most two bi-sets.*

Proof. Since two comparable bi-sets are not independent, the second statement of the lemma is immediate.

Construct an undirected graph $B = (U, A)$ whose nodes correspond to the elements of $\mathcal{F}$ and two nodes are connected by an undirected edge if the corresponding members X, Y of $\mathcal{F}$ can be covered by an edge of D^*, that is, if they are not independent. Since $\mathcal{F}$ consists of two chains, B is the complement of a bipartite graph, and hence B is perfect.

Claim 3.6. *Let $Q \subseteq U$ be the node-set of a clique of graph B and let $\mathcal{F}_Q$ denote the members of $\mathcal{F}$ corresponding to the elements of Q. Then there is an edge of D^* covering all members of $\mathcal{F}_Q$.*

Proof. Assume first that $\mathcal{F}_Q$ is a chain. Let t be any node in the inner set of the smallest member of $\mathcal{F}_Q$ while s any node outside the outer set of the largest member of $\mathcal{F}_Q$. Then st covers all members of $\mathcal{F}_Q$. Therefore $\mathcal{F}_Q$ may be assumed to be the union of two non-empty chains $\mathcal{C}'_1$ and $\mathcal{C}'_2$. Let X_1 and X_2 be the smallest members of $\mathcal{C}'_1$ and $\mathcal{C}'_2$.

As Q is a clique, X_1 and X_2 are not independent, so there is a node $t \in V$ in the intersection of their inner sets. Similarly, let Y_1 and Y_2 be the largest members of $\mathcal{C}'_1$ and $\mathcal{C}'_2$. They are not independent either so there is a node $s \in V$ outside the union of their outer sets. Then st covers all the members of $\mathcal{C}'_1$ and $\mathcal{C}'_2$. ∎

The claim and the hypothesis of the lemma imply that the largest clique of B has at most h elements. Since B is perfect, its node set can be partitioned into h stable sets. Therefore $\mathcal{F}$ can be partitioned into h independent sets of bi-sets families. ∎ ∎

Let us turn to the proof of the non-trivial inequality $\tau(p) \le \nu(p)$ in the theorem. We proceed by induction on $\sum \left[p(X) \colon X \in \mathcal{P}_2\right]$. If this sum is zero, then the digraph $(V, \emptyset)$ with no edge will cover p. Suppose now that this sum is positive. For an edge $e \in A^*$, let $p_e(X) := \big(p(X) - 1\big)^+$ if e enters X and $p_e(X) := p(X)$ otherwise. Since the in-degree function (on bi-sets) of a digraph is fully submodular, $p_e(X)$ is 2/3-supermodular.

Lemma 3.7. *If $p(Z) > 0$ for a bi-set Z, then there is an edge $e \in A^*$ entering Z such that $\nu(p_e) < \nu(p)$.*

Proof. Let A denote the set of edges entering Z and suppose on the contrary that $\nu(p_e) = \nu(p)$ for each element e of A. That is, there is an independent pair $\mathcal{F}_e := \{X, Y\}$ of bi-sets for which e enters neither X nor Y and $p(X) + p(Y) = \nu(p)$

Let $\mathcal{F}'$ consist of bi-set Z plus all of the bi-sets which are members of some $\mathcal{F}_e$ in the sense that each bi-set X is taken into $\mathcal{F}'$ in as many copies as the number of pairs $\mathcal{F}_e$ containing X. Note that $(*)$ every edge of D^* enters at most $h := |A|$ members of $\mathcal{F}'$. The uncrossing procedure consists of finding two non-comparable elements X, Y of $\mathcal{F}'$ for which the supermodular inequality holds and replacing them by their intersection and union. Apply the uncrossing procedure as long as possible. Because the sum $\sum \left[|X_I|^2 + |X_O|^2 \colon X \in \mathcal{F}' \right]$ strictly increases at each uncrossing step, the procedure terminates after a finite number of steps. Discard all members with p-value zero and let $\mathcal{F}$ denote the resulting family. Clearly $|\mathcal{F}| \leq |\mathcal{F}'|$, $p(Z) + h\nu(p) = p(\mathcal{F}') \leq p(\mathcal{F})$, and $(*)$ holds for $\mathcal{F}$, too. $\mathcal{F}$ cannot contain three pairwise non-comparable bi-sets for otherwise, by the 2/3-supermodularity of p, two of them would satisfy the supermodular inequality, and then they could have been uncrossed. If a partially ordered set contains no three pairwise uncomparable elements, then, by Dilworth's theorem, there are two disjoint chains covering the ground-set of the poset. Therefore the members of $\mathcal{F}$ can be partitioned into two chains $\mathcal{C}_1$ and $\mathcal{C}_2$. By Lemma 3.5 the members of $\mathcal{F}$ can be partitioned into h independent parts $\mathcal{I}_i$ $(i = 1, \ldots, h)$. So for one of these we must have $p(\mathcal{I}_i) \geq \lfloor p(\mathcal{F})/h \rfloor > \nu(p)$, a contradiction. ■

For the edge e provided by Lemma 3.7, we have by induction $\tau(p_e) - 1 \leq \tau(p) \leq \nu(p) \leq \nu(p_e) - 1 \leq \tau(p_e) - 1$ from which equality holds throughout, and in particular $\tau(p) = \nu(p)$. ■ ■ ■

Theorem 3.4 has a self-refining nature as it gives rise to its own extension. Let S and T be two non-empty subsets of V. We call a directed edge st an ST**-edge** if $s \in S$ and $t \in T$. Two bi-sets X and Y are ST**-independent** if there is no ST-edge covering both (or, equivalently, at least one of the sets $T \cap X_I \cap Y_I$ and $S - (X_O \cup Y_O)$ is empty).

Theorem 3.8. *Let q be a positively 2/3-supermodular bi-set function so that $q(X)$ can be positive only if there is an ST-edge covering X. Then q can be covered by γ (possibly parallel) ST-edges if and only if $q(X) + q(Y) \leq \gamma$ for every pair of ST-independent bi-sets X, Y.*

Proof. The necessity of the condition is evident since an ST-edge cannot cover two ST-independent bi-sets. For the sufficiency, define a bi-set function p on $\mathcal{P}_2$, as follows.

(15)

$$p(X) := \begin{cases} \max\big\{q(X')\colon X' \in \mathcal{P}_2(V),\ X_I = X'_I \cap T, \\ \quad X_O = \big(X'_O \cup (V-S)\big)\big\} & \text{if } X_I \subseteq T \text{ and } V = X_O \cup S, \\ 0 & \text{otherwise.} \end{cases}$$

Proposition 3.9. *p is positively $2/3$-supermodular.*

Proof. Let X, Y, and Z be bi-sets for which $p(X) > 0$, $p(Y) > 0$, and $p(Z) > 0$. By the definition of p, there is a biset X' for which $p(X) = q(X')$ and $X_I = X'_I \cap T$, $X_O = \big(X'_O \cup (V-S)\big)$, and similarly there are bi-sets Y' and Z' with analogous properties.

It follows that

$$X_I \cap Y_I = (X'_I \cap Y'_I) \cap T \quad \text{and} \quad X_O \cap Y_O = (X'_O \cap Y'_O) \cup (V-S)$$

from which $q(X' \cap Y') \le p(X \cap Y)$, and analogously,

$$X_I \cup Y_I = (X'_I \cup Y'_I) \cap T \quad \text{and} \quad X_O \cup Y_O = (X'_O \cup Y'_O) \cup (V-S)$$

from which $q(X' \cup Y') \le p(X \cup Y)$.

Since q is positively $2/3$-supermodular, among the three bi-sets X', Y', Z', there are two, say X' and Y' satisfying the supermodular inequality. Hence $p(X)+p(Y) = q(X')+q(Y') \le q(X' \cap Y')+q(X' \cup Y') \le p(X \cap Y)+p(X \cup Y)$, as required. ■

If $p(X) > 0$, then $p(X) = q(X')$ for some X' and hence $p(X) = q(X') \le \gamma$. If $p(X) > 0$ and $p(Y) > 0$ for independent X and Y, then there are bi-sets X' and Y' for which $p(X) = q(X')$ and $p(X) = q(X')$. The definition of p implies that X' and Y' are ST-independent and hence $p(X) + p(Y) = q(X') + q(Y') \le \gamma$. Therefore Theorem 3.4 implies the existence of a set of γ edges covering p. The definition of p implies that every edge covering a bi-set X with $p(X) > 0$ is necessarily an ST-edge, moreover any set covering p also covers q, and hence the theorem follows. ■ ■

As a corollary, we have the following extension of Theorem 3.3.

Theorem 3.10. *Let q_1 and q_2 be two positively supermodular bi-set functions for which $q_i(X)$ can be positive only if there is an ST-edge covering X. Let $q := \max\{q_1, q_2\}$. Then q can be covered by γ ST-edges if and only if $q_1(X) + q_2(Y) \le \gamma$ for every pair of ST-independent bi-sets X, Y.* ■

We will point out in Subsection 3.3 that positively supermodular functions do not behave well from an algorithmic point of view. In typical applications, however, one encounters with fully supermodular functions defined on a ring-family of bi-sets that may take negative values. For this case, Theorem 3.10 specializes as follows.

Theorem 3.11. *For $i = 1, 2$ let p_i be a supermodular function on a ring-family $\mathcal{R}_i$ of bi-sets and assume that $p_i(X)$ may be positive only if there is an ST-edge covering X. There is a set of γ ST-edges covering both p_1 and p_2 if and only if*

$$p_i(X) \le \gamma \quad \text{for every} \quad X \in \mathcal{R}_i \quad (i = 1, 2) \tag{16}$$

and

$$p_1(X') + p_2(X'') \le \gamma \quad \text{for every} \quad ST\text{-independent} \quad X' \in \mathcal{R}_1, \quad X'' \in \mathcal{R}_2. \tag{17}$$

Equivalently, the minimum number of (possibly parallel) *ST-edges covering p_1 and p_2 is equal to $\nu := \max\{\nu_1, \nu_2, \nu_3, \nu_4\}$ where*

$$\nu_1 = \max\{p_1(X') \colon X' \in \mathcal{R}_1\}$$

$$\nu_2 = \max\{p_2(X'') \colon X'' \in \mathcal{R}_2\},$$

$$\nu_3 = \max\{p_1(X') + p_2(X'') \colon X' \in \mathcal{R}_1, \ X'' \in \mathcal{R}_2, \ X'_I \cap X''_I \cap T = \emptyset\},$$

$$\nu_4 = \max\{p_1(X') + p_2(X') \colon X' \in \mathcal{R}_1, \ X'' \in \mathcal{R}_2, \ S - (X'_O \cup X''_O) = \emptyset\},$$

where the maximum on the empty set is defined to be zero. ■

This implies the following equivalent version of Edmonds' polymatroid intersection theorem [1] (which was originally formulated for submodular functions).

Theorem 3.12 (Edmonds). *Let p_1 and p_2 be supermodular functions on a common ground-set T. Then*

$$\min\left\{z(T)\colon z\colon T \to \mathbf{Z}_+,\ z(X) \ge \max\left\{p_1(X), p_2(X)\right\} \text{ for every } X \subseteq T\right\}$$
$$= \max\left\{p_1(X) + p_2(Y)\colon X \cap Y = \emptyset\right\}.$$

Proof. Let s be a new element, $V := T + s$ and $S := \{s\}$. Apply Theorem 3.11 for the special case when $\mathcal{R}_i := \left\{(X_O, X_I)\colon X_O = X_I \subseteq T\right\}$ and observe that in this case $\nu = \nu_3$ and the ST-edges can be identified with the elements of T. ■

It should be noted that Edmonds extended the theorem for the more general case as well when p_i is intersecting supermodular only (that is, the supermodular inequality is required only for intersecting sets). In this case the maximum formula in the theorem is more complicated as it includes partitions rather then sets only. No extension of Theorem 3.11 is known to cover this form. The difficulty is indicated by the fact that the following natural-looking statement is **false**: *For $i = 1, 2$ let p_i be a non-negative integer-valued intersecting supermodular function on an intersecting family $\mathcal{R}_i$ of sets so that $p_i(X)$ may be positive only if there is an ST-edge covering X. The minimum number of (possibly parallel) ST-edges covering p_1 and p_2 is equal to* $\max\left\{\sum\left[p_1(X)\colon X \in \mathcal{F}_1\right] + \sum\left[p_2(X)\colon X \in \mathcal{F}_2\right]\right\}\colon \mathcal{F}_i \subseteq \mathcal{R}_i$ *is laminar, $\mathcal{F}_1 \cup \mathcal{F}_2$ is ST-independent.*$\}$ (A family $\mathcal{R}$ of sets is intersecting if it contains $X \cap Y$ and $X \cup Y$ whenever $X, Y \in \mathcal{R}, X \cap Y \neq \emptyset$. $\mathcal{R}$ is laminar if one of the sets $X - Y$, $Y - X$, $X \cap Y$ is empty for every two members X, Y.) Let $V = \{v_1, v_2, v_3, v_4\}$, $\mathcal{R}_1 := \left\{\{v_1\}, \{v_2, v_3\}, \{v_1, v_2, v_3\}\right\}$, $\mathcal{R}_2 := \left\{\{v_3, v_4\}, \{v_1, v_3, v_4\}\right\}$, let p_i be identically one on $\mathcal{R}_i$, and let $S := T := V$. Then the minimum value in the statement is 3 while the maximum is only 2.

Edmonds' intersection theorem extends to the weighted case, as well, asserting, in a concise form, that the linear system $\left\{z \ge 0,\ z(X) \ge \max\left\{p_1(X), p_2(X)\right\} \text{ for every } X \subseteq T\right\}$ is totally dual integral (TDI). The min-cost version of Theorem 3.10 includes NP-complete connectivity augmentation problems so it is unlikely to have a TDI-ness result concerning Theorem 3.10. In other connectivity augmentation problems however [8] the special case of node-induced costs were nicely solvable (where node-induced means that the cost of an edge st arises as the sum of the given node-costs of s and t). The corresponding problem in the enviroment of Theorem 3.10 remains open.

3.2. Applications to bipartite graphs and digraphs

Let us derive a graphical consequence concerning bipartite matchings.

Theorem 3.13. *For $i = 1, 2$, let $G_i = (S, T; E_i)$ be a bitpartite graph with $n = |S| = |T|$. There is a set F of at most γ (undirected) ST-edges so that both $G_1 + F$ and $G_2 + F$ has a perfect matching if and only if*

$$q_1(Z') + q_2(Z'') \le \gamma \tag{18}$$

holds for every two disjoint subsets Z', Z'' of S and for every two disjoint subsets Z', Z'' of T. Here $q_i(Z) := |Z| - \big|\Gamma_i(Z)\big|$ where $\Gamma_i(Z)$ (for $Z \subseteq S$ or $Z \subseteq T$) denotes the set of nodes having at least one neighbour in Z in the graph G_i $(i = 1, 2)$.

Proof. The necessity of (18) is evident, we prove its sufficiency. For $i = 1, 2$, define ring-families $\mathcal{R}_i$ of bi-sets as follows. $\mathcal{R}_i := \{X = (X_O, X_I)\colon X_I = X_O \cap T,\ X_O \supseteq \big(X_I \cup \Gamma_i(X_I)\big\}$. For simplicity we will not distinguish between the directed ST-edges and the undirected edges connecting S and T. Since an edge of G_i, when considered to be oriented toward T, cannot cover any member of $\mathcal{R}_i$, it follows that $\mathcal{R}_i$ is indeed a ring-family. For $X \in \mathcal{R}_i$, let

$$p_i(X) := 2|X_I| - |X_O|.$$

Clearly, p_i is supermodular on $\mathcal{R}_i$. For $X \in \mathcal{R}_i$ let $Z := S - X_O$. Since $\Gamma_i(X_I) \subseteq X_O - X_I$ we have $p_i(X) = |X_I| - \big(|X_O| - |X_I|\big) \le |X_I| - \big|\Gamma_i(X_I)\big| = q_i(X_I)$. Since $\Gamma_i(Z) \subseteq T - X_I$, we have $|X_I| + \big|\Gamma_i(Z)\big| \le |T| = |S| = |Z| + |X_O - X_I| = q_i(Z)$. Based on these, (18) implies (17). Since the bi-set $(\emptyset, \emptyset)$ belongs to $\mathcal{R}_i$ and $p_i(\emptyset, \emptyset) = 0$, we conclude that (17) implies (16).

By Theorem 3.11, there is a set F of γ ST-edges covering both p_1 and p_2. We claim that $G_i^+ := G_i + F$ satisfies the Hall condition. Indeed, if the set Y' of neighbours of a subset $Y \subseteq T$ in G_i^+ had fewer elements than $|Y|$, then $p_i(Y \cup Y', Y) > 0$ and F would not cover the bi-set $(Y \cup Y', Y)$. By Hall's theorem, G_i^+ has a perfect matching, as required. ∎

It should be noted that requiring (18) only for the subsets of T is not sufficient (unlike the situation in Hall's theorem on perfect matching in bipartite graphs where Hall's criterion $|X| \le \big|\Gamma(X)\big|$ is violated by a subset of S if and only if it is violated by a subset of T). Given the simple condition (18) in Theorem 3.13, one may feel tempted to derive the result from classical matching theory or matroid intersection, and, indeed, J. Pap [20] found a short, elegant way to derive Theorem 3.13 directly from Edmonds' matroid intersection theorem [3].

As another corollary, we exhibit a connectivity augmentation result concerning simultaneous augmentations of two digraphs. In order to handle edge-disjoint and node-disjoint paths uniformly, the following common generalization was introduced in [10].

Let $D = (V, F)$ be a digraph and $g\colon V \to \mathbf{Z}_+$ a function. A set of edge-disjoint st-paths is said to be **g-bounded** if each node $v \in V - \{s,t\}$ is used by at most $g(v)$ of these paths. We stress that g-boundedness automatically means that the paths are edge-disjoint. Let $\lambda_g(s,t;D)$ denote the maximum number of g-bounded st-paths. Note that for large g (say, $g \equiv |F|$) $\lambda_g(s,t;D)$ is the maximum number of edge-disjoint st-paths, while for $g \equiv 1$, $\lambda_g(s,t;D)$ is the maximum number of openly disjoint st-paths.

We will need the bi-set function μ_g defined by

$$\mu_g(X) := \sum \big[g(v)\colon v \in X_O - X_I\big]. \tag{19}$$

It is easily seen that for bi-sets X and Y

$$\mu_g(X) + \mu_g(Y) = \mu_g(X \cap Y) + \mu_g(X \cup Y). \tag{20}$$

The following characterization can be easily derived from the edge-version of Menger's theorem (and was done in [10]).

Proposition 3.14 (Variation of Menger's theorem). *In a digraph $D = (V, F)$ there are k g-bounded st-paths if and only if*

$$\varrho_F(X) \ge k - \mu_g(X) \text{ holds for every bi-set } X = (X_O, X_I) \text{ with } t \in X_I \text{ and } X_O \subseteq V - s. \tag{21}$$ ■

We say that D is **(k,g)-connected** from s to t if there are k g-bounded paths from s to t.

Suppose now that $D_i = (V, A_i)$ are digraphs for $i = 1, 2$ on the same node set V in which s_i and t_i are designated source and sink nodes. Moreover, let $g_i\colon V \to \mathbf{Z}_+$ be a function and k_i positive integer. Consider the ring-family

$\mathcal{R}_i := \{X \in \mathcal{P}_2(V) \colon t_i \in X_I,\ X_O \subseteq V - s_i\}$ of bi-sets and define a bi-set function p_i on $\mathcal{R}_i$ by

$$p_i(X) := k_i - \varrho_{D_i}(X) - \mu_{g_i}(X).$$

Since ϱ_{D_i} is submodular and μ_{g_i} is modular, p_i is supermodular on $\mathcal{R}_i$. Let S and T be two non-empty subsets of V so that there is an ST-edge covering each bi-set with positive p_1- or p_2-value. By Theorem 3.11, we get the following.

Theorem 3.15. *Given D_i, s_i, t_i, $\mathcal{R}_i$ g_i, k_i, p_i S, T for $i = 1, 2$ as above, there is a set F of γ ST-edges whose addition to D_i results in a digraph which is (k_i, g_i)-connected from s_i to t_i if and only if*

$$p_i(X) \le \gamma \ \text{ for every } \ X \in \mathcal{R}_i, \ \ (i = 1, 2), \ \text{ and}$$

$$p_1(X') + p_2(X'') \le \gamma \ \text{ for every } \ ST\text{-independent } \ X' \in \mathcal{R}_1$$

$$\text{and } \ X'' \in \mathcal{R}_2. \qquad \blacksquare$$

3.3. Algorithmic aspects

Before sketching an algorithmic approach, we make some observations on classes of supermodular functions.

Claim 3.16. *If a positively 2/3-supermodular function is given by an evaluation oracle, then its maximum cannot be computed in polynomial time.*

Proof. Let p be a set function which takes positive value on exactly one subset and zero otherwise. This is positively 2/3-supermodular and to find out its maximum one must, in worst case, call for the value of all subsets. ■

Therefore there is no polynomial algorithm for computing the extrema in Theorem 3.4 if the 2/3-submodular function is given by an evaluation oracle. The question arises whether the problem in Theorem 3.4 is more general at all than the one in Theorem 3.11. The next claim shows that the answer is yes.

Claim 3.17. *Not every 2/3-supermodular function arises from two positively supermodular functions as their maximum.*

Proof. Let the ground-set $V = \{v_1, v_2, \ldots, v_5, s\}$ have six elements so that the first five elements are arranged around a circle according to their subscripts. Define $p(X)$ to be 1 if $s \in X$ and the elements of $X - s$ are consecutive around the circle (in particular, if $X = V$ or $X = \{s\}$), otherwise let $p(X) = 0$. Then easy case-checking shows that p is 2/3-supermodular but there cannot be two positively supermodular functions p_1 and p_2 so that $p(X) = \max\{p_1(X), p_2(X)\}$. Indeed, $p(V_i) = 1$ for $V_i := V - v_i$, $1 \leq i \leq 5$. Since the non-consecutive pairs form a five-gon, one of p_1 and p_2, say p_1, must take value one on two sets V_i, V_j with non-consecutive v_i, v_j. But then p_1 cannot be positively supermodular since $p_1(V_i \cap V_j) \leq p(V_i \cap V_j) = 0$ by definition and $p_1(V_i \cup V_j) \leq p(V_i \cup V_j) \leq 1$. ■

An analogous question concerning positively supermodular functions was answered by T. Király [15]:

Claim 3.18. *Not all positively supermodular functions arise as the non-negative part of a fully supermodular function.*

Proof. Let X_1, X_2, X_3 be three subsets of a ground-set V in general position. Let $p(X_i) = 1, p(X_i \cup X_j) = 2 \ \ (i \neq j)$, $p(X_1 \cup X_2 \cup X_3) = 4$ and $p(X) = 0$ on the remaining sets. Then p is positively supermodular and a simple argument shows that it cannot be the nonnegative part of a supermodular function. ■

The only general construction we know for positively supermodular functions is taking the non-negative part of a supermodular function on a ring-family, and likewise, we do not know any general class, let alone applications, of 2/3-supermodular functions which are not the maximum of two supermodular ones. On one hand, these function classes gave rise to formally more general results and their use made the proofs technically simpler, on the other hand they are not convenient for algorithmic handling. This is why we formulated separately Theorem 3.11: there is a strongly polynomial algorithm for computing the extrema in that theorem.

The very nature of the theorem makes it possible to compute a digraph H covering p_1 and p_2 with a minimum number of edges, provided that a subroutine is available for computing ν given in the theorem. With some work, such a subroutine can indeed be constructed by making use of an existing algorithm for maximizing supermodular functions [13, 21] (and in the special case of Theorem 3.15 even a Max-flow Min-cut subroutine suffices). So suppose that such a subroutine is available. The digraph H with a minimum number of edges that covers p_1 and p_2 will be defined with the help of a function $z\colon A^* \to \mathbf{Z}_+$ which tells us the number $z(a) \geq 0$ of parallel copies of every possible ST-edge a to be taken into H. The digraph defined by z covers p_i if $\varrho_z \geq p_i$ for $i = 1, 2$.

For a given z, let $\nu(z)$ denote the optimum in Theorem 3.11 with respect to the revised bi-set functions $p_1 - \varrho_z$ and $p_2 - \varrho_z$. Call a function $z\colon A^* \to \mathbf{Z}_+$ **good** if

$$\nu = \nu(z) + z(A^*). \tag{22}$$

By definition $z \equiv 0$ is good and the problem of finding a minimum z is equivalent to construct a good z covering of p.

Consider the elements of A^* in an arbitrary order $a_1, \ldots, a_m$. At the beginning $z \equiv 0$. At a general step, suppose that the values of $z(a_1), \ldots, z(a_{i-1})$ have already been computed in such a way that the vector $z = \big(z(a_1), \ldots, z(a_{i-1}), 0, \ldots, 0\big)$ is good. Compute $\nu(z)$. If this number is zero, then z is a covering of p and the algorithm terminates by returning z. Suppose now that $\nu(z) > 0$. Let z' be a vector arising from z by setting $z(a_i)$ to be a big enough number M and compute $\nu(z')$. It follows from Theorem 3.11 that by setting $z(a_i)$ to be $\nu(z) - \nu(z')$ the revised vector keeps to be good and the algorithm may proceed to the next index $i+1$.

References

[1] J. Edmonds, Submodular functions, matroids and certain polyhedra, in: *Combinatorial Structures and their Applications* (R. Guy, H. Hanani, N. Sauer and J. Schönheim, eds.), Gordon and Breach, New York (1970), pp. 69–87.

[2] J. Edmonds, Edge-disjoint branchings, in: *Combinatorial Algorithms* (B. Rustin, ed.), Acad. Press, New York, (1973), 91–96.

[3] J. Edmonds, Matroid intersection, *Annals of Discrete Math.*, **4** (1979), 39–49.

[4] J. Edmonds and R. Giles, A min-max relation for submodular functions on graphs, *Annals of Discrete Mathematics,* **1** (1977), 185–204.

[5] A. Frank, Kernel systems of directed graphs, *Acta Scientiarum Mathematicarum (Szeged),* **41, 1–2** (1979), 63–76.

[6] A. Frank, Submodular flows, in: *Progress in Combinatorial Optimization* (ed. W. Pulleyblank), Academic Press (1984), 147–165.

[7] A. Frank and É. Tardos, An application of submodular flows, *Linear Algebra and its Applications,* **114/115** (1989), 329–348.

[8] A. Frank, Augmenting graphs to meet edge-connectivity requirements, *SIAM J. on Discrete Mathematics* (1992 February), Vol. 5, No. 1, pp. 22–53.

[9] A. Frank, Connectivity augmentation problems in network design, in: *Mathematical Programming: State of the Art 1994* (J. R. Birge and K. G. Murty, eds.), The University of Michigan, pp. 34–63.

[10] A. Frank, Rooted k-connections in digraphs, *Discrete Applied Mathematics,* to appear.

[11] A. Frank and T. Jordán, Minimal edge-coverings of pairs of sets, *J. Combinatorial Theory, Ser. B.,* Vol. 65, No. 1 (1995, September), pp. 73–110.

[12] A. Frank, T. Király and M. Kriesell, On decomposing a hypergraph into k connected sub-hypergraphs, in: *Submodularity* (guest editor S. Fujishige), Discrete Applied Mathematics, Vol. 131, Issue 2. (September 2003), pp. 373–383.

[13] S. Iwata, L. Fleischer and S. Fujishige, A combinatorial, strongly polynomial-time algorithm for minimizing submodular functions, *J. of the ACM,* **48, 4** (2001), pp. 761–777. A preliminary version appeared in: Proceedings of the 32'nd STOC 2000, held in Portland, Oregon USA, pp. 97–106.

[14] N. Kamiyama, N. Katoh and A. Takizawa, Arc-disjoint in-trees in directed graphs, in: *Proc. the nineteenth Annual ACM-SIAM Symposium on Discrete Mathematics* (SODA 2008), pp. 218–526, 2008. The journal version is going to appear in *Combinatorica.*

[15] T. Király, *Edge-connectivity of undirected and directed hypergraphs,* Ph.D. Dissertation (2003), Eötvös University, Budapest, see: p. 126.

[16] L. Lovász, Solution to Problem 11, see pp. 168–169 in: Report on the Memorial Mathematical Contest Miklós Schweitzer of the year 1968, (in Hungarian), *Matematikai Lapok,* **20** (1969), pp. 145–171.

[17] L. Lovász, A generalization of Kőnig's theorem, *Acta. Math. Acad. Sci. Hungar.,* **21** (1970), 443–446.

[18] L. Lovász, On two minimax theorems in graph theory, *J. Combinatorial Theory (B),* **21** (1976), 96–103.

[19] C. L. Lucchesi and D. H. Younger, A minimax relation for directed graphs, *J. London Math. Soc.* (2), **17** (1978), 369–374.

[20] J. Pap, *A note on making two bipartite graphs perfectly matchable,* EGRES quickproof series QP-2006-04, (http://www.cs.elte.hu/egres/www/quickproof.html).

[21] A. Schrijver, A combinatorial algorithm minimizing submodular functions in strongly polynomial time, *Journal of Combinatorial Theory, Ser. B.,* Vol. 80 (2000), pp. 575–588.

[22] L. Szegő, Note on covering intersecting set-systems by digraphs, *Discrete Mathematics,* **234** (2001), 187–189.

[23] R. E. Tarjan, A good algorithm for edge-disjoint branching, *Information Processing Letters,* **3** (1974), 52–53.

Kristóf Bérczi
Egerváry Research Group of MTA-ELTE
Department of Operations Research
Eötvös University
Pázmány P. s. 1/c
Budapest
Hungary, H-1117
e-mail: berkri@cs.elte.hu

András Frank
Egerváry Research Group of MTA-ELTE
Department of Operations Research
Eötvös University
Pázmány P. s. 1/c
Budapest
Hungary, H-1117
e-mail: frank@cs.elte.hu

BOLYAI SOCIETY
MATHEMATICAL STUDIES, 19

Building Bridges
pp. 165–203.

Random Walks, Arrangements, Cell Complexes, Greedoids, and Self-organizing Libraries

ANDERS BJÖRNER

To László Lovász on his 60th birthday

The starting point is the known fact that some much-studied random walks on permutations, such as the Tsetlin library, arise from walks on real hyperplane arrangements. This paper explores similar walks on complex hyperplane arrangements. This is achieved by involving certain cell complexes naturally associated with the arrangement. In a particular case this leads to walks on libraries with several shelves.

We also show that interval greedoids give rise to random walks belonging to the same general family. Members of this family of Markov chains, based on certain semigroups, have the property that all eigenvalues of the transition matrices are non-negative real and given by a simple combinatorial formula.

Background material needed for understanding the walks is reviewed in rather great detail.

1. Introduction

The following random walk, called *Tsetlin's library*, is a classic in the theory of combinatorial Markov chains. Consider books labeled by the integers $1, 2, \ldots, n$ standing on a shelf in some order. A book is withdrawn according to some probability distribution w and then placed at the beginning of the shelf. Then another book is withdrawn according to w and placed at the beginning of the shelf, and so on. This Markov chain is of interest also for computer science, where it goes under names such as *dynamic file management* and *cache management*.

Much is known about the Tsetlin library, for instance good descriptions of its stationary distribution, good estimates of the rate of convergence to stationarity, exact formulas for the eigenvalues of its transition matrix P_w, and more. These eigenvalues are nonnegative real and their indexing and multiplicities, as well as their value, are given by very explicit combinatorial data.

The Tsetlin library is the simplest of a class of Markov chains on permutations that can be described in terms of books on a shelf. Instead of one customer visiting the library to borrow one book which when returned is placed at the beginning of the shelf, we can picture several customers who each borrows several books. When the books are returned, the books of the first borrower are placed at the beginning of the shelf in the induced order (i.e. the order they had before being borrowed). Then the books of the second borrower are placed in their induced order, and so on. Finally, the remaining books that noone borrowed stand, in the induced order, at the end of the shelf.

The analysis of such a "dynamic library" became part of a vastly more general theory through the work of Bidigare, Hanlon and Rockmore [2], continued and expanded by Brown and Diaconis [13, 14, 15, 16]. They created an attractive theory of random walks on hyperplane arrangements $\mathcal{A}$ in $\mathbb{R}^d$, for which the states of the Markov chain are the regions making up the complement of $\cup\mathcal{A}$ in $\mathbb{R}^d$. When adapted to the braid arrangement, whose regions are in bijective correspondence with the permutations of $\{1, 2, \ldots, n\}$, their theory specializes precisely to the "self-organizing", or "dynamic", one-shelf library that we just described. The theory was later further generalized by Brown [13, 14] to a class of semigroups.

The genesis of this paper is the question: *what about random walks on complex hyperplane arrangements?* It is of course not at all clear what is meant. The complement in $\mathbb{C}^d$ of the union of a finite collection of hyperplanes is a $2d$-dimensional manifold, so what determines a finite Markov chain?

The idea is to consider not the complement itself, but rather a certain finite cell complex determining the complement up to homotopy type. In addition, we need that this complex extends to a cell complex for the whole singularity link, since much of the probability mass is typically placed in that extension. Such complexes were introduced by Ziegler and the author in [11]. The construction and basic properties partly run parallel to a similar construction in the real case, well-known from the theory of oriented matroids.

The complex hyperplane walks take place on such cell complexes in a manner that will be described in Section 4.3. These cell complexes have a semigroup structure to which the theory of Brown [13] applies. Thus we get results for complex hyperplane walks analogous to those for the real case.

As mentioned, when specialized to the real braid arrangement the general theory of walks on *real* arrangements leads to the *one-shelf dynamic library*. What happens when we similarly specialize random walks on complex arrangements to the *complex* braid arrangement? The pleasant answer is that we are led to Markov chains modelling *dynamic libraries with several shelves*. These are self-organizing libraries where the books are placed on different shelves according to some classification (combinatorics books, geometry books, etc.), and not only the books on each shelf but also the shelves themselves are permuted in the steps of the Markov chain. Depending on the distribution of probability mass there are different versions.

Here is one. Say that a customer withdraws a subset $E \subseteq [n]$ of books from the library. The books are replaced in the following way. Permute the shelves so that the ones that contain one of the books from E become the top ones, maintaining the induced order among them and among the remaining shelves, which are now at the bottom. Then, on each shelf move the books from E to the beginning of the shelf, where they are placed in the induced order.

The exact description is given in Section 4.4. These Markov chains may be of interest also for file management applications in computer science.

In this paper we take a somewhat leisurely walk through the territory leading to complex hyperplane walks, recalling and assembling results along the way that in the end lead to the desired conclusions. We are not seeking the greatest generality, the aim is rather for simplicity of statements and illuminating ideas through special cases. Some proofs that would interfere with this aspiration are banished to an appendix.

Several topics touched upon in this paper relate to joint work with László Lovász. This is the case for the k-equal arrangements [8] in Section 2.4 and for the greedoids [6] in Section 4.5. It is a pleasure to thank Laci for all the pleasant collaborations and interesting discussions over many years.

Also, I am grateful to Persi Diaconis for inspiration and encouragement, and to Jakob Jonsson for helpful remarks.

2. Real Hyperplane Arrangements

We review the basic facts about real hyperplane arrangements. This material is described in greater detail in many places, for instance in [7] and [19], to where we refer for more detailed information. Also, we adhere to the notation for posets and lattices in [20].

2.1. Basics

Let $\ell_1, \ldots, \ell_t$ be linear forms on $\mathbb{R}^d$, and $H_i = \{x \colon \ell_i(x) = 0\} \subseteq \mathbb{R}^d$ the corresponding hyperplanes. We call $\mathcal{A} = \{H_1, \ldots, H_t\}$ a *real hyperplane arrangement*. The arrangement is *essential* if $\cap H_i = \{0\}$, and we usually assume that this is the case.

The *complement* $M_{\mathcal{A}} = \mathbb{R}^d \setminus \cup\mathcal{A}$ consists of a collection $C_{\mathcal{A}}$ of open convex cones R_i called *regions*. They are the connected components of the decomposition $M_{\mathcal{A}} = \biguplus R_i$ into contractible pieces.

With $\mathcal{A}$ we associate its *intersection lattice* $L_{\mathcal{A}}$, consisting of all intersections of subfamilies of hyperplanes H_i ordered by set inclusion. Each subspace belonging to $L_{\mathcal{A}}$ can be represented by the set of hyperplanes from $\mathcal{A}$ whose intersection it is. In this way the elements of $L_{\mathcal{A}}$ can be viewed either as subsets of $\mathbb{R}^d$ or as subsets of $\mathcal{A}$. The latter is for simplicity encoded as subsets of $[n]$ via the labeling $i \leftrightarrow H_i$.

Let $L_{\mathcal{A}}^{\mathrm{op}}$ denote $L_{\mathcal{A}}$ with the opposite partial order, so in $L_{\mathcal{A}}^{\mathrm{op}}$ the subspaces of $\mathbb{R}^d$ are ordered by *reverse inclusion*. This is a geometric lattice, whose atoms are the hyperplanes H_i.

The number of regions of $\mathcal{A}$ is determined by $L_{\mathcal{A}}$ via its Möbius function in the following way.

Theorem 2.1 (Zaslavsky [22]). $|C_{\mathcal{A}}| = \sum_{x \in L_{\mathcal{A}}} \left|\mu(x, \hat{1})\right|.$

There is a useful way to encode the position of a point $x \in \mathbb{R}^d$ with respect to $\mathcal{A}$. Define the *sign vector (position vector)* $\sigma(x) = \{\sigma_1, \ldots, \sigma_t\} \in \{0, +, -\}^t$ by

$$\sigma_i \stackrel{\text{def}}{=} \begin{cases} 0, & \text{if } \ell_i(x) = 0 \\ +, & \text{if } \ell_i(x) > 0 \\ -, & \text{if } \ell_i(x) < 0 \end{cases}$$

In words, the ith entry σ_i of the sign vector $\sigma(x)$ tells us whether the point x is on the hyperplane H_i, or on its positive resp. negative side.

Let $F_{\mathcal{A}} \stackrel{\text{def}}{=} \sigma(\mathbb{R}^d) \subseteq \{+,-,0\}^t$ and make this collection of sign vectors into a poset by componentwise ordering via

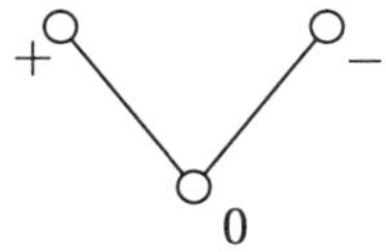

Thus, we have a surjective map $\sigma\colon \mathbb{R}^d \to F_{\mathcal{A}}$. Note that $F_{\mathcal{A}}$, called the *face semilattice*, has minimum element $(0,\ldots,0)$ and its maximal elements $F_{\mathcal{A}} \cap \{+,-\}^t$ are in bijective correspondence with the regions, as is illustrated in Figure 1.

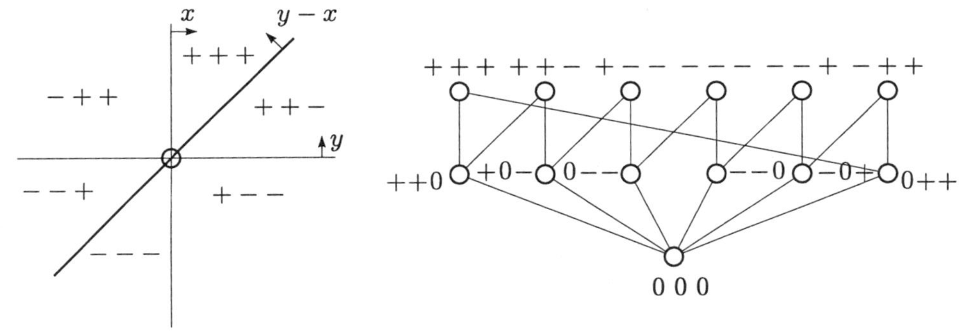

Fig. 1. Face semilattice of an arrangement of three lines in $\mathbb{R}^2$

The *composition* $X \circ Y$ of two sign vectors $X, Y \in \{0,+,-\}^t$ is defined by

$$(X \circ Y)_i \stackrel{\text{def}}{=} \begin{cases} X_i, \text{if } \; X_i \neq 0 \\ Y_i, \text{if } \; X_i = 0 \end{cases}$$

This operation on $\{0,+,-\}^t$ is associative, idempotent, and has unit element $(0,\ldots,0)$. Furthermore, $F_{\mathcal{A}}$ forms a closed subsystem: if $X, Y \in F_{\mathcal{A}}$ then $X \circ Y \in F_{\mathcal{A}}$. Here is the geometric reason: choose points $x, y \in \mathbb{R}^d$ such that $\sigma(x) = X$ and $\sigma(y) = Y$. Move a small distance from x along the straight line segment from x to y. The point z reached has the position $\sigma(z) = X \circ Y$.

Hence,

(2.1) $(F_{\mathcal{A}}, \circ)$ is an idempotent semigroup.

The combinatorics of sign vectors is systematically developed in *oriented matroid theory*, where the elements of $F_{\mathcal{A}}$ are called "covectors" and the system $(F_{\mathcal{A}}, \circ)$ is the basis for one of the fundamental axiom systems, see [7, Section 3.7].

There is an important *span map*

$$\text{span}\colon F_{\mathcal{A}} \to L_{\mathcal{A}} \tag{2.2}$$

which can be characterized in two ways. Combinatorially, it sends the sign-vector X to the set of positions of its non-zero components (a subset of $[n]$). Geometrically, it sends the cone $\sigma^{-1}(X)$ to its linear span.

The span map is a rank-preserving and order-preserving semigroup map, meaning that

$$\text{rk}_{F_{\mathcal{A}}}(X) = \text{rk}_{L_{\mathcal{A}}}(\text{span}\,(X)) \tag{2.3}$$

$$X \leq Y \Rightarrow \text{span}\,(X) \leq \text{span}\,(Y) \tag{2.4}$$

$$\text{span}\,(X \circ Y) = \text{span}\,(X) \vee \text{span}\,(Y) \tag{2.5}$$

Also, we have that

$$X \circ Y = Y \Leftrightarrow X \leq Y \tag{2.6}$$

$$X \circ Y = X \Leftrightarrow \text{span}\,(Y) \leq \text{span}\,(X) \tag{2.7}$$

2.2. The braid arrangement

The *braid arrangement* $\mathcal{B}_n = \{x_i - x_j \mid 1 \leq i < j \leq n\}$ in $\mathbb{R}^n$ plays an important role in this paper, due to its close connections with the combinatorics of permutations and partitions. The hyperplanes in $\mathcal{B}_n$ all contain the diagonal line $(t, t, \ldots, t)$. By intersecting with the hyperplane orthogonal to this line we get an essential arrangement, now in $\mathbb{R}^{d-1}$.

The intersection lattice $L_{\mathcal{B}_n}$ is isomorphic to the partition lattice Π_n, i.e. the partitions of the set $[n]$ ordered by reverse refinement. The correspondence between a set partition and a subspace obtained by intersecting some hyperplanes $x_i - x_j$ is easily understood from examples, such as

$$(134 \mid 27 \mid 5 \mid 6) \;\leftrightarrow\; \begin{cases} x_1 = x_3 = x_4 \\ x_2 = x_7 \end{cases}$$

and

$$(1345 \mid 267) < (134 \mid 27 \mid 5 \mid 6).$$

The face semilattice $F_{\mathcal{B}_n}$ is isomorphic to the meet-semilattice of *ordered* set partitions Π_n^{ord} (so, the order of the blocks matters), ordered by reverse refinement. For instance,

$$\langle 134 \mid 6 \mid 27 \mid 5 \rangle \leftrightarrow \begin{cases} x_1 = x_3 = x_4 \\ x_4 < x_6 < x_2 \\ x_2 = x_7 \\ x_7 < x_5 \end{cases}$$

and

$$\langle 1346 \mid 257 \rangle < \langle 134 \mid 6 \mid 27 \mid 5 \rangle.$$

Under this correspondence the regions of $\mathbb{R}^{n-1} \setminus \cup\mathcal{B}_n$ are in bijection with the ordered partitions into singleton sets, or in other words, with the permutations of the set $[n]$. The span map (2.2) is the map $\Pi_n^{\mathrm{ord}} \to \Pi_n$ that sends an ordered partition $\langle \dots \rangle$ to an unordered partition $(\dots)$ by forgetting the ordering of its blocks.

Composition in $F_{\mathcal{B}_n}$ has the following description. If $X = \langle X_1, \dots, X_p \rangle$ and $Y = \langle Y_1, \dots, Y_q \rangle$ are ordered partitions of $[n]$, then $X \circ Y = \langle X_i \cap Y_j \rangle$ with the blocks ordered lexicographically according to the pairs of indices (i, j). For instance,

$$\langle 257 \mid 3 \mid 146 \rangle \circ \langle 17 \mid 25 \mid 346 \rangle = \langle 7 \mid 25 \mid 3 \mid 1 \mid 46 \rangle,$$

as can conveniently be seen from the computation table

(2.8)

$\circ$	$1,7$	$2,5$	$3,4,6$
$2,5,7$	7	$2,5$	
3			3
$1,4,6$	1		$4,6$

2.3. Cell complexes and zonotopes

The whole idea of random walks on complex hyperplane arrangements rests on the idea of walking on the cells of an associated cell complex. We therefore review the construction used in [11] of such cell complexes. The basic idea is given together with two applications. The first one is the construction of cell complexes for the complement of a linear subspace arrangement in $\mathbb{R}^d$ at the end of this section. The other is the construction of cell complexes for hyperplane arrangements in $\mathbb{C}^d$, to which we return in Section 3.2. See e.g. [4] for topological terminology.

A regular cell decomposition Γ of the unit sphere S^{d-1} is said to be *PL* if its barycentric subdivision (equivalently, the order complex of its face poset) is a piecewise linear triangulation of S^{d-1}. Here is a simple combinatorial procedure for producing regular cell complexes of certain specific homotopy types from posets.

Proposition 2.2 [11, Prop. 3.1]. *Suppose that Γ is a PL regular cell decomposition of S^{d-1}, with face poset F_Γ. Let $T \subseteq S^{d-1}$ be a subspace of the sphere such that $T = \cup_{\tau \in G} \tau$ for some order ideal $G \subseteq F_\Gamma$. Then the poset $(F_\Gamma \setminus G)^{\mathrm{op}}$ is the face poset of a regular cell complex having the homotopy type of the complement $S^{d-1} \setminus T$.*

Now, let $\mathcal{A}$ be an essential hyperplane arrangement in $\mathbb{R}^d$. For a general sign vector $X \in F_{\mathcal{A}}$ the set $\sigma^{-1}(X)$ is a convex cone in $\mathbb{R}^d$ which is open in its linear span. Let $\tau_X \stackrel{\text{def}}{=} \sigma^{-1}(X) \cap S^{d-1}$. The sets τ_X, for $X \in F_{\mathcal{A}} \setminus \widehat{0}$, partition the the unit sphere and are in fact the open cells of a regular CW decomposition of S^{d-1}. Furthermore, the inclusion relation of their closures $\overline{\tau_X}$ coincides with the partial order we have defined on $F_{\mathcal{A}}$. Thus, $F_{\mathcal{A}} \setminus \widehat{0}$ is the face poset of a regular cell decomposition $\Gamma_{\mathcal{A}}$ of the unit sphere in $\mathbb{R}^d$, namely the cell decomposition naturally cut out by the hyperplanes.

The cell complex $\Gamma_{\mathcal{A}} = \{\tau_X\}_{X \in F_{\mathcal{A}}}$ induced by a hyperplane arrangement $\mathcal{A}$ is PL. Thus, via Proposition 2.2 we can construct cell complexes determining the complement of a subcomplex up to homotopy type. Combinatorially the description is simple: erase from the face poset $F_{\mathcal{A}}$ all the cells that belong to the given subcomplex and then turn the remaining subposet upside down. Done!

The cell complexes constructed this way from a hyperplane arrangement $\mathcal{A}$ can be geometrically realized on the boundary of an associated convex polytope. Namely, with $\mathcal{A}$ is associated its *zonotope* $\mathbf{Z}_{\mathcal{A}} = [-e_1, e_1] \oplus \cdots \oplus$

$[-e_t, e_t]$. Here e_i is a normal vector in $\mathbb{R}^d$ to the hyperplane H_i and the right-hand side denotes Minkowski sum of centrally symmetric line segments. Thus, $\mathbf{Z}_{\mathcal{A}}$ is a centrally symmetric convex polytope, determined this way up to combinatorial equivalence. A key property of $\mathbf{Z}_{\mathcal{A}}$ is that there exists an order-reversing bijection between the faces on its boundary and the cells of $\Gamma_{\mathcal{A}}$. In other words, the poset of proper faces of $\mathbf{Z}_{\mathcal{A}}$ is isomorphic to the opposite of the face poset of $\mathcal{A}$:

$$F_{\mathbf{Z}_{\mathcal{A}}} \cong \left(F_{\mathcal{A}} \setminus \widehat{0}\right)^{\mathrm{op}} \tag{2.9}$$

Suppose that $\mathcal{A}$ is an arrangement of linear subspaces of arbitrary dimensions in $\mathbb{R}^d$. Say that we want to construct a cell complex having the homotopy type of its complement $\mathbb{R}^d \setminus \cup\mathcal{A}$. This complement is by radial projection homotopy equivalent to its intersection with the unit sphere S^{d-1}. Therefore the preceding construction is applicable. We just have to choose an auxiliary hyperplane arrangement $\mathcal{H}$ into which $\mathcal{A}$ *embeds*, meaning that each subspace in $\mathcal{A}$ is the intersection of some of the hyperplanes from $\mathcal{H}$. This is clearly always possible. Putting the various pieces of information together and applying Proposition 2.2 we obtain the following description.

Theorem 2.3 [11]. *Let $\mathcal{A}$ be an arrangement of linear subspaces in $\mathbb{R}^d$. Choose a hyperplane arrangement $\mathcal{H}$ into which $\mathcal{A}$ embeds. Then the complement $\mathbb{R}^d \setminus \cup\mathcal{A}$ has the homotopy type of a subcomplex $\mathbf{Z}_{\mathcal{H},\mathcal{A}}$ of the boundary of the zonotope $\mathbf{Z}_{\mathcal{H}}$. The complex $\mathbf{Z}_{\mathcal{H},\mathcal{A}}$ is obtained by deleting from the boundary of $\mathbf{Z}_{\mathcal{H}}$ all faces that correspond to cells τ_X contained in $\cup\mathcal{A}$.*

2.4. The permutohedron and the k-equal arrangements

We illustrate the general constructions of the preceding section by applying them to the *k-equal arrangements* $\mathcal{A}_{n,k} = \{x_{i_1} = x_{i_2} = \cdots = x_{i_k} : 1 \le i_1 < i_2 < \cdots < i_k \le n\}$ in $\mathbb{R}^n$. The topology of their complements plays a crucial role in the solution of a complexity-theoretic problem in joint work with Lovász and Yao [8, 9]. See also [10], where their homology groups were computed. The k-equal arrangements embed into the braid arrangement (the $k = 2$ case), so Theorem 2.3 is applicable. It tells us that, up to homotopy type, the topology of the complement of the k-equal arrangement $\mathcal{A}_{n,k}$ is realized by some subcomplex of the zonotope of the braid arrangement. This subcomplex can be very explicitly described.

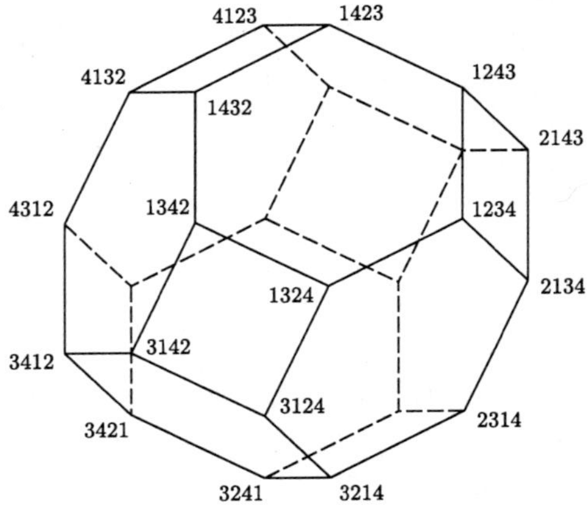

Fig. 2. The permutohedron $\mathbf{Z}_4^{\text{perm}}$

The zonotope of the braid arrangement $\mathcal{B}_n$ is the *permutohedron* $\mathbf{Z}_n^{\text{perm}}$, that is, the convex hull of the $n!$ points in $\mathbb{R}^n$ whose coordinates are given by a permutation of the numbers $1, 2, \ldots, n$. Its $n!$ vertices are in bijection with the $n!$ regions of $\mathcal{B}_n$, in accordance with the duality (2.9).

We want to describe the subcomplex $\mathbf{Z}_{\mathcal{A}_{n,k}}$ of the boundary of $\mathbf{Z}_n^{\text{perm}}$ which is homotopy equivalent to the complement $\mathcal{M}_{n,k}$ of $\mathcal{A}_{n,k}$.

For this one argues as follows, keeping Section 2.2 in fresh memory. Let $f\colon \Pi_n^{\text{ord}} \to \Pi_n$ be the span map, i.e., the forgetful map that sends an ordered partition of $[n]$ to the corresponding unordered partition. The set $\Pi_n^{\text{ord}} \setminus \widehat{0}$ ordered by refinement is the poset of proper faces of the permutohedron $\mathbf{Z}_n^{\text{perm}}$, whereas the set Π_n ordered by refinement is the opposite of intersection lattice of the braid arrangement. The image $f(\pi)$ for $\pi \in \Pi_n^{\text{ord}}$ is a partition determining the span of the corresponding cell (i.e., the smallest intersection subspace of the braid arrangement in which the cell is contained). More precisely, the span of π is the subspace obtained by setting $x_{i_1} = x_{i_2} = \cdots = x_{i_j}$ for each block $\{i_1, i_2 \ldots, i_j\}$ of π. Thus, a cell $\pi \in \Pi_n^{\text{ord}}$ lies in the union of the k-equal arrangement if and only if some block has size at least k.

It follows that the complex $\mathbf{Z}_{\mathcal{B}_n, \mathcal{A}_{n,k}}$ consists of those cells on the boundary of the permutohedron $\mathbf{Z}_n^{\text{perm}}$ that correspond to ordered partitions with all blocks of size less than k. If an ordered partition has blocks of sizes $b_1, \ldots, b_e$, then the corresponding face of $\mathbf{Z}_n^{\text{perm}}$ is the product of smaller permutohedra of dimensions $b_1 - 1, \ldots, b_e - 1$. Therefore, the final description of the cell complex $\mathbf{Z}_{\mathcal{B}_n \mathcal{A}_{n,k}}$ is that one should delete from $\mathbf{Z}_n^{\mathrm{perm}}$ all faces that contain a q-dimensional permutohedron, for $q \geq k-1$, in its decomposition.

We are led to the following result, obtained independently by E. Babson for $k = 3$ (see [1]) and the author [5].

Theorem 2.4. *Delete from the boundary of the permutohedron $\mathbf{Z}_n^{\mathrm{perm}}$ every face that contains a d-dimensional permutohedron, $d \geq k-1$, in its decomposition. Then the remaining subcomplex has the homotopy type of the complement of the k-equal arrangement.*

Thus, for $k = 2$ one deletes everything but the vertices, for $k = 3$ one deletes all cells except those that are products of edges (equivalently, keep only the cubical faces), for $k = 4$ one deletes all cells except those that are products of edges (1-dimensional zonotope) and hexagons (2-dimensional zonotope), and so on.

The case $k = 3$ is especially interesting. The complex is in that case cubical. In particular, the fundamental group of $\mathcal{M}_{n,3}$ is the same as the fundamental group of the cell complex obtained from the graph (1-skeleton) of $\mathbf{Z}_n^{\mathrm{perm}}$ by gluing a 2-cell (membrane) into every 4-cycle.

Remark 2.5. What was just said is part of a more general result about gluing 2-cells into 4-cycles of a zonotopal graph.

Let $\mathcal{H}$ be an arbitrary central and essential hyperplane arrangement, and let $\mathcal{A}$ be the subspace arrangement consisting of codimension 2 intersections of 3 or more planes from $\mathcal{H}$ (assuming that there are such).

Next, let G be the 1-skeleton of the zonotope $\mathbf{Z}_{\mathcal{H}}$. The 2-cells of $\mathbf{Z}_{\mathcal{H}}$ are $2m$-gons (corresponding to codimension 2 subspaces where m planes meet). Let $\Gamma_{\mathcal{A}}$ be the cell complex obtained by gluing 2-cells into the 4-cycles of the graph G. Then the general construction above shows (since fundamental groups live on 2-skeleta) that the fundamental group of $\Gamma_{\mathcal{A}}$ is isomorphic to that of the complement $\mathcal{M}_{\mathcal{A}}$.

One can go on and describe the higher-dimensional cells needed to obtain a cell complex having the homotopy type of the complement of such a codimension 2 arrangement $\mathcal{A}$. They are all the cubes in the boundary of $\mathbf{Z}_{\mathcal{H}}$, just like for the special case of the 3-equal arrangement.

Remark 2.6. The two-dimensional faces of $\mathbf{Z}_n^{\text{perm}}$ are either 4-gons or 6-gons. What happens if we take the graph of the permutohedron and glue in only the hexagonal 2-cells? The answer is that we get a two-dimensional cell complex whose fundamental group is isomorphic to that of the complement of another subspace arrangement, namely the arrangement $\mathcal{A}_{[2,2]}$ consisting of codimension 2 subspaces of $\mathbb{R}^n$ obtained as intersections of pairs of hyperplanes $x_i = x_j$ and $x_k = x_l$, for all distinct i, j, k, l. Actually, for $\mathcal{A}_{[2,2]}$ a stronger statement is true: the 2-dimensional cell complex described (i.e. the permutohedron graph plus all hexagonal 2-cells) has the homotopy type of the complement of $\mathcal{A}_{[2,2]}$.

It is an interesting fact that the codimension 2 arrangements $\mathcal{A}_{n,3}$ and $\mathcal{A}_{[2,2]}$, corresponding to the two ways of gluing 2-cells into the permutohedron graph, share a significant topological property, namely that their complements are $K(\pi, 1)$ spaces. See Khovanov [17].

3. Complex Hyperplane Arrangements

We now move the discussion to complex space. To begin with many of the concepts and results are parallel to the real case. But new interesting features soon start to appear. This whole chapter summarizes material from [11].

3.1. Basics

We call $\mathcal{A} = \{H_1, \ldots, H_t\}$ a *complex hyperplane arrangement* if $H_i = \{z : \ell_i(z) = 0\} \subseteq \mathbb{C}^d$ for some linear forms $\ell_1, \ldots, \ell_t$ on $\mathbb{C}^d$. A particular choice of defining linear forms is assumed throughout, so we can also write $\mathcal{A} = \{\ell_1, \ldots, \ell_t\}$. The arrangement is *essential* if $\cap H_i = \{0\}$, and we usually assume that this is the case. The real and imaginary parts of $w = x+iy \in \mathbb{C}$ are denoted, respectively, by $\Re(w) = x$ and $\Im(w) = y$.

The position of a point $z \in \mathbb{C}^d$ with respect to $\mathcal{A}$ is combinatorially encoded in the following way. Define the *sign vector (position vector)* $\sigma(z) = \{\sigma_1, \ldots, \sigma_t\} \in \{0, +, -, i, j\}^t$ by

$$\sigma_i = \begin{cases} 0, & \text{if } \ell_i(z) = 0 \\ +, & \text{if } \Im\big(\ell_i(z)\big) = 0,\ \Re(\ell_i(x) > 0 \\ -, & \text{if } \Im\big(\ell_i(z)\big) = 0,\ \Re(\ell_i(x) < 0 \\ i, & \text{if } \Im\big(\ell_i(z)\big) > 0 \\ j, & \text{if } \Im\big(\ell_i(z)\big) < 0 \end{cases}$$

Let $F_{\mathcal{A}} \stackrel{\text{def}}{=} \sigma(\mathbb{C}^d) \subseteq \{0, +, -, i, j\}^t$ and make this collection of sign vectors into a poset, called the *face poset*, by componentwise ordering via

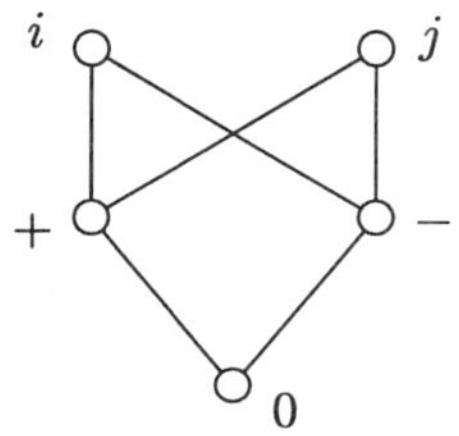

Proposition 3.1 [11].

(1) *$F_{\mathcal{A}}$ is a ranked poset of length $2d$. Its unique minimal element is* 0.

(2) *The maximal elements of $F_{\mathcal{A}}$ are the sign vectors in $F_{\mathcal{A}} \cap \{i, j\}^t$.*

(3) $\mu(Z, W) = (-1)^{\mathrm{rk}(W) - \mathrm{rk}(Z)}$, *for all $Z \leq W$ in $F_{\mathcal{A}} \cup \widehat{1}$.*

Figure 3 (borrowed from [11]), shows the face poset of $\mathcal{A} = \{z, w, w - z\}$ in $\mathbb{C}^2$. The reason for marking the elements not containing any zero with filled dots becomes clear in Section 3.2

The *composition* of two complex sign vectors $Z \circ W \in \{0, +, -, i, j\}^t$ is defined by

$$(Z \circ W)_i = \begin{cases} Z_i, & \text{if } W_i \not> Z_i \\ W_i, & \text{if } W_i > Z_i \end{cases} \tag{3.1}$$

Just as in the corresponding real case this operation on $\{0, +, -, i, j\}^t$ is associative, idempotent, and has unit element $(0, \ldots, 0)$. Also, for geometric

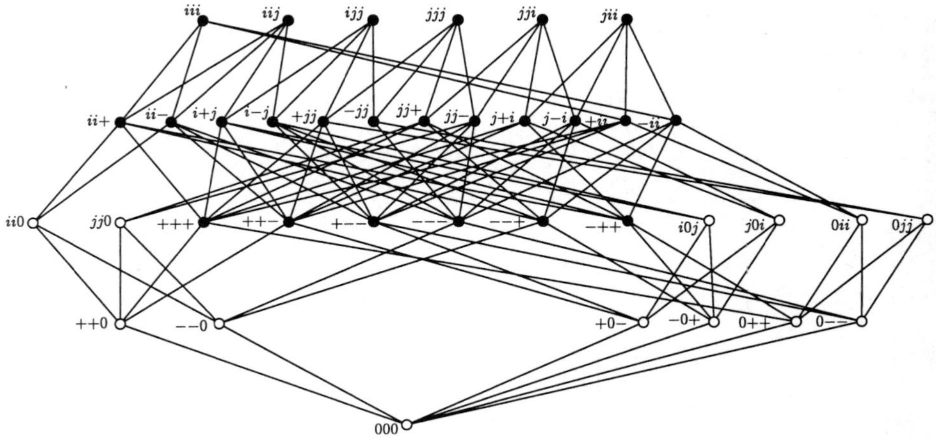

Fig. 3. Face poset of an arrangement of three lines in $\mathbb{C}^2$

reasons (analogous to the ones in the real case) $X, Y \in F_{\mathcal{A}}$ implies that $X \circ Y \in F_{\mathcal{A}}$. Hence,

(3.2) $$(F_{\mathcal{A}}, \circ) \text{ is an idempotent semigroup.}$$

For complex arrangements the notion of intersection lattice splits into two.

1. The *intersection lattice* $L_{\mathcal{A}}$ consists of all intersections of subfamilies of hyperplanes H_i ordered by set inclusion.

2. The *augmented intersection lattice* $L_{\mathcal{A},\text{aug}}$ is the collection of all intersections of subfamilies of the *augmented arrangement*

$$\mathcal{A}_{\text{aug}} = \{H_1, \ldots, H_t, H_1^{\mathbb{R}}, \ldots, H_t^{\mathbb{R}}\}$$

ordered by set inclusion. Here, $H_i^{\mathbb{R}} \stackrel{\text{def}}{=} \big\{z \in \mathbb{C}^d \colon \Im\big(\ell_i(z)\big) = 0\big\}$ is a $(2d-1)$-dimensional real hyperplane in $\mathbb{C}^d \cong \mathbb{R}^{2d}$ containing H_i.

Again as in the real case, we denote by $L_{\mathcal{A}}^{\text{op}}$ and $L_{\mathcal{A},\text{aug}}^{\text{op}}$ the opposite lattices, obtained by reversing the partial order.

Proposition 3.2. (1) *$L_{\mathcal{A}}^{\text{op}}$ is a geometric lattice of length d.*

(2) *$L_{\mathcal{A},\text{aug}}^{\text{op}}$ is a semimodular lattice of length $2d$.*

There is a *span map*

(3.3) $$\text{span} \colon F_{\mathcal{A}} \to L_{\mathcal{A},\text{aug}}$$

defined by sending the convex cone $\sigma^{-1}(Z)$, for $Z \in F_{\mathcal{A}}$, to the intersection of all subspaces in $\mathcal{A}_{\text{aug}}$ that contain $\sigma^{-1}(Z)$. This map preserves poset and semigroup structure as well as poset rank.

Proposition 3.3 [11].

$$\operatorname{rk}_{F_{\mathcal{A}}}(Z) = \operatorname{rk}_{L_{\mathcal{A},\text{aug}}}\big(\operatorname{span}(Z)\big) \tag{3.4}$$

$$Z \le W \Rightarrow W \circ Z = W \;\Leftrightarrow\; \operatorname{span}(Z) \le \operatorname{span}(W) \tag{3.5}$$

$$\operatorname{span}(Z \circ W) = \operatorname{span}(Z) \vee \operatorname{span}(W) \tag{3.6}$$

3.2. Cell complexes

The *complement* $M_{\mathcal{A}} = \mathbb{C}^d \setminus \cup\mathcal{A}$ is a complex manifold of real dimension $2d$. There is a huge literature on the topology of such spaces, see e.g. [19]. Among the basic results we mention that the Betti numbers of $M_{\mathcal{A}}$ are determined by $L_{\mathcal{A}}$ via its Möbius function in the following way.

Theorem 3.4 [19, p. 20]. $\displaystyle \beta_i(M_{\mathcal{A}}) = \sum_{x \in L_{\mathcal{A}}:\ \operatorname{rk}(x) = d-i} \big|\mu(x, \widehat{1})\big|.$

Let $\mathcal{A}$ be an essential complex hyperplane arrangement in $\mathbb{C}^d$, as before. For every sign vector $Z \in F_{\mathcal{A}} \setminus 0$ the inverse image $\sigma^{-1}(Z)$ is a relative-open convex cone in $\mathbb{C}^d$. The intersections of these cones with the unit sphere S^{2d-1} in $\mathbb{C}^d$ are the open cells of a PL regular cell decomposition of S^{2d-1} whose face poset is isomorphic to $F_{\mathcal{A}}$. Hence, as an application of Proposition 2.2 we get part (3) of the following result. Part (2) can be seen from the fact that x is an $\operatorname{rk}(x)$-dimensional linear subspace, so $x \cap S^{2d-1}$ is an $\big(\operatorname{rk}(x) - 1\big)$-dimensional sphere, for all $x \in L_{\mathcal{A},\text{aug}} \setminus \widehat{0}$, where "rk" denotes poset rank in $L_{\mathcal{A},\text{aug}}$.

Theorem 3.5 [11].

1. *The poset $F_{\mathcal{A}}$ is the face poset of a regular cell decomposition of the unit sphere in $\mathbb{R}^{2d} \cong \mathbb{C}^d$.*
2. *The subposet $\operatorname{span}^{-1}\big((L_{\mathcal{A},\text{aug}})_{\le x}\big)$ is the face poset of a regular cell decomposition of the sphere $S^{\operatorname{rk}(x)-1}$, for all $x \in L_{\mathcal{A},\text{aug}} \setminus \widehat{0}$.*

3. *The subposet* $C_{\mathcal{A}} \stackrel{\text{def}}{=} F_{\mathcal{A}} \cap \{+,-,i,j\}^t$, *with opposite order, is the face poset of a regular cell complex having the homotopy type of the complement* $M_{\mathcal{A}}$.

For an example, have a look at Figure 3. The sign vectors in $F_{\mathcal{A}}$ that lack a zero component are shown by filled dots. Hence, the cell complex $C_{\mathcal{A}}$ can be viewed by turning the page upside-down and looking at the subposet of filled dots only.

Combining some of this topological information with Theorem 5.1 of the Appendix we obtain the following analogue of Zaslavsky's theorem 2.1 for the number of maximal cells in the complex case.

Theorem 3.6. $\left|\max(F_{\mathcal{A}})\right| = \sum_{x \in L_{\mathcal{A},\text{aug}}} \left|\mu(x,\hat{1})\right|.$

Proof. We apply Theorem 5.1 to the span map: $F_{\mathcal{A}} \to L_{\mathcal{A},\text{aug}}$. There are six conditions to verify. With the exception of (5), they all follow from Propositions 3.1 and 3.2. Condition (5) is the consequence for the Euler characteristic of Theorem 3.5(2). ■

3.3. Complexified ℝ-arrangements

This section concerns the special case when all the linear forms $\ell_i(z)$ have real coefficients. The forms then define both a real arrangement $\mathcal{A}^{\mathbb{R}}$ in $\mathbb{R}^d$ and a complex arrangement $\mathcal{A}^{\mathbb{C}}$ in $\mathbb{C}^d$. These are of course related, and we here summarize what expression this relation takes for the combinatorial structures of interest.

First a few observations about complex sign vectors. A sign vector Z is called *real* if all its entries come from $\{0,+,-\}$. Every complex sign vector Z can be obtained as a composition $Z = X \circ iY$ [1] for two real sign vectors X and Y. Only the vector Y is unique in this decomposition.

For any poset P, let $\text{Int}(P)$ denote the set of its closed intervals. In the case of the face poset $F_{\mathcal{A}^{\mathbb{R}}}$ of a real arrangement $\mathcal{A}^{\mathbb{R}}$ we make $\text{Int}(F_{\mathcal{A}^{\mathbb{R}}})$ into a poset by introducing the following partial order:

$$[Y,X] \le [R,S] \ \leftrightarrow \ \begin{cases} Y \le R \\ R \circ X \le S \end{cases} \tag{3.7}$$

[1] here $i \cdot 0 = 0$, $i \cdot + = i$, $i \cdot - = j$.

Proposition 3.7 [11]. *The map* $\phi\colon \operatorname{Int}(F_{\mathcal{A}^{\mathbb{R}}}) \to F_{\mathcal{A}^{\mathbb{C}}}$ *given by* $[Y, X] \mapsto X \circ iY$ *is a poset isomorphism.*

For example,

$$\phi\colon \big[(0 - +0\,0 -), (- - +0 + -)\big] \;\mapsto\; (-j\, i 0 + j)$$

Hence, the entire structure of the complex face poset $F_{\mathcal{A}^{\mathbb{C}}}$ can be dealt with in terms of intervals in the real face poset $F_{\mathcal{A}^{\mathbb{R}}}$. In particular, the cells in the complement of $\mathcal{A}$, being the sign vectors without any zero coordinate, get this description.

$$C_{\mathcal{A}^{\mathbb{C}}} \;\overset{\phi}{\leftrightarrow}\; \text{intervals } [Y, X] \text{ with } X \in \max\big(F_{\mathcal{A}^{\mathbb{R}}}\big)$$

Composition of complex sign vectors (3.1) takes the following form when translated to intervals:

$$[Y, X] \circ [R, S] = [Y \circ R,\ Y \circ R \circ X \circ S] \tag{3.8}$$

The augmented intersection lattice $L_{\mathcal{A}^{\mathbb{C}},\text{aug}}$ is similarly determined by the intervals of $L_{\mathcal{A}^{\mathbb{R}}}$, namely

$$L_{\mathcal{A}^{\mathbb{C}},\text{aug}} \cong \operatorname{Int}(L_{\mathcal{A}^{\mathbb{R}}}), \tag{3.9}$$

this time with the partial order defined by

$$(x, y) \le (x', y') \quad \text{if and only if} \quad x \le x' \text{ and } y \le y'.$$

The span map is the natural one

$$\operatorname{Int}(F_{\mathcal{A}^{\mathbb{R}}}) \cong F_{\mathcal{A}^{\mathbb{C}}} \to L_{\mathcal{A}^{\mathbb{C}},\text{aug}} \cong \operatorname{Int}(L_{\mathcal{A}^{\mathbb{R}}}) \tag{3.10}$$

sending $[Y, X]$ to $\big[\operatorname{span}(Y), \operatorname{span}(X)\big]$. The Möbius function of $\operatorname{Int}(L_{\mathcal{A}^{\mathbb{R}}})$ is described in terms of the Möbius function of the lattice $L_{\mathcal{A}^{\mathbb{R}}}$ in Appendix 5.2.

Example 3.8. The braid arrangement $\mathcal{B}_n^{\mathbb{C}} = \{z_i - z_j \mid 1 \le i < j \le n\}$ in $\mathbb{C}^n$ is the complexification of the real braid arrangement, discussed in Section 2.2. Hence we can translate its combinatorics into the language of intervals, as outlined in this section.

We obtain that $\mathcal{B}_n^{\mathbb{C}}$ has face semilattice

$$F_{\mathcal{B}_n^{\mathbb{C}}} \cong \operatorname{Int}\big(F_{\mathcal{B}_n^{\mathbb{R}}}\big) \cong \operatorname{Int}(\Pi_n^{\text{ord}})$$

and augmented intersection lattice

$$L_{\mathcal{B}_n^{\mathbb{C}},\mathrm{aug}} \cong \mathrm{Int}\left(L_{\mathcal{B}_n^{\mathbb{R}}}\right) \cong \mathrm{Int}\,(\Pi_n).$$

Thus, the complex sign vectors of $\mathcal{B}_n^{\mathbb{C}}$ are encoded into pairs $[Y, X]$ of ordered partitions, where X is an refinement of Y. The composition (3.8) is illustrated in this computation table:

(3.11)

$\circ$	1	3	5	4	7	6	2
3, 7		3			7		
1	1						
2, 5, 6			5			6	2
4				4			

from which we read that

$$\langle 37 \mid 1 \mid\mid 256 \mid 4\rangle \circ \langle 1 \mid 3 \mid 5 \mid\mid 4 \mid 7 \mid\mid 6 \mid 2\rangle = \langle 3 \mid 1 \mid\mid 7 \mid\mid 5 \mid\mid 4 \mid\mid 6 \mid 2\rangle$$

Here single bars denote the separation of the ground set $[7]$ into ordered blocks according to X, and double bars the coarser partition Y. The rule is to read off the coarser partition of the composition by ordering the double bar boxes lexicographically, and then read off the refinement by ordering the single bar boxes within each double bar box lexicographically (empty boxes are skipped).

Notice that the cells in the complement of the complex braid arrangement, cf. Theorem 3.5 (3), correspond to block-divided permutations:

$$\begin{aligned} C_{\mathcal{B}_n^{\mathbb{C}}} &\leftrightarrow \text{sign vectors } X \circ iY \text{ without zero coordinates} \\ &\leftrightarrow \text{intervals } [Y, X],\ X \text{ maximal} \\ &\leftrightarrow \text{permutations } X \text{ divided into ordered blocks } Y \end{aligned}$$

4. Random Walks

This chapter begins with a summary of Brown's theory for random walks on a class of semigroups [13]. The motivating example, namely walks on real hyperplane arrangements, is then recalled. After that comes a sequence of applications.

4.1. Walks on semigroups

A *semigroup* is a set with an associative composition. We also assume the existence of an identity element, denoted "e", and we write the composition in multiplicative notation.

Definition 4.1. An *LRB semigroup* is a finite semigroup Σ with identity satisfying

(1) $x^2 = x$ for all $x \in \Sigma$,

(2) $xyx = xy$ for all $x, y \in \Sigma$.

A *left ideal* of Σ is a subset $I \subseteq \Sigma$ such that $x \in \Sigma$, $y \in I \ \Rightarrow \ xy \in I$.

The acronym LRB stands for "Left-Regular Band", a name by which this class of semigroups is sometimes known in the literature. Brown [13] defined a class of random walks on semigroups of this type. This section summarizes some material from [13], to where we refer for more information, background and references.

Definition 4.2. Let I be a left ideal of Σ, and let w be a probability distribution on Σ. A random walk on I is defined in the following way. If the current position of the walk is at an element $y \in I$, then choose $x \in \Sigma$ according to the distribution w and move to xy.

Brown's main theorem gives surprisingly exact information about such random walks. In order to be able to state it we need to first introduce two related poset structures.

Let Σ be an LRB semigroup. We define a relation "$\leq$" on Σ by

$$x \leq y \quad \Leftrightarrow \quad xy = y \tag{4.1}$$

This turns out to be a partial order relation, so we may think of an LRB semigroup also as a poset. The identity element e is the unique minimal element. The set $\max(\Sigma)$ of maximal elements is a left ideal in Σ.

There is also another partial order significantly related to Σ.

Proposition 4.3 [13]. *Let Σ be an LRB semigroup. Then there exists a unique finite lattice Λ and an order-preserving and surjective map*

$$\text{supp}\colon \Sigma \to \Lambda \tag{4.2}$$

such that for all $x, y \in \Sigma$:

(1) $\operatorname{supp}(xy) = \operatorname{supp}(x) \vee \operatorname{supp}(y)$

(2) $\operatorname{supp}(x) \leq \operatorname{supp}(y) \quad \Leftrightarrow \quad yx = y$

We call Λ the *support lattice* and supp the *support map.* Observe that

$$\operatorname{supp}^{-1}(\widehat{0}) = \{e\} \quad \text{and} \quad \operatorname{supp}^{-1}(\widehat{1}\,) = \max(\Sigma),$$

where $\widehat{0}$ and $\widehat{1}$ denote the bottom and top elements of Λ. In fact, the following conditions on an element $c \in \Sigma$ are equivalent:

(1) $\operatorname{supp}(c) = \widehat{1}$,

(2) $c \in \max(\Sigma)$,

(3) $cx = c$, for all $x \in \Sigma$.

Here is the main result on the random walks of Definition 4.2.

Theorem 4.4 (Brown [13]). *Let Σ be an LRB semigroup and Λ its support lattice. Furthermore, let $\{w_x\}$ be a probability distribution on Σ and P_w the transition matrix of the induced random walk on the ideal* $\max(\Sigma)$*:*

$$P_w(c,d) = \sum_{x\,:\; xc=d} w_x$$

for $c, d \in \max(\Sigma)$. Then,

(1) *The matrix P_w is diagonalizable.*

(2) *For each $X \in \Lambda$ there is an eigenvalue $\varepsilon_X = \sum_{y\,:\;\operatorname{supp}(y) \leq X} w_y$.*

(3) *The multiplicity of the eigenvalue ε_X is $m_X = \sum_{Y\,:\; Y \geq X} \mu_\Lambda(X,Y) c_Y$, where $c_Y \stackrel{\text{def}}{=} \bigl|\max(\Sigma_{\geq y})\bigr|$, for any $y \in \operatorname{supp}^{-1}(Y)$.*

(4) *These are all the eigenvalues of P.*

(5) *Suppose that Σ is generated by $\{x \in \Sigma\colon w_x > 0\}$. Then the random walk on* $\max(\Sigma)$ *has a unique stationary distribution π.*

By Möbius inversion the multiplicities can be determined also from the relations

$$c_X = \sum_{Y\,:\; Y \geq X} m_Y. \tag{4.3}$$

Theorem 4.4 is a generalization from the special case of face semigroups of real hyperplane arangements, to be briefly reviewed in the following section. In that case the theorem emanates from the work of Bidigare, Hanlon and Rockmore [2] and was expanded by Brown and Diaconis [15]. The generalization to LRB semigroups was given by Brown [13, 14].

The cited papers also contain information about the rate of convergence to stationarity and various descriptions of the stationary distribution, e.g. via sampling techniques, see [2, 13, 14, 15, 16] for such information.

The following proposition describes two ways in which smaller LRB semigroups are induced.

Proposition 4.5 [13]. *Let Σ be an LRB semigroup with support lattice Λ. Suppose that $x \in \Sigma$ and $X \in \Lambda$. Then*

(1) $\Sigma_{\geq x} \overset{\text{def}}{=} \{y \in \Sigma \colon y \geq x\}$ *is an LRB semigroup whose support lattice is the interval* $\big[\operatorname{supp}(x), \widehat{1}\big]$ *in* Λ.

(2) *If* $\operatorname{supp}(x) = \operatorname{supp}(y)$ *then* $\Sigma_{\geq x} \cong \Sigma_{\geq y}$.

(3) $\operatorname{Fib}_\Lambda(X) \overset{\text{def}}{=} \big\{y \in \Sigma \colon \operatorname{supp}(y) \leq X\big\}$ *is an LRB semigroup (we call it the* fiber semigroup *at X), whose support lattice is the interval* $[\widehat{0}, X]$ *in* Λ.

4.2. Walks on $\mathbb{R}$-arrangements

Let $\mathcal{A}$ be an essential hyperplane arrangement in $\mathbb{R}^d$ with face semilattice $F_{\mathcal{A}}$ and intersection lattice $L_{\mathcal{A}}$. The following is easily seen from observations (2.1)–(2.5).

Proposition 4.6. *$(F_{\mathcal{A}}, \circ)$ is an LRB semigroup with support lattice $L_{\mathcal{A}}$ and support map* span.

Let $C_{\mathcal{A}}$ be the set of regions in the complement of $\mathcal{A}$. There is a one-to-one correspondence $C_{\mathcal{A}} \leftrightarrow \max(F_{\mathcal{A}})$. Thus the general theory produces a class of random walks on $C_{\mathcal{A}}$ to which Theorem 4.4 is applicable. The description of this case is as follows.

> *Random walk on $C_{\mathcal{A}}$: Fix a probability distribution w on $F_{\mathcal{A}}$. If the walk is currently in region $C \in C_{\mathcal{A}}$, then choose a face $X \in F_{\mathcal{A}}$ according to w and move to the region $X \circ C$.*

Let P_w be the transition matrix

$$P_w(C, D) = \sum_{F:\ F \circ C = D} w_F$$

Theorem 4.4 specializes to the following, where part (3) relies on Zaslavsky's formula (Theorem 2.1) together with relation (4.3).

Theorem 4.7 (Bidigare–Hanlon–Rockmore [2], Brown-Diaconis [15]).

(1) *P_w is diagonalizable.*

(2) *For each $X \in L_{\mathcal{A}}$ there is an eigenvalue $\varepsilon_X = \sum_{F:\ \mathrm{span}\,(F) \subseteq X} w_F$.*

(3) *The multiplicity of ε_X is $\left| \mu_{L_{\mathcal{A}}}(X, \widehat{1}\,) \right|$.*

(4) *These are all the eigenvalues.*

(5) *Assume that the probability mass w is not concentrated on any single hyperplane H_i. Then there is a unique stationary distribution π.*

Remark 4.8. The following interesting result appears in [3]. Let w be the uniform distribution on the set of vertices (minimal elements of $F_{\mathcal{A}} \setminus \{0\}$) of an arrangement in $\mathbb{R}^3$. Then the probability (according to π) of being in a region with k sides is proportional to $k - 2$. It is an open problem to give any such geometric characterization of the stationary distribution for arrangements in $\mathbb{R}^d$, $d \geq 4$.

4.3. Walks on $\mathbb{C}$-arrangements

Let $\mathcal{A}$ be an essential hyperplane arrangement in $\mathbb{C}^d$ with face semilattice $F_{\mathcal{A}}$ and intersection lattices $L_{\mathcal{A}}$ and $L_{\mathcal{A},\mathrm{aug}}$. The following strengthening of observation (3.2) is immediate.

Proposition 4.9. *$(F_{\mathcal{A}}, \circ)$ is an LRB semigroup with support lattice $L_{\mathcal{A},\mathrm{aug}}$ and support map* span.

Applying the general theory directly to $F_{\mathcal{A}}$ and the ideal $\max\,(F_{\mathcal{A}})$ we get a walk on the maximal complex sign vectors which is a direct analogue of the real walks in Section 4.2.

Theorem 4.10. *The statements of Theorem 4.7 are valid for the complex walks, with the following replacements for items* (2) *and* (3) :

(2) *For each* $X \in L_{\mathcal{A},\mathrm{aug}}$ *such that* $\mu_{L_{\mathcal{A},\mathrm{aug}}}(X, \widehat{1}\,) \neq 0$ *there is an eigenvalue* $\varepsilon_X = \sum_{F:\ \mathrm{span}\,(F) \subseteq X} w_F$.

(3) *The multiplicity of* ε_X *is* $\left|\mu_{L_{\mathcal{A},\mathrm{aug}}}(X, \widehat{1}\,)\right|$.

The proof of part (3) relies here on the generalized Zaslavsky formula (Theorem 3.6) together with relation (4.3). Note that in the formulation of Theorem 4.7 we need not demand that $\mu(X, \widehat{1}\,) \neq 0$, since that is automatically true for geometric lattices. However, in Theorem 4.10 all we know is that the lattice is lower semimodular, which implies that the Möbius function alternates in sign but not that it is nonzero.

Specializing in various directions there are several semigroup-induced random walks coming out of this situation. We describe two of them.

Case 1. Suppose that the probability mass w is concentrated on the real sign vectors and let $Z = X \circ iY \in F_{\mathcal{A}}$, for real sign vectors X and Y. Choose $W \in F_{\mathcal{A}} \cap \{0, +, -\}^t$ according to w and move to $W \circ Z = (W \circ X) \circ iY$. Then Z and $W \circ Z$ have the same imaginary part iY. It can be checked that the subset of $F_{\mathcal{A}}$ consisting of sign vectors with fixed imaginary part iY is an LRB semigroup. Note that it doesn't come from a filter of a fiber, as in Proposition 4.5.

For complexified real arrangements, where sign vectors correspond to intervals, we have in this case that

$$[0, X] \circ [R, S] = [0 \circ R, 0 \circ R \circ X \circ S] = [R, R \circ X \circ S]$$

So, probability mass concentrated on elements $[0, X]$ (real sign vectors) gives a random walk on the set of intervals $[R, S]$, S maximal, for any fixed element R.

Case 2. Let $\mathcal{A}^{\mathbb{C}}$ be the complexification of a real arrangement $\mathcal{A}^{\mathbb{R}}$. We have that $L_{\mathcal{A}^{\mathbb{C}},\mathrm{aug}} \cong \mathrm{Int}\left(L_{\mathcal{A}^{\mathbb{R}}}\right)$. The purpose here is to determine the transition matrix eigenvalues for the fiber semigroup $\mathrm{Fib}\,(X) = \left\{y \in F_{\mathcal{A}^{\mathbb{C}}}\colon \mathrm{supp}\,(y) \leq X\right\}$, for $X = [\pi, \widehat{1}\,] \in \mathrm{Int}\left(L_{\mathcal{A}^{\mathbb{R}}}\right)$. The support lattice of $\mathrm{Fib}\,(X)$ is the interval $[\widehat{0}, X]$ in $L_{\mathcal{A}^{\mathbb{C}},\mathrm{aug}}$, cf. Proposition 4.5.

Theorem 4.4 shows that the eigenvalues are indexed by intervals $[\alpha, \beta] \in \left[[\widehat{0}, \widehat{0}\,], [\pi, \widehat{1}\,]\right]$, i.e., intervals $[\alpha, \beta]$ such that $\alpha \leq \pi$. Furthermore, the multiplicity of such an eigenvalue is, according to Theorems 4.10, 5.1 and 5.2, the absolute value of

$$\mu_{\mathrm{Int}\,(L)}\big([\alpha,\beta],[\pi,\widehat{1}\,]\big) = \begin{cases} \mu_L(\alpha,\pi)\,\mu_L(\beta,\widehat{1}\,), & \text{if } \pi \le \beta \\ 0, & \text{otherwise.} \end{cases}$$

Thus, eigenvalues of positive multiplicity occur only when $\alpha \le \pi \le \beta$, and we have proved the following.

Theorem 4.11. *The statements of Theorem* 4.4 *are valid for the complex hyperplane walks induced on fibers* $\mathrm{Fib}\,(X)$, *as explained, with the following replacements for items* (2) *and* (3) :

(2) *For each* $(\alpha,\beta) \in [\widehat{0},\pi] \times [\pi,\widehat{1}\,]$ *there is an eigenvalue* $\varepsilon_{(\alpha,\beta)}$.

(3) *The multiplicity of* $\varepsilon_{(\alpha,\beta)}$ *is* $\big|\mu_L(\alpha,\pi)\,\mu_L(\beta,\widehat{1}\,)\big|$.

The exact value of $\varepsilon_{(\alpha,\beta)}$ can of course be stated as a special case of Theorem 4.4, but we leave this aside.

4.4. Walks on libraries

This section concerns the walks produced by the braid arrangements, both real and complex. By translating from permutation and partition structures we can interpret the states of such walks as distributions of books on shelves. This library terminology also provides a convenient image for picturing and explaining these walks.

Real case. Here one obtains random walks on permutations governed by probability distributions w on ordered partitions. This case is thoroughly discussed and exemplified in the literature, see [2, 3, 13, 15, 16]. We mention just two examples.

First, suppose that the probability mass is concentrated on the two-block ordered partitions whose first block is a singleton. That is,

$$\text{probability} = \begin{cases} w_i, & \text{for the partition } \{i\} \mid [n] \setminus \{i\} \\ 0, & \text{for all other ordered partitions.} \end{cases}$$

Then the random walk is precisely the Tsetlin library, for which book i is chosen with probability w_i and moved to the beginning of the shelf.

Second, more generally allow non-zero probability for all two-block ordered partitions:

$$\text{probability} \;=\; \begin{cases} w_E, & \text{for the partition } E \mid [n] \setminus E \\ 0, & \text{for all other ordered partitions.} \end{cases}$$

Then the steps of the random walk consist of removing the books belonging to the subset E with probability w_E and then replacing them in the induced order at the beginning of the shelf.

In the general case, when non-zero probability is allowed for arbitrary ordered partitions, we obtain the one-shelf dynamic library with several borrowers described in the Introduction.

Complex case. Let us now see what happens in the case of the complex braid arrangement. We work out the case of a particular fiber LRB semigroup, namely the one determined by choosing $X = [\pi, \widehat{1}]$, where π is a partition $(B_1, \ldots, B_k) \in \Pi_n$ and $\widehat{1}$, is the partition into singletons.

In our library there are n books labeled by the integers 1 through n, and k shelves labeled by the integers 1 through k. Think of π as a division of the books into k groups corresponding to the blocks B_i. For instance, B_1 could be the set of books on combinatorics, B_2 the set of algebra books, and so on. We are going to consider placements of these n books on the k shelves so that the books in any particular class B_i stand (in some order) on some particular shelf dedicated to that class.

The inverse image $\operatorname{supp}^{-1}(X)$ consists of pairs $[p, s]$, where p is an ordered partition of the given blocks, $p = \langle B_{p_1}, \ldots, B_{p_k} \rangle$, and s is a permutation of $[n]$ refining p. We interpret such an element $[p, s]$ as a particular placement of the books: the books in B_{p_1} stand on the top shelf in the order assigned by s, then the books in B_{p_2} stand on the next shelf in the order assigned by s, and so on.

The fiber semigroup $\operatorname{Fib}(X) = \operatorname{supp}^{-1}(\Lambda_{\leq X})$ consists of pairs $[q, t]$, where q is an ordered partition such that $\operatorname{supp}(q)$ is a coarsening of the given partition $\pi = \{B_1, \ldots, B_k\}$, and t is an ordered partition refining q.

A step in the Markov chain is of the form $[p, s] \mapsto [q, t] \circ [p, s] = [q \circ p, q \circ p \circ t \circ s]$. What is its combinatorial meaning? Well, $q \circ p$ is an ordered partition with blocks $B_1, \ldots, B_k$, and $q \circ p \circ t \circ s$ is a permutation refining $q \circ p$. Hence, the combinatorial meaning of such a step in the Markov chain is that we permute the shelf assignments for the blocks B_i according to $q \circ p$, and then permute the books on each shelf as induced by the permutation $q \circ p \circ t \circ s$.

Here is a concrete example. Say we have 14 books of 4 types, namely the algebra books $B_{\mathrm{alg}} = \{1, 4, 5, 7\}$, the combinatorics books $B_{\mathrm{comb}} = \{2, 8, 11, 12, 14\}$, the geometry books $B_{\mathrm{geom}} = \{6, 13\}$, and the topology books $B_{\mathrm{top}} = \{3, 9, 10\}$. Furthermore, say that the present state of the Markov chain is this library configuration:

(4.4)

11	14	2	12	8
6	13			
4	7	5	1	
10	9	3		

So, in particular, we have the combinatorics books on the top shelf, the geometry books on the next shelf, and so on.

Now, let

$$q = \langle B_{\mathrm{alg}} \mid B_{\mathrm{comb}} \cup B_{\mathrm{top}} \mid B_{\mathrm{geom}} \rangle$$

and

$$t = \langle 4, 5 \mid 1, 7 \mid 8, 9, 12 \mid 14 \mid 2, 3, 10, 11 \mid 6, 13 \rangle$$

Then, $[q, t]$ acting on the state (4.4) leads to the following configuration

(4.5)

4	5	7	1	
12	8	14	11	2
9	10	3		
6	13			

From now on we specialize the discussion to what seems like a "realistic" special case, in which the Markov chain is driven by choices of subsets $E \subseteq [n]$ of the books. This walk has the following description in words.

Library walk: A borrower enters the library and borrows a subset $E \subseteq [n]$ of the books with probability w_E. These books may come from several shelves. When returned the books are put back in the following way. Permute the shelves so that the ones that contained one of the borrowed books become the top ones, maintaining the induced order among them and among the remaining shelves, which are now at the bottom. Then, on each shelf place the books belonging to E at the beginning of the *shelf, in the induced order, followed by the remaining books in their induced order.*

For example, if this procedure is carried out on the library configuration (4.4) for the choice $E = \{1, 2, 3, 4\}$ we obtain the new configuration (4.6).

(4.6)

2	11	14	12	8
4	1	7	5	
3	10	9		
6	13			

In mathematical language, the following is going on. For the subset $E \subseteq [n]$ let $K_E \stackrel{\text{def}}{=} \cup_{i\colon B_i \cap E \neq \emptyset} B_i$ and

$$q_E \stackrel{\text{def}}{=} \big\langle K_E \mid [n] \setminus K_E \big\rangle \quad \text{and} \quad t_E \stackrel{\text{def}}{=} \big\langle E \mid K_E \setminus E \mid [n] \setminus K_E \big\rangle.$$

The mathematical description of the library walk is that we assign the following distribution

$$\text{probability} = \begin{cases} w_E, & \text{for the partition interval } [q_E, t_E], \text{ all } E \subseteq [n] \\ 0, & \text{for all other intervals of ordered partitions} \end{cases}$$

to the elements of the fiber semigroup $\mathrm{Fib}\big([\pi, \widehat{1}\,]\big)$, and then we refer to Theorem 4.11 for the consequences.

To exemplify how the interval $[q_E, t_E]$ acts on a library configuration we return once more to the configuration (4.4). Suppose that $E = \{1, 2, 3, 4\}$ and let the interval $[q_E, t_E]$ act on (4.4). This leads to the library configuration (4.6), as shown by the following computation table

$\circ$	11	14	2	12	8	6	13	4	7	5	1	10	9	3
$1, 2, 3, 4$			2					4			1			3
$5, 7, 8, 9, 10, 11, 12, 14$	11	14		12	8				7	5		10	9	
$6, 13$						6	13							

Summing up the discussion we obtain the following result.

Theorem 4.12. *The statements of Theorem 4.11 are valid for the library walk, with the following replacements for parts* (2) *and* (3) :

(2) *For each pair of unordered partitions* (α, β) *such that* $\alpha \leq \pi \leq \beta$ *(i.e.,* β *refines* π *and* π *refines* α*) there is an eigenvalue* $\varepsilon_{(\alpha,\beta)}$. *Furthermore,*

$$\varepsilon_{(\alpha,\beta)} = \sum w_E,$$

the sum extending over all $E \subseteq [n]$ *such that* E *is a union of blocks from* β *and the totality of books standing on the shelves containing some element of* E *is a union of blocks from* α.

(3) *The multiplicity of* $\varepsilon_{(\alpha,\beta)}$ *is* $\prod(p_i - 1)! \prod(q_j - 1)!$, *where* $(p_1, p_2, \ldots)$ *are the block sizes of* β *and* $(q_1, q_2, \ldots)$ *the block sizes of* α *modulo* π.

Here part (3) uses the well-known formula for the Möbius function of the partition lattice Π_n in terms of factorials, see e.g. [20, p. 128]

Example 4.13. We exemplify the preceding with a worked-out example. Let $n = 3$ and $\pi = (1, 2 \mid 3)$. Then there are four library configurations indexing the rows and columns of the transition matrix P_w:

	1 2 3	2 1 3	3 1 2	3 2 1
1 2 3	$w_1 + w_{1,2} + w_{1,3}$	$w_1 + w_{1,3}$	$w_1 + w_{1,2}$	w_1
2 1 3	$w_2 + w_{2,3}$	$w_2 + w_{1,2} + w_{2,3}$	w_2	$w_2 + w_{1,2}$
3 1 2	w_3	0	$w_3 + w_{1,3}$	$w_{1,3}$
3 2 1	0	w_3	$w_{2,3}$	$w_3 + w_{2,3}$

We ignore the trivial choices $E = \emptyset$ and $E = \{1, 2, 3\}$, so six elementary probabilities w_E are assigned. For instance, the entry $w_2 + w_{1,2}$ records that if books E are removed from the library configuration $\begin{array}{cc} 3 & \\ \hline 2 & 1 \\ \hline \end{array}$ and replaced

according to the rules, then configuration $\begin{array}{cc} 2 & 1 \\ \hline 3 & \\ \hline \end{array}$ is obtained precisely if

$E = \{2\}$ or $E = \{1, 2\}$.

We have that $\widehat{0} \lessdot \pi \lessdot \widehat{1}$ ($\cdot\lessdot\cdot$ indicates coverings), so according to Theorem 4.12 there are four pairs (α, β) indexing the eigenvalues, all of which have multiplicity one, and these eigenvalues are $\begin{cases} \varepsilon_{(\widehat{0},\pi)} = 0 \\ \varepsilon_{(\widehat{0},\widehat{1})} = w_{1,3} + w_{2,3} \\ \varepsilon_{(\pi,\pi)} = w_3 + w_{1,2} \\ \varepsilon_{(\pi,\widehat{1})} = 1 \end{cases}$

It is instructive to also check how the elementary probabilities w_E contribute to the various eigenvalues $\varepsilon_{(\alpha,\beta)}$ in terms of the associated intervals:

E	$[q_E, t_E]$	contributes to $\varepsilon_{(\alpha,\beta)}$
1	$\big[\langle 12 \mid 3\rangle, \langle 1 \mid 2 \mid 3\rangle\big]$	$[\alpha,\beta] = [\pi, \widehat{1}]$
2	$\big[\langle 12 \mid 3\rangle, \langle 2 \mid 1 \mid 3\rangle\big]$	$[\alpha,\beta] = [\pi, \widehat{1}]$
3	$\big[\langle 3 \mid 12\rangle, \langle 3 \mid 12\rangle\big]$	$[\alpha,\beta] = [\pi, \widehat{1}]$ or $[\pi, \pi]$
$1, 2$	$\big[\langle 12 \mid 3\rangle, \langle 12 \mid 3\rangle\big]$	$[\alpha,\beta] = [\pi, \widehat{1}]$ or $[\pi, \pi]$
$1, 3$	$\big[\langle 123\rangle, \langle 13 \mid 2\rangle\big]$	$[\alpha,\beta] = [\pi, \widehat{1}]$ or $[\widehat{0}, \widehat{1}]$
$2, 3$	$\big[\langle 123\rangle, \langle 23 \mid 1\rangle\big]$	$[\alpha,\beta] = [\pi, \widehat{1}]$ or $[\widehat{0}, \widehat{1}]$

4.5. Walks on greedoids

Denote by E^* the set of repetition-free words $\alpha = x_1 x_2 \ldots x_k$ in letters $x_i \in E$. A *greedoid* is a language $\mathcal{L} \subseteq E^*$ such that

(G1) if $\alpha\beta \in \mathcal{L}$ then $\alpha \in \mathcal{L}$, for all $\alpha, \beta \in E^*$,

(G2) if $\alpha, \beta \in \mathcal{L}$ and $|\alpha| > |\beta|$, then α contains a letter x such that $\beta x \in \mathcal{L}$.

The words in $\mathcal{L}$ are called *feasible* and the longest feasible words are called *basic.* All basic words have the same length, and $\mathcal{L}$ is determined by the basic words as the collection of all their prefixes.

Greedoids were introduced in the early 1980s by Korte and Lovász, see the accounts in [12] and [18]. The concept can equivalently be formulated in terms of set systems, but only the (ordered) language version will concern us here.

Important examples of greedoids are provided by matroids (abstraction of linear hull) and antimatroids (abstraction of convex hull). Other examples come from branchings in rooted directed graphs and various optimization procedures (involving some versions of "the greedy algorithm").

If $\alpha, \beta \in \mathcal{L}$ and $|\alpha| > |\beta|$, then repeated use of the exchange property (G2) shows that β can be augmented to a feasible word $\beta x_1 x_2 \dots x_j$ with $j = |\alpha| - |\beta|$ letters x_i drawn from α. But the letters x_i might not occur in $\beta x_1 x_2 \dots x_j$ in the "right" order, i.e., in the order induced by their placement in α. This motivates defining an important subclass of greedoids.

Definition 4.14. An *interval greedoid* is a language $\mathcal{L} \subseteq E^*$ satisfying (G1) and the following strong exchange property

(G3) if $\alpha, \beta \in \mathcal{L}$ and $|\alpha| > |\beta|$, then α contains a subword γ of length $|\gamma| = |\alpha| - |\beta|$ such that $\beta\gamma \in \mathcal{L}$.

By *subword* we mean what can be obtained by erasing some letters and then closing the gaps. Matroids, antimatroids and branchings are examples of interval greedoids.

Let $\mathcal{L}$ be a greedoid on the finite alphabet E. We define an equivalence relation on $\mathcal{L}$ by

$$\alpha \sim \beta \quad \Leftrightarrow \quad \{\gamma \in E^* \colon \alpha\gamma \in \mathcal{L}\} = \{\gamma \in E^* \colon \beta\gamma \in \mathcal{L}\}. \tag{4.7}$$

So, α and β are equivalent if and only if they have the same set of feasible continuations. The equivalence classes $[\alpha] \in \mathcal{L}/\sim$ are the *flats* of the greedoid, and the *poset of flats*

$$\Phi \stackrel{\text{def}}{=} (\mathcal{L}/\sim, \leq)$$

consists of these classes ordered by

$$[\alpha] \leq [\beta] \quad \Leftrightarrow \quad \alpha\gamma \sim \beta, \text{ for some } \gamma \in E^*.$$

For instance, the poset of flats of a matroid defined in this way is easily seen to be isomorphic to the usual geometric "lattice of flats" of matroid theory.

The feasible words of a greedoid can be composed in the following manner. If $x_1 x_2 \dots x_j \in \mathcal{L}$ and $y_1 y_2 \dots y_k \in \mathcal{L}$ then

$$x_1 x_2 \dots x_j \circ y_1 y_2 \dots y_k \stackrel{\text{def}}{=} x_1 x_2 \dots x_j y_{i_1} y_{i_2} \dots y_{i_e} \tag{4.8}$$

where $i_1 < i_2 < \ldots < i_e$ is the lexicographically first non-extendable increasing sequence such that $x_1x_2\ldots x_jy_{i_1}y_{i_2}\cdots y_{i_e} \in \mathcal{L}$. Letting $\alpha = x_1x_2\ldots x_j$ it is equivalent to say that $\alpha \circ y_1y_2\ldots y_k = \alpha y_{i_1}y_{i_2}\cdots y_{i_e}$ is the word obtained, starting from α, by processing the letters y_i of $y_1y_2\ldots y_k$ from left to right and adding at the end of the word being formed only those letters y_i whose inclusion at that stage preserves feasibility.

For instance, consider the greedoid on $E = \{x, y, z, w\}$ whose 14 basic words are the words in E^* of length 3 that do not begin with a permutation of $\{x, y, z\}$ or $\{z, w\}$. This greedoid is discussed on pp. 290–291 of [12]. Here are two sample computations:

$$x \circ yzw = xyw \quad \text{and} \quad (x \circ z) \circ w = xzw \neq xz = x \circ (z \circ w)$$

This example shows that the composition (4.8) is not associative, and hence does not in general produce a semigroup. For this reason we must limit the discussion to a smaller class of greedoids.

Theorem 4.15. *Let $\mathcal{L}$ be an interval greedoid. Then $\mathcal{L}$ with the composition* (4.8) *is an LRB semigroup. Its support lattice is the lattice of flats Φ, and its support map $\mathcal{L} \to \Phi$ sends a feasible word α to its class $[\alpha]$.*

That matroids give rise to LRB semigroups in this way was mentioned by Brown [13, p. 891]. In the matroid case the result is quite obvious, whereas for the general case some details turn out to be a little more tricky. The proof is deferred to Appendix 5.3.

Being an LRB semigroup means that Brown's theory of random walks, summarized in Section 4.1, is applicable. What can be said about the eigenvalue distribution when specialized to greedoid walks?

There is an eigenvalue ε_X for each $X \in \Phi$ whose value and multiplicity m_X are determined according to parts (2) and (3) of Theorem 4.4. However, as Example 4.16 shows, for greedoids the multiplicities do not depend only on the structure of the interval $[X, \widehat{1}]$ in Φ, as was the case in the corresponding situation for real and complex hyperplane walks.

We now illustrate greedoid walks for the important case of branchings. Let G be a directed rooted graph with node set $\{r, 1, 2, \ldots, n\}$ and edge set E. A *branching* is a tree directed away from the root node r. A subset $R \subseteq \{1, 2, \ldots, n\}$ is *reachable* if it is the set of nodes of some branching.

The *branching greedoid* $\mathcal{L}_G$ consists of ordered strings of edges such that each initial segment is a branching. It models common search procedures on graphs. See [12] and [18] for more information.

The poset of flats of $\mathcal{L}_G$ is the lattice Φ_G of reachable sets ordered by inclusion. This is, in fact, a join-distributive lattice, see the cited references. The support map sends a branching to the reachable set of its nodes.

According to Theorem 4.4 there is an eigenvalue ε_X associated with every reachable node set X. Its value is the sum of the probabilities for the branchings that reach a subset of X, and its multiplicity is given by

$$m_X = \sum_{Y:\, Y \geq X} \mu(X,Y) c_Y$$

Here c_X is the number of ordered edge sequences feasibly extending (any branching reaching) X to a maximal branching.

Since Φ_G is join-distributive its Möbius function takes the simple form

$$\mu(X,Y) = \begin{cases} (-1)^{|Y|-|X|}, & \text{if the interval is Boolean,} \\ 0, & \text{otherwise.} \end{cases}$$

For each reachable set X, let $\operatorname{dom}(X)$ denote the superset of all nodes that are either in X or else can be reached from $X \cup \{r\}$ along a single edge of G. It is clear that every set of nodes contained between X and $\operatorname{dom}(X)$ is reachable, and that the *domination set* $\operatorname{dom}(X)$ is maximal with this property. Hence, we get the following simplified expression for the eigenvalue multiplicity at X:

$$m_X = \sum_{X \leq Y \leq \operatorname{dom}(X)} (-1)^{|Y|-|X|} c_Y \tag{4.9}$$

Example 4.16. The rooted directed graph in Figure 4 gives a branching greedoid of rank 3 with 9 basic words: *abc*, *abd*, *acb*, *ace*, *aec*, *aed*, *bac*, *bad*, *bda*. All subsets of $\{1, 2, 3\}$ except $\{2\}$ are reachable.

Assign probabilities w_α to the seven feasible words (ordered branchings) of rank one and two: *a*, *b*, *ab*, *ac*, *ae*, *ba*, *bd*. A step in the random walk on the nine ordered maximal branchings consists in choosing one of these words α according to the given probabilities w_α and then extending α to a maximal branching by adding edges in sequence from the currently visited maximal branching according to the composition rule (4.8).

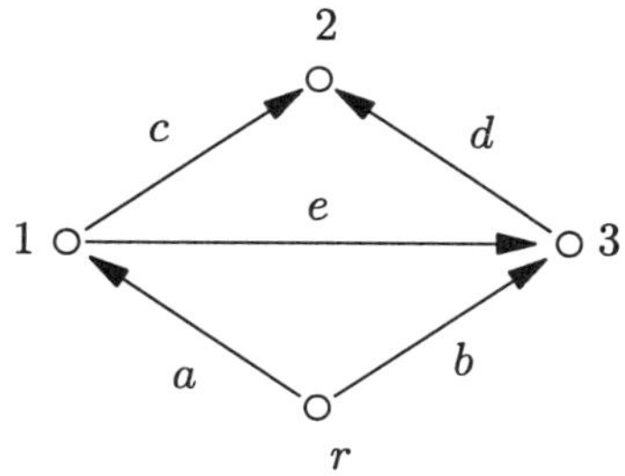

Fig. 4. Branching greedoid

Here are the eigenvalues for the walk on this branching greedoid:

X	c_X	dom (X)	m_X	ε_X
123	1	123	1	1
12	2	123	1	$w_a + w_{ac}$
13	2	123	1	$w_a + w_b + w_{ab} + w_{ae} + w_{ba}$
23	1	123	0	$w_b + w_{bd}$
1	6	123	3	w_a
3	3	123	1	w_b
$\emptyset$	9	13	2	0

Remark 4.17. By copying the procedure that leads from the sign vector system of a real hyperplane arrangement to that of its complexification (Section 3.3) we can formally introduce the complexification of any LRB semigroup. Namely, let Σ be an LRB semigroup with support lattice Λ. Define $\Sigma^{\mathbb{C}}$ to be the set of intervals $\{ [x,y] \colon x \leq y \text{ in } \Sigma \}$ with the composition

$$[x,y][z,w] \stackrel{\text{def}}{=} [xz, xzyw]$$

One readily verifies that this is an LRB semigroup and that its support lattice is $Int\,(\Lambda)$, with the partial order defined in Appendix 5.2.

This way one can complexify e.g. the greedoids walks.

5. Appendix

In this section we gather some proofs. Familiarity with the Möbius function is assumed, a good reference is [20].

5.1. A generalized Zaslavsky formula

A ranked poset R with $\widehat{0}$ and $\widehat{1}$ is said to be *Eulerian* if $\mu_R(x,y) = (-1)^{\mathrm{rk}\,(y)-\mathrm{rk}\,(x)}$ for all $x < y$ in R. Denote by $\max(P)$ the set of maximal elements of a poset P.

Theorem 5.1. *Suppose that $f\colon P \to Q$ satisfies the following conditions:*

(1) *the posets P and Q are ranked and of the same length r,*

(2) *Q has a unique maximal element $\widehat{1}_Q$,*

(3) *$\widehat{P} \stackrel{\mathrm{def}}{=} P \uplus \{\widehat{0}_P, \widehat{1}_P\}$ is Eulerian,*

(4) *f is an order-preserving, rank-preserving and surjective map,*

(5) *$\mu_P\big(f^{-1}(Q_{\le x})\big) = (-1)^{\mathrm{rk}\,(x)}$, for all $x \in Q$,*

(6) *$(-1)^{r-\mathrm{rk}\,(x)}\mu_Q(x,\widehat{1}_Q) \ge 0$, for all $x \in Q$.*

Then,

$$\big|\max(P)\big| = \sum_{x\in Q\uplus\widehat{0}} \big|\mu(x,\widehat{1}_Q)\big|$$

Proof. According to the "Möbius-theoretic Alexander duality" formula [20, p. 137] condition (3) implies that

$$\mu(R) = (-1)^{r-1}\,\mu(P \setminus R)$$

for all subsets $R \subseteq P$. In particular,

$$\big|\max(P)\big| = \mu\big(\max(P)\big) + 1 = (-1)^{r-1}\,\mu\big(P\setminus\max(P)\big) + 1. \tag{5.1}$$

On the other hand, according to the "Möbius-theoretic fiber formula" [21, p. 377] applied to the map $f\colon P\setminus\max(P) \to Q\setminus\widehat{1}$ we have that

$$\mu\big(P\setminus\max(P)\big) = \mu(Q\setminus\widehat{1}) - \sum_{x\in Q\setminus\widehat{1}} \mu\big(f^{-1}(Q_{\le x})\big)\,\mu(x,\widehat{1}_Q). \tag{5.2}$$

Thus,

$$\begin{aligned}\big|\max(P)\big| &= 1 + (-1)^{r-1}\Big[\mu(Q\setminus\widehat{1}) - \sum_{x\in Q\setminus\widehat{1}} \mu\big(f^{-1}(Q_{\le x})\big)\,\mu(x,\widehat{1}_Q)\Big] \\ &= (-1)^r \sum_{x\in Q\uplus\widehat{0}} (-1)^{\operatorname{rk}(x)}\mu(x,\widehat{1}_Q)) = \sum_{x\in Q\uplus\widehat{0}} \big|\mu(x,\widehat{1}_Q)\big|. \qquad\blacksquare\end{aligned}$$

Applying this result to the span map $F_{\mathcal{A}} \to L_{\mathcal{A}}$ of a real hyperplane arrangement $\mathcal{A}$ we obtain Zaslavsky's theorem 2.1. Applying it to the span map $F_{\mathcal{A}} \to L_{\mathcal{A},\mathrm{aug}}$ of a complex hyperplane arrangement $\mathcal{A}$ we obtain Theorem 3.6.

5.2. Lattice of intervals

Let L be a lattice and $\operatorname{Int}(L) \stackrel{\text{def}}{=} \big\{(x,y)\colon x\le y\big\}$ the set of its intervals partially ordered by

$$(x,y)\le(x',y') \quad \text{if and only if} \quad x\le x' \text{ and } y\le y'.$$

The poset $\operatorname{Int}(L)$ is itself a lattice with componentwise operations

$$(x,y)\vee(x',y') = (x\vee x', y\vee y') \quad\text{and}\quad (x,y)\wedge(x',y') = (x\wedge x', y\wedge y').$$

Its Möbius function is related to that of L in the following way.

Theorem 5.2.

$$\mu_{\operatorname{Int}(L)}\big((x,y),(x',y')\big) = \begin{cases} \mu_L(x,x')\,\mu_L(y,y'), & \textit{if } x'\le y \\ 0, & \textit{otherwise.}\end{cases}$$

Proof. If $x'\le y$ then $\big[(x,y),(x',y')\big] \cong [x,x']\times[y,y']$, so this case follows from the product property of the Möbius function.

Assume that $x'\not\le y$. We claim that the element $[x'\wedge y, x'\vee y]$ lacks a lattice-theoretic complement in the interval $\big[(x,y),(x',y')\big]$. For, say that $[s,t]$ is such a complement. This means that

$$\begin{aligned} s\vee(x'\wedge y) &= x' \qquad & t\vee(x'\vee y) &= y' \\ s\wedge(x'\wedge y) &= x & t\wedge(x'\vee y) &= y \end{aligned}$$

Then: $$s \le x' \wedge t \le (x' \vee y) \wedge t = y$$

$$\Rightarrow s \le x' \wedge y$$
$$\Rightarrow s = s \wedge (x' \wedge y) = x$$
$$\Rightarrow x' = x \vee (x' \wedge y) = x' \wedge y$$
$$\Rightarrow x' \le y,$$

contradicting the assumption. Thus, the interval $\big[(x,y),(x',y')\big]$ is not complemented, so by Crapo's complementation theorem [20, p. 160] its Möbius function is zero. ■

5.3. Interval greedoids

The lattice-theoretical structure of semimodularity is closely related to interval greedoids.

Theorem 5.3 [12, Thm. 8.8.7]. *The poset of flats Φ of an interval greedoid is a semimodular lattice. Conversely, every finite semimodular lattice arises from some interval greedoid in this way.*

This will be used in the proof of Theorem 4.15, to which we now turn. For economy of presentation we assume familiarity with the notation, conventions and results on pp. 332–334 of [12]. See particularly the proof of Theorem 8.2.5 on p. 334.

Proof. Let $\alpha = x_1 \dots x_j$ and $\beta = y_1 \dots y_k$ be feasible words of an interval greedoid $\mathcal{L}$. By letting $X_i = [x_1 \dots x_i]$ and $Y_i = [y_1 \dots y_i]$, these words correspond to edge-labeled unrefinable chains $\emptyset \lessdot X_1 \lessdot \cdots \lessdot X_j$ and $\emptyset \lessdot Y_1 \lessdot \cdots \lessdot Y_k$ in the semimodular lattice Φ. In the same manner (cf. Lemma 8.8.8 of [12]) the composition $x_1x_2 \dots x_j \circ y_1y_2 \dots y_k$ corresponds to the edge-labeled unrefinable chain

$$\emptyset \lessdot X_1 \lessdot \cdots \lessdot X_j \le X_j \vee Y_1 \le \cdots \le X_j \vee Y_k.$$

Here, due to semimodularity, the relation $X_j \vee Y_i \le X_j \vee Y_{i+1}$ is either a covering $X_j \vee Y_i \lessdot X_j \vee Y_{i+1}$ or an equality $X_j \vee Y_i = X_j \vee Y_{i+1}$, in which case we omit it from the chain. This shows that

$$[\alpha \circ \beta] = X_j \vee Y_k = [\alpha] \vee [\beta] \tag{5.3}$$

which in turn is used to see that

$$(5.4) \qquad [\beta] \leq [\alpha] \Leftrightarrow [\alpha] \vee [\beta] = [\alpha] \Leftrightarrow [\alpha \circ \beta] = [\alpha] \Leftrightarrow \alpha \circ \beta = \alpha$$

Thus, once associativity of the composition of feasible words has been established the proof will be complete. The other identities required of an LRB semigroup are trivially fulfilled, since feasible words lack repeated letters. Relations (5.3) and (5.4) then show, in view of Proposition 4.3, that Φ is indeed the support lattice of $\mathcal{L}$ as an LRB semigroup.

To deal with associativity, let γ be a third feasible word. We want to show that

$$(5.5) \qquad (\alpha \circ \beta) \circ \gamma = \alpha \circ (\beta \circ \gamma)$$

By definition

$$(\alpha \circ \beta) \circ \gamma = \alpha\beta'\gamma' \quad \text{and} \quad \alpha \circ (\beta \circ \gamma) = \alpha\beta'\gamma''$$

where β' is a subword of β and γ' and γ'' are subwords of γ. Thus it remains to convince ourselves that $\gamma' = \gamma''$. A crucial first step is to show that they are of equal length.

Let $\emptyset \lessdot Z_1 \lessdot \cdots \lessdot Z_l$ be the edge-labelled chain in Φ corresponding to γ. Then $(\alpha \circ \beta) \circ \gamma$ corresponds to the chain

$$\emptyset \lessdot X_1 \lessdot \cdots \lessdot X_j \leq X_j \vee Y_1 \leq \cdots \leq X_j \vee Y_k \leq (X_j \vee Y_k) \vee Z_1$$
$$\leq \cdots \leq (X_j \vee Y_k) \vee Z_l$$

and $\alpha \circ (\beta \circ \gamma)$ corresponds to

$$\emptyset \lessdot X_1 \lessdot \cdots \lessdot X_j \leq X_j \vee Y_1 \leq \cdots \leq X_j \vee Y_k \leq X_j \vee (Y_k \vee Z_1)$$
$$\leq \cdots \leq X_j \vee (Y_k \vee Z_l).$$

Due to associativity of the lattice join operation $\cdot \vee \cdot$ these chains are identical, and by construction the induced edge-labelings yield the words $\alpha\beta'\gamma'$ and $\alpha\beta'\gamma''$. Hence, being related to the same segment of the common chain, γ' and γ'' are of the same length.

We now prove (5.5) by induction on the length of the word γ. Suppose that $\gamma = t$ is a single letter. Then $\gamma' = \gamma''$ since the subwords of t of length 0 and 1 are unique. Hence,

$$(\alpha \circ \beta) \circ t = \alpha \circ (\beta \circ t)$$

Suppose now that $\gamma = \delta t$, meaning that the last letter of γ is t. Using the induction assumption and the length one case we obtain

$$\begin{aligned}(\alpha \circ \beta) \circ \gamma &= \big((\alpha \circ \beta) \circ \delta)\big) \circ t = \big(\alpha \circ (\beta \circ \delta)\big) \circ t \\ &= \alpha \circ \big((\beta \circ \delta) \circ t\big) = \alpha \circ \big(\beta \circ (\delta \circ t)\big) \\ &= \alpha \circ (\beta \circ \gamma). \qquad \blacksquare\end{aligned}$$

References

[1] H. Barcelo, and R. Laubenbacher, Perspectives on A-homotopy theory and its applications, *Discrete Math.,* **298** (2005), 36–91.

[2] P. Bidigare, P. Hanlon and D. Rockmore, A combinatorial description of the spectrum for the Tsetlin library and its generalization to hyperplane arrangements, *Duke Math J.,* **99** (1999), 135–174.

[3] L. J. Billera, K. S. Brown and P. Diaconis, Random walks and plane arrangements in three dimensions, *Amer. Math. Soc.,* **106** (1999), 502–524.

[4] A. Björner, Topological Methods, in: *Handbook of Combinatorics* (eds. R. Graham, M. Grötschel and L. Lovász), North-Holland, Amsterdam, 1995, pp. 1819-1872.

[5] A. Björner, letter to R. Laubenbacher, Aug. 7, 2001.

[6] A. Björner, B. Korte and L. Lovász, Homotopy properties of greedoids, *Advances in Appl. Math.,* **6** (1985), 447–494.

[7] A. Björner, M. Las Vergnas, B. Sturmfels, N. White and G. M. Ziegler, *Oriented Matroids,* Cambridge Univ. Press, 1993. (Second edition 1999)

[8] A. Björner and L. Lovász, Linear decision trees, subspace arrangements and Möbius functions, *J. Amer. Math. Soc.,* **7** (1994), 677–706.

[9] A. Björner, L. Lovász and A. Yao, Linear decision trees: volume estimates and topological bounds, in: *Proc. 24th ACM Symp. on Theory of Computing (May 1992),* ACM Press, N.Y. (1992), pp. 170–177.

[10] A. Björner and V. Welker, The homology of "k-equal" manifolds and related partition lattices, *Advances in Math.,* **110** (1995), 277–313.

[11] A. Björner and G. M. Ziegler, Combinatorial stratification of complex arrangements, *J. Amer. Math. Soc.,* **5** (1992), 105–149.

[12] A. Björner and G. M. Ziegler, Introduction to greedoids, in: *Matroid Applications* (ed. N. White), Cambridge Univ. Press (1992), pp. 284–357.

[13] K. S. Brown, Semigroups, rings and Markov chains, *J. Theoret. Probab.,* **13** (2000), 871–938.

[14] K. S. Brown, Semigroup and ring theoretical methods in probability, *Fields Inst. Commun.*, **40** (2004), 3–26.

[15] K. S. Brown and P. Diaconis, Random walks and hyperplane arrangements, *Ann. Probab.*, **26** (1998), 1813–1854.

[16] P. Diaconis, From shuffling cards to walking around the building: An introduction to modern Markov chain theory, in: *Proceedings of the International Congress of Mathematicians, Berlin, 1998,* Vol. 1, 187–204.

[17] M. Khovanov, Real $K(\pi, 1)$ arrangements from finite root systems, *Math. Res. Lett.*, **3** (1996), 261–274.

[18] B. Korte, L. Lovász and R. Schrader, *Greedoids,* Springer-Verlag, Berlin, 1991.

[19] P. Orlik and H. Terao, *Arrangements of Hyperplanes,* Springer, Berlin, 1992.

[20] R. P. Stanley, *Enumerative Combinatorics, Vol. 1,* Cambridge Univ. Press, 1997.

[21] J. W. Walker, Homotopy type and Euler characteristic of partially ordered sets, *Europ. J. Combinatorics,* **2** (1981), 373–384.

[22] T. Zaslavsky, Facing up to arrangements: face count formulas for partitions of space by hyperplanes, *Mem. Amer. Math, Soc.,* **1** (1976), 154.

Anders Björner

Royal Institute of Technology
Department of Mathematics
S-100 44 Stockholm
Sweden

e-mail: `bjorner@math.kth.se`

BOLYAI SOCIETY
MATHEMATICAL STUDIES, 19

Building Bridges
pp. 205–218.

The Finite Field Kakeya Problem

AART BLOKHUIS and FRANCESCO MAZZOCCA

A Besicovitch set in $AG(n,q)$ is a set of points containing a line in every direction. The Kakeya problem is to determine the minimal size of such a set. We solve the Kakeya problem in the plane, and substantially improve the known bounds for $n > 4$.

1. Introduction

We denote by π_q the projective plane $PG(2,q)$ over the Galois field $GF(q)$ with q elements, $q > 2$ a prime power.

Let ℓ be a line in π_q and, for every point P on ℓ, let ℓ_P be a line on P other than ℓ. The set

$$K = \Big(\bigcup_{P\in\ell} \ell_P\Big) \setminus \ell \tag{1}$$

is called a *Kakeya set,* or a *minimal Besicovitch set.* The *finite plane Kakeya problem* asks for the smallest size $k(q)$ of a Kakeya set; it is the two-dimensional version of the *finite field Kakeya problem* posed by *T. Wolff* in his influential paper [11] of 1996.

In the following, unless explicitly mentioned otherwise, we will use the same notation of (1) for the lines defining a Kakeya set K.

Let Ω be a set of $q+2$ points in π_q. A point $P \in \Omega$ is said to be an *internal nucleus* of Ω if every line through P meets Ω in exactly one other point. Internal nuclei of $(q+2)-$sets were first considered by *A. Bichara* and *G. Korchmáros* in [1]; here they proved the following result.

Proposition 1 (1982). *Let q be an odd prime-power. Every set of $q+2$ points in π_q has at most two internal nuclei.*

The $q+2$ lines defining a Kakeya set in π_q can be viewed as a set of $q+2$ points with an internal nucleus in the dual plane π_q^*. More precisely, if K is a Kakeya set in π_q, the lines ℓ and ℓ_P, $P \in \ell$, give rise in π_q^* to a set $\Omega(K)$ of $q+2$ points with ℓ as an internal nucleus. Vice versa, every set of $q+2$ points with an internal nucleus in π_q defines in an obvious way a Kakeya set in π_q^*. Thanks to this duality, the finite plane Kakeya problem is equivalent to ask for *the smallest number $k^*(q)$ of lines in π_q meeting a set of $q+2$ points with an internal nucleus;* to be precise, we have

$$k^*(q) = 1 + q + k(q).$$

2. Old and New Results in the Plane

Let us start by recalling that the first author and *A. A. Bruen* studied in [2] the smallest number of lines intersecting a set of $q+2$ points in π_q; here no assumption on the existence of internal nuclei is made. Nevertheless the dual of the theorem 1.3 of [2] contains the following result as a special case.

Proposition 2 (1989). *If $q \geq 7$ is odd, then*

$$|K| \geq \frac{q(q+1)}{2} + \frac{q+2}{3},$$

for every Kakeya set K.

Example 1. Assume q is even and consider in π_q a dual hyperoval $\mathcal{H}$, i.e. a $(q+2)$−set of lines, no three of which are concurrent. Fix a line $\ell \in \mathcal{H}$ and, for every point $P \in \ell$, let ℓ_P the line of $\mathcal{H}$ on P other than ℓ. Then the Kakeya set

$$K(\mathcal{H}, \ell) = \Big(\bigcup_{P\in\ell} \ell_P\Big)\Big\backslash \ell$$

is said to be associated to $\mathcal{H}$ and ℓ and it is of size

$$\big|K(\mathcal{H}, \ell)\big| = \frac{q(q+1)}{2}. \qquad \blacksquare$$

Example 2. Assume q is odd and consider in π_q a dual oval $\mathcal{O}$, i.e. a $(q+1)-$set of lines, no three concurrent. Let ℓ be a fixed line in $\mathcal{O}$. Every point P on ℓ, but one, belongs to a second line $\ell_P \in \mathcal{O}$ other than ℓ. If A is this remaining point on ℓ, let ℓ_A be a(ny) line through it different from ℓ. Then the Kakeya set

$$K(\mathcal{O},\ell,\ell_A) = \Big(\bigcup_{P\in\ell} \ell_P\Big)\Big\backslash \ell$$

is said to be associated to $\mathcal{H}$, ℓ and ℓ_A; moreover it is of size

$$\big|K(\mathcal{O},\ell,\ell_A)\big| = \frac{q(q+1)}{2} + \frac{q-1}{2}. \qquad \blacksquare$$

For any point A of a Kakeya set K, we denote by m_A the number of lines ℓ_P, $P \in \ell$, on A and we set

$$\sigma(K) = \sum_{A\in K} \frac{(m_A-1)(m_A-2)}{2}. \tag{2}$$

In [7], *X. W. C. Faber* described special cases of *Examples* 1 and 2 and, by a counting argument, proved the following result.

Proposition 3 (Incidence formula, 2006). *The size of a Kakeya set K is given by*

$$|K| = \frac{q(q+1)}{2} + \sigma(K). \tag{3}$$

Since $\sigma(K) \geq 0$, for every Kakeya set K, a first consequence of (3) is that

$$|K| \geq \frac{q(q+1)}{2}. \tag{4}$$

Let us note that *T. Wolff* in [11] proved that $|K| \geq q^2/2$; in fact his method gives inequality (4). Equality in (4) is actually attained in *Example* 1 and it is easy to see that this happens only in this case. So, when q is even, our problem is quite simple: *every Kakeya set K in π_q, q even, satisfies inequality* (4) *and equality holds iff K is associated to a dual hyperoval and one of its lines.* When q is odd the plane π_q contains no hyperovals and

$\sigma(K) > 0$, for every Kakeya set K. In this case the Kakeya set closest to that of *Example* 1 is the set $K(\mathcal{O}, \ell, \ell_A)$ described in *Example* 2. This is the reason for the following conjecture recently raised and studied by *X. W. C. Faber* in [7].

Conjecture 1 (2006). *If q is odd, then*

$$|K| \geq \frac{q(q+1)}{2} + \frac{q-1}{2},$$

for every Kakeya set K.

We remark that the *Blokhuis-Bruen* inequality in *Proposition* 2 is not so far from that of the conjecture. Moreover in [7], *X. W. C. Faber* obtained the following two results; the second one is a slight improvement of *Proposition* 2.

Proposition 4 (Triple point lemma, 2006). *Let K be a Kakeya set in π_q, q odd. Then, for every point $P \in \ell$, except possibly one, there exists a point $A \in \ell_P$ with $m_A \geq 3$.*

Proposition 5 (2006). *If q is odd, then*

$$|K| \geq \frac{q(q+1)}{2} + \frac{q}{3}, \tag{5}$$

for every Kakeya set K.

The *triple point lemma* is the the main tool in the proof of *Proposition* 5 and it is worth to remark that it is just the dual of *Proposition* 1. Actually it was proved by the same argument of *Bichara* and *Korchmáros:* the celebrated *Segre's lemma of tangents,* that was the key ingredient in his famous characterization of the $q+1$ rational points of an irreducible conic in π_q with q odd ([9]).

Let Ω be a $(q+2)$–set in π_q with an internal nucleus and let ℓ_∞ a line through this nucleus. Then, in the affine plane $AG(2,q) = \pi_q \setminus \ell_\infty$, the point set $\Omega \setminus \ell_\infty$ can be arranged as the graph $\{(a, f(a)) : a \in GF(q)\}$ of a function f, f being either a permutation or a semipermutation (i.e. a function whose range has size $q-1$) of $GF(q)$. This graph has been recently introduced and studied by *J. Cooper* in [6] and the following improvement to the *Faber's* inequality (5) has been obtained.

Proposition 6 (2006). *If q is odd, then*

$$|K| \geq \frac{q(q+1)}{2} + \frac{5q}{14} - \frac{1}{14}, \tag{6}$$

for every Kakeya set K.

Finally, we can settle Faber's conjecture, also characterizing the unique example realizing it. Actually we have the following sharp result.

Proposition 7. *If q is odd, then*

$$|K| \geq \frac{q(q+1)}{2} + \frac{q-1}{2},$$

for every Kakeya set K. Equality holds if and only if K is of type $K(\mathcal{O}, \ell, \ell_A)$, as in Example 2.

The essential ingredients in the proof are the *Segre's* lemma of tangents and the *Jamison–Brouwer–Schrijver* bound on the size of blocking sets in desarguesian affine planes ([3], [8]).

3. Solution of Kakeya's Problem in the Plane

We will give the proof of Proposition 7. It is more convenient however to phrase it in its dual form.

Proposition 8. *Let Ω be a set of $q+2$ points in $PG(2,q)$, with an internal nucleus. Then the number of lines intersecting Ω is at least*

$$k^*(q) = \frac{(q+1)(q+2)}{2} + \frac{q-1}{2}.$$

Equality implies that Ω consists of the points of an irreducible conic together with an external point.

Proof. Let a_i be the number of lines in $AG(2,q)$ intersecting Ω in i points. Then:

$$\begin{cases} \sum a_i = q^2+q+1 \\ \sum i\, a_i = (q+2)(q+1) \\ \sum \binom{i}{2} a_i = (q+2)(q+1)/2 \end{cases}$$

The first equation counts the total number of lines in the affine plane. In the second we count incident point-line pairs (P,ℓ), where P is a point of Ω. Finally in the third we count ordered triples (P,Q,ℓ), where P and Q are different points from Ω (and ℓ the unique line joining them). It follows that

$$a_0 + a_3 + 3a_4 + \cdots + \binom{q}{2} a_{q+1} = (q^2-q)/2.$$

Also, for later use we note that:

$$a_1 = 3a_3 + 8a_4 + \cdots = \sum_{n>2} (n^2-2n) a_n.$$

We aim for the situation where Ω is a conic together with an external point. In that case $a_1 = (q-1)+(q-1)/2$, $a_2 = (q^2+5)/2$, $a_3 = (q-1)/2$ and $a_0 = (q-1)^2/2$ (and the number of intersecting lines is $(q^2+4q+1)/2$).

Let the number of intersecting lines be $(q+2)(q+1)/2+f$ for some f, so that $a_0 = (q^2-q)/2 - f$. This gives us for f the equation

$$a_3 + 3a_4 + \cdots + \binom{q}{2} a_{q+1} = f,$$

and we would like to show that $f \geq (q-1)/2$.

We know from *Bichara–Korchmáros* result (Prop.1), that there are at most 2 internal nuclei (in the example exactly 2) and by assumption there is at least one. Every other point is therefore on at least one tangent, and hence also on at least one (≥ 3)-secant. In particular $f \geq q/3$, with equality if every other point is on exactly one tangent and one three-secant (this does happen if $q=3$).

Every point, with the exception of the internal nucleus (nuclei), is on an odd intersector. So the odd intersectors form a blocking set of the dual affine plane if there is just one nucleus (this should maybe be called a dual blocking set, but we will use this term with a different meaning later). In this case:

$$a_1 + a_3 + a_5 + \cdots \geq 2q - 1,$$

and therefore

$$4a_3 + 8a_4 + 15a_5 + \cdots \geq 2q - 1,$$

and hence $f \geq (2q-1)/4$, more than we want.

From now on we assume that there are two internal nuclei, N_1 and N_2. Adding a random line on one of the internal nuclei, but not containing the other one, we again get a blocking set of the dual affine plane, and we obtain

$$4a_3 + 8a_4 + 15a_5 + \cdots \geq 2q - 2,$$

and hence $f \geq (2q-2)/4$ with equality if $a_k = 0$ for $k > 3$. So we have proved our lower bound, and we proceed to characterize the case of equality.

If $f = (q-1)/2$ then we have $(q-1)/2$ three-secants, and $3(q-1)/2$ tangents. Now if a point Q, is on exactly one tangent, and this happens often, then also on a unique three-secant, and we will show, that their intersection points with ℓ are related: if one is $(1 : \lambda)$ the other is $(1 : -\lambda)$), where coordinates are chosen such that $N_1 = (1 : 0)$ and $N_2 = (0 : 1)$.

Consider a three-secant containing two points on a unique tangent. Then these two tangents intersect in a point on the line joining the two internal nuclei (ℓ). This is true in the example and follows from a *Segre*-type computation: if the three secant intersects the line ℓ in $(1 : \lambda : 0)$ then the unique tangents go through $(1 : -\lambda : 0)$), where the coordinates are set up in such a way that the two internal nuclei are $(1 : 0 : 0)$ and $(0 : 1 : 0)$.

We will use *Segre*-type computations a lot in the sequel. The general setup is the following. Consider three points $E_1 = (1 : 0 : 0)$, $E_2 = (0 : 1 : 0)$, $E_3 = (0 : 0 : 1)$. Let X be any set of points such that no point of X is on one of the coordinate lines E_iE_j. For $x = (x_1 : x_2 : x_3)$ write down the triple $x' = (x'_1, x'_2, x'_3) := (x_2/x_1, x_3/x_2, x_1/x_3)$. It is clear from the definition that $\prod_{x \in X} x'_1 x'_2 x'_3 = 1$. On the other hand, it is sometimes possible, because of geometric properties of X to say something about $p_i = \prod_{x \in X} x'_i$. Applying this together with $p_1p_2p_3 = 1$ is called *Segre*'s lemma of tangents or a *Segre* computation. In our case the argument runs as follows. Let U be a point on a unique three-secant, further choose coordinates such that $U = (0 : 0 : 1)$, and some random fourth point equals $(1 : 1 : 1)$. Recall that $N_1 = (1 : 0 : 0)$ and $N_2 = (0 : 1 : 0)$. Let the three-secant through U intersect ℓ in $(1 : \lambda : 0)$ and let the unique tangent intersect ℓ in $(1 : \mu : 0)$. The remaining $q - 1$

points of Ω (other than N_1, N_2 and U) have (homogeneous) coordinates $(a_i : b_i : c_i)$ with $a_i b_i c_i \neq 0$. We associate to such a point the triple $(b_i/a_i, c_i/b_i, a_i/c_i)$. Taking the product of all the entries in all triples we clearly get 1, because that is the contribution of each triple. On the other hand we have $\prod_i c_i/b_i = -1$, because on each line through N_1 we have a unique point of Ω so we just have the product of all non-zero field elements. In the same way $\prod a_i/c_i = -1$ by considering lines through N_2. To compute $\prod b_i/a_i$ we consider the lines through $U = (0 : 0 : 1)$. The three secant gives the value $b_i/a_i = \lambda$ twice, but the value $b_i/a_i = \mu$ is absent. All other nonzero field elements occur exactly once in the product, so for this product we end up with $-\lambda/\mu$, so $(-1)(-1)(-\lambda/\mu) = 1$ and we conclude that $\mu = -\lambda$.

We will show that, unless $q = 3$, the three points of Ω on a three-secant cannot all be points with a unique tangent, by applying again a *Segre* computation.

Apart from the 2 internal nuclei our set has q points, and all of them are on at least one tangent. The total number of tangents is

$$3(q-1)/2 = q + (q-3)/2$$

hence at least $(q+3)/2$ points are on exactly one tangent (and one three-secant). So we certainly find a three-secant with (at least) two unique-tangent points on it.

Let $N_1 = (1 : 0 : 0)$ and $N_2 = (0 : 1 : 0)$ (as before) be the internal nuclei.

Let $U_1 = (0 : 0 : 1)$ and $U_2 = (1 : 1 : 1)$ be two one-tangent points on a common three-secant, and let $V = (a : b : 1)$ be a one-tangent point not on the line U_1U_2, so a and b are nonzero, and $a \neq b$.

Note that in our example we have that N_1, N_2, U_1 and U_2 are on a conic, and the tangents at $U_{1,2}$ are also known. So the conic has to be: $-2x_1x_2 + x_2x_3 + x_3x_1 = 0$. So we should expect that $-2ab + a + b = 0$ for $V = (a : b : 1)$.

The three-secant U_1U_2 meets N_1N_2 in $(1 : 1 : 0) = N_1 + N_2$, so the tangents at U_1 and U_2 meet in $N_1 - N_2 = (1 : -1 : 0)$. Let the tangent at V pass through $(1 : \lambda : 0)$, then the three-secant on V passes through the point $(1 : -\lambda : 0)$.

First we consider the triangle U_1N_1V. The tangent at V_1 intersects U_1N_1 in $U_1 + \big((\lambda a - b)/\lambda\big)\, N_1$, the three-line in $U_1 + \big((\lambda a + b)/\lambda\big)\, N_1$. The tangent through U_1 intersects N_1V in $N_1+\big(-1/(a+b)\big)V$, the three-line in $N_1+\big(1/(b-a)\big)V$. On VU_1 there are no special 'missing' or 'extra' points. *Segre* gives:

$$(a+b)(\lambda a+b)=(b-a)(\lambda a-b).$$

And we get the important fact $\lambda=-b^2/a^2$.

Next we consider the triangle $N_1U_2U_1$. Let the third point of Ω on U_1U_2 be $U_2+\mu U_1$. On N_1U_2 we 'miss' the point $(-1:1:1)=N_1+(-1/2)U_2$. On U_2U_1 we 'miss' the point $U_2+\mu U_1$, and finally on U_1N_1 the point $(2:0:1)=U_1+2N_1$. Here we used that since the three-line on U_1 goes through $(1:1:0)$, the tangent passes through $(1:-1:0)$. It follows from the *Segre* product that $\mu=1$.

We now turn to the triangle $U_1U_2V_1$. On U_1U_2 we find the 'extra' point, the intersection with the three line through V:

$$U_1+\frac{(b+a\lambda)/(1+\lambda)}{1-(b+a\lambda)/(1+\lambda)}U_2.$$

and 'missing' points U_1+U_2 (the third point of Ω on U_1U_2) and the intersection of the tangent through V with U_1U_2:

$$U_1+\frac{(b-a\lambda)/(1-\lambda)}{1-(b-a\lambda)/(1-\lambda)}U_2.$$

This is of course just the expression for the three-secant with $-\lambda$ instead of λ. On U_2V and VU_1 we find 'missing' coordinates $-2/(a+b)$ and $-1+(a+b)/2$. The *Segre* computation gives us

$$(a+b)(b+a\lambda)(1-b+(a-1)\lambda)=(a+b-2)(b-a\lambda)(1-b-(a-1)\lambda).$$

This we may rewrite as

$$a(a-1)\lambda^2+(a+b-1)(a-b)\lambda+b(1-b)=0.$$

Now substitute $\lambda=-b^2/a^2$, multiply by a^3 and divide by b. We get:

$$(a-b)(a+b)(2ab-a-b)=0.$$

We already remarked that $a\neq b$, but also $a\neq -b$ because otherwise V would be on the tangent through U_1. Hence $2ab-a-b=0$ and V is a point on the conic we are aiming for. A direct computation shows that also the tangent is 'right' and that the three-secant through V passes through the 'special point' $(1:1:2) = U_1 + U_2$.

Some counting to end the story. Let there be k points on a unique tangent. This means that our special point $U_1 + U_2$ is on at least $k/2$ three-secants, and hence on at least $k/2$ tangents. What is left in Ω (apart from the internal nuclei, the special point and the unique tangent points) is a set of $q-1-k$ points on at least 2 tangents, and a set of at most $3(q-1)/2 - k - k/2$ tangents. So

$$3(q-1)/2 - k - k/2 \geq 2(q-1-k).$$

This means $k \geq q-1$, so all other points are on the conic, and we finished the proof.

4. Applications to Dual Blocking Sets

A blocking set B in $\pi_q = PG(2,q)$ is a point set meeting every line and containing none.

Definition 1. A dual blocking set S in π_q is a point set meeting every blocking set and containing no lines.

Example 3. A Kakeya set $K = \left(\bigcup_{P\in\ell} \ell_P\right) \setminus \ell$ in π_q contains no lines. Moreover, for every blocking set B of π_q, a point P exists on $\ell \setminus B$ and so K meets B in a point of $\ell_P \setminus \ell$. It follows that K *is a dual blocking set.* ■

Example 4. The complement $S = \pi_q \setminus (\ell \cup m)$ of the union of two distinct lines ℓ and m in π_q contains no lines. Moreover, no blocking set is contained in the union of two lines and so S meets every blocking set. It follows that S *is a dual blocking set.* ■

Dual blocking sets were introduced by *P. Cameron, F. Mazzocca* and *R. Meshulam* in [4]; the first of the two main results of this paper is the following.

Proposition 9 (1988). *Let S be a dual blocking set in π_q. Then*

$$|S| \geq \frac{q(q+1)}{2}.$$

Equality holds if and only if either

(i) *S is the Kakeya set associated to a dual hyperoval and one of its lines; or*

(ii) *$q = 3$ and S is the complement of the union of two distinct lines.*

The argument in the proof of this proposition implicitly shows that every minimal (with respect to inclusion) dual blocking set in π_q is of one of types described in examples (3) and (4). For the sake of completeness we give an explicit proof of this result.

Proposition 10. *Let S be a minimal dual blocking set in π_q. Then one of the two following possibilities occur:*

(i) *$S = \big(\bigcup_{P\in\ell} \ell_P\big) \setminus \ell$ is a Kakeya set;*

(ii) *$S = \pi_q \setminus (\ell \cup m)$ is the complement of the union of two distinct lines ℓ and m.*

Proof. First of all we observe that there is a line ℓ disjoint from S, for if not, then, since S does not contain a line, S and its complement are blocking sets; a contradiction as S must meet every blocking set. Now we distinguish the following two cases.

Case 1. Assume that S is disjoint from exactly one line ℓ, and let P be a point of this line. If, for every line $m \neq \ell$ through P, there is a point $Q \neq P$ on m but not in S, then

$$B = (\ell \setminus \{P\}) \cup \left(\bigcup_{P\in m\neq\ell} m \right)$$

is a blocking set disjoint from S; a contradiction. Hence, for every point $P \in \ell$, there exists a line ℓ_P through P with $\ell_P \setminus \{P\} \subseteq S$. Then S contains the Kakeya set $K = \big(\bigcup_{P\in\ell} \ell_P\big) \setminus \ell$, which is a dual blocking set. From the minimality of S it follows that $S = K$.

Case 2. Assume that there are two lines ℓ and m disjoint from S. For any point $P \notin \ell \cup m$, let n be a line on P meeting $\ell \setminus m$ and $m \setminus \ell$ in the points L and M, respectively. Then $\big(\ell \cup m \cup \{P\}\big) \setminus \{L, M\}$ is a blocking set contained in $\ell \cup m \cup \{P\}$. It follows that P must belong to S and S is the complement of $\ell \cup m$. ■

By *Propositions* 9 and 10 we can conclude that all the bounds previously shown for the size of a Kakeya set give, in the case that q is odd, corresponding new bounds for the size of a minimal dual blocking set, improving the result of *Proposition* 9. In fact, as a corollary of *Proposition* 7, we have the following sharp result.

Proposition 11. *Let S be a dual blocking set in π_q, q odd. Then*

$$|S| \geq \frac{q(q+1)}{2} + \frac{q-1}{2}$$

and equality holds if and only if S is a Kakeya set of type described in Example 2.

5. Old and New Results in Higher Dimensions

In contrast to the plane case we only have bounds and conjectures for higher dimensions. In [11] it is shown that the number of points in a Kakeya set in $AG(n,q)$ is at least $c \cdot q^{(n+2)/2}$, which is good for $n=2$ but probably not for any larger n. The case $n=3$ is the first open problem, but for $n=4$ *T. Tao* has shown ([10]) that the exponent 3 can be improved to $3+\frac{1}{16}$. In what follows we will show that for general n we get the lower bound $c \cdot q^{n-1}$, where $c = 1/(n-1)!$, so this improves the previous bounds when n is at least 5 and comes close to the conjectured $c_n q^n$. Unfortunately our ideas are for several reasons very unlikely to lead to improvements in the case of the 'real' Kakeya problem.

Very recently however, *Zeev Dvir* [5] has proved the finite field Kakeya problem, by showing that the number of points of a Kakeya set in $AG(n,q)$ is at least $\binom{q+n-1}{n}$.

Since our result and proof are similar in nature but still slightly different, we will include it for historical reasons, and with the hope that an improved argument will give a bound equivalent or even slightly better than that of *Dvir.* To improve the bound in higher dimensions we use a bound on the dimension of a certain geometric codes.

Consider the line-point incidence matrix of $PG(n,q)$. Number the points (so the columns): first the points in the hyperplane at infinity, then the points not in the Kakeya set, and finally the points in the Kakeya set. As usual we denote the number of points (and hyperplanes) in $PG(n,q)$ by $\theta_n = \theta_n(q) = (q^{n+1}-1)/(q-1)$.

Let the first θ_{n-1} rows be labeled by the lines defining the Kakeya set, in the right order. The top consisting of the first θ_{n-1} rows of the incidence matrix now looks like this:

$$\mathbf{T} = (\mathbf{I}\,;\,\mathbf{O}\,;\,\mathbf{K}).$$

Here $\mathbf{I}$ is the identity matrix, and $\mathbf{K}$ is the θ_{n-1} by $|K|$ line-point incidence matrix of Kakeya-lines versus Kakeya-points Let $d = d_{n-1}$ be the dimension of C_{n-1}, the $GF(p)$-code (where $q = p^t$) spanned by the lines of $PG(n-1,q)$ (the hyperplane at infinity). Then there is a subset C of the points, of size $\theta_{n-1}-d_{n-1}$ that does not contain the support of a codeword (this is obvious: after normalization a generator matrix for this code has the form $(I\,;\,A)$ and every nonzero codeword has a nonzero coordinate in one of the first d_{n-1} positions, so no codeword has its support contained in the 'tail' of length $\theta_{n-1}-d_{n-1}$). It follows that the set of Kakeya points has at least this size: Consider the $\theta_{n-1}-d_{n-1}$ rows of $\mathbf{T}$ corresponding to the Kakeya lines having a direction in C. Suppose the corresponding rows of $\mathbf{K}$ are dependent (over $GF(p)$). Then this dependency would produce a codeword in the line-point code of $PG(n,q)$ with support contained in the set C in the hyperplane at infinity. But such a word is already in the point-line code of this hyperplane. To see this, let C_n stand for the line code of $PG(n,q)$, and C_{n-1} for the line code of the hyperplane H. Clearly $C_n^{\perp}|_H \subseteq C_{n-1}^{\perp}$. We show that in fact equality holds, for let u be a word in $C_{n-1}^{\perp}$, and now take a point $P \notin H$ and form the cone with top P over u, but remove P. This defines in an obvious way a word $\tilde{u}$ in $C_n^{\perp}$ whose restriction to H is u.

So we find $|K| \geq \dim C_{n-1}^{\perp}$. The dimension of C_{n-1} is known, and equal to something complicated. For us the bound

$$|K| \geq \dim C_{n-1}^{\perp} \geq \binom{q+n-2}{n-1} \geq q^{n-1}/(n-1)!$$

suffices. In fact, if q is prime we have equality, if not we have a little improvement, but not an essential one.

Acknowledgments. This research was done when the first author was visiting the *Seconda Università degli Studi di Napoli* at Caserta. The authors wish to thank for their supports the research group GNSAGA of Italian

Istituto Nazionale di Alta Matematica and the *Mathematics Department* of the *Seconda Università degli Studi di Napoli.*

References

[1] A. Bichara and G. Korchmáros, Note on $(q+2)-$Sets in a Galois Plane of Orde, q, *Ann. Discrete Math.,* **14** (1982), 117–121.

[2] A. Blokhuis and A. A. Bruen, The Minimal Number of Lines Intersected by a Set of $q+2$ Points, Blocking Sets, and Intersecting Circles, *J. Combin. Theory* Ser. A, **50** (1989), 308–315.

[3] A. E. Brouwer and A. Schrijver, The Blocking Number of an Affine Space, *J. Combin. Theory* Ser. A, **24** (1978), 251-253.

[4] P. J. Cameron, F. Mazzocca and R. Meshulam, Dual Blocking Sets in Projective and Affine Planes, *Geom. Dedicata,* **27** (1988), 203–207.

[5] Z. Dvir, On the Size of Kakeya Sets in Finite Fields, *J. of the AMS* (to appear), (2008).

[6] J. Cooper, Collinear Triple Hypergraphs and the Finite Plane Kakeya Problem, *Arxiv preprint,* math.CO/0607734, (2006) - arxiv.org.

[7] X. W. C. Faber, On the Finite Field Kakeya Problem in Two Dimensions, *J. Number Theory,* **117** (2006), 471–481.

[8] R. Jamison, Covering Finite Fields with Cosets of Subspaces, *J. Combin. Theory* Ser. A, **22** (1977), 253–266.

[9] B. Segre, Ovals in Finite Projective Planes, *Canad. J. Math.,* **7** (1955), 414–416.

[10] T. Tao, A New Bound for Finite Field Besicovitch Sets in Four Dimensions, *Pacific J. Math.,* **222** (2005), 337–364.

[11] T. Wolff, Recent Work Connected with the Kakeya Problem, *Prospects in Matematics* (Princeton, NJ, 1996), Amer. Math. Soc., Providence, RI, (1999), 129–162.

Aart Blokhuis
Eindhoven University of Technology
Department of Mathematics and Computing Science
P.O. Box 513
5600 MB Eindhoven
The Nederlands

e-mail: a.blokhuis@tue.nl

Francesco Mazzocca
Seconda Università degli Studi di Napoli
Dipartimento di Matematica
via Vivaldi 43
I-81100 Caserta
Italy

e-mail: francesco.mazzocca@unina2.it

BOLYAI SOCIETY
MATHEMATICAL STUDIES, 19

Building Bridges
pp. 219–240.

An Abstract Szemerédi Regularity Lemma

BÉLA BOLLOBÁS* and VLADIMIR NIKIFOROV

We extend Szemerédi's Regularity Lemma to abstract measure spaces.

1. Introduction

Szemerédi's Regularity Lemma is one of the few truly universal tools in modern combinatorics, with numerous important applications. In particular, this lemma is the cornerstone of the theory of convergent sequences of dense graphs launched recently by Lovász and Szegedy [15], Borgs, Chayes, Lovász, Sós and Vesztergombi [3], [4] and Borgs, Chayes and Lovász [5]. The germ of a similar theory for sparse graphs, started by Bollobás and Riordan [2], relies on variants of this lemma. The so-called weak version of this lemma was proved by Frieze and Kannan [8], and Lovász and Szegedy [16] gave a beautiful and suprising proof of a similar result. A number of its extensions to sparse graphs were proved by Kohayakawa [12], Kohayakawa and Rödl [13], Gerke, Kohayakawa, Rödl and Steger [9], and others; in a different direction, Rödl and Skokan [17] and Gowers [10] proved deep hypergraph variants. In several recent papers, including those by Lovász and Szegedy [16], Tao [19], Ishigami [11], and Elek and Szegedy [7], the original lemma was deduced from more abstract assertions. Our main result here is close to that of Tao but is simpler an more axiomatic: distilling the essential components of the original proof of Szemerédi, we obtain a regularity lemma for abstract measure spaces, having a multitude of concrete applications well beyond graph theory. Some of these results can be ob-

*Research supported in part by NSF grants DMS-0505550, CNS-0721983 and CCF-0728928

tained in other ways, but others seem to be tailor made for our approach: in particular, we are not aware of other ways of proving them.

Our notation follows [1].

1.1. Measure triples

A finitely additive measure triple or, briefly, *a measure triple* $(X, \mathbb{A}, \mu)$ consists of a set X, an algebra $\mathbb{A} \subset 2^X$, and a complete, nonnegative, finitely additive measure μ on $\mathbb{A}$ with $\mu(X) = 1$. Thus, $\mathbb{A}$ contains X and is closed under finite intersections, unions and differences; the elements of $\mathbb{A}$ are called *measurable subsets* of X.

Letting $[n] = \{1, \ldots, n\}$, $2^{[n]}$ be the power set of $[n]$ and $\mu(A) = |A|/n$ for every $A \subset [n]$, we see that $\left([n], 2^{[n]}, \mu\right)$ is a measure triple. For later reference we first outline two specializations of this simple example.

Example 1. Let $k, n \geq 1$, write $2^{[n]^k}$ for the power set of $[n]^k$, and define μ^k by $\mu^k(A) = |A|/n^k$ for every $A \subset [n]^k$. Then $\left([n]^k, 2^{[n]^k}, \mu^k\right)$ is a measure triple.

Note that there is a bijection between undirected k-graphs on the vertex set $[n]$ and subsets $G \subset 2^{[n]^k}$ such that if $(v_1, \ldots, v_k) \in G$, then G contains every permutation of $(v_1, \ldots, v_k)$. In view of this, we shall consider subsets of $2^{[n]^k}$ as labelled directed k-graphs (with loops) on the vertex set $[n]$.

Example 2. Let $k \geq 2$, and let $X_1, \ldots, X_k$ be finite nonempty disjoint sets. Write $2^{X_1 \times \cdots \times X_k}$ for the power set of $X_1 \times \cdots \times X_k$, and define μ^k by $\mu^k(A) = |A|/\left(|X_1| \cdots |X_k|\right)$ for every $A \subset X_1 \times \cdots \times X_k$. Then $(X_1 \times \cdots \times X_k, 2^{X_1 \times \cdots \times X_k}, \mu^k)$ is a measure triple.

Note that a subset of $2^{X_1 \times \cdots \times X_k}$ is naturally identified with a k-partite k-graph with vertex classes $X_1, \ldots, X_k$.

Another essential, but less trivial example of a measure triple is the k-dimensional unit cube.

Example 3. Let $k \geq 1$, and let $\mathbb{B}^k$ be the algebra of the Borel subsets of the unit cube $[0, 1]^k$; write λ^k for the Lebesgue measure on $\mathbb{B}^k$. Then $\left([0, 1]^k, \mathbb{B}^k, \lambda^k\right)$ is a measure triple.

1.2. SR-systems

We now introduce the main objects of our study, SR-systems: measure triples with a suitably chosen semi-ring. Here SR stands for "Szemerédi regularity" rather than "semi-ring".

Recall that a set system $\mathbb{S}$ is a *semi-ring* if it is closed under intersection and for all $A, B \in \mathbb{S}$, the difference $A \backslash B$ is a disjoint union of a finite number of members of $\mathbb{S}$.

A semi-ring $\mathbb{S}$ is called *r-built* if for all $A, B \in \mathbb{S}$, the difference $A \backslash B$ is a disjoint union of at most r members of $\mathbb{S}$; we say that $\mathbb{S}$ is *boundedly built* if it is r-built for some r.

An *SR-system* is a quadruple $(X, \mathbb{A}, \mu, \mathbb{S})$, where $(X, \mathbb{A}, \mu)$ is a measure triple and $\mathbb{S} \subset \mathbb{A}$ is a boundedly built semi-ring.

Clearly, the quadruple $(X, \mathbb{A}, \mu, \mathbb{A})$ is an SR-system based on the measure triple $(X, \mathbb{A}, \mu)$. For the rest of the section, let us fix an SR-system $(X, \mathbb{A}, \mu, \mathbb{S})$.

Given a set system $\mathbb{Z}$ and $k \geq 1$, let $\mathbb{Z}^{\langle k \rangle}$ be the collection of products of k elements of $\mathbb{Z}$ any two of which are either disjoint or coincide, i.e.,

(1)
$$\mathbb{Z}^{\langle k \rangle} = \left\{ Z_1 \times \cdots \times Z_k \colon Z_i \in \mathbb{Z} \text{ and } Z_i \cap Z_j = \emptyset \text{ or } Z_i = Z_j \text{ for all } i, j \in [k] \right\}.$$

The proof of the following lemma is given in Section 4.

Lemma 4. *Given a boundedly built semi-ring $\mathbb{S} \subset \mathbb{A}$, the set system $\mathbb{S}^{\langle k \rangle}$ is a boundedly built semi-ring.*

Using this assertion, we can construct the following general SR-system.

Example 5. Let $\mathbb{A}$ be a set algebra, and for $k \geq 1$, set

$$\mathbb{A}^k = \left\{ A_1 \times \cdots \times A_k \colon A_i \in \mathbb{A} \text{ for all } i \in [k] \right\}.$$

Write $\mathcal{A}(\mathbb{A}^k)$ for the algebra generated by the set system $\mathbb{A}^k$, and μ^k for the product measure on $\mathcal{A}(\mathbb{A}^k)$. The quadruple $\left(X^k, \mathcal{A}(\mathbb{A}^k), \mu^k, \mathbb{A}^{\langle k \rangle}\right)$ is an SR-system.

Let us outline three particular cases of the above construction.

Example 6. For $k \geq 1$ set $\mathcal{G}^k(n) = \left([n]^k, 2^{[n]^k}, \mu^k, \left(2^{[n]}\right)^{\langle k \rangle}\right)$, where $\left([n]^k, 2^{[n]^k}, \mu^k\right)$ is the measure triple defined in Example 1, and $\left(2^{[n]}\right)^{\langle k \rangle}$

is the set of all products of k subsets of $[n]$ any two of which are either disjoint or coincide.

Example 7. For $k \geq 1$ set $\mathcal{B}^k = \left([0,1]^k, \mathbb{B}^k, \lambda^k, \mathbb{B}^{\langle k \rangle}\right)$, where $\left([0,1]^k, \mathbb{B}^k, \lambda^k\right)$ is the measure triple defined in Example 3, and $\mathbb{B}^{\langle k \rangle}$ is the set of all products of k Borel subsets of $[0,1]$ any two of which are either disjoint or coincide.

Example 8. For $k \geq 1$ set $\mathcal{BI}^k = \left([0,1]^k, \mathbb{B}^k, \lambda^k, \mathbb{I}^{\langle k \rangle}\right)$, where $\left([0,1]^k, \mathbb{B}^k, \lambda^k\right)$ is the measure triple defined in Example 3, and $\mathbb{I}^{\langle k \rangle}$ is the set of all products of k intervals $[a,b) \subset [0,1]$ any two of which are either disjoint or coincide.

The construction $\mathbb{Z}^{\langle k \rangle}$ given in (1) is important in some applications, but is otherwise nonessential to our general approach. Here is a general SR-system, where this construction is not used.

Example 9. Suppose $k \geq 2$ and $X_1, \ldots, X_k$ are finite nonempty disjoint sets. Set

$$\mathcal{PG}(X_1, \ldots, X_k) = \left(X_1 \times \cdots \times X_k, 2^{X_1 \times \cdots \times X_k}, \mu^k, \mathbb{P}(X_1, \ldots, X_k)\right),$$

where

$$\mathbb{P}(X_1, \ldots, X_k) = \left\{ A_1 \times \cdots \times A_k \colon A_i \subset X_i \text{ for all } i \in [k] \right\}$$

and $(X_1 \times \cdots \times X_k, 2^{X_1 \times \cdots \times X_k}, \mu^k)$ is the measure triple defined in Example 2. Then $\mathcal{PG}(X_1, \ldots, X_k)$ is an SR-system.

1.2.1. Using SR-systems to define ε-regularity. The primary goal of introducing SR-systems is to extend the concept of ε-regular pairs of Szemerédi [18] (for definitions and background see also [1] and [14]).

Suppose $(X, \mathbb{A}, \mu, \mathbb{S})$ is a fixed SR-system. For every $A, V \in \mathbb{A}$ set

$$d(A, V) = \frac{\mu(A \cap V)}{\mu(V)}$$

if $\mu(V) > 0$, and $d(A, V) = 0$ if $\mu(V) = 0$.

Definition 10. Suppose that $0 < \varepsilon < 1$ and $V \in \mathbb{S}$ satisfies $\mu(V) > 0$. A set $A \in \mathbb{A}$ is called **ε-regular** in V if

$$\left| d(A, U) - d(A, V) \right| < \varepsilon$$

for every $U \in \mathbb{S}$ such that $U \subset V$ and $\mu(U) > \varepsilon\mu(V)$.

Let us first see what Definition 10 says about k-partite k-graphs. Take the SR-system $\mathcal{PG}(V_1, \ldots, V_k)$ from Example 9. The edge set $E(G)$ of a k-partite k-graph G with vertex classes $V_1, \ldots, V_k$ is just a subset of $V_1 \times \cdots \times V_k$. Given $U_1 \subset V_1, \ldots, U_k \subset V_k$, write $e(U_1, \ldots, U_k)$ for the number of edges $(v_1, \ldots, v_k) \in E(G)$ such that $v_i \in U_i$ for $i = 1, \ldots, k$.

Now if G is $\varepsilon^{1/k}$-regular in $V_1 \times \cdots \times V_k$, then, for every ordered k-tuple $(U_1, \ldots, U_k)$ such that $U_i \subset V_i$ and $|U_i| > \varepsilon|V_i|$ for $i = 1, \ldots, k$, we obtain

$$\left| \frac{e(V_1, \ldots, V_k)}{|V_1| \cdots |V_k|} - \frac{e(U_1, \ldots, U_k)}{|U_1| \cdots |U_k|} \right| < \varepsilon.$$

Note that for $k = 2$ this condition is equivalent to the traditional "ε-regular pair" up to the choice of ε (see [1], [14], and [18]).

The ε-regularity for directed k-graphs is easily understood too. Indeed, take the SR-system $\mathcal{G}^k(n)$ from Example 6. As suggested above, the edge set $E(G)$ of a directed k-graph G with $V(G) = [n]$ is just a subset of $[n]^k$. Let $(V_1, \ldots, V_k)$ be an ordered k-tuple of disjoint nonempty subsets of $[n]$. If G is $\varepsilon^{1/k}$-regular in $V_1 \times \cdots \times V_k$, then, for every ordered k-tuple $(U_1, \ldots, U_k)$ such that $U_i \subset V_i$ and $|U_i| > \varepsilon|V_i|$ for $i = 1, \ldots, k$, we obtain

$$\left| \frac{e(V_1, \ldots, V_k)}{|V_1| \cdots |V_k|} - \frac{e(U_1, \ldots, U_k)}{|U_1| \cdots |U_k|} \right| < \varepsilon.$$

For convenience we shall extend ε-regularity to whole partitions.

Definition 11. Let $0 < \varepsilon < 1$ and $\mathcal{P}$ be a partition of X into sets belonging to $\mathbb{S}$. We call a set $A \in \mathbb{A}$ **ε-regular** in $\mathcal{P}$ if

$$\sum \left\{ \mu(P) \colon P \in \mathcal{P},\ A \text{ is not } \varepsilon\text{-regular in } P \right\} < \varepsilon.$$

1.3. Partitions in measure triples

Given a collection $\mathbb{Z}$ of subsets of X, we write $\Pi(\mathbb{Z})$ for the family of finite partitions of X into sets belonging to $\mathbb{Z}$. We shall be mainly interested in $\Pi(\mathbb{S})$. As usual, we write $|\mathcal{P}|$ for the number of elements of a partition $\mathcal{P}$.

Let $\mathcal{P}$, $\mathcal{Q}$ be partitions of X, and $A \subset X$. We say that $\mathcal{P}$ refines A (in notation $\mathcal{P} \succ A$) if A is a union of members of $\mathcal{P}$, and that $\mathcal{P}$ refines $\mathcal{Q}$ (in notation $\mathcal{P} \succ \mathcal{Q}$) if $\mathcal{P}$ refines each $Q \in \mathcal{Q}$. We write $\mathcal{P} \cap \mathcal{Q}$ for the partition consisting of all nonempty intersections $P \cap Q$, where $P \in \mathcal{P}$ and $Q \in \mathcal{Q}$. Finally, define the partition $\mathcal{P}^k$ of X^k as

$$\mathcal{P}^k = \left\{ P_{i_1} \times \cdots \times P_{i_k} \colon P_{i_j} \in \mathcal{P}, \text{ for all } j \in [k] \right\}.$$

1.3.1. Bounding families of partitions. We say that a family of partitions $\Phi \subset \Pi(\mathbb{S})$ *bounds* $\Pi(\mathbb{S})$ if for every $\mathcal{P} \in \Pi(\mathbb{S})$, there exists $\mathcal{Q} \in \Phi$ such that $\mathcal{Q} \succ \mathcal{P}$ and $|\mathcal{Q}| \le \varphi\big(|\mathcal{P}|\big)$, where $\varphi \colon \mathbb{N} \to \mathbb{N}$ is a fixed increasing function, called the *rate* of Φ.

Here is an example of a bounding family: let $\big(X^k, \mathcal{A}(\mathbb{A}^k), \mu^k, \mathbb{S}^{\langle k \rangle}\big)$ be the SR-system in Example 5, and let the family of partitions $\Phi^k \subset \Pi\big(\mathbb{S}^{\langle k \rangle}\big)$ be defined as

$$\Phi^k = \left\{ \mathcal{P}^k \colon \mathcal{P} \in \Pi(\mathbb{S}) \right\}.$$

Lemma 12. *If $\mathbb{S}$ is r-built, then Φ^k bounds $\Pi\big(\mathbb{S}^{\langle k \rangle}\big)$ with rate $\varphi(p) \le (kpr)^{k^2 pr}$.*

Lemma 12 is proved in Section 4.

2. The main result

We are ready now to state our main theorem. Its proof is presented in 4.1.

Theorem 13. *Suppose that $0 < \varepsilon < 1$, l, p and r are positive integers and $\varphi \colon \mathbb{N} \to \mathbb{N}$ is an increasing function. Then there exists an integer $q = q(\varepsilon, l, p, r, \varphi)$ such that the following assertion holds.*

Let $(X, \mathbb{A}, \mu, \mathbb{S})$ be an SR-system, where $\mathbb{S}$ is r-built, and let Φ be a family of partitions bounding $\Pi(\mathbb{S})$ with rate φ. For every collection $\mathcal{L} \subset \mathbb{A}$

of l sets and every partition $\mathcal{P} \in \Pi(\mathbb{S})$ into p sets, there exists a partition $\mathcal{Q} \in \Phi$ such that:

(i) $\mathcal{Q} \succ \mathcal{P}$;

(ii) every $A \in \mathcal{L}$ *is* ε*-regular in* $\mathcal{Q}$;

(iii) $|\mathcal{Q}| \leq q$.

Our next goal is to show that Theorem 13 implies various types of regularity lemmas. Let us emphasize the three steps that are necessary for its application:

(1) select a measure triple $(X, \mathbb{A}, \mu)$;

(2) introduce ε-regularity by fixing a boundedly built semi-ring $\mathbb{S} \subset \mathbb{A}$;

(3) select a bounding family of partitions $\Phi \subset \Pi(\mathbb{S})$ by demonstrating an upper bound on its rate $\varphi(\cdot)$.

We turn now to specific applications.

3. Applications

To obtain more familiar versions of the Regularity Lemma, we extend the concept of "ε-equitable partitions" and investigate when such partitions form bounding families.

3.1. Equitable partitions

Given $\varepsilon > 0$ and a measure triple $(X, \mathbb{A}, \mu)$, a partition $\mathcal{P} = \{P_0, \ldots, P_p\} \in \Pi(\mathbb{A})$ is called *ε-equitable*, if $\mu(P_0) \leq \varepsilon$ and $\mu(P_1) = \cdots = \mu(P_p) \leq \varepsilon$.

Let $k \geq 2$, take the SR-system $\left(X^k, \mathcal{A}(\mathbb{A}^k), \mu^k, \mathbb{S}^{\langle k \rangle}\right)$, and define a family of partitions $\Phi^k(\varepsilon) \subset \Pi\left(\mathbb{S}^{\langle k \rangle}\right)$ as

$$\Phi^k(\varepsilon) = \left\{\mathcal{P}^k \colon \mathcal{P} \in \Pi(\mathbb{A}) \text{ and } \mathcal{P} \text{ is } \varepsilon\text{-equitable}\right\}. \tag{2}$$

It is possible to prove that under some mild conditions on $(X, \mathbb{A}, \mu)$ the family $\Phi^k(\varepsilon)$ bounds $\Pi\left(\mathbb{S}^{\langle k \rangle}\right)$. To avoid technicalities, we shall illustrate this claim for the SR-system $\mathcal{G}^k(n)$ in Example 6.

Lemma 14. *Let $0 < \varepsilon < 1$ and $n > 1/\varepsilon$. Suppose that the family of partitions $\Phi^k(n,\varepsilon)$ is defined by (2) for the SR-system $\mathcal{G}^k(n) = \Big([n]^k, 2^{[n]^k}, \mu^k, \big(2^{[n]}\big)^{\langle k\rangle}\Big)$. Then $\Phi^k(n,\varepsilon)$ bounds $\Pi\Big(\big(2^{[n]}\big)^{\langle k\rangle}\Big)$ with rate*

$$\varphi(p) = \big(\lceil 2/\varepsilon\rceil + 1\big)^k 2^{pk^2}.$$

Lemma 14 is proved in 4.2. We mention without a proof a similar statement for the SR-system $\mathcal{B}^k$ in Example 7.

Lemma 15. *Let $0 < \varepsilon < 1$. Suppose that the family of partitions $\Phi^k\big([0,1],\varepsilon\big)$ is defined by (2) for the SR-system $\mathcal{B}^k = \big([0,1]^k, \mathbb{B}^k, \lambda^k, \mathbb{B}^{\langle k\rangle}\big)$. Then $\Phi^k\big([0,1],\varepsilon\big)$ bounds $\Pi\big(\mathbb{B}^{\langle k\rangle}\big)$ with rate*

$$\varphi(p) = \big(\lceil 1/\varepsilon\rceil + 1\big)^k 2^{pk^2}. \qquad \blacksquare$$

3.2. Regularity lemmas for k-graphs

We first state a regularity lemma for directed k-graphs, which, as mentioned above, we represent as subsets of $2^{[n]^k}$. Defining ε-regularity in terms of the SR-system $\mathcal{G}^k(n)$ in Example 6, we obtain the following

Theorem 16. *For all $0 < \varepsilon < 1$ and positive integers k, l, there exist $n_0(k,\varepsilon)$ and $q(k,l,\varepsilon)$ such that if $n > n_0(k,\varepsilon)$ and $\mathcal{L}$ is a collection of l directed k-graphs on the vertex set $[n]$, then there exists a partition $\mathcal{Q} = \{Q_0, \ldots, Q_q\}$ of $[n]$ satisfying*

(i) $q \le q(k,l,\varepsilon)$;

(ii) $|Q_0| < \varepsilon n$, $|Q_1| = \cdots = |Q_q| < \varepsilon n$;

(iii) every graph $G \in \mathcal{L}$ is ε-regular in at least $(1-\varepsilon)q^k$ sets $Q_{i_1} \times \cdots \times Q_{i_k}$, where $(i_1, \ldots, i_k)$ is a k-tuple of distinct elements of $[q]$.

As a consequence, we obtain a regularity lemma for undirected k-graphs. For $k = 2$ this is the result of Szemerédi, for $k > 2$ this is the result of Chung [6]. Recall that undirected k-graphs are subsets $G \subset 2^{[n]^k}$ such that if $(v_1, \ldots, v_k) \in G$, then $\{v_1, \ldots, v_k\}$ is a set of size k and G contains each permutation of $(v_1, \ldots, v_k)$.

Theorem 17. *For all $0 < \varepsilon < 1$ and positive integers k, l, there exist $n_0(k,\varepsilon)$ and $q(k,l,\varepsilon)$ such that if $n > n_0(k,\varepsilon)$ and $\mathcal{L}$ is a collection of l undirected k-graphs on the vertex set $[n]$, then there exists a partition $\mathcal{Q} = \{Q_0, \ldots, Q_q\}$ of $[n]$ satisfying:*

(i) $q \le q(k,l,\varepsilon)$;

(ii) $|Q_0| < \varepsilon n$, $|Q_1| = \cdots = |Q_q| < \varepsilon n$;

(iii) for every graph $G \in \mathcal{L}$, there exist at least $(1-\varepsilon)\binom{q}{k}$ sets $\{i_1, \ldots, i_k\}$ of distinct elements of $[q]$ such that G is ε-regular in $Q_{j_1} \times \cdots \times Q_{j_k}$ for every permutation $(j_1, \ldots, j_k)$ of $\{i_1, \ldots, i_k\}$.

3.2.1. A regularity lemma for k-partite k-graphs.

Using the SR-system $\mathcal{PG}(X_1, \ldots, X_k)$ in Example 9, we obtain a regularity lemma for k-partite k-graphs.

Theorem 18. *Let $X_1, \ldots, X_k$ be disjoint sets with $|X_1| = \cdots = |X_k| = n$. For all $0 < \varepsilon < 1$ and positive integers k, l, there exist $n_0(k,\varepsilon)$ and $q(k,l,\varepsilon)$ such that if $n > n_0(k,\varepsilon)$ and $\mathcal{L}$ is a collection of l distinct k-partite k-graphs with vertex classes $X_1, \ldots, X_k$, then for each $i \in [k]$, there exists a partition $\mathcal{Q}_i = \{Q_{i0}, \ldots, Q_{iq}\}$ of X_i, satisfying:*

(i) $q \le q(k,l,\varepsilon)$;

(ii) $|Q_{i,0}| < \varepsilon n$, $|Q_{i,1}| = \cdots = |Q_{i,q}| < \varepsilon n$;

(iii) for every graph $G \in \mathcal{L}$, there exist at least $(1-\varepsilon)q^k$ vectors $(i_1, \ldots, i_k) \in [q]^k$ such that G is ε-regular in $Q_{1,i_1} \times \cdots \times Q_{k,i_k}$.

3.3. Regularity lemmas for the cube $[0,1]^k$

Using the SR-system $\mathcal{B}^k = \left([0,1]^k, \mathbb{B}^k, \lambda^k, \mathbb{B}^{\langle k \rangle}\right)$, we obtain a regularity lemma for the k-dimensional unit cube.

Theorem 19. *For all $0 < \varepsilon < 1$ and positive integers k, l, there exists $q(k,l,\varepsilon)$ such that if $\mathcal{L}$ is a collection of l measurable subsets of the cube $[0,1]^k$, then there exists a partition $\mathcal{Q} = \{Q_0, \ldots, Q_q\}$ of $[0,1]$ into measurable sets satisfying:*

(i) $q \le q(k,l,\varepsilon)$;

(ii) $\mu(Q_0) < \varepsilon$, $\mu(Q_1) = \cdots = \mu(Q_q) < \varepsilon$;

(iii) every set $L \in \mathcal{L}$ is ε-regular in at least $(1-\varepsilon)q^k$ sets $Q_{i_1} \times \cdots \times Q_{i_k}$, where $(i_1, \ldots, i_k)$ is a k-tuple of distinct elements of $[q]$.

Finally, using the SR-system $\mathcal{BI}^k = \left([0,1]^k, \mathbb{B}^k, \lambda^k, \mathbb{I}^{\langle k \rangle}\right)$ in Example 8, we obtain a result that seems to illustrate our approach particularly well.

Theorem 20. *For all $0 < \varepsilon < 1$ and positive integers k, l, there exists $q(k,l,\varepsilon)$ such that if $\mathcal{L}$ is a collection of l measurable subsets of the cube $[0,1]^k$ then there exists a partition $\mathcal{Q} = \{Q_0, \ldots, Q_q\}$ of $[0,1]$ satisfying:*

(i) $q \le q(k,l,\varepsilon)$;

(ii) $\lambda(Q_0) < \varepsilon$, and the sets $Q_1, \ldots, Q_q$ are intervals of equal length less than ε;

(iii) Every set $L \in \mathcal{L}$ is ε-regular in at least $(1-\varepsilon)q^k$ sets $Q_{i_1} \times \cdots \times Q_{i_k}$, where $(i_1, \ldots, i_k)$ is a k-tuple of distinct elements of $[q]$.

3.4. Regularity lemmas for functions and matrices

Suppose $(X, \mathbb{A}, \mu, \mathbb{S})$ is an SR-system and let $\int$ be the integral defined in $(X, \mathbb{A}, \mu)$. As usual, write $L_1(X)$ be the space of real measurable functions defined in X and with integrable absolute value. Write χ_U for the indicator function of a set $U \subset X$.

Definition 21. Suppose $\varepsilon > 0$, $V \in \mathbb{S}$, and $\mu(V) > 0$. A function $f \in L_1(X)$ is said to be **ε-regular** in V if

$$\left| \frac{\int f \cdot \chi_U}{\mu(U)} - \frac{\int f \cdot \chi_V}{\mu(V)} \right| < \varepsilon$$

for every $U \in \mathbb{S}$ such that $U \subset V$ and $\mu(U) > \varepsilon\mu(V)$.

As with sets, we extend ε-regularity of functions to partitions.

Definition 22. Let $0 < \varepsilon < 1$ and $\mathcal{P} \in \Pi(\mathbb{S})$. We call a function $f \in L_1(X)$ **ε-regular** in $\mathcal{P}$ if

$$\sum \left\{ \mu(P) \colon P \in \mathcal{P},\ f \text{ is not } \varepsilon\text{-regular in } P \right\} < \varepsilon.$$

Remarks:

– if a function f is constant in $V \in \mathbb{S}$, then f is ε-regular in V;

– a set $A \in \mathbb{A}$ is ε-regular in $V \in \mathbb{S}$ iff χ_A is ε-regular in V;

– a set $A \in \mathbb{A}$ is ε-regular in a partition $\mathcal{P} \in \Pi(\mathbb{S})$ iff χ_A is ε-regular in $\mathcal{P}$.

Having extended all necessary definitions, we give next a regularity lemma for L_1-functions; the proof is in 4.3.

Theorem 23. *Suppose that* $0 < \varepsilon < 1$, r *is a positive integer and* $\varphi\colon \mathbb{N} \to \mathbb{N}$ *is an increasing function. There exists* $q = q(\varepsilon, r, \varphi)$ *such that the following assertion holds.*

Let $(X, \mathbb{A}, \mu, \mathbb{S})$ *be an SR-system, where* $\mathbb{S}$ *is* r*-built, and let* Φ *be a family of partitions bounding* $\Pi(\mathbb{S})$ *with rate* φ. *For every* $f \in L_1(X)$ *satisfying* $|f| \le 1$, *there exists a partition* $\mathcal{Q} \in \Phi$ *with* $|\mathcal{Q}| \le q$ *such that* f *is* ε*-regular in* $\mathcal{Q}$.

3.4.1. A regularity lemma for matrices. Let $\mathcal{PG}\big([m], [n]\big)$ be the SR-system in Example 9 with $X_1 = [m]$, $X_2 = [n]$. Then the set of L_1 functions defined in $[m] \times [n]$ is just the set of all real $m \times n$ matrices. It is worth spelling out the definition for ε-regularity for this particular case.

Let $P \subset [m]$, $Q \subset [n]$. An $m \times n$ matrix $A = (a_{ij})$ is ε-regular in $P \times Q$ if for every $U \subset P$, $V \subset Q$ such that $|U|\,|V| > \varepsilon |P|\,|Q|$, we have

$$\left| \frac{\sum_{i\in U,\ j\in V} a_{ij}}{|U|\,|V|} - \frac{\sum_{i\in P,\ j\in Q} a_{ij}}{|P|\,|Q|} \right| < \varepsilon.$$

Thus, we obtain the following regularity lemma for real $m \times n$ matrices.

Theorem 24. *For all* $0 < \varepsilon < 1$ *and positive integers* k *there exist* $n_0(k, \varepsilon)$ *and* $q(k, \varepsilon)$ *such that if* $m > n_0(k, \varepsilon)$, $n > n_0(k, \varepsilon)$, *and* $A = (a_{ij})$ *is a real* $m \times n$ *matrix, satisfying* $|a_{ij}| \le 1$, *then there exist partitions* $\mathcal{P} = \{P_0, \dots, P_p\}$ *of* $[m]$ *and* $\mathcal{Q} = \{Q_0, \dots, Q_q\}$ *of* $[n]$ *such that* $p \le q(k, \varepsilon)$, $q \le q(k, \varepsilon)$, *and:*

(i) $|P_0| < \varepsilon m$, $|P_1| = \cdots = |P_p| < \varepsilon m$;

(ii) $|Q_0| < \varepsilon n$, $|Q_1| = \cdots = |Q_q| < \varepsilon n$;

(iii) A *is* ε*-regular in at least* $(1 - \varepsilon)pq$ *sets* $P_i \times Q_j$, $i \in [p]$, $j \in [q]$.

Remark. Along the same lines we can obtain a regularity lemma for multidimensional matrices, which can be thought also as weighted, directed, multipartite, uniform hypergraphs.

4. Proofs

Proof of Lemma 4. Let

$$A = A_1 \times \cdots \times A_k,\ A_i \in \mathbb{S} \text{ and } A_i \cap A_j = \emptyset \text{ or } A_i = A_j \text{ for all } i,j \in [k]$$

$$B = B_1 \times \cdots \times B_k,\ B_i \in \mathbb{S} \text{ and } B_i \cap B_j = \emptyset \text{ or } A_i = A_j \text{ for all } i,j \in [k]$$

Since

$$A \cap B = (A_1 \cap B_1) \times \cdots \times (A_k \cap B_k),$$

we see that $\mathbb{S}^{\langle k \rangle}$ is closed under intersections.

Suppose now that $\mathbb{S}$ is r-built. We shall show that $\mathbb{S}^{\langle k \rangle}$ is $(2r+1)^{k^2}$-built. For every $i,j \in [k]$, there is a partition $\mathcal{P}_{ij}$ of $A_i \cup B_j$ into at most $2r+1$ members of $\mathbb{S}$. Define the family of sets $\mathcal{R}$ as

$$\mathcal{R} = \left\{ R \cap S \colon R \in \mathcal{P}_{ij}, S \in \mathcal{P}_{gh},\ i,j,g,h \in [k] \right\}.$$

Clearly the sets in $\mathcal{R}$ belong to $\mathbb{S}$. Since $\mathcal{R}$ partitions A_i and B_i for all $i \in [k]$, we see that $\mathcal{R}^{\langle k \rangle}$ partitions A, B and $A \cap B$, and so $\mathcal{R}^{\langle k \rangle}$ partitions $A \backslash B$. The members of $\mathcal{R}^{\langle k \rangle}$ belong to $\mathbb{S}^{\langle k \rangle}$; in view of $|\mathcal{R}| \le (2r+1)^{k^2}$, we see that $A \backslash B$ is a union of at most $(2r+1)^{k^2}$ members of $\mathbb{S}^{\langle k \rangle}$, completing the proof. ■

Proof of Lemma 12. Suppose that $\mathcal{P} = \{P_1, \ldots, P_p\} \in \Pi\big(\mathbb{S}^{\langle k \rangle}\big)$, and for every $i \in [p]$, let

$$P_i = Q_{i1} \times \cdots \times Q_{ik}, \quad \text{where} \quad Q_{ij} \in \mathbb{S} \quad \text{for} \quad j = 1, \ldots, k.$$

Observe that $X = \cup_{i,j} Q_{ij}$; thus for every $i \in [p]$, $j \in [k]$, there exist disjoint sets $R_{ij1}, \ldots, R_{ijq} \in \mathbb{S}$ such that

$$X \backslash Q_{ij} = \bigcup_{s=1}^{q} R_{ijs}$$

and $q \le (kp-1)r$. Now setting $\mathcal{R}_{ij} = \{Q_{ij}, R_{ij1}, \ldots, R_{ijq}\}$, we see that $\mathcal{R}_{ij} \in \Pi(\mathbb{S})$ and $|\mathcal{R}_{ij}| < kpr$ for all $i \in [p]$, $j \in [k]$. Finally, set

$$\mathcal{Q} = \bigcap_{i \in [p], j \in [k]} \mathcal{R}_{ij}$$

and note that $\mathcal{Q}$ refines each set Q_{ij}, $\mathcal{Q}$ belongs to $\Pi(\mathbb{S})$ and $|\mathcal{Q}| < (kpr)^{kp}$. Since $\mathcal{Q}^k$ refines P_i for all $i \in [p]$ and $\mathcal{Q}^k \in \Phi^k$, we see that the rate of Φ^k is at most $|\mathcal{Q}^k| < (kpr)^{k^2pr}$, completing the proof of Lemma 12. ■

4.1. Proof of Theorem 13

Our proof is an adaptation of the original proof of Szemerédi [18] (see also [1] and [14]). The following basic lemma is known as the "defect form of the Cauchy–Schwarz inequality"; for a proof see [1].

Lemma 25. *Let x_i and c_i be positive numbers for $i = 1, \ldots, n$. Then*

$$\sum_{i=1}^{n} c_i \sum_{i=1}^{n} c_i x_i^2 \geq \left(\sum_{i=1}^{n} c_i x_i \right)^2 .$$

If J is a proper subset of $[n]$ and $\gamma > 0$ is such that

$$\sum_{i=1}^{n} c_i \sum_{i \in J} c_i x_i \geq \sum_{i=1}^{n} c_i x_i \sum_{i \in J} c_i + \gamma,$$

then

$$\sum_{i=1}^{n} c_i \sum_{i=1}^{n} c_i x_i^2 \geq \left(\sum_{i=1}^{n} c_i x_i \right)^2 + \gamma^2 \Big/ \left(\sum_{i \in J} c_i \sum_{i \in [n] \setminus J} c_i \right). \qquad ■$$

Let $\mathcal{P} = \{P_1, \ldots, P_p\} \in \Pi(\mathbb{S})$, and $A \in \mathbb{A}$. Define the *index* of $\mathcal{P}$ with respect to A as

$$\operatorname{ind}_A \mathcal{P} = \sum_{P_i \in \mathcal{P}} \mu(P_i) d^2(A \cap P_i).$$

Note that for every $A \in \mathbb{A}$,

(3)
$$\operatorname{ind}_A \mathcal{P} = \sum_{P_i \in \mathcal{P}} \mu(P_i) d^2(A \cap P_i) \leq \sum_{P_i \in \mathcal{P},\ \mu(P_i) > 0} \frac{\mu(A \cap P_i)\mu(P_i)}{\mu(P_i)} = \mu(A) \leq 1.$$

Lemma 26. *If $\mathcal{P}, \mathcal{Q} \in \Pi(\mathbb{S})$, $A \in \mathbb{A}$, and $\mathcal{Q} \succ \mathcal{P}$ then $\operatorname{ind}_A \mathcal{Q} \geq \operatorname{ind}_A \mathcal{P}$.*

Proof. For simplicity we shall assume that $\mathcal{P}$ and $\mathcal{Q}$ consist only of sets of positive measure. Fix $P \in \mathcal{P}$ and for every $Q_i \subset P$, set

$$c_i = \mu(Q_i) \quad \text{and} \quad x_i = d(A \cap Q_i).$$

Note that

$$\sum_{Q_i \subset P} c_i = \sum_{Q_i \subset P} \mu(Q_i) = \mu(P) \quad \text{and} \quad \sum_{Q_i \subset P} c_i x_i = \mu(A \cap P).$$

The Cauchy–Schwarz inequality (the first part of Lemma 25) implies that

$$\sum_{Q_i \subset P} \mu(Q_i) d^2(A \cap Q_i) = \sum_{Q_i \subset P} c_i x_i^2 \geq \frac{1}{\mu(P)} \left(\sum_{Q_i \subset P} c_i x_i \right)^2 = \frac{\mu^2(A \cap P)}{\mu(P)}.$$

Summing over all sets $P \in \mathcal{P}$, the desired inequality follows. ∎

The next lemma supports the proof of Lemma 28.

Lemma 27. *Suppose $A, S, T \in \mathbb{A}$, $T \subset S$ and $\mu(T) > 0$. If*

$$\big| d(A \cap T) - d(A \cap S) \big| \geq \varepsilon \tag{4}$$

then every partition $\mathcal{U} = \{U_1, \ldots, U_p\} \in \Pi(\mathbb{A})$ such that $\mathcal{U} \succ S$ and $\mathcal{U} \succ T$, satisfies

$$\sum_{U_i \subset S} \mu(U_i) d^2(A, U_i) \geq \mu(S) d^2(A, S) + \varepsilon^2 \mu(T).$$

Proof. Let the partition $\mathcal{U} = \{U_1, \ldots, U_p\} \in \Pi(\mathbb{A})$ be such that $\mathcal{U} \succ S$ and $\mathcal{U} \succ T$. For every $U_i \subset S$, set

$$c_i = \mu(U_i), \qquad x_i = d(A, U_i),$$

and observe that

$$\sum_{U_i \subset S} c_i = \sum_{U_i \subset S} \mu(U_i) = \mu(S) \quad \text{and} \quad \sum_{U_i \subset S} c_i x_i = \sum_{U_i \subset S} \mu(A \cap U_i) = \mu(A \cap S).$$

Similarly, we have

$$\sum_{U_i \subset T} c_i = \mu(T) \quad \text{and} \quad \sum_{U_i \subset T} c_i x_i = \sum_{U_i \subset T} \mu(A \cap U_i) = \mu(A \cap T).$$

Inequality (4) implies that either

$$d(A,T) > d(A,S) + \varepsilon \tag{5}$$

or

$$d(A,S) > d(A,T) + \varepsilon.$$

The arguments in the two cases are identical; thus assume that (5) holds. Hence, $\mu(T) \neq \mu(S)$, so $T \subset S$ implies that $\mu(T) < \mu(S)$. Furthermore,

$$\sum_{U_i \subset T} c_i x_i = d(A,T)\mu(T) > \big(d(A,S)+\varepsilon\big)\mu(T) = \big(d(A,S)+\varepsilon\big)\sum_{U_i \subset T} c_i$$

$$= \frac{\sum_{U_i \subset S} c_i x_i}{\sum_{U_i \subset S} c_i} \sum_{U_i \subset T} c_i + \varepsilon \sum_{U_i \subset T} c_i.$$

By the definition of c_i and x_i, we have

$$\sum_{U_i \subset S} \mu(U_i) d^2(A,U_i) = \sum_{U_i \subset S} c_i x_i^2.$$

Therefore, setting

$$\lambda = \varepsilon \sum_{U_i \subset S} c_i \sum_{U_i \subset T} c_i = \varepsilon \mu(T)\mu(S),$$

and applying the second part of Lemma 25, we find that

$$\mu(S) \sum_{U_i \subset S} \mu(U_i) d^2(A,U_i) \geq \Big(\sum_{U_i \subset S} c_i x_i\Big)^2 + \lambda^2 \Big/ \Big(\sum_{U_i \subset T} c_i \sum_{U_i \subset S \setminus T} c_i\Big)$$

$$= \mu^2(A \cap S) + \varepsilon^2 \frac{\mu^2(S)\mu(T)}{\mu(S) - \mu(T)}.$$

Hence,

$$\sum_{U_i \subset S} \sum_{U_i \subset S} \mu(U_i) d^2(A,U_i) \geq \mu(S) d^2(A,S) + \varepsilon^2 \frac{\mu(T)}{\mu(S)\big(\mu(S) - \mu(T)\big)}$$

$$> \mu(S) d^2(A,S) + \varepsilon^2 \mu(T)$$

and this is exactly the desired inequality. ■

The following lemma gives a condition for an absolute increase of $\operatorname{ind}_A \mathcal{P}$ resulting from refining.

Lemma 28. *Let $0 < \varepsilon < 1$ and $\mathbb{S}$ be r-built. If $\mathcal{P} \in \Pi(\mathbb{S})$ and $A \in \mathbb{A}$ is not ε-regular in $\mathcal{P}$ then there exists $\mathcal{Q} \in \Pi(\mathbb{S})$ satisfying $\mathcal{Q} \succ \mathcal{P}$, $|\mathcal{Q}| \leq (r+1)|\mathcal{P}|$, and*

$$\operatorname{ind}_A \mathcal{Q} \geq \operatorname{ind}_A \mathcal{P} + \varepsilon^4. \tag{6}$$

Proof. Let $\mathcal{P} = \{P_1, \ldots, P_p\}$ and $\mathcal{N}$ be the set of all P_i for which A is not ε-regular in P_i. Since A is not ε-regular in $\mathcal{P}$, by definition, we have

$$\sum_{P_i \in \mathcal{N}} \mu(P_i) \geq \varepsilon.$$

For every $P_i \in \mathcal{N}$, since A is not ε-regular in P_i, there is a set $T_i \subset P_i$ such that $T_i \in \mathbb{S}$, $\mu(T_i) > \varepsilon\mu(P_i)$, and

$$\left| d(A, P_i) - d(A, T_i) \right| \geq \varepsilon.$$

Since $\mathbb{S}$ is r-built, for every $P_i \in \mathcal{N}$, there is a partition of $P_i \backslash T_i$ into r disjoint sets $A_{i1}, \ldots A_{is} \in \mathbb{S}$; hence $\{A_{i1}, \ldots A_{is}, T_i\}$ is a partition of P_i into at most $(r+1)$ sets belonging to $\mathbb{S}$. Let $\mathcal{Q}$ be the collection of all sets $A_{i1}, \ldots A_{is}, T_i$, where $P_i \in \mathcal{N}$, together with all sets $P_j \in \mathcal{P} \backslash \mathcal{N}$. Clearly $\mathcal{Q} \in \Pi(\mathbb{S})$; also, $\mathcal{Q} \succ T_i$ and $\mathcal{Q} \succ P_i$ for every $P_i \in \mathcal{N}$, and

$$|\mathcal{Q}| \leq (r+1)|\mathcal{N}| + |\mathcal{P}| - |\mathcal{N}| \leq (r+1)|\mathcal{P}|.$$

Thus, to finish the proof, we have to prove (6). Let $\mathcal{Q} = \{Q_1, \ldots, Q_q\}$. For every $P_k \in \mathcal{N}$, Lemma 27 implies that

$$\begin{aligned} \sum_{Q_i \subset P_k} \mu(Q_i) d^2(A, Q_i) &\geq \mu(P_k) d^2(A, P_k) + \varepsilon^2 \mu(T_k) \\ &\geq \mu(P_k) d^2(A, P_k) + \varepsilon^3 \mu(P_k). \end{aligned} \tag{7}$$

For any $P_k \in \mathcal{P}$, Lemma 26 implies that

$$\sum_{Q_i \subset P_k} \mu(Q_i) d^2(A, Q_i) \geq \mu(P_k) d^2(A, P_k). \tag{8}$$

Now, by (7) and (8), we obtain

$$\operatorname{ind}_A \mathcal{Q} = \sum_{Q_i \subset \mathcal{Q}} \mu(Q_i) d^2(A, Q_i) \geq \sum_{P_i \in \mathcal{P}} \mu(P_k) d^2(A, P_i) + \varepsilon^3 \sum_{P_i \in \mathcal{N}} \mu(P_k)$$

$$\geq \sum_{P_i \in \mathcal{P}} \mu(P_k) d^2(A, P_i) + \varepsilon^4 = \operatorname{ind}_A \mathcal{P} + \varepsilon^4,$$

completing the proof. ■

Proof of Theorem 13. Suppose that $\mathbb{S}$ is r-built, Φ bounds $\Pi(\mathbb{S})$ with rate $\varphi(\cdot)$ and set $p = |\mathcal{P}|$. Define a function $\psi\colon \mathbb{N} \to \mathbb{N}$ by

(9) $$\psi(1, p) = p;$$

$$\psi(s+1, p) = (r+1)\varphi\big(\psi(s, p)\big), \quad \text{for every } s > 1.$$

We shall show that the partition $\mathcal{Q} \in \Phi$ may be selected so that $|\mathcal{Q}| \leq \psi\big(l\lfloor\varepsilon^{-4}\rfloor, p\big)$.

Select first a partition $\mathcal{P}_0 \in \Phi$ such that $\mathcal{P}_0 \succ \mathcal{P}$ and $|\mathcal{P}_0| \leq \varphi\big(|\mathcal{P}|\big)$. We build recursively a sequence of partitions $\mathcal{P}_1, \mathcal{P}_2, \ldots$ satisfying

(10) $$\mathcal{P}_{i+1} \succ \mathcal{P}_i,$$

(11) $$|\mathcal{P}_{i+1}| \leq \varphi\big((r+1)|\mathcal{P}_i|\big),$$

(12) $$\exists A_i \in \mathcal{G}\colon \operatorname{ind}_{A_i} \mathcal{P}_{i+1} \geq \operatorname{ind}_{A_i} \mathcal{P}_i + \varepsilon^4$$

for every $i = 0, 1, \ldots$ The sequence $\mathcal{P}_1, \mathcal{P}_2, \ldots$ is built according the following rule: if all $A \in \mathcal{L}$ are ε-regular in $\mathcal{P}_i$, then we stop; otherwise select a set $A_i \in \mathcal{L}$ that is not ε-regular in $\mathcal{P}_i$. Then, by Lemma 28, there is a partition $\mathcal{P}_i' \in \Pi(\mathbb{S})$ such that

$$\mathcal{P}_i' \succ \mathcal{P}_i,$$

$$|\mathcal{P}_i'| \leq (r+1)|\mathcal{P}_i|,$$

$$\operatorname{ind}_{A_i} \mathcal{P}_i' \geq \operatorname{ind}_{A_i} \mathcal{P}_i + \varepsilon^4.$$

Since Φ bounds $\mathbb{S}$ with rate φ, we can select a partition $\mathcal{P}_{i+1} \in \Phi$ such that $\mathcal{P}_{i+1} \succ \mathcal{P}_i'$ and $|\mathcal{P}_{i+1}| \leq \varphi\big(|\mathcal{P}_i'|\big)$. Hence, (10), (11), and (12) hold.

Set $k = \lfloor \varepsilon^{-4} \rfloor$. If the sequence $\mathcal{P}_0, \mathcal{P}_1, \ldots$ has more than lk terms then, by the pigeonhole principle, there exist a set $A \in \mathcal{L}$ and a sequence $\mathcal{P}_{i_1}, \ldots, \mathcal{P}_{i_{k+1}}$ such that

$$\operatorname{ind}_A \mathcal{P}_{i_{j+1}} \geq \operatorname{ind}_A \mathcal{P}_{i_j} + \varepsilon^4$$

for every $j = 1, \ldots, k$. Hence we find that

$$\operatorname{ind}_A \mathcal{P}_{i_{k+1}} \geq \operatorname{ind}_A \mathcal{P}_{i_1} + k\varepsilon^4 > k\varepsilon^4 \geq 1,$$

contradicting (3). Therefore, all $A \in \mathcal{L}$ are ε-regular in some partition $\mathcal{Q} = \mathcal{P}_i$. By (11), $|\mathcal{Q}| \leq \psi\left(l \lfloor \varepsilon^{-4} \rfloor, p\right)$, completing the proof. ■

4.2. Proof of Lemma 14

Proof. Select a partition $\mathcal{P} = \{P_1, \ldots, P_p\} \in \Pi\left(\left(2^{[n]}\right)^{\langle k \rangle}\right)$; for every $i \in [p]$, let

$$P_i = R_{i1} \times \cdots \times R_{ik}, \; R_{ij} \subset [n], \text{ for } j \in [k].$$

Let

$$\mathcal{R} = \bigcap_{i \in [p], \; j \in [k]} \{R_{ij}, X \backslash R_{ij}\}.$$

and set $r = |\mathcal{R}|$. Clearly, $r \leq 2^{pk}$. Our first goal is to find an ε-equitable partition $\mathcal{Q} \succ \mathcal{R}$ with

$$|\mathcal{Q}| \leq \left(\frac{2}{\varepsilon} + 1\right) 2^{pk}.$$

Suppose first that $n \geq 2r/\varepsilon$. Partition every $R \in \mathcal{R}$ into sets of size $\lfloor \varepsilon n/r \rfloor$ and a smaller residual set; write $\mathcal{Q}$ for the resulting partition. The measure of each member of $\mathcal{Q}$ that is not residual is at most $\lfloor \varepsilon n/r \rfloor / n \leq \varepsilon$. The total measure of all residual sets is less than

$$\frac{\lfloor \varepsilon n/r \rfloor}{n} r \leq \varepsilon,$$

thus, $\mathcal{Q}$ is an ε-equitable partition refining $\mathcal{R}$. Since

$$|\mathcal{Q}| \leq \frac{n}{\lfloor \varepsilon n/r \rfloor} + r \leq \frac{2n}{\varepsilon n/r} + r = \left(\frac{2}{\varepsilon} + 1\right) r \leq \left(\frac{2}{\varepsilon} + 1\right) 2^{pk},$$

$\mathcal{Q}$ has the required properties.

Let now $n < 2r/\varepsilon$ and $\mathcal{Q}$ be the partition of $[n]$ into n sets of size 1. Since $\varepsilon > 1/n$, the partition $\mathcal{Q}$ is ε-equitable and refines $\mathcal{R}$. Since

$$|\mathcal{Q}| = n < \frac{2}{\varepsilon} r \leq \left(\frac{2}{\varepsilon} + 1\right) 2^{pk},$$

$\mathcal{Q}$ has the required properties.

To complete the proof observe that $\mathcal{Q}^k \in \Phi^k(n, \varepsilon)$, $\mathcal{Q}^k \succ \mathcal{R}^k \succ \mathcal{P}$, and

$$|\mathcal{Q}^k| \leq |\mathcal{Q}|^k \leq \left(\frac{2}{\varepsilon} + 1\right)^k 2^{pk^2}. \qquad \blacksquare$$

4.3. Proof of Theorem 23

Proof. Set $l = \lceil 4/\varepsilon \rceil$, $\delta = \varepsilon/(2l+1)$. For every $k \in [l]$, define the sets

$$A_k = \left\{x \colon \frac{k-1}{l} < f(x) \leq \frac{k}{l}\right\}, \qquad A_{-k} = \left\{x \colon -\frac{k}{l} \leq f(x) < -\frac{k-1}{l}\right\}$$

and let $A_0 = \{x \colon f(x) = 0\}$. Since f is measurable, all sets $A_{-l}, \ldots, A_l$ belong to $\mathbb{A}$. Let q be the function defined in Theorem 13; applying Theorem 13 with $\varepsilon = \delta$, $\mathcal{P} = \{X\}$, and $\mathcal{L} = \{A_{-l}, \ldots, A_l\}$, we see that there exists a partition $\mathcal{Q} \in \Phi$ such that $\mathcal{Q} \succ X$, $|\mathcal{Q}| \leq q(\delta, 2l+1, 1, r, \varphi)$, and every A_k is δ-regular in $\mathcal{Q}$. We shall deduce that f is ε-regular in $\mathcal{Q}$. Set

$$\widehat{f} = \sum_{k=1}^{l} \frac{k-1}{l} \chi_{A_k} - \sum_{k=1}^{l} \frac{k-1}{l} \chi_{A_{-k}}$$

and note that $\left|f(x) - \widehat{f}(x)\right| < 1/l$ for every $x \in X$. Let $Q \in \mathcal{Q}$ be such that all $A_{-l}, \dots, A_l$ are δ-regular in Q, and let $U \in \mathbb{S}$ satisfies $U \subset Q$ and $\mu(U) > \varepsilon\mu(Q)$. Then, we have

$$\begin{aligned}
&\left|\frac{\int f \cdot \chi_U}{\mu(U)} - \frac{\int f \cdot \chi_Q}{\mu(Q)}\right| \\
&\leq \left|\frac{\int \widehat{f} \cdot \chi_U}{\mu(U)} - \frac{\int \widehat{f} \cdot \chi_Q}{\mu(Q)}\right| + \left|\frac{\int |f - \widehat{f}| \cdot \chi_U}{\mu(U)}\right| + \left|\frac{\int |f - \widehat{f}| \cdot \chi_Q}{\mu(Q)}\right| \\
&< \sum_{k=1}^{l} \frac{k-1}{l} \left|\frac{\int \chi_{A_k} \cdot \chi_U}{\mu(U)} - \frac{\int \chi_{A_k} \cdot \chi_Q}{\mu(Q)}\right| \\
&+ \sum_{k=1}^{l} \frac{k}{l} \left|\frac{\int \chi_{A_{-k}} \cdot \chi_U}{\mu(U)} - \frac{\int \chi_{A_{-k}} \cdot \chi_Q}{\mu(Q)}\right| + 2\int |f - \widehat{f}| \\
&= 2\sum_{k=1}^{l} \frac{k-1}{l}\delta + \frac{2}{l} = (l-1)\delta + \frac{2}{l} = (l-1)\frac{\varepsilon}{2l+1} + \frac{2}{4/\varepsilon} < \varepsilon.
\end{aligned}$$

This implies that f is ε-regular in every set $Q \in \mathcal{Q}$ such that all $A_{-l}, \dots, A_l$ are δ-regular in Q. On the other hand, all sets $A_{-l}, \dots, A_l$ are δ-regular in $\mathcal{Q}$; hence, for every s, such that $-l \leq s \leq l$, we have

$$\sum \left\{ \mu(P) \colon P \in \mathcal{Q},\ A_s \text{ is not } \delta\text{-regular in } P \right\} < \delta.$$

Therefore,

$$\begin{aligned}
&\sum \left\{ \mu(P) \colon P \in \mathcal{Q},\ f \text{ is not } \varepsilon\text{-regular in } P \right\} \\
&\leq \sum \{ \mu(P) \colon P \in \mathcal{Q}, \text{ there exists } s,\ -l \leq s \leq l, \\
&\qquad \text{such that } A_s \text{ is not } \delta\text{-regular in } P \} \\
&< (2l+1)\delta = \varepsilon.
\end{aligned}$$

Hence, $f(x)$ is ε-regular in $\mathcal{Q}$, completing the proof. ■

REFERENCES

[1] B. Bollobás, *Modern Graph Theory,* Graduate Texts in Mathematics **184,** Springer-Verlag, New York (1998), xiv + 394 pp.

[2] B. Bollobás and O. Riordan, Sparse graphs: metrics and random models, to appear; preprint available at arXiv:0708.1919

[3] C. Borgs, J. T. Chayes, L. Lovász, V.T. Sós and K. Vesztergombi, Convergent sequences of dense graphs I: Subgraph frequencies, metric properties and testing. Preprint, May 2006 (revised Jan 2007).

[4] C. Borgs, J. T. Chayes, L. Lovász, V.T. Sós and K. Vesztergombi, Convergent sequences of dense graphs II: Multiway cuts and statistical physics. Preprint, March 2007.

[5] C. Borgs, J. T. Chayes and L. Lovász, Unique limits of dense graph sequences (in preparation).

[6] F. Chung, Regularity lemmas for hypergraphs and quasi-randomness, *Random Structures and Algorithms,* **2** (1991), 241–252.

[7] G. Elek and B. Szegedy, Limits of hypergraphs, removal and regularity lemmas. A non-standard approach, preprint available at *arXiv:0705.2179v1*

[8] A. Frieze and R. Kannan, Quick approximation to matrices and applications, *Combinatorica,* **19** (1999), 175–220.

[9] S. Gerke, Y. Kohayakawa, V. Rödl and A. Steger, Small subsets inherit sparse ε-regularity, *J. Combin. Theory Ser. B.,* **97** (2007), 34–56.

[10] T. Gowers, Hypergraph regularity and the multidimensional Szemerédi theorem, *Ann. of Math.,* **166** (2007), 897–946.

[11] Y. Ishigami, A simple regularization of hypergraphs, preprint available at *arXiv:math/0612838v1*

[12] Y. Kohayakawa, Szemerédi's regularity lemma for sparse graphs, in *Foundations of Computational Mathematics* (Rio de Janeiro, 1997), Springer, Berlin, 1997, pp. 216–230.

[13] Y. Kohayakawa and V. Rödl, Regular pairs in sparse random graphs I, *Random Structures Algorithms,* **22** (2003), 359–434.

[14] J. Komlós and M. Simonovits, Szemerédi's regularity lemma and its applications to graph theory, in: *Combinatorics: Paul Erdős is Eighty,* Vol. 2, (D. Miklós, V. T. Sós and T. Szőnyi, eds.), Bolyai Society Math. Studies, Keszthely, Hungary, 1996, pp. 295–352.

[15] L. Lovász and B. Szegedy, Limits of dense graph sequences, *J. Combin. Theory B.,* **96** (2006), 933–957.

[16] L. Lovász and B. Szegedy, Szemerédi's lemma for the analyst, *Geometric and Functional Analysis,* **17** (2007), 252–270.

[17] V. Rödl and J. Skokan, Regularity lemma for k-uniform hypergraphs, *Random Structures and Algorithms,* **25** (2004), 1–42.

[18] E. Szemerédi, Regular partitions of graphs, in: *Colloques Internationaux C.N.R.S. No. 260 – Problèmes Combinatoires et Théorie des Graphes*, Orsay, 1976, pp. 399–401.

[19] T. Tao, Szemerédi's regularity lemma revisited, *Contrib. Discrete Math.*, **1** (2006), 8–28.

Béla Bollobás
Department of Pure Mathematics and Mathematical Statistics
University of Cambridge
Cambridge CB3 0WB
U.K.

Béla Bollobás and
Vladimir Nikiforov
Department of Mathematical Sciences
University of Memphis
Memphis TN 38152
U.S.A.

BOLYAI SOCIETY
MATHEMATICAL STUDIES, 19

Building Bridges
pp. 241–281.

Isotropic PCA and Affine-Invariant Clustering

S. CHARLES BRUBAKER and SANTOSH S. VEMPALA

We present an extension of Principal Component Analysis (PCA) and a new algorithm for clustering points in R^n based on it. The key property of the algorithm is that it is affine-invariant. When the input is a sample from a mixture of two arbitrary Gaussians, the algorithm correctly classifies the sample assuming only that the two components are separable by a hyperplane, i.e., there exists a halfspace that contains most of one Gaussian and almost none of the other in probability mass. This is nearly the best possible, improving known results substantially [15, 10, 1]. For $k > 2$ components, the algorithm requires only that there be some $(k-1)$-dimensional subspace in which the *overlap* in every direction is small. Here we define overlap to be the ratio of the following two quantities: 1) the average squared distance between a point and the mean of its component, and 2) the average squared distance between a point and the mean of the mixture. The main result may also be stated in the language of linear discriminant analysis: if the standard Fisher discriminant [9] is small enough, labels are not needed to estimate the optimal subspace for projection. Our main tools are isotropic transformation, spectral projection and a simple reweighting technique. We call this combination *isotropic PCA*.

1. Introduction

We present an extension to Principal Component Analysis (PCA), which is able to go beyond standard PCA in identifying "important" directions. When the covariance matrix of the input (distribution or point set in R^n) is a multiple of the identity, then PCA reveals no information; the second moment along any direction is the same. Such inputs are called isotropic. Our extension, which we call *isotropic PCA*, can reveal interesting information in such settings. We use this technique to give an affine-invariant clustering algorithm for points in R^n. When applied to the problem of unraveling mixtures of arbitrary Gaussians from unlabeled samples, the algorithm yields substantial improvements of known results.

To illustrate the technique, consider the uniform distribution on the set $X = \{(x,y) \in \mathbb{R}^2 : x \in \{-1,1\},\ y \in [-\sqrt{3}, \sqrt{3}]\}$, which is isotropic. Suppose this distribution is rotated in an unknown way and that we would like to recover the original x and y axes. For each point in a sample, we may project it to the unit circle and compute the covariance matrix of the resulting point set. The x direction will correspond to the greater eigenvector, the y direction to the other. See Figure 1 for an illustration. Instead of projection onto the unit circle, this process may also be thought of as importance weighting, a technique which allows one to simulate one distribution with another. In this case, we are simulating a distribution over the set X, where the density function is proportional to $(1+y^2)^{-1}$, so that points near $(1,0)$ or $(-1,0)$ are more probable.

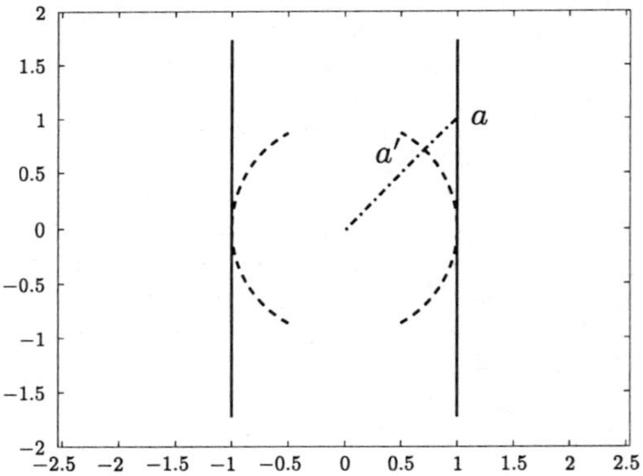

Fig. 1. Mapping points to the unit circle and then finding the direction of maximum variance reveals the orientation of this isotropic distribution.

In this paper, we describe how to apply this method to mixtures of arbitrary Gaussians in $\mathbb{R}^n$ in order to find a set of directions along which the Gaussians are well-separated. These directions span the Fisher subspace of the mixture, a classical concept in Pattern Recognition. Once these directions are identified, points can be classified according to which component of the distribution generated them, and hence all parameters of the mixture can be learned.

What separates this paper from previous work on learning mixtures is that our algorithm is affine-invariant. Indeed, for every mixture distribution that can be learned using a previously known algorithm, there is a linear transformation of bounded condition number that causes the algorithm to fail. For $k = 2$ components our algorithm has nearly the best possible guarantees (and subsumes all previous results) for clustering Gaussian mixtures. For $k > 2$, it requires that there be a $(k-1)$-dimensional subspace where the *overlap* of the components is small in every direction (See section 1.2). This condition can be stated in terms of the Fisher discriminant, a quantity commonly used in the field of Pattern Recognition with labeled data. Because our algorithm is affine invariant, it makes it possible to unravel a much larger set of Gaussian mixtures than had been possible previously.

The first step of our algorithm is to place the mixture in isotropic position (see Section 1.2) via an affine transformation. This has the effect of making the $(k-1)$-dimensional Fisher subspace, i.e., the one that minimizes the Fisher discriminant, the same as the subspace spanned by the means of the components (they only coincide in general in isotropic position), for *any* mixture. The rest of the algorithm identifies directions close to this subspace and uses them to cluster, without access to labels. Intuitively this is hard since after isotropy, standard PCA reveals no additional information. Before presenting the ideas and guarantees in more detail, we describe relevant related work.

1.1. Previous work

A mixture model is a convex combination of distributions of known type. In the most commonly studied version, a distribution F in R^n is composed of k unknown Gaussians. That is,

$$F = \mathrm{w}_1 N(\mu_1, \Sigma_1) + \ldots + \mathrm{w}_k N(\mu_k, \Sigma_k),$$

where the mixing weights w_i, means μ_i, and covariance matrices Σ_i are all unknown. Typically, $k \ll n$, so that a concise model explains a high dimensional phenomenon. A random sample is generated from F by first choosing a component with probability equal to its mixing weight and then picking a random point from that component distribution. In this paper, we study the classical problem of unraveling a sample from a mixture, i.e., labeling each point in the sample according to its component of origin.

Heuristics for classifying samples include "expectation maximization" [6] and "k-means clustering" [12]. These methods can take a long time and can get stuck with suboptimal classifications. Over the past decade, there has been much progress on finding polynomial-time algorithms with rigorous guarantees for classifying mixtures, especially mixtures of Gaussians

[5, 16, 15, 18, 10, 1]. Starting with Dasgupta's paper [5], one line of work uses the concentration of pairwise distances and assumes that the components' means are so far apart that distances between points from the same component are likely to be smaller than distances from points in different components. Arora and Kannan [15] establish nearly optimal results for such distance-based algorithms. Unfortunately their results inherently require separation that grows with the dimension of the ambient space and the largest variance of each component Gaussian.

To see why this is unnatural, consider k well-separated Gaussians in $\mathbb{R}^k$ with means $e_1, \ldots, e_k$, i.e. each mean is 1 unit away from the origin along a unique coordinate axis. Adding extra dimensions with arbitrary variance does not affect the separability of these Gaussians, but these algorithms are no longer guaranteed to work. For example, suppose that each Gaussian has a maximum variance of $\varepsilon \ll 1$. Then, adding $O^*(k\varepsilon^{-2})$ extra dimensions with variance ε will violate the necessary separation conditions.

To improve on this, a subsequent line of work uses spectral projection (PCA). Vempala and Wang [18] showed that for a mixture of *spherical* Gaussians, the subspace spanned by the top k principal components of the mixture contains the means of the components. Thus, projecting to this subspace has the effect of shrinking the components while maintaining the separation between their means. This leads to a nearly optimal separation requirement of

$$\|\mu_i - \mu_j\| \geq \tilde{\Omega}(k^{1/4}) \max\{\sigma_i, \sigma_j\}$$

where μ_i is the mean of component i and σ_i^2 is the variance of component i along any direction. Note that there is no dependence on the dimension of the distribution. Kannan et al. [10] applied the spectral approach to arbitrary mixtures of Gaussians (and more generally, logconcave distributions) and obtained a separation that grows with a polynomial in k and the largest variance of each component:

$$\|\mu_i - \mu_j\| \geq \text{poly}(k) \max\{\sigma_{i,\max}, \sigma_{j,\max}\}$$

where $\sigma_{i,\max}^2$ is the maximum variance of the ith component in any direction. The polynomial in k was improved in [1] along with matching lower bounds for this approach, suggesting this bound, in particular the dependence on the maximum component variances, to be the limit of spectral methods. Going beyond this "spectral threshold" for arbitrary Gaussians has been a major open problem.

The representative hard case is the special case of two parallel "pancakes", i.e., two Gaussians that are spherical in $n-1$ directions and narrow in the last direction, so that a hyperplane orthogonal to the last direction separates the two. The spectral approach requires a separation that grows with their largest standard deviation which is unrelated to the distance between the pancakes (their means). Other examples can be generated by starting with Gaussians in k dimensions that are separable and then adding other dimensions, one of which has large variance. Because there is a subspace where the Gaussians are separable, the separation requirement should depend only on the dimension of this subspace and the components' variances in it.

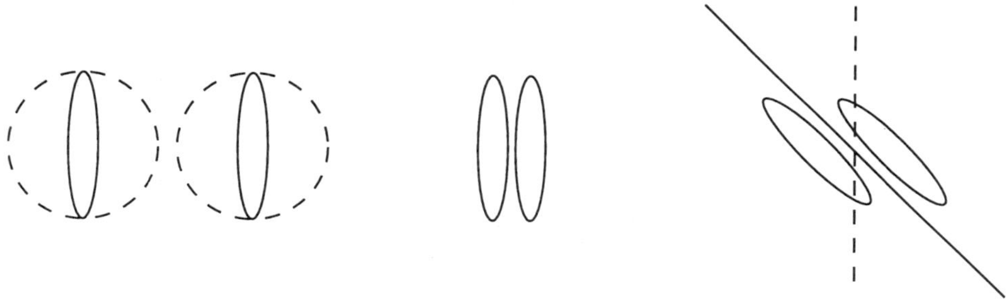

(a) Distance Concentration Separability (b) Hyperplane Separability (c) Intermean Hyperplane and Fisher Hyperplane

Fig. 2. Previous work requires distance concentration separability which depends on the maximum directional variance (a). Our results require only hyperplane separability, which depends only on the variance in the separating direction (b). For non-isotropic mixtures the best separating direction may not be between the means of the components (c).

A related line of work considers learning symmetric product distributions, where the coordinates are independent. Feldman et al [7] have shown that mixtures of axis-aligned Gaussians can be approximated without any separation assumption at all in time exponential in k. Chaudhuri and Rao [3] have recently given a polynomial-time algorithm for clustering mixtures of product distributions (axis-aligned Gaussians) under mild separation conditions. A. Dasgupta et al [4] and later Chaudhuri and Rao [2] gave algorithms for clustering mixtures of heavy-tailed distributions.

1.2. Results

We assume we are given a lower bound w on the minimum mixing weight and k, the number of components. With high probability, our algorithm UNRAVEL returns a partition of space by hyperplanes so that each part (a polyhedron) encloses almost all of the probability mass of a single component and almost none of the other components. The error of such a set of polyhedra is the total probability mass that falls outside the correct polyhedron.

We first state our result for two Gaussians in a way that makes clear the relationship to previous work that relies on separation.

Theorem 1. *Let* w_1, μ_1, Σ_1 *and* w_2, μ_2, Σ_2 *define a mixture of two Gaussians. There is an absolute constant* C *such that, if there exists a direction* v *such that*

$$\left|\operatorname{proj}_v(\mu_1 - \mu_2)\right| \geq C\left(\sqrt{v^T\Sigma_1 v} + \sqrt{v^T\Sigma_2 v}\,\right)\mathrm{w}^{-2}\log^{1/2}\left(\frac{1}{\mathrm{w}\delta} + \frac{1}{\eta}\right),$$

then with probability $1-\delta$ *algorithm* UNRAVEL *returns two complementary halfspaces that have error at most* η *using time and a number of samples that is polynomial in* n, w^{-1}, $\log(1/\delta)$.

So the separation required between the means is comparable to the standard deviation in *some direction*. This separation condition of Theorem 1 is affine-invariant and much weaker than conditions of the form $\|\mu_1 - \mu_2\| \gtrsim \max\{\sigma_{1,\max}, \sigma_{2,\max}\}$ used in previous work. See Figure 2. The dotted line shows how previous work effectively treats every component as spherical. We also note that the separating direction does not need to be the intermean direction as illustrated in Figure 2(c). The dotted line illustrates hyperplane induced by the intermean direction, which may be far from the optimal separating hyperplane shown by the solid line.

It will be insightful to state this result in terms of the Fisher discriminant, a standard notion from Pattern Recognition [9, 8] that is used with labeled data. In words, the Fisher discriminant along direction p is

$$J(p) = \frac{\text{the intra-component variance in direction } p}{\text{the total variance in direction } p}.$$

Mathematically, this is expressed as

$$J(p) = \frac{E\left[\left\|\text{proj}_p\left(x - \mu_{\ell(x)}\right)\right\|^2\right]}{E\left[\left\|\text{proj}_p(x)\right\|^2\right]} = \frac{p^T(\mathrm{w}_1\Sigma_1 + \mathrm{w}_2\Sigma_2)p}{p^T\left(\mathrm{w}_1(\Sigma_1 + \mu_1\mu_1^T) + \mathrm{w}_2(\Sigma_2 + \mu_2\mu_2^T)\right)p}$$

for x distributed according to a mixture distribution with means μ_i and covariance matrices Σ_i. We use $\ell(x)$ to indicate the component from which x was drawn.

Theorem 2. *There is an absolute constant C for which the following holds. Suppose that $\mathcal{F}$ is a mixture of two Gaussians such that there exists a direction p for which*

$$J(p) \leq C\mathrm{w}^3 \log^{-1}\left(\frac{1}{\delta \mathrm{w}} + \frac{1}{\eta}\right).$$

With probability $1-\delta$, algorithm Unravel *returns a halfspace with error at most η using time and sample complexity polynomial in n, w^{-1}, $\log(1/\delta)$.*

There are several ways of generalizing the Fisher discriminant for $k = 2$ components to greater k [8]. These generalizations are most easily understood when the distribution is isotropic. An isotropic distribution has the identity matrix as its covariance and the origin as its mean. An isotropic mixture therefore has

$$\sum_{i=1}^{k} \mathrm{w}_i\mu_i = 0 \quad \text{and} \quad \sum_{i=1}^{k} \mathrm{w}_i(\Sigma_i + \mu_i\mu_i^T) = I.$$

It is well known that any distribution with bounded covariance matrix (and therefore any mixture) can be made isotropic by an affine transformation. As we will see shortly, for $k = 2$, for an isotropic mixture, the line joining the means is the direction that minimizes the Fisher discriminant.

Under isotropy, the denominator of the Fisher discriminant is always 1. Thus, the discriminant is just the expected squared distance between the projection of a point and the projection of its mean, where projection is onto some direction p. The generalization to $k > 2$ is natural, as we may simply replace projection onto direction p with projection onto a $(k-1)$-dimensional subspace S. For convenience, let

$$\Sigma = \sum_{i=1}^{k} \mathrm{w}_i\Sigma_i.$$

Let the vector $p_1, \ldots, p_{k-1}$ be an orthonormal basis of S and let $\ell(x)$ be the component from which x was drawn. We then have under isotropy

$$J(S) = E\Big[\big\|\mathrm{proj}_S\big(x - \mu_{\ell(x)}\big)\big\|^2\Big] = \sum_{j=1}^{k-1} p_j^T \Sigma p_j$$

for x distributed according to a mixture distribution with means μ_i and covariance matrices Σ_i. As Σ is symmetric positive definite, it follows that the smallest $k-1$ eigenvectors of the matrix are optimal choices of p_j and S is the span of these eigenvectors.

This motivates our definition of the Fisher subspace for *any* mixture with bounded second moments (not necessarily Gaussians).

Definition 1. Let $\{\mathrm{w}_i, \mu_i, \Sigma_i\}$ be the weights, means, and covariance matrices for an isotropic[1] mixture distribution with mean at the origin and where $\dim\big(\mathrm{span}\{\mu_1, \ldots, \mu_k\}\big) = k-1$. Let $\ell(x)$ be the component from which x was drawn. The *Fisher subspace* F is defined as the $(k-1)$-dimensional subspace that minimizes

$$J(S) = E\Big[\big\|\mathrm{proj}_S\big(x - \mu_{\ell(x)}\big)\big\|^2\Big].$$

over subspaces S of dimension $k-1$.

Note that $\dim\big(\mathrm{span}\{\mu_1, \ldots, \mu_k\}\big)$ is only $k-1$ because isotropy implies $\sum_{i=1}^k \mathrm{w}_i\mu_i = 0$. The next lemma provides a simple alternative characterization of the Fisher subspace as the span of the means of the components (after transforming to isotropic position). The proof is given in Section 3.2.

Lemma 1. *Suppose* $\{\mathrm{w}_i, \mu_i, \Sigma_i\}_{i=1}^k$ *defines an isotropic mixture in* $\mathbb{R}^n$. *Let* $\lambda_1 \geq \ldots \geq \lambda_n$ *be the eigenvalues of the matrix* $\Sigma = \sum_{i=1}^k \mathrm{w}_i\Sigma_i$ *and let*

[1]For non-isotropic mixtures, the Fisher discriminant generalizes to

$$\sum_{j=1}^{k-1} p_j^T \Big(\sum_{i=1}^{k} \mathrm{w}_i(\Sigma_i + \mu_i\mu_i^T)\Big)^{-1} \Sigma p_j$$

and the overlap to

$$p^T \Big(\sum_{i=1}^{k} \mathrm{w}_i(\Sigma_i + \mu_i\mu_i^T)\Big)^{-1} \Sigma p.$$

$v_1, \ldots, v_n$ *be the corresponding eigenvectors. If the dimension of the span of the means of the components is* $k-1$, *then the Fisher subspace*

$$F = \text{span}\{v_{n-k+1}, \ldots, v_n\} = \text{span}\{\mu_1, \ldots, \mu_k\}.$$

Our algorithm attempts to find the Fisher subspace (or one close to it) and succeeds in doing so, provided the discriminant is small enough.

The next definition will be useful in stating our main theorem precisely.

Definition 2. The *overlap* of a mixture given as in Definition 1 is

$$\phi = \min_{S:\ \dim(S)=k-1} \max_{p \in S} p^T \Sigma p. \tag{1}$$

It is a direct consequence of the Courant-Fisher min-max theorem that ϕ is the $(k-1)$th smallest eigenvalue of the matrix Σ and the subspace achieving ϕ is the Fisher subspace, i.e.,

$$\phi = \left\| E\left[\text{proj}_F\left(x - \mu_{\ell(x)}\right)\text{proj}_F\left(x - \mu_{\ell(x)}\right)^T\right]\right\|_2.$$

We can now state our main theorem for $k > 2$.

Theorem 3. *There is an absolute constant* C *for which that following holds. Suppose that* $\mathcal{F}$ *is a mixture of* k *Gaussian components where the overlap satisfies*

$$\phi \leq C \text{w}^3 k^{-3} \log^{-1}\left(\frac{nk}{\delta \text{w}} + \frac{1}{\eta}\right)$$

with probability $1-\delta$, *algorithm* UNRAVEL *returns a set of* k *polyhedra that have error at most* η *using time and a number of samples that is polynomial in* n, w^{-1}, $\log(1/\delta)$.

In words, the algorithm successfully unravels arbitrary Gaussians provided there exists a $(k-1)$-dimensional subspace in which along every direction, the expected squared distance of a point to its component mean is smaller than the expected squared distance to the overall mean by roughly a $\text{poly}(k, 1/\text{w})$ factor. There is no dependence on the largest variances of the individual components, and the dependence on the ambient dimension is logarithmic. This means that the addition of extra dimensions (even where the distribution has large variance) as discussed in Section 1.1 has little impact on the success of our algorithm.

2. Algorithm

The algorithm has three major components: an initial affine transformation, a reweighting step, and identification of a direction close to the Fisher subspace and a hyperplane orthogonal to this direction which leaves each component's probability mass almost entirely in one of the halfspaces induced by the hyperplane. The key insight is that the reweighting technique will either cause the mean of the mixture to shift in the intermean subspace, or cause the top $k-1$ principal components of the second moment matrix to approximate the intermean subspace. In either case, we obtain a direction along which we can partition the components.

We first find an affine transformation W which when applied to $\mathcal{F}$ results in an isotropic distribution. That is, we move the mean to the origin and apply a linear transformation to make the covariance matrix the identity. We apply this transformation to a new set of m_1 points $\{x_i\}$ from $\mathcal{F}$ and then reweight according to a spherically symmetric Gaussian $\exp\big(-\|x\|^2/(2\alpha)\big)$ for $\alpha = \Theta(n/\mathrm{w})$. We then compute the mean $\hat{u}$ and second moment matrix $\hat{M}$ of the resulting set.[2]

After the reweighting, the algorithm chooses either the new mean or the direction of maximum second moment and projects the data onto this direction h. By bisecting the largest gap between points, we obtain a threshold t, which along with h defines a hyperplane that separates the components. Using the notation $H_{h,t} = \{x \in \mathbb{R}^n : h^T x \geq t\}$, to indicate a halfspace, we then recurse on each half of the mixture. Thus, every node in the recursion tree represents an intersection of half-spaces. To make our analysis easier, we assume that we use different samples for each step of the algorithm. The reader might find it useful to read Section 2.1, which gives an intuitive explanation for how the algorithm works on parallel pancakes, before reviewing the details of the algorithm.

[2]This practice of transforming the points and then looking at the second moment matrix can be viewed as a form of kernel PCA; however the connection between our algorithm and kernel PCA is superficial. Our transformation does not result in any standard kernel. Moreover, it is dimension-preserving (it is just a reweighting), and hence the "kernel trick" has no computational advantage.

Algorithm Unravel
Input: Integer k, scalar w.
Initialization: $P = \mathbb{R}^n$.

1. (Isotropy) Use samples lying in P to compute an affine transformation W that makes the distribution nearly isotropic (mean zero, identity covariance matrix).
2. (Reweighting) Use m_1 samples in P and for each compute a weight $e^{-\|x\|^2/2(n/\mathrm{w})}$.
3. (Separating Direction) Find the mean of the reweighted data $\hat{\mu}$. If $\|\hat{\mu}\| > \sqrt{\mathrm{w}}/(32\alpha)$ (where $\alpha > n/\mathrm{w}$), let $h = \hat{\mu}$. Otherwise, find the covariance matrix $\hat{M}$ of the reweighted points and let h be its top principal component.
4. (Recursion) Project m_2 sample points to h and find the largest gap between points in the interval $[-1/2, 1/2]$. If this gap is less than $1/4(k-1)$, then return P. Otherwise, set t to be the midpoint of the largest gap, recurse on $P \cap H_{h,t}$ and $P \cap H_{-h,-t}$, and return the union of the polyhedra produces by these recursive calls.

2.1. Parallel pancakes

The following special case, which represents the open problem in previous work, will illuminate the intuition behind the new algorithm. Suppose $\mathcal{F}$ is a mixture of two spherical Gaussians that are well-separated, i.e. the intermean distance is large compared to the standard deviation along any direction. We consider two cases, one where the mixing weights are equal and another where they are imbalanced.

After isotropy is enforced, each component will become thin in the intermean direction, giving the density the appearance of two parallel pancakes. When the mixing weights are equal, the means of the components will be equally spaced at a distance of $1 - \phi$ on opposite sides of the origin. For imbalanced weights, the origin will still lie on the intermean direction but will be much closer to the heavier component, while the lighter component will be much further away. In both cases, this transformation makes the variance of the mixture 1 in every direction, so the principal components give us no insight into the inter-mean direction.

Consider next the effect of the reweighting on the mean of the mixture. For the case of equal mixing weights, symmetry assures that the mean does not shift at all. For imbalanced weights, however, the heavier component, which lies closer to the origin will become heavier still. Thus, the reweighted mean shifts toward the mean of the heavier component, allowing us to detect the intermean direction.

Finally, consider the effect of reweighting on the second moments of the mixture with equal mixing weights. Because points closer to the origin are weighted more, the second moment in every direction is reduced. However, in the intermean direction, where part of the moment is due to the displacement of the component means from the origin, it shrinks less. Thus, the direction of maximum second moment is the intermean direction.

2.2. Overview of analysis

To analyze the algorithm, in the general case, we will proceed as follows. Section 3 shows that under isotropy the Fisher subspace coincides with the intermean subspace (Lemma 1), gives the necessary sampling convergence and perturbation lemmas and relates overlap to a more conventional notion of separation (Prop. 3). Section 3.3 gives approximations to the first and second moments. Section 4 then combines these approximations with the perturbation lemmas to show that the vector h (either the mean shift or the largest principal component) lies close to the intermean subspace. Finally, Section 5 shows the correctness of the recursive aspects of the algorithm.

3. Preliminaries

3.1. Matrix properties

For a matrix Z, we will denote the ith largest eigenvalue of Z by $\lambda_i(Z)$ or just λ_i if the matrix is clear from context. Unless specified otherwise all norms are the 2-norm. For symmetric matrices, this is $\|Z\|_2 = \lambda_1(Z) = \max_{x \in \mathbb{R}^n} \|Zx\|_2 / \|x\|_2$.

The following two facts from linear algebra will be useful in our analysis.

Fact 1. *Let $\lambda_1 \geq \ldots \geq \lambda_n$ be the eigenvalues for an n-by-n symmetric positive definite matrix Z and let $v_1, \ldots v_n$ be the corresponding eigenvectors.*

Then

$$\lambda_n + \ldots + \lambda_{n-k+1} = \min_{S:\ \dim(S)=k} \sum_{j=1}^{k} p_j^T Z p_j,$$

where $\{p_j\}$ *is any orthonormal basis for* S. *If* $\lambda_{n-k} > \lambda_{n-k+1}$, *then* span $\{v_n, \ldots, v_{n-k+1}\}$ *is the unique minimizing subspace.*

Recall that a matrix Z is positive semi-definite if $x^T Z x \geq 0$ for all non-zero x.

Fact 2. *Suppose that the matrix*

$$Z = \begin{bmatrix} A & B^T \\ B & D \end{bmatrix}$$

is symmetric positive semi-definite and that A *and* D *are square submatrices. Then* $\|B\| \leq \sqrt{\|A\|\,\|D\|}$.

Proof. Let y and x be the top left and right singular vectors of B, so that $y^T B x = \|B\|$. Because Z is positive semi-definite, we have that for any real γ,

$$0 \leq \left[\gamma x^T\ y^T\right] Z \left[\gamma x^T\ y^T\right]^T = \gamma^2 x^T A x + 2\gamma y^T B x + y^T D y.$$

This is a quadratic polynomial in γ that can have only one real root. Therefore the discriminant must be non-positive:

$$0 \geq 4\left(y^T B x\right)^2 - 4\left(x^T A x\right)\left(y^T D y\right).$$

We conclude that

$$\|B\| = y^T B x \leq \sqrt{(x^T A x)(y^T D y)} \leq \sqrt{\|A\|\,\|D\|}. \qquad \blacksquare$$

3.2. The Fisher criterion and isotropy

We begin with the proof of the lemma that for an isotropic mixture the Fisher subspace is the same as the intermean subspace.

Proof of Lemma 1. By definition for an isotropic distribution, the Fisher subspace minimizes

$$J(S) = E\left[\left\|\mathrm{proj}_S\left(x - \mu_{\ell(x)}\right)\right\|^2\right] = \sum_{j=1}^{k-1} p_j^T \Sigma p_j,$$

where $\{p_j\}$ is an orthonormal basis for S.

By Fact 1 one minimizing subspace is the span of the smallest $k-1$ eigenvectors of the matrix Σ, i.e. $v_{n-k+2}, \ldots, v_n$. Because the distribution is isotropic,

$$\Sigma = I - \sum_{i=1}^{k} \mathrm{w}_i \mu_i \mu_i^T,$$

and these vectors become the largest eigenvectors of $\sum_{i=1}^{k} \mathrm{w}_i \mu_i \mu_i^T$. Clearly, $\mathrm{span}\{v_{n-k+2}, \ldots, v_n\} \subseteq \mathrm{span}\{\mu_1, \ldots, \mu_k\}$, but both spans have dimension $k-1$ making them equal.

Since v_{n-k+1} must be orthogonal to the other eigenvectors, it follows that $\lambda_{n-k+1} = 1 > \lambda_{n-k+2}$. Therefore, $\mathrm{span}\{v_{n-k+2}, \ldots, v_n\} \subseteq \mathrm{span}\{\mu_1, \ldots, \mu_k\}$ is the unique minimizing subspace. ■

It follows directly that under the conditions of Lemma 1, the overlap may be characterized as

$$\phi = \lambda_{n-k+2}(\Sigma) = 1 - \lambda_{k-1}\left(\sum_{i=1}^{k} \mathrm{w}_i \mu_i \mu_i^T\right).$$

For clarity of the analysis, we will assume that Step 1 of the algorithm produces a perfectly isotropic mixture. Theorem 4 gives a bound on the required number of samples to make the distribution nearly isotropic, and as our analysis shows, our algorithm is robust to small estimation errors.

We will also assume for convenience of notation that the unit vectors along the first $k-1$ coordinate axes $e_1, \ldots e_{k-1}$ span the intermean (i.e.

Fisher) subspace. That is, $F = \text{span}\{e_1, \ldots, e_{k-1}\}$. When considering this subspace it will be convenient to be able to refer to projection of the mean vectors to this subspace. Thus, we define $\tilde{\mu_i} \in R^{k-1}$ to be the first $k-1$ coordinates of μ_i; the remaining coordinates are all zero. In other terms,

$$\tilde{\mu_i} = \begin{bmatrix} I_{k-1} & 0 \end{bmatrix} \mu_i.$$

In this coordinate system the covariance matrix of each component has a particular structure, which will be useful for our analysis. For the rest of this paper we fix the following notation: an isotropic mixture is defined by $\{\mathrm{w}_i, \mu_i, \Sigma_i\}$. We assume that $\text{span}\{e_1, \ldots, e_{k-1}\}$ is the intermean subspace and A_i, B_i, and D_i are defined such that

$$\mathrm{w}_i \Sigma_i = \begin{bmatrix} A_i & B_i^T \\ B_i & D_i \end{bmatrix} \tag{2}$$

where A_i is a $(k-1) \times (k-1)$ submatrix and D_i is a $(n-k+1) \times (n-k+1)$ submatrix.

Lemma 2 (Covariance Structure). *Using the above notation,*

$$\|A_i\| \leq \phi, \quad \|D_i\| \leq 1, \quad \|B_i\| \leq \sqrt{\phi}$$

for all components i.

Proof of Lemma 2. Because $\text{span}\{e_1, \ldots, e_{k-1}\}$ is the Fisher subspace

$$\phi = \max_{v \in \mathbb{R}^{k-1}} \frac{1}{\|v\|^2} \sum_{i=1}^{k} v^T A_i v = \left\| \sum_{i=1}^{k} A_i \right\|_2.$$

Also $\sum_{i=1}^k D_i = I$, so $\left\| \sum_{i=1}^k D_i \right\| = 1$. Each matrix $\mathrm{w}_i \Sigma_i$ is positive definite, so the principal minors A_i, D_i must be positive definite as well. Therefore, $\|A_i\| \leq \phi$, $\|D_i\| \leq 1$, and $\|B_i\| \leq \sqrt{\|A_i\| \|D_i\|} = \sqrt{\phi}$ using Fact 2. ■

For small ϕ, the covariance between intermean and non-intermean directions, i.e. B_i, is small. For $k = 2$, this means that all densities will have a "nearly parallel pancake" shape. In general, it means that $k-1$ of the principal axes of the Gaussians will lie close to the intermean subspace.

We conclude this section with a proposition connecting, for $k = 2$, the overlap to a standard notion of separation between two distributions, so that Theorem 1 becomes an immediate corollary of Theorem 2.

Proposition 3. *If there exists a unit vector p such that*

$$\left|p^T(\mu_1-\mu_2)\right| > t\left(\sqrt{p^T \mathrm{w}_1\Sigma_1 p} + \sqrt{p^T \mathrm{w}_2\Sigma_2 p}\right),$$

then the overlap $\phi \le J(p) \le (1+\mathrm{w}_1\mathrm{w}_2 t^2)^{-1}$.

Proof of Proposition 3. Since the mean of the distribution is at the origin, we have $\mathrm{w}_1 p^T\mu_1 = -\mathrm{w}_2 p^T \mu_2$. Thus,

$$\begin{aligned}\left|p^T\mu_1 - p^T\mu_2\right|^2 &= \left(p^T\mu_1\right)^2 + \left(p^T\mu_2\right)^2 + 2\left|p^T\mu_1\right|\left|p^T\mu_2\right| \\ &= \left(\mathrm{w}_1 p^T\mu_1\right)^2\left(\frac{1}{\mathrm{w}_1^2}+\frac{1}{\mathrm{w}_2^2}+\frac{2}{\mathrm{w}_1\mathrm{w}_2}\right),\end{aligned}$$

using $\mathrm{w}_1+\mathrm{w}_2=1$. We rewrite the last factor as

$$\frac{1}{\mathrm{w}_1^2}+\frac{1}{\mathrm{w}_2^2}+\frac{2}{\mathrm{w}_1\mathrm{w}_2} = \frac{\mathrm{w}_1^2+\mathrm{w}_2^2+2\mathrm{w}_1\mathrm{w}_2}{\mathrm{w}_1^2\mathrm{w}_2^2} = \frac{1}{\mathrm{w}_1^2\mathrm{w}_2^2} = \frac{1}{\mathrm{w}_1\mathrm{w}_2}\left(\frac{1}{\mathrm{w}_1}+\frac{1}{\mathrm{w}_2}\right).$$

Again, using the fact that $\mathrm{w}_1 p^T\mu_1 = -\mathrm{w}_2 p^T\mu_2$, we have that

$$\begin{aligned}\left|p^T\mu_1 - p^T\mu_2\right|^2 &= \frac{\left(\mathrm{w}_1 p^T\mu_1\right)^2}{\mathrm{w}_1\mathrm{w}_2}\left(\frac{1}{\mathrm{w}_1}+\frac{1}{\mathrm{w}_2}\right) \\ &= \frac{\mathrm{w}_1\left(p^T\mu_1\right)^2 + \mathrm{w}_2\left(p^T\mu_2\right)^2}{\mathrm{w}_1\mathrm{w}_2}.\end{aligned}$$

Thus, by the separation condition

$$\begin{aligned}\mathrm{w}_1\left(p^T\mu_1\right)^2 + \mathrm{w}_2\left(p^T\mu_2\right)^2 &= \mathrm{w}_1\mathrm{w}_2\left|p^T\mu_1 - p^T\mu_2\right|^2 \\ &\ge \mathrm{w}_1\mathrm{w}_2 t^2\left(p^T\mathrm{w}_1\Sigma_1 p + p^T\mathrm{w}_2\Sigma_2 p\right).\end{aligned}$$

To bound $J(p)$, we then argue

$$\begin{aligned}J(p) &= \frac{p^T\mathrm{w}_1\Sigma_1 p + p^T\mathrm{w}_2\Sigma_2 p}{\mathrm{w}_1\left(p^T\Sigma_1 p + (p^T\mu_1)^2\right) + \mathrm{w}_2\left(p^T\Sigma_2 p + (p^T\mu_2)^2\right)} \\ &= 1 - \frac{\mathrm{w}_1(p^T\mu_1)^2 + \mathrm{w}_2(p^T\mu_2)^2}{\mathrm{w}_1\left(p^T\Sigma_1 p + (p^T\mu_1)^2\right) + \mathrm{w}_2\left(p^T\Sigma_2 p + (p^T\mu_2)^2\right)}\end{aligned}$$

$$\le 1 - \frac{w_1 w_2 t^2 (w_1 p^T \Sigma_1 p + w_2 p^T \Sigma_2 p)}{w_1\left(p^T \Sigma_1 p + (p^T \mu_1)^2\right) + w_2\left(p^T \Sigma_2 p + (p^T \mu_2)^2\right)}$$

$$\le 1 - w_1 w_2 t^2 J(p),$$

and $J(p) \le 1/(1 + w_1 w_2 t^2)$. ■

3.3. Approximation of the reweighted moments

Our algorithm works by computing the first and second reweighted moments of a point set from $\mathcal{F}$. In this section, we examine how the reweighting affects the second moments of a single component and then give some approximations for the first and second moments of the entire mixture.

3.3.1. Single Component. The first step is to characterize how the reweighting affects the moments of a single component. Specifically, we will show for any function f (and therefore x and xx^T in particular) that for $\alpha > 0$,

$$E\left[f(x)\exp\left(-\frac{\|x\|^2}{2\alpha}\right)\right] = \sum_i w_i \rho_i E_i\left[f(y_i)\right],$$

Here, $E_i[\cdot]$ denotes expectation taken with respect to the component i, the quantity $\rho_i = E_i\left[\exp\left(-\frac{\|x\|^2}{2\alpha}\right)\right]$, and y_i is a Gaussian variable with parameters slightly perturbed from the original ith component.

Claim 4. If $\alpha = n/w$, the quantity $\rho_i = E_i\left[\exp\left(-\frac{\|x\|^2}{2\alpha}\right)\right]$ is at least 1/2.

Proof. Because the distribution is isotropic, for any component i, $w_i E_i\left[\|x\|^2\right] \le n$. Therefore,

$$\rho_i = E_i\left[\exp\left(-\frac{\|x\|^2}{2\alpha}\right)\right] \ge E_i\left[1 - \frac{\|x\|^2}{2\alpha}\right] \ge 1 - \frac{1}{2\alpha}\frac{n}{w_i} \ge \frac{1}{2}. \quad ■$$

Lemma 5 (Reweighted Moments of a Single Component). *For any $\alpha > 0$, with respect to a single component i of the mixture*

$$E_i\left[x\exp\left(-\frac{\|x\|^2}{2\alpha}\right)\right] = \rho_i\left(\mu_i - \frac{1}{\alpha}\Sigma_i\mu_i + f\right)$$

and

$$E_i\left[xx^T\exp\left(-\frac{\|x\|^2}{2\alpha}\right)\right]$$

$$=\rho\left(\Sigma_i+\mu_i\mu_i^T-\frac{1}{\alpha}(\Sigma_i\Sigma_i+\mu_i\mu_i^T\Sigma_i+\Sigma_i\mu_i\mu_i^T)+F\right)$$

where $\|f\|, \|F\| = O(\alpha^{-2})$.

We first establish the following claim.

Claim 6. Let x be a random variable distributed according to the normal distribution $N(\mu,\Sigma)$ and let $\Sigma = Q\Lambda Q^T$ be the singular value decomposition of Σ with $\lambda_1,\ldots,\lambda_n$ being the diagonal elements of Λ. Let $W=\mathrm{diag}\big(\alpha/(\alpha+\lambda_1),\ldots,\alpha/(\alpha+\lambda_n)\big)$. Finally, let y be a random variable distributed according to $N\big(QWQ^T\mu, QW\Lambda Q^T\big)$. Then for any function $f(x)$,

$$E\left[f(x)\exp\left(-\frac{\|x\|^2}{2\alpha}\right)\right]=\det(W)^{1/2}\exp\left(-\frac{\mu^TQWQ^T\mu}{2\alpha}\right)E\big[f(y)\big].$$

Proof of Claim 6. We assume that $Q=I$ for the initial part of the proof. From the definition of a Gaussian distribution, we have

$$E\left[f(x)\exp\left(-\frac{\|x\|^2}{2\alpha}\right)\right]$$

$$=\det(\Lambda)^{-1/2}(2\pi)^{-n/2}\int_{\mathbb{R}^n} f(x)\exp\left(-\frac{x^Tx}{2\alpha}-\frac{(x-\mu)^T\Lambda^{-1}(x-\mu)}{2}\right).$$

Because Λ is diagonal, we may write the exponents on the right hand side as

$$\sum_{i=1}^n x_i^2\alpha^{-1}+(x_i-\mu_i)^2\lambda_i^{-1}=\sum_{i=1}^n x_i^2(\lambda^{-1}+\alpha^{-1})-2x_i\mu_i\lambda_i^{-1}+\mu_i^2\lambda_i^{-1}.$$

Completing the square gives the expression

$$\sum_{i=1}^n\left(x_i-\mu_i\frac{\alpha}{\alpha+\lambda_i}\right)^2\left(\frac{\lambda_i\alpha}{\alpha+\lambda_i}\right)^{-1}+\mu_i^2\lambda_i^{-1}-\mu_i^2\lambda_i^{-1}\frac{\alpha}{\alpha+\lambda_i}.$$

The last two terms can be simplified to $\mu_i^2/(\alpha + \lambda_i)$. In matrix form the exponent becomes

$$(x - W\mu)^T (W\Lambda)^{-1} (x - W\mu) + \mu^T W \mu \alpha^{-1}.$$

For general Q, this becomes

$$(x - QWQ^T\mu)^T Q(W\Lambda)^{-1} Q^T (x - QWQ^T\mu) + \mu^T QWQ^T \mu\alpha^{-1}.$$

Now recalling the definition of the random variable y, we see

$$E\left[f(x)\exp\left(-\frac{\|x\|^2}{2\alpha}\right)\right] = \det(\Lambda)^{-1/2}(2\pi)^{-n/2}\exp\left(-\frac{\mu^T QWQ^T\mu}{2\alpha}\right)$$

$$\int_{\mathbb{R}^n} f(x)\exp\left(-\frac{1}{2}(x - QWQ^T\mu)^T Q(W\Lambda)^{-1}Q^T(x - QWQ^T\mu)\right)$$

$$= \det(W)^{1/2}\exp\left(-\frac{\mu^T QWQ^T\mu}{2\alpha}\right) E[f(y)]. \qquad \blacksquare$$

The proof of Lemma 5 is now straightforward.

Proof of Lemma 5. For simplicity of notation, we drop the subscript i from ρ_i, μ_i, Σ_i with the understanding that all statements of expectation apply to a single component. Using the notation of Claim 6, we have

$$\rho = E\left[\exp\left(-\frac{\|x\|^2}{2\alpha}\right)\right] = \det(W)^{1/2}\exp\left(-\frac{\mu^T QWQ^T\mu}{2\alpha}\right).$$

A diagonal entry of the matrix W can expanded as

$$\frac{\alpha}{\alpha + \lambda_i} = 1 - \frac{\lambda_i}{\alpha + \lambda_i} = 1 - \frac{\lambda_i}{\alpha} + \frac{\lambda_i^2}{\alpha(\alpha + \lambda_i)},$$

so that

$$W = I - \frac{1}{\alpha}\Lambda + \frac{1}{\alpha^2}W\Lambda^2.$$

Thus,

$$E\left[x\exp\left(-\frac{\|x\|^2}{2\alpha}\right)\right] = \rho(QWQ^T\mu)$$

$$= \rho(QIQ^T\mu - \frac{1}{\alpha}Q\Lambda Q^T\mu + \frac{1}{\alpha^2}QW\Lambda^2Q^T\mu)$$

$$= \rho(\mu - \frac{1}{\alpha}\Sigma\mu + f),$$

where $\|f\| = O(\alpha^{-2})$.

We analyze the perturbed covariance in a similar fashion.

$$E\left[xx^T\exp\left(-\frac{\|x\|^2}{2\alpha}\right)\right] = \rho\big(Q(W\Lambda)Q^T + QWQ^T\mu\mu^TQWQ^T\big)$$

$$= \rho\Bigg(Q\Lambda Q^T - \frac{1}{\alpha}Q\Lambda^2Q^T + \frac{1}{\alpha^2}QW\Lambda^3Q^T$$

$$+ \left(\mu - \frac{1}{\alpha}\Sigma\mu + f\right)\left(\mu - \frac{1}{\alpha}\Sigma\mu + f\right)^T\Bigg)$$

$$= \rho\left(\Sigma + \mu\mu^T - \frac{1}{\alpha}(\Sigma\Sigma + \mu\mu^T\Sigma + \Sigma\mu\mu^T) + F\right),$$

where $\|F\| = O(\alpha^{-2})$. ∎

3.3.2. Mixture moments. The second step is to approximate the first and second moments of the entire mixture distribution. Let ρ be the vector where $\rho_i = E_i\big[\exp\big(-\frac{\|x\|^2}{2\alpha}\big)\big]$ and let $\bar{\rho}$ be the average of the ρ_i. We also define

$$(3)\qquad u \equiv E\left[x\exp\left(-\frac{\|x\|^2}{2\alpha}\right)\right] = \sum_{i=1}^{k}\mathrm{w}_i\rho_i\mu_i - \frac{1}{\alpha}\sum_{i=1}^{k}\mathrm{w}_i\rho_i\Sigma_i\mu_i + f$$

$$(4) \quad M \equiv E\left[xx^T \exp\left(-\frac{\|x\|^2}{2\alpha}\right)\right]$$
$$= \sum_{i=1}^{k} \mathrm{w}_i \rho_i \left(\Sigma_i + \mu_i\mu_i^T - \frac{1}{\alpha}(\Sigma_i\Sigma_i + \mu_i\mu_i^T\Sigma_i + \Sigma_i\mu_i\mu_i^T)\right) + F$$

with $\|f\| = O(\alpha^{-2})$ and $\|F\| = O(\alpha^{-2})$. We denote the estimates of these quantities computed from samples by $\hat{u}$ and $\hat{M}$ respectively.

Lemma 7. *Let $v = \sum_{i=1}^k \rho_i \mathrm{w}_i \mu_i$. Then*

$$\|u - v\|^2 \le \frac{4k^2}{\alpha^2 \mathrm{w}}\phi.$$

Proof of Lemma 7. We argue from Eqn. (2) and Eqn. (3) that

$$\|u - v\| = \frac{1}{\alpha}\left\|\sum_{i=1}^{k} \mathrm{w}_i\rho_i\Sigma_i\mu_i\right\| + O(\alpha^{-2})$$
$$\le \frac{1}{\alpha\sqrt{\mathrm{w}}}\sum_{i=1}^{k}\rho_i\left\|(\mathrm{w}_i\Sigma_i)\left(\sqrt{\mathrm{w}_i}\mu_i\right)\right\| + O(\alpha^{-2})$$
$$\le \frac{1}{\alpha\sqrt{\mathrm{w}}}\sum_{i=1}^{k}\rho_i\left\|\left[A_i, B_i^T\right]^T\right\|\left\|\left(\sqrt{\mathrm{w}_i}\mu_i\right)\right\| + O(\alpha^{-2}).$$

From isotropy, it follows that $\left\|\sqrt{\mathrm{w}_i}\mu_i\right\| \le 1$. To bound the other factor, we argue

$$\left\|\left[A_i, B_i^T\right]^T\right\| \le \sqrt{2}\max\left\{\|A_i\|, \|B_i\|\right\} \le \sqrt{2\phi}.$$

Therefore,

$$\|u - v\|^2 \le \frac{2k^2}{\alpha^2\mathrm{w}}\phi + O(\alpha^{-3}) \le \frac{4k^2}{\alpha^2\mathrm{w}}\phi,$$

for sufficiently large n, as $\alpha \ge n/\mathrm{w}$. ■

Lemma 8. *Let*

$$\Gamma = \begin{bmatrix} \sum_{i=1}^k \rho_i(\mathrm{w}_i\tilde{\mu}_i\tilde{\mu}_i^T + A_i) & 0 \\ 0 & \sum_{i=1}^k \rho_i D_i - \frac{\rho_i}{\mathrm{w}_i\alpha}D_i^2 \end{bmatrix}.$$

If $\|\rho - 1\bar{\rho}\|_\infty < 1/(2\alpha)$, *then*

$$\|M - \Gamma\|_2^2 \leq \frac{16^2 k^2}{\mathrm{w}^2 \alpha^2} \phi.$$

Before giving the proof, we summarize some of the necessary calculation in the following claim.

Claim 9. The matrix of second moments

$$M = E\left[xx^T \exp\left(-\frac{\|x\|^2}{2\alpha}\right)\right] = \begin{bmatrix} \Gamma_{11} & 0 \\ 0 & \Gamma_{22} \end{bmatrix} + \begin{bmatrix} \Delta_{11} & \Delta_{21}^T \\ \Delta_{21} & \Delta_{22} \end{bmatrix} + F,$$

where

$$\Gamma_{11} = \sum_{i=1}^k \rho_i \big(\mathrm{w}_i \tilde{\mu}_i \tilde{\mu}_i{}^T + A_i\big)$$

$$\Gamma_{22} = \sum_{i=1}^k \rho_i D_i - \frac{\rho_i}{\mathrm{w}_i \alpha} D_i^2$$

$$\Delta_{11} = -\sum_{i=1}^k \frac{\rho_i}{\mathrm{w}_i \alpha} B_i^T B_i + \frac{\rho_i}{\mathrm{w}_i \alpha}\big(\mathrm{w}_i \tilde{\mu}_i \tilde{\mu}_i{}^T A_i + \mathrm{w}_i A_i \tilde{\mu}_i \tilde{\mu}_i{}^T + A_i^2\big)$$

$$\Delta_{21} = \sum_{i=1}^k \rho_i B_i - \frac{\rho_i}{\mathrm{w}_i \alpha}\big(B_i\big(\mathrm{w}_i \tilde{\mu}_i \tilde{\mu}_i{}^T\big) + B_i A_i + D_i B_i\big)$$

$$\Delta_{22} = -\sum_{i=1}^k \frac{\rho_i}{\mathrm{w}_i \alpha} B_i B_i^T,$$

and $\|F\| = O(\alpha^{-2})$.

Proof. The calculation is straightforward. ■

Proof of Lemma 8. We begin by bounding the 2-norm of each of the blocks. Since $\left\| \mathrm{w}_i \tilde{\mu_i} \tilde{\mu_i}^T \right\| < 1$ and $\|A_i\| \leq \phi$ and $\|B_i\| \leq \sqrt{\phi}$, we can bound

$$\|\Delta_{11}\| = \max_{\|y\|=1} \sum_{i=1}^{k} \frac{\rho_i}{\mathrm{w}_i \alpha} y^T B_i^T B_i y^T$$

$$- \frac{\rho_i}{\mathrm{w}_i \alpha} y^T \left(\mathrm{w}_i \tilde{\mu_i} \tilde{\mu_i}^T A_i + \mathrm{w}_i A_i \tilde{\mu_i} \tilde{\mu_i}^T + A_i^2 \right) y + O(\alpha^{-2})$$

$$\leq \sum_{i=1}^{k} \frac{\rho_i}{\mathrm{w}_i \alpha} \|B_i\|^2 + \frac{\rho_i}{\mathrm{w}_i \alpha} \left(2\|A\| + \|A\|^2 \right) + O(\alpha^{-2})$$

$$\leq \frac{4k}{\mathrm{w}\alpha} \phi + O(\alpha^{-2}).$$

By a similar argument, $\|\Delta_{22}\| \leq k\phi/(\mathrm{w}\alpha) + O(\alpha^{-2})$. For Δ_{21}, we observe that $\sum_{i=1}^{k} B_i = 0$. Therefore,

$$\|\Delta_{21}\| \leq \left\| \sum_{i=1}^{k} (\rho_i - \bar{\rho}) B_i \right\|$$

$$+ \left\| \sum_{i=1}^{k} \frac{\rho_i}{\mathrm{w}_i \alpha} \left(B_i (\mathrm{w}_i \tilde{\mu_i} \tilde{\mu}_i^T) + B_i A_i + D_i B_i \right) \right\| + O(\alpha^{-2})$$

$$\leq \sum_{i=1}^{k} |\rho_i - \bar{\rho}| \, \|B_i\| + \sum_{i=1}^{k} \frac{\rho_i}{\mathrm{w}_i \alpha} \Big(\| B_i (\mathrm{w}_i \tilde{\mu_i} \tilde{\mu}_i^T) \|$$

$$+ \|B_i A_i\| + \|D_i B_i\| \Big) + O(\alpha^{-2})$$

$$\leq k\|\rho - 1\bar{\rho}\|_\infty \sqrt{\phi} + \sum_{i=1}^{k} \frac{\rho_i}{\mathrm{w}_i \alpha} \left(\sqrt{\phi} + \phi\sqrt{\phi} + \sqrt{\phi} \right) + O(\alpha^{-2})$$

$$\leq k\|\rho - 1\bar{\rho}\|_\infty \sqrt{\phi} + \frac{3k\bar{\rho}}{\mathrm{w}\alpha} \sqrt{\phi}$$

$$\leq \frac{7k}{2\mathrm{w}\alpha} \sqrt{\phi} + O(\alpha^{-2}).$$

Thus, we have $\max\{\|\Delta_{11}\|, \|\Delta_{22}\|, \|\Delta_{21}\|\} \leq 4k\sqrt{\phi}/(\mathrm{w}\alpha)+O(\alpha^{-2})$, so that

$$\begin{aligned}\|M-\Gamma\| &\leq \|\Delta\| + O(\alpha^{-2}) \leq 2\max\{\|\Delta_{11}\|, \|\Delta_{22}\|, \|\Delta_{21}\|\} \\ &\leq \frac{8k}{\mathrm{w}\alpha}\sqrt{\phi} + O(\alpha^{-2}) \leq \frac{16k}{\mathrm{w}\alpha}\sqrt{\phi}.\end{aligned}$$

for sufficiently large n, as $\alpha \geq n/\mathrm{w}$. ■

3.4. Sample convergence

In this section, we collect some bounds on the convergence of the sample mean u and sample matrix of second moments M to their expectations $\hat{u}$ and $\hat{M}$. For the convergence of second moment matrices, we use the following lemma due to Rudelson [13], which was presented in this form in [14].

Lemma 10. *Let y be a random vector from a distribution D in R^n, with $\sup_D \|y\| = A$ and $\|\mathsf{E}(yy^T)\| \leq 1$. Let $y_1, \ldots, y_m$ be independent samples from D. Let*

$$\eta = CA\sqrt{\frac{\log m}{m}}$$

where C is an absolute constant. Then,

(i) *If $\eta < 1$, then*

$$\mathsf{E}\left(\left\|\frac{1}{m}\sum_{i=1}^{m} y_i y_i^T - \mathsf{E}(yy^T)\right\|\right) \leq \eta.$$

(ii) *For every $t \in (0,1)$,*

$$\mathsf{P}\left(\left\|\frac{1}{m}\sum_{i=1}^{m} y_i y_i^T - \mathsf{E}(yy^T)\right\| > t\right) \leq 2e^{-ct^2/\eta^2}.$$

This lemma can be used to show that a mixture of k logconcave densities can be made nearly isotropic using only $O^*(kn)$ samples. This is already known for a single logconcave density [13, 11] and the extension to a mixture is straightforward.

Theorem 4. *There is an absolute constant C such that for an isotropic mixture of k logconcave distributions, with probability at least $1-\delta$, a sample of size*

$$m > C\frac{kn\log^2(n/\delta)}{\varepsilon^2}$$

gives a sample mean $\hat{u}$ and sample covariance $\hat{M}$ so that

$$\|\hat{u}\| \le \varepsilon \quad \text{and} \quad \|\hat{M} - I\| \le \varepsilon.$$

The isotropic transformation is computed simply by estimating the mean and covariance matrix of a sample, and computing the affine transformation that puts the sample in isotropic position.

Lemma 11. *Let $\varepsilon, \delta > 0$ and let $\hat{u}$ be the mean of a set of m points drawn from an isotropic mixture of k Gaussians in n dimensions, where*

$$m \ge \frac{2n\alpha}{\varepsilon^2}\log\frac{2n}{\delta}.$$

Then

$$\mathsf{P}\big[\,\|\hat{u} - u\| > \varepsilon\big] \le \delta$$

Proof. We first consider only a single coordinate of the vector $\hat{u}$. Let $y = x_1\exp\big(-\|x\|^2/(2\alpha)\big) - u_1$. We observe that

$$\left| x_1 \exp\left(-\frac{\|x\|^2}{2\alpha}\right)\right| \le |x_1|\exp\left(-\frac{x_1^2}{2\alpha}\right) \le \sqrt{\frac{\alpha}{e}} < \sqrt{\alpha}.$$

Thus, each term in the sum $m\hat{u}_1 = \sum_{j=1}^m y_j$ falls the range $\big[-\sqrt{\alpha} - u_1, \sqrt{\alpha} - u_1\big]$. We may therefore apply Hoeffding's inequality to show that

$$\mathsf{P}\big[|\hat{u}_1 - u_1| \ge \varepsilon/\sqrt{n}\big] \le 2\exp\left(-\frac{2m^2\big(\varepsilon/\sqrt{n}\big)^2}{m\cdot\big(2\sqrt{\alpha}\big)^2}\right) \le 2\exp\left(-\frac{m\varepsilon^2}{2\alpha n}\right) \le \frac{\delta}{n}.$$

Taking the union bound over the n coordinates, we have that with probability $1-\delta$ the error in each coordinate is at most $\varepsilon/\sqrt{n}$, which implies that $\|\hat{u} - u\| \le \varepsilon$. ■

Lemma 12. *Let $\varepsilon, \delta > 0$ and let $\hat{M}$ be the sample covariance matrix of m points drawn from an isotropic mixture of k Gaussians in n dimensions, where*

$$m \geq C_1 \frac{n\alpha}{\varepsilon^2} \log \frac{n\alpha}{\delta}.$$

and C_1 is an absolute constant. Then

$$\mathsf{P}\left[\|\hat{M} - I\| > \varepsilon\right] < \delta.$$

Proof. We will apply Lemma 10. Define $y = x \exp\left(-\|x\|^2/(2\alpha)\right)$. Then,

$$y_i^2 \leq x_i^2 \exp\left(-\frac{\|x\|^2}{\alpha}\right) \leq x_i^2 \exp\left(-\frac{x_i^2}{\alpha}\right) \leq \frac{\alpha}{e} < \alpha.$$

Therefore $\|y\| \leq \sqrt{2\alpha n}$.

Next, since the mixture is in isotropic position, we have for any unit vector v,

$$\mathsf{E}\left((v^T y)^2\right) \leq \mathsf{E}\left((v^T x)^2\right) \leq 1$$

and so $\left\|\mathsf{E}(yy^T)\right\| \leq 1$.

Now we apply the second part of Lemma 10 with $\eta = \varepsilon\sqrt{c/\ln(2/\delta)}$ and $t = \eta\sqrt{\ln(2/\delta)/c}$. This requires that

$$\eta = \frac{c\varepsilon}{\ln(2/\delta)} \leq C\sqrt{2\alpha n}\sqrt{\frac{\log m}{m}}$$

which is satisfied for our choice of m. ■

Lemma 13. *Let X be a collection of m points drawn from a Gaussian with mean μ and variance σ^2. With probability $1 - \delta$,*

$$|x - \mu| \leq \sigma\sqrt{2\log m/\delta}.$$

for every $x \in X$.

3.5. Perturbation lemma

We will use the following key lemma due to Stewart [17] to show that when we apply the spectral step, the top $k-1$ dimensional invariant subspace will be close to the Fisher subspace.

Lemma 14 (Stewart's Theorem). *Suppose A and $A+E$ are n-by-n symmetric matrices and that*

$$A = \begin{array}{c} \begin{bmatrix} D_1 & 0 \\ 0 & D_2 \end{bmatrix} \begin{array}{c} r \\ n-r \end{array} \\ \begin{array}{cc} r & n-r \end{array} \end{array} \qquad E = \begin{array}{c} \begin{bmatrix} E_{11} & E_{21}^T \\ E_{21} & E_{22} \end{bmatrix} \begin{array}{c} r \\ n-r \end{array} \\ \begin{array}{cc} r & n-r \end{array} \end{array}.$$

Let the columns of V be the top r eigenvectors of the matrix $A+E$ and let P_2 be the matrix with columns $e_{r+1}, \ldots, e_n$. If $d = \lambda_r(D_1) - \lambda_1(D_2) > 0$ and

$$\|E\| \le \frac{d}{5},$$

then

$$\|V^T P_2\| \le \frac{4}{d}\|E_{21}\|_2.$$

4. Finding a Vector near the Fisher Subspace

In this section, we combine the approximations of Section 3.3 and the perturbation lemma of Section 3.5 to show that the direction h chosen by step 3 of the algorithm is close to the intermean subspace. Section 5 argues that this direction can be used to partition the components. Finding the separating direction is the most challenging part of the classification task and represents the main contribution of this work.

We first assume zero overlap and that the sample reweighted moments behave exactly according to expectation. In this case, the mean shift $\hat{u}$ becomes

$$v \equiv \sum_{i=1}^{k} \mathrm{w}_i \rho_i \mu_i.$$

We can intuitively think of the components that have greater ρ_i as gaining mixing weight and those with smaller ρ_i as losing mixing weight. As long as the ρ_i are not all equal, we will observe some shift of the mean in the intermean subspace, i.e. Fisher subspace. Therefore, we may use this direction to partition the components. On the other hand, if all of the ρ_i are equal, then $\hat{M}$ becomes

$$\Gamma \equiv \begin{bmatrix} \sum_{i=1}^k \rho_i \left(\mathrm{w}_i \tilde{\mu_i} \tilde{\mu_i}^T + A_i\right) & 0 \\ 0 & \sum_{i=1}^k \rho_i D_i - \frac{\rho_i}{\mathrm{w}_i \alpha} D_i^2 \end{bmatrix}$$

$$= \bar{\rho} \begin{bmatrix} I & 0 \\ 0 & I - \frac{1}{\alpha} \sum_{i=1}^k \frac{1}{\mathrm{w}_i} D_i^2 \end{bmatrix}.$$

Notice that the second moments in the subspace span $\{e_1, \ldots, e_{k-1}\}$ are maintained while those in the complementary subspace are reduced by poly $(1/\alpha)$. Therefore, the top eigenvector will be in the intermean subspace, which is the Fisher subspace.

We now argue that this same strategy can be adapted to work in general, i.e., with nonzero overlap and sampling errors, with high probability. A critical aspect of this argument is that the norm of the error term $\hat{M} - \Gamma$ depends only on ϕ and k and not the dimension of the data. See Lemma 8 and the supporting Lemma 2 and Fact 2.

Since we cannot know directly how imbalanced the ρ_i are, we choose the method of finding a separating direction according the norm of the vector $\|\hat{u}\|$. Recall that when $\|\hat{u}\| > \sqrt{\mathrm{w}}/(32\alpha)$ the algorithm uses $\hat{u}$ to determine the separating direction h. Lemma 15 guarantees that this vector is close to the Fisher subspace. When $\|\hat{u}\| \leq \sqrt{\mathrm{w}}/(32\alpha)$, the algorithm uses the top eigenvector of the covariance matrix $\hat{M}$. Lemma 16 guarantees that this vector is close to the Fisher subspace.

Lemma 15 (Mean Shift Method). *Let $\varepsilon > 0$. There exists a constant C such that if $m_1 \geq Cn^4 \mathrm{poly}\,(k, \mathrm{w}^{-1}, \log n/\delta)$, then the following holds with probability $1 - \delta$. If $\|\hat{u}\| > \sqrt{\mathrm{w}}/(32\alpha)$ and*

$$\phi \leq \frac{\mathrm{w}^2 \varepsilon}{2^{14} k^2},$$

then

$$\frac{\|\hat{u}^T v\|}{\|\hat{u}\| \, \|v\|} \geq 1 - \varepsilon.$$

Lemma 16 (Spectral Method). *Let $\varepsilon > 0$. There exists a constant C such that if $m_1 \geq Cn^4 \text{poly}(k, \mathrm{w}^{-1}, \log n/\delta)$, then the following holds with probability $1-\delta$. Let $v_1, \ldots, v_{k-1}$ be the top $k-1$ eigenvectors of $\hat{M}$. If $\|\hat{u}\| \leq \sqrt{\mathrm{w}}/(32\alpha)$ and*

$$\phi \leq \frac{\mathrm{w}^2 \varepsilon}{640^2 k^2}$$

then

$$\min_{v \in \text{span}\{v_1, \ldots, v_{k-1}\},\ \|v\|=1} \left\| \text{proj}_F(v) \right\| \geq 1 - \varepsilon.$$

4.1. Mean shift

Proof of Lemma 15. We will make use of the following claim.

Claim 17. For any vectors $a, b \neq 0$,

$$\frac{|a^T b|}{\|a\| \, \|b\|} \geq \left(1 - \frac{\|a-b\|^2}{\max\left\{ \|a\|^2, \|b\|^2 \right\}} \right)^{1/2}.$$

By the triangle inequality, $\|\hat{u} - v\| \leq \|\hat{u} - u\| + \|u - v\|$. By Lemma 7,

$$\|u - v\| \leq \sqrt{\frac{4k^2}{\alpha^2 \mathrm{w}} \phi} = \sqrt{\frac{4k^2}{\alpha^2 \mathrm{w}} \cdot \frac{\mathrm{w}^2 \varepsilon}{2^1 4k^2}} \leq \sqrt{\frac{\mathrm{w}\varepsilon}{2^{12}\alpha^2}}.$$

By Lemma 11, for large m_1 we obtain the same bound on $\|\hat{u} - u\|$ with probability $1 - \delta$. Thus,

$$\|\hat{u} - v\| \leq \sqrt{\frac{\mathrm{w}\varepsilon}{2^{10}\alpha^2}}.$$

Applying the claim gives

$$\begin{aligned} \frac{\|\hat{u}^T v\|}{\|\hat{u}\| \, \|v\|} &\geq 1 - \frac{\|\hat{u} - v\|^2}{\|\hat{u}\|^2} \\ &\geq 1 - \frac{\mathrm{w}\varepsilon}{2^{10}\alpha^2} \cdot \frac{32^2 \alpha^2}{\mathrm{w}} \\ &= 1 - \varepsilon. \qquad \blacksquare \end{aligned}$$

Proof of Claim 17. Without loss of generality, assume $\|u\| \geq \|v\|$ and fix the distance $\|u-v\|$. In order to maximize the angle between u and v, the vector v should be chosen so that it is tangent to the sphere centered at u with radius $\|u-v\|$. Hence, the vectors u, v, $(u-v)$ form a right triangle where $\|u\|^2 = \|v\|^2 + \|u-v\|^2$. For this choice of v, let θ be the angle between u and v so that

$$\frac{u^T v}{\|u\|\,\|v\|} = \cos\theta = (1-\sin^2\theta)^{1/2} = \left(1 - \frac{\|u-v\|^2}{\|u\|^2}\right)^{1/2}. \qquad \blacksquare$$

4.2. Spectral method

We first show that the smallness of the mean shift $\hat{u}$ implies that the coefficients ρ_i are sufficiently uniform to allow us to apply the spectral method.

Claim 18 (Small Mean Shift Implies Balanced Second Moments). If $\|\hat{u}| \leq \sqrt{\mathrm{w}}/(32\alpha)$ and

$$\sqrt{\phi} \leq \frac{\mathrm{w}}{64k},$$

then

$$\|\rho - 1\bar{\rho}\|_2 \leq \frac{1}{8\alpha}.$$

Proof. Let $q_1, \ldots, q_k$ be the right singular vectors of the matrix $U = [\mathrm{w}_1\mu_1, \ldots, \mathrm{w}_k\mu_k]$ and let $\sigma_i(U)$ be the ith largest singular value. As $\sum_{i=1}^k \mathrm{w}_i\mu_i = 0$, we have that $\sigma_k(U) = 0$ and $q_k = 1/\sqrt{k}$. Recall that ρ is the k vector of scalars $\rho_1, \ldots, \rho_k$ and that $v = U\rho$. Then

$$\begin{aligned}
\|v\|^2 &= \|U\rho\|^2 \\
&= \sum_{i=1}^{k-1} \sigma_i(U)^2 (q_i^T\rho)^2 \\
&\geq \sigma_{k-1}(U)^2 \left\|\rho - q_k(q_k^T\rho)\right\|_2^2 \\
&= \sigma_{k-1}(U)^2 \|\rho - 1\bar{\rho}\|_2^2.
\end{aligned}$$

Since $q_{k-1} \in \text{span}\{\mu_1, \ldots, \mu_k\}$, we have that $\sum_{i=1}^k \mathrm{w}_i q_{k-1}^T \mu_i \mu_i^T q_{k-1} \geq 1 - \phi$. Therefore,

$$\begin{aligned} \sigma_{k-1}(U)^2 &= \|U q_{k-1}\|^2 \\ &= q_{k-1}^T \left(\sum_{i=1}^k \mathrm{w}_i^2 \mu_i \mu_i^T \right) q_{k-1} \\ &\geq \mathrm{w} q_{k-1}^T \left(\sum_{i=1}^k \mathrm{w}_i \mu_i \mu_i^T \right) q_{k-1} \\ &\geq \mathrm{w}(1 - \phi). \end{aligned}$$

Thus, we have the bound

$$\|\rho - 1\bar{\rho}\|_\infty \leq \frac{1}{\sqrt{(1-\phi)\mathrm{w}}} \|v\| \leq \frac{2}{\sqrt{\mathrm{w}}} \|v\|.$$

By the triangle inequality $\|v\| \leq \|\hat{u}\| + \|\hat{u} - v\|$. As argued in Lemma 7,

$$\|\hat{u} - v\| \leq \sqrt{\frac{4k^2}{\alpha^2 \mathrm{w}} \phi} = \sqrt{\frac{4k^2}{\alpha^2 \mathrm{w}} \cdot \frac{\mathrm{w}^2}{64^2 k^2}} = \leq \frac{\sqrt{\mathrm{w}}}{32\alpha}.$$

Thus,

$$\begin{aligned} \|\rho - 1\bar{\rho}\|_\infty &\leq \frac{2\bar{\rho}}{\sqrt{\mathrm{w}}} \|v\| \\ &\leq \frac{2\bar{\rho}}{\sqrt{\mathrm{w}}} \left(\frac{\sqrt{\mathrm{w}}}{32\alpha} + \frac{\sqrt{\mathrm{w}}}{32\alpha} \right) \\ &\leq \frac{1}{8\alpha}. \qquad \blacksquare \end{aligned}$$

We next show that the top $k - 1$ principal components of Γ span the intermean subspace and put a lower bound on the spectral gap between the intermean and non-intermean components.

Lemma 19 (Ideal Case). *If* $\|\rho - 1\bar{\rho}\|_\infty \leq 1/(8\alpha)$, *then*

$$\lambda_{k-1}(\Gamma) - \lambda_k(\Gamma) \geq \frac{1}{4\alpha},$$

and the top $k - 1$ eigenvectors of Γ span the means of the components.

Proof of Lemma 19. We first bound $\lambda_{k-1}(\Gamma_{11})$. Recall that

$$\Gamma_{11} = \sum_{i=1}^{k} \rho_i \left(\mathrm{w}_i \tilde{\mu_i} \tilde{\mu_i}^T + A_i \right).$$

Thus,

$$\lambda_{k-1}(\Gamma_{11}) = \min_{\|y\|=1} \sum_{i=1}^{k} \rho_i y^T \left(\mathrm{w}_i \tilde{\mu_i} \tilde{\mu_i}^T + A_i \right) y$$

$$\geq \bar{\rho} - \max_{\|y\|=1} \sum_{i=1}^{k} (\bar{\rho} - \rho_i) y^T \left(\mathrm{w}_i \tilde{\mu_i} \tilde{\mu_i}^T + A_i \right) y.$$

We observe that $\sum_{i=1}^{k} y^T \left(\mathrm{w}_i \tilde{\mu_i} \tilde{\mu_i}^T + A_i \right) y = 1$ and each term is non-negative. Hence the sum is bounded by

$$\sum_{i=1}^{k} (\bar{\rho} - \rho_i) y^T \left(\mathrm{w}_i \tilde{\mu_i} \tilde{\mu_i}^T + A_i \right) y \leq \|\rho - 1\bar{\rho}\|_\infty,$$

so,

$$\lambda_{k-1}(\Gamma_{11}) \geq \bar{\rho} - \|\rho - 1\bar{\rho}\|_\infty.$$

Next, we bound $\lambda_1(\Gamma_{22})$. Recall that

$$\Gamma_{22} = \sum_{i=1}^{k} \rho_i D_i - \frac{\rho_i}{\mathrm{w}_i \alpha} D_i^2$$

and that for any $n-k$ vector y such that $\|y\| = 1$, we have $\sum_{i=1}^{k} y^T D_i y = 1$. Using the same arguments as above,

$$\lambda_1(\Gamma_{22}) = \max_{\|y\|=1} \bar{\rho} + \sum_{i=1}^{k} (\rho_i - \bar{\rho}) y^T D_i y - \frac{\rho_i}{\mathrm{w}_i \alpha} y^T D_i^2 y$$

$$\leq \bar{\rho} + \|\rho - 1\bar{\rho}\|_\infty - \min_{\|y\|=1} \sum_{i=1}^{k} \frac{\rho_i}{\mathrm{w}_i \alpha} y^T D_i^2 y.$$

To bound the last sum, we observe that $\rho_i - \bar{\rho} = O(\alpha^{-1})$. Therefore

$$\sum_{i=1}^{k} \frac{\rho_i}{\mathrm{w}_i \alpha} y^T D_i^2 y \geq \frac{\bar{\rho}}{\alpha} \sum_{i=1}^{k} \frac{1}{\mathrm{w}_i} y^T D_i^2 y + O(\alpha^{-2}).$$

Without loss of generality, we may assume that $y = e_1$ by an appropriate rotation of the D_i. Let $D_i(\ell, j)$ be element in the ℓth row and jth column of the matrix D_i. Then the sum becomes

$$\begin{aligned}\sum_{i=1}^{k} \frac{1}{\mathrm{w}_i} y^T D_i^2 y &= \sum_{i=1}^{k} \frac{1}{\mathrm{w}_i} \sum_{j=1}^{n} D_j(1,j)^2 \\ &\geq \sum_{i=1}^{k} \frac{1}{\mathrm{w}_i} D_j(1,1)^2.\end{aligned}$$

As $\sum_{i=1}^{k} D_i = I$, we have $\sum_{i=1}^{k} D_i(1,1) = 1$. From the Cauchy–Schwartz inequality, it follows

$$\left(\sum_{i=1}^{k} \mathrm{w}_i\right)^{1/2} \left(\sum_{i=1}^{k} \frac{1}{\mathrm{w}_i} D_i(1,1)^2\right)^{1/2} \geq \sum_{i=1}^{k} \sqrt{\mathrm{w}_i} \frac{D_i(1,1)}{\sqrt{\mathrm{w}_i}} = 1.$$

Since $\sum_{i=1}^{k} \mathrm{w}_i = 1$, we conclude that $\sum_{i=1}^{k} \frac{1}{\mathrm{w}_i} D_i(1,1)^2 \geq 1$. Thus, using the fact that $\bar{\rho} \geq 1/2$, we have

$$\sum_{i=1}^{k} \frac{\rho_i}{\mathrm{w}_i \alpha} y^T D_i^2 y \geq \frac{1}{2\alpha}$$

Putting the bounds together

$$\lambda_{k-1}(\Gamma_{11}) - \lambda_1(\Gamma_{22}) \geq \frac{1}{2\alpha} - 2\|\rho - 1\bar{\rho}\|_\infty \geq \frac{1}{4\alpha}. \qquad \blacksquare$$

Proof of Lemma 16. To bound the effect of overlap and sample errors on the eigenvectors, we apply Stewart's Lemma (Lemma 14). Define $d = \lambda_{k-1}(\Gamma) - \lambda_k(\Gamma)$ and $E = \hat{M} - \Gamma$.

We assume that the mean shift satisfies $\|\hat{u}\| \leq \sqrt{\mathrm{w}}/(32\alpha)$ and that ϕ is small. By Lemma 19, this implies that

$$d = \lambda_{k-1}(\Gamma) - \lambda_k(\Gamma) \geq \frac{1}{4\alpha}. \tag{5}$$

To bound $\|E\|$, we use the triangle inequality $\|E\| \leq \|\Gamma - M\| + \|M - \hat{M}\|$. Lemma 8 bounds the first term by

$$\|M - \Gamma\| \leq \sqrt{\frac{16^2 k^2}{\mathrm{w}^2 \alpha^2} \phi} = \sqrt{\frac{16^2 k^2}{\mathrm{w}^2 \alpha^2} \cdot \frac{\mathrm{w}^2 \varepsilon}{640^2 k^2}} \leq \frac{1}{40\alpha} \sqrt{\varepsilon}. \tag{6}$$

By Lemma 12, we obtain the same bound on $\|M - \hat{M}\|$ with probability $1 - \delta$ for large enough m_1. Thus,

$$\|E\| \leq \frac{1}{20\alpha} \sqrt{\varepsilon}.$$

Combining the bounds of Eqn. (5) and (6), we have

$$\sqrt{1 - (1 - \varepsilon)^2} d - 5\|E\| \geq \sqrt{1 - (1 - \varepsilon)^2} \frac{1}{4\alpha} - 5 \frac{1}{20\alpha} \sqrt{\varepsilon} \geq 0,$$

as $\sqrt{1 - (1 - \varepsilon)^2} \geq \sqrt{\varepsilon}$. This implies both that $\|E\| \leq d/5$ and that $4\|E_{21}\|/d < \sqrt{1 - (1 - \varepsilon)^2}$, enabling us to apply Stewart's Lemma to the matrix pair Γ and $\hat{M}$.

By Lemma 19, the top $k-1$ eigenvectors of Γ, i.e. $e_1, \ldots, e_{k-1}$, span the means of the components. Let the columns of P_1 be these eigenvectors. Let the columns of P_2 be defined such that $[P_1, P_2]$ is an orthonormal matrix and let $v_1, \ldots, v_k$ be the top $k-1$ eigenvectors of $\hat{M}$. By Stewart's Lemma, letting the columns of V be $v_1, \ldots, v_{k-1}$, we have

$$\|V^T P_2\|_2 \leq \sqrt{1 - (1 - \varepsilon)^2},$$

or equivalently,

$$\min_{v \in \operatorname{span}\{v_1, \ldots, v_{k-1}\},\ \|v\|=1} \|\operatorname{proj}_F v\| = \sigma_{k-1}(V^T P_1) \geq 1 - \varepsilon. \qquad \blacksquare$$

5. Recursion

In this section, we show that for every direction h that is close to the intermean subspace, the "largest gap clustering" step produces a pair of complementary halfspaces that partitions $\mathbb{R}^n$ while leaving only a small part of the probability mass on the wrong side of the partition, small enough that with high probability, it does not affect the samples used by the algorithm.

Lemma 20. *Let $\delta, \delta' > 0$, where $\delta' \leq \delta/(2m_2)$, and let m_2 satisfy $m_2 \geq n/k \log(2k/\delta)$. Suppose that h is a unit vector such that*

$$\|\mathrm{proj}_F(h)\| \geq 1 - \frac{\mathrm{w}}{2^{10}(k-1)^2 \log \frac{1}{\delta'}}.$$

Let $\mathcal{F}$ be a mixture of $k > 1$ Gaussians with overlap

$$\phi \leq \frac{\mathrm{w}}{2^9(k-1)^2} \log^{-1} \frac{1}{\delta'}.$$

Let X be a collection of m_2 points from $\mathcal{F}$ and let t be the midpoint of the largest gap in set $\{h^T x \colon x \in X\}$. With probability $1-\delta$, the halfspace $H_{h,t}$ has the following property. For a random sample y from $\mathcal{F}$ either

$$y, \mu_{\ell(y)} \in H_{h,t} \text{ or } y, \mu_{\ell(y)} \notin H_{h,t}$$

with probability $1 - \delta'$.

Proof of Lemma 20. The idea behind the proof is simple. We first show that two of the means are at least a constant distance apart. We then bound the width of a component along the direction h, i.e. the maximum distance between two points belonging to the same component. If the width of each component is small, then clearly the largest gap must fall between components. Setting t to be the midpoint of the gap, we avoid cutting any components.

We first show that at least one mean must be far from the origin in the direction h. Let the columns of P_1 be the vectors $e_1, \ldots, e_{k-1}$. The span of these vectors is also the span of the means, so we have

$$\begin{aligned}\max_i \left(h^T\mu_i\right)^2 &= \max_i \left(h^T P_1 P_1^T \mu_i\right)^2 \\ &= \| P_1^T h\|^2 \max_i \left(\frac{\left(P_1^T h\right)^T}{\|P_1 h\|}\tilde{\mu}_i\right)^2 \\ &\geq \| P_1^T h\|^2 \sum_{i=1}^k \mathrm{w}_i \left(\frac{\left(P_1^T h\right)^T}{\|P_1 h\|}\tilde{\mu}_i\right)^2 \\ &\geq \| P_1^T h\|^2 (1-\phi) \\ &> \frac{1}{2}.\end{aligned}$$

Since the origin is the mean of the means, we conclude that the maximum distance between two means in the direction h is at least $1/2$. Without loss of generality, we assume that the interval $[0, 1/2]$ is contained between two means projected to h.

We now show that every point x drawn from component i falls in a narrow interval when projected to h. That is, x satisfies $h^T x \in b_i$, where $b_i = \left[h^T\mu_i - \left(8(k-1)\right)^{-1}, h^T\mu_i + \left(8(k-1)\right)^{-1}\right]$. We begin by examining the variance along h. Let $e_k, \ldots, e_n$ be the columns of the matrix n-by-$(n-k+1)$ matrix P_2. Recall from Eqn. (2) that $P_1^T \mathrm{w}_i \Sigma_i P_1 = A_i$, that $P_2^T \mathrm{w}_i \Sigma_i P_1 = B_i$, and that $P_2^T \mathrm{w}_i \Sigma_i P_2 = D_i$. The norms of these matrices are bounded according to Lemma 2. Also, the vector $h = P_1 P_1^T h + P_2 P_2^T h$. For convenience of notation we define ε such that $\|P_1^T h\| = 1 - \varepsilon$. Then $\|P_2^T h\|^2 = 1 - (1-\varepsilon)^2 \leq 2\varepsilon$. We now argue

$$\begin{aligned}h^T \mathrm{w}_i \Sigma_i h &\leq \left(h^T P_1 A_i P_1^T h + 2h^T P_2 B_i P_1 h + h^T P_2^T D_i P_2 h\right) \\ &\leq 2\left(h^T P_1 A_i P_1^T h + h^T P_2 D_i P_2^T h\right) \\ &\leq 2\left(\| P_1^T h\|^2 \|A_i\| + \| P_2^T h|^2 \|D_i\|\right) \\ &\leq 2(\phi + 2\varepsilon).\end{aligned}$$

Using the assumptions about ϕ and ε, we conclude that the maximum variance along h is at most

$$\max_i h^T \Sigma_i h \leq \frac{2}{\mathrm{w}} \left(\frac{\mathrm{w}}{2^9(k-1)^2} \log \frac{1}{\delta'} + 2 \frac{\mathrm{w}}{2^{10}(k-1)^2} \log \frac{1}{\delta'} \right)$$
$$\leq \left(2^7 (k-1)^2 \log 1/\delta' \right)^{-1}.$$

We now translate these bounds on the variance to a bound on the difference between the minimum and maximum points along the direction h. By Lemma 13, with probability $1 - \delta/2$

$$\left| h^T \left(x - \mu_{\ell(x)} \right) \right| \leq \sqrt{2 h^T \Sigma_i h \log \left(2 m_2 / \delta \right)}$$
$$\leq \frac{1}{8(k-1)} \cdot \frac{\log \left(2 m_2 / \delta \right)}{\log \left(1/\delta' \right)} \leq \frac{1}{8(k-1)}.$$

Thus, with probability $1 - \delta/2$, every point from X falls into the union of intervals $b_1 \cup \ldots \cup b_k$ where $b_i = \left[h^T \mu_i - \left(8(k-1) \right)^{-1}, h^T \mu_i + \left(8(k-1) \right)^{-1} \right]$. Since these intervals are centered about the means, at least the equivalent of one interval must fall outside the range $[0, 1/2]$, which we assumed was contained between two projected means. Thus, the measure of subset of $[0, 1/2]$ that does not fall into one of the intervals is

$$\frac{1}{2} - (k-1) \frac{1}{4(k-1)} = \frac{1}{4}.$$

This set can be cut into at most $k-1$ intervals, so the smallest possible gap between these intervals is $\left(4(k-1) \right)^{-1}$, which is exactly the width of an interval.

Since we have $m_2 = k/\mathrm{w} \log \left(2k/\delta \right)$, the set X contains at least one sample from every component with probability $1 - \delta/2$. Overall, with probability $1 - \delta$ every component has at least one sample and all samples from component i fall in b_i. Thus, the largest gap between the sampled points will not contain one of the intervals $b_1, \ldots, b_k$. Moreover, the midpoint t of this gap must also fall outside of $b_1 \cup \ldots \cup b_k$, ensuring that no b_i is cut by t.

By the same argument given above, any single point y from $\mathcal{F}$ is contained in $b_1 \cup \ldots \cup b_k$ with probability $1 - \delta'$ proving the Lemma. ∎

In the proof of the main theorem for large k, we will need to have every point sampled from $\mathcal{F}$ in the recursion subtree classified correctly by the halfspace, so we will assume δ' considerably smaller than m_2/δ.

The second lemma shows that all submixtures have smaller overlap to ensure that all the relevant lemmas apply in the recursive steps.

Lemma 21. *The removal of any subset of components cannot induce a mixture with greater overlap than the original.*

Proof of Lemma 21. Suppose that the components $j+1, \ldots k$ are removed from the mixture. Let $\omega = \sum_{i=1}^{j} \mathrm{w}_i$ be a normalizing factor for the weights. Then if $c = \sum_{i=1}^{j} \mathrm{w}_i \mu_i = -\sum_{i=j+1}^{k} \mathrm{w}_i \mu_i$, the induced mean is $\omega^{-1} c$. Let T be the subspace that minimizes the maximum overlap for the full k component mixture. We then argue that the overlap $\tilde{\phi}^2$ of the induced mixture is bounded by

$$\begin{aligned}
\tilde{\phi} &= \min_{\dim(S)=j-1} \max_{v \in S} \frac{\omega^{-1} v^T \Sigma v}{\omega^{-1} \sum_{i=1}^{j} \mathrm{w}_i v^T (\mu_i \mu_i^T - cc^T + \Sigma_i) v} \\
&\leq \max_{v \in \mathrm{span}\,\{e_1, \ldots, e_{k-1}\} \setminus \mathrm{span}\,\{\mu_{j+1}, \ldots, \mu_k\}} \frac{\sum_{i=1}^{j} \mathrm{w}_i v^T \Sigma_i v}{\sum_{i=1}^{j} \mathrm{w}_i v^T (\mu_i \mu_i^T - cc^T + \Sigma_i) v}.
\end{aligned}$$

Every $v \in \mathrm{span}\,\{e_1, \ldots, e_{k-1}\} \setminus \mathrm{span}\,\{\mu_{j+1}, \ldots, \mu_k\}$ must be orthogonal to every μ_ℓ for $j+1 \leq \ell \leq k$. Therefore, v must be orthogonal to c as well. This also enables us to add the terms for $j+1, \ldots, k$ in both the numerator and denominator, because they are all zero.

$$\begin{aligned}
\tilde{\phi} &\leq \max_{v \in \mathrm{span}\,\{e_1, \ldots, e_{k-1}\} \setminus \mathrm{span}\,\{\mu_{j+1}, \ldots, \mu_k\}} \frac{v^T \Sigma v}{\sum_{i=1}^{k} \mathrm{w}_i v^T (\mu_i \mu_i^T + \Sigma_i) v} \\
&\leq \max_{v \in \mathrm{span}\,\{e_1, \ldots, e_{k-1}\}} \frac{v^T \Sigma v}{\sum_{i=1}^{k} \mathrm{w}_i v^T (\mu_i \mu_i^T + \Sigma_i) v} \\
&= \phi. \qquad \blacksquare
\end{aligned}$$

The proofs of the main theorems are now apparent. Consider the case of $k = 2$ Gaussians first. As argued in Section 3.4, using $m_1 = \omega\big(kn^4 \mathrm{w}^{-3} \log (n/\delta \mathrm{w})\big)$ samples to estimate $\hat{u}$ and $\hat{M}$ is sufficient to guarantee that the estimates are accurate. For a well-chosen constant C, the condition

$$\phi \leq J(p) \leq C \mathrm{w}^3 \log^{-1} \left(\frac{1}{\delta \mathrm{w}} + \frac{1}{\eta} \right)$$

of Theorem 2 implies that

$$\sqrt{\phi} \leq \frac{\mathrm{w}\sqrt{\varepsilon}}{640 \cdot 2},$$

where

$$\varepsilon = \frac{\mathrm{w}}{2^9} \log^{-1}\left(\frac{2m_2}{\delta} + \frac{1}{\eta}\right).$$

The arguments of Section 4 then show that the direction h selected in step 3 satisfies

$$\|P_1^T h\| \geq 1 - \varepsilon = 1 - \frac{\mathrm{w}}{2^9} \log^{-1}\left(\frac{m_2}{\delta} + \frac{1}{\eta}\right).$$

Already, for the overlap we have

$$\sqrt{\phi} \leq \frac{\mathrm{w}\sqrt{\varepsilon}}{640 \cdot 2} \leq \sqrt{\frac{\mathrm{w}}{2^9(k-1)^2}} \log^{-1/2} \frac{1}{\delta'}.$$

so we may apply Lemma 20 with $\delta' = (m_2/\delta + 1/\eta)^{-1}$. Thus, with probability $1 - \delta$ the classifier $H_{h,t}$ is correct with probability $1 - \delta' \geq 1 - \eta$.

We follow the same outline for $k > 2$, with the quantity $1/\delta' = m_2/\delta + 1/\eta$ being replaced with $1/\delta' = m/\delta + 1/\eta$, where m is the total number of samples used. This is necessary because the half-space $H_{h,t}$ must classify every sample point taken below it in the recursion subtree correctly. This adds the n and k factors so that the required overlap becomes

$$\phi \leq C\mathrm{w}^3 k^{-3} \log^{-1}\left(\frac{nk}{\delta \mathrm{w}} + \frac{1}{\eta}\right)$$

for an appropriate constant C. The correctness in the recursive steps is guaranteed by Lemma 21. Assuming that all previous steps are correct, the termination condition of step 4 is clearly correct when a single component is isolated.

6. Conclusion

We have presented an affine-invariant extension of principal components. We expect that this technique should be applicable to a broader class of problems. For example, mixtures of distributions with some mild properties such as center symmetry and some bounds on the first few moments might be solvable using isotropic PCA. It would be nice to characterize the full scope of the technique for clustering and also to find other applications, given that standard PCA is widely used.

Acknowledgements. We are indebted to Ravi Kannan and Dan Spielman for thoughtful comments. An extended abstract of this paper appears in the proceedings of FOCS 2008. This research was supported in part by NSF award CCF-0721503, the Algorithms and Randomness Center at Georgia Tech and a Raytheon fellowship.

References

[1] D. Achlioptas and F. McSherry, On spectral learning of mixtures of distributions, in: *Proc. of COLT*, 2005.

[2] K. Chaudhuri and S. Rao, *Beyond gaussians: Spectral methods for learning mixtures of heavy-tailed distributions* (2008).

[3] K. Chaudhuri and S. Rao, *Learning mixtures of product distributions using correlations and independence* (2008).

[4] A. Dasgupta, J. Hopcroft, J. Kleinberg and M. Sandler, On learning mixtures of heavy-tailed distributions, in: *Proc. of FOCS* (2005).

[5] S. DasGupta, Learning mixtures of gaussians, in: *Proc. of FOCS* (1999).

[6] A. P. Dempster, N. M. Laird and D. B. Rubin, Maximum likelihood from incomplete data via the em algorithm, *Journal of the Royal Statistical Society B,* **39** (1977), 1–38.

[7] Jon Feldman, Rocco A. Servedio and Ryan O'Donnell, Pac learning axis-aligned mixtures of gaussians with no separation assumption, in: *COLT* (2006), pp. 20–34.

[8] K. Fukunaga, *Introduction to Statistical Pattern Recognition,* Academic Press (1990).

[9] R. O. Duda, P. E. Hart and D. G. Stork, *Pattern Classification,* John Wiley & Sons (2001).

[10] R. Kannan, H. Salmasian and S. Vempala, The spectral method for general mixture models, in: *Proceedings of the 18th Conference on Learning Theory,* University of California Press (2005).

[11] L. Lovász and S. Vempala, The geometry of logconcave functions and and sampling algorithms, *Random Strucures and Algorithms,* **30(3)** (2007), 307–358.

[12] J. B. MacQueen, Some methods for classification and analysis of multivariate observations, in: *Proceedings of 5-th Berkeley Symposium on Mathematical Statistics and Probability,* volume 1. University of California Press (1967), pp. 281–297.

[13] M. Rudelson, Random vectors in the isotropic position, *Journal of Functional Analysis,* **164** (1999), 60–72.

[14] M. Rudelson and R. Vershynin, Sampling from large matrices: An approach through geometric functional analysis, *J. ACM,* **54(4)** (2007).

[15] R. Kannan and S. Arora, Learning mixtures of arbitrary gaussians, *Ann. Appl. Probab.,* **15(1A)** (2005), 69–92.

[16] L. Schulman and S. DasGupta, A two-round variant of em for gaussian mixtures, in: *Sixteenth Conference on Uncertainty in Artificial Intelligence* (2000).

[17] G. W. Stewart and Ji guang Sun, *Matrix Perturbation Theory,* Academic Press, Inc. (1990).

[18] S. Vempala and G. Wang, A spectral algorithm for learning mixtures of distributions, *Proc. of FOCS 2002; JCCS,* **68(4)** (2004), 841–860.

S. Charles Brubaker

College of Computing
Georgia Tech.

e-mail: brubaker@cc.gatech.edu

Santosh S. Vempala

College of Computing
Georgia Tech.

e-mail: vempala@cc.gatech.edu

BOLYAI SOCIETY
MATHEMATICAL STUDIES, 19

Building Bridges
pp. 283–307.

Small Linear Dependencies for Binary Vectors of Low Weight

URIEL FEIGE

We show that every set of $m \simeq cn\sqrt{n \log\log n}$ vectors in $\{0,1\}^n$ in which every vector has Hamming weight 3 contains a subset of $O(\log n)$ vectors that form a linear dependency. Our proof is based on showing that in every graph of average degree at least $c \log\log n$, every legal edge coloring produces a cycle in which one of the colors appears either once or twice. (In both results, c is some constant.) The results proved are used (in a companion work) in refutation algorithms for semirandom 3CNF formulas.

1. Introduction

The problem studied in this paper can be viewed either as a problem involving linear dependencies among binary vectors, or as a problem on hypergraphs. We present here the hypergraph formulation.

Definition 1.1. An *even cover* in a hypergraph is a nonempty set of hyperedges that contains each vertex an even number of times (either not at all, or twice, or four times, etc.). The *size* of an even cover is the number of hyperedges in the even cover.

It is not hard to see that every hypergraph on n vertices with more than n hyperedges has an even cover of size at most $n+1$. This follows by viewing each hyperedge as an indicator vector for its variables, noting that this gives a vector space of dimension at most n, and that every minimal set of linearly dependent binary vectors (addition performed modulo 2) corresponds to an even cover. As the number of hyperedges in a hypergraph increases, smaller even covers may appear. For r-uniform hypergraphs with

$r \geq 3$, it is reasonable to conjecture the following relation between number of hyperedges and size of even covers. (The $\tilde{O}$ notation is meant to suppress an $O(\log n)$ multiplicative term, though the author would be happy to settle also for somewhat larger low order multiplicative terms.)

Conjecture 1.2. Let c be sufficiently large. Then every r-uniform hypergraph on n vertices and $m = c\beta n$ hyperedges (with $1 < \beta \leq O(n^{(r-2)/2})$) has an even cover of size at most $\tilde{O}\big(n/\beta^{2/(r-2)}\big)$.

For graphs ($r = 2$), minimal even covers are simply cycles. The natural analog of the conjecture for graphs would be that every graph of sufficiently high constant average degree has a cycle of length $O(\log n)$, which is well known to be true. For general r, the conjecture is not known to hold, except of course at the very low density case when $\beta < (\log n)^{(r-2)/2}$, in which case the conjecture is trivially true. When r is even, the conjecture is known to be true also at the very high density case, say, when $\beta = 2n^{(r-2)/2}$ (see Proposition 2.2). The current work addresses the very high density case of Conjecture 1.2 when r is odd, and comes closer to proving the conjecture in this case.

The case that motivates the current work is that of $r = 3$. In this case, β can range from 1 to $\sqrt{n}$. As we shall explain in Section 1.1, this case comes up in refutation of random 3CNF formulas. Some of the results in this work are stated only for this special case, but easily extend to all odd r.

The following theorem is implicit in the work of Naor and Vastreate [11]. If not for the term $\log n$ in the value of β, it would prove the conjecture for the very high density case.

Theorem 1.3. *Every 3-uniform hypergraph with n vertices and βn hyperedges contains an even cover of size at most $\log n$. Here $\beta = c \log n \sqrt{n}$ for some sufficiently large universal constant c.*

In our work, we improve over the value of β and show:

Theorem 1.4. *Let H be an arbitrary 3-uniform hypergraph with n vertices and $m = \beta n$ hyperedges, and let c be a sufficiently large universal constant. Then:*

1. *If $\beta \geq c\sqrt{n \log n / \log\log n}$ then H contains an even cover of size $O(\log n / \log\log n)$.*
2. *If $\beta \geq c\sqrt{n \log\log n}$ then H contains an even cover of size $O(\log n)$.*

Moreover, in both cases there is a polynomial time algorithm that finds the respective even cover.

Our proof of Theorem 1.4 produces even covers which are of a special form (correspond to linear dependencies over any field). See Corollary 4.5. Our proof technique reduces the problem of even covers in hypergraphs to an extremal problem in graphs.

Definition 1.5. Given a graph G and an arbitrary legal coloring of its edges (incident edges have different colors), a simple cycle (namely, a cycle that does not visit any vertex more than once) is called a 1-cycle (2-cycle, respectively) if some color is used in order to color exactly one (two, respectively) of its edges. We say that a cycle in an edge colored graph is *distinguished* if it is either a 1-cycle or a 2-cycle.

The extremal problem is as follows: what is the maximum number of edges that a legally colored n-vertex graph can have and still not contain a distinguished cycle of length at most t? Observe that a graph may have arbitrarily many edges and still not have a 2-cycle (if every edge is colored by a different color). It may have $\Omega(n \log n)$ edges and still not have a 1-cycle (e.g, color edges of the hypercube by the name of the coordinate that is flipped). However, once both 1-cycles and 2-cycles are forbidden, we show that the number of edges is at most $O(n \log \log n)$.

Theorem 1.6. *For a sufficiently large constant c,*

1. *For every graph on n vertices and average degree at least $c\frac{\log n}{\log\log n}$, every legal edge coloring creates a distinguished cycle of length $O\left(\frac{\log n}{\log\log n}\right)$.*
2. *For every graph on n vertices and average degree at least $c \log\log n$, every legal edge coloring creates a distinguished cycle of length $O(\log n \log\log n)$.*

Moreover, in both cases there is a polynomial time algorithm that finds the respective distinguished cycle.

As we shall see in the proof of Corollary 2.12, the correspondence between Theorem 1.6 and Theorem 1.4 is as follows. A degree of d in Theorem 1.6 gives a density of $\beta = O(\sqrt{dn})$ in Theorem 1.4 for even covers that are twice as large as the corresponding distinguished cycles. Hence item 1 of Theorem 1.4 follows from item 1 of Theorem 1.6. Likewise, item 2 of Theorem 1.4 almost follows from item 2 of Theorem 1.6, except for a $\log\log n$

term in the size of the even cover. To remove this $\log \log n$ term, we appeal to some elementary properties of cycle bases in graphs.

1.1. Motivation and related work

The author's motivation for studying small even covers comes from a sequence of works on refuting random 3CNF formulas. The goal of these works is to design algorithms that when given a nonsatisfiable 3CNF Boolean formula (conjunction of clauses, where each clause is a disjunction of three literals such as $x_1 \vee \bar{x_2} \vee x_3$) certifies that no satisfying assignment exists. In general, this problem is coNP-hard, but it turns out that for sufficiently dense random (or semirandom) formulas efficient refutation algorithms exist (with high probability over the choice of the input formula). The methodology developed in [8, 4, 7, 6, 5] to design such algorithms reduces the problem of refuting 3CNF formulas to a stronger form of refutation but for an easier problem, max-3LIN2. Namely, given an inconsistent system of linear equations with three Boolean variables per equation (such as $x_1 + x_2 + x_3 = 1$ modulo 2), certify that the system is "far" from being satisfiable (in the sense that every assignment leaves "many" equations not satisfied). We call this *strong* refutation (though in a sense not as strong as that of [3]). Refuting satisfiability of 3LIN2 is easy (by Gaussian elimination), but strong refutation is in general NP-hard (by a rephrasing of the known hardness of approximation results [9] for max-3LIN2). However, the max-3LIN2 systems that are obtained from the reduction from random 3CNF formulas are random rather than worst case instances, and hence there is hope for strongly refuting them. Here is the approach developed in the above works.

Given a 3LIN2 system ϕ, let H_ϕ be the following 3-uniform hypergraph. The vertices of H_ϕ are the variables of ϕ. The hyperedges are the equations of ϕ. For example, the clause $x_1 + x_2 + x_3 = 0$ gives rise to the hyperedge (x_1, x_2, x_3). The hypergraph does not represent the right hand side of the equations. Assume that H_ϕ has an even cover of size 2ℓ (observe that an even cover always has an even number of hyperedges, because every hyperedge contains three vertices, and every vertex appears an even number of times). Consider the 2ℓ linear equations that correspond to the hyperedges of the even cover. Summing up all equations, the left hand side gives 0 (since every variable appears an even number of times and addition is performed modulo 2). As to the right hand side, if there is some randomness in the equation in the sense that for every equation independently there is some small probability δ that its right hand side is random, then with probability $\Omega(\delta\ell)$ (or at least 1/4 if $\delta\ell > 1$) the right hand side will sum up to 1, leading to a contradiction. Moreover, if H_ϕ has many disjoint even covers that can be found efficiently, this gives many disjoint subsystems, and if the original 3LIN system is somewhat random in the above sense, then many of them are likely not to be satisfiable. This is exactly what we want to achieve by strong refutation. Observe that a theorem such as Theorem 1.4 implies the existence of many disjoint even covers (and not just one), because after a small even cover is found, it can be removed from the hypergraph without substantially changing the number of hyperedges, and then the theorem can be applied again.

Theorem 1.4 plays a central role in the refutation algorithm presented in [5]. Its proof was only sketched in [5] and is presented in full in the current paper. (Remark: the bounds proved in the current paper are stronger than the corresponding bounds claimed in [5].)

Results in [6] support Conjecture 1.2. There, the special case of $r = 3$ and $\beta = n^{0.4}$ was considered. If was shown that if the hypergraph is random, then indeed it is likely to have even covers of size $O(n/\beta^2) = O(n^{0.2})$. The proof works for other values of β as well (except possibly for very small values of β – this needs to be checked). We remark that random hypergraphs serve as examples showing that Conjecture 1.2 cannot be improved upon. This follows from a simple expectation computation. See for example [6].

The "super-high" density case of $\beta = n^{\delta+(r-2)/2}$ for some $\delta > 0$ was studied in [11]. The motivation there comes from studying the Hamming distance of codes that have low density parity check matrices (see more details in [11]). When $\delta > 0$ there are even covers of constant size, and the goal is to figure out how this constant depends on δ. In our work we slightly improve over the bounds proven in [11] when r is odd and large (see the end of Section 2). More importantly, our proof technique, though sharing some features with that of [11], works for a wider range of parameters than the proof technique of [11]. Hence we start getting meaningful results when $\beta \geq n^{(r-2)/2}\sqrt{\log\log n}$, whereas the techniques of [11] require $\beta \geq n^{(r-2)/2}\log n$.

Some nontrivial upper bounds on the size of even covers can be obtained by proving the existence of subhypergraphs that contain more hyperedges than vertices, and on such a hypergraph invoking the argument that follows Definition 1.1. For example, in [1] it is shown that every 3-uniform hypergraph with βn hyperedges has a set of $\ell = O(n \log n/\beta)$ vertices that induce at least $\ell + 1$ hyperedges. It follows that even covers of size $O(n \log n/\beta)$

always exist. This line of work suggest the following conjecture as an intermediate step towards proving Conjecture 1.2.

Conjecture 1.7. Let c be sufficiently large. Then every 3-uniform hypergraph on n vertices and $m = c\beta n$ hyperedges (with $1 < \beta \leq O(\sqrt{n})$) has a set of $n' \leq \tilde{O}(n/\beta^2)$ vertices that induce at least $2n'/3$ hyperedges.

2. Distinguished Cycles Shorter than $\log n$

This section contains the proof of item (1) of Theorem 1.4. It is based on a simplification of the proof technique of [11], and works when $\beta > \sqrt{n \log n / \log\log n}$. In passing, we also improve some other results of [11]. For this reason, the presentation will be for r-uniform hypergraphs for general r, even though 3-uniform hypergraphs suffice for Theorem 1.4.

It will be convenient for us to view hyperedges of H as r-tuples of vertices, rather than as sets of vertices. Hence we shall use the convention that vertices of H are sorted in some arbitrary order, and likewise, the r vertices in a hyperedge are sorted according to the same order. The r-tuple corresponding to a hyperedge is this sorted list of vertices.

The following lemma is a result taken from [2]. (The result in [2] is slightly stronger. Weaker results that would also suffice for the purpose of our paper were known previous to [2].)

Lemma 2.1. *In any n-vertex graph of average degree $d > 2$ there is a cycle of length no longer than 2ℓ, if $d(d-1)^{\ell-1} > n$.*

Proof. If the graph is d-regular, the proof follows easily by performing breadth first search, starting from an arbitrary vertex. The (known) proof for nonregular graphs is not as simple. See [2] for details. ■

The difficulty in proving Theorem 1.4 stems from the fact that it deals with r-uniform hypergraphs with odd $r = 3$. It is instructive to first see how a corresponding (in fact, stronger) theorem can be proved when r is even. The following proposition is taken from [11].

Proposition 2.2. *For even r and $d > 1$, every r-uniform hypergraph H with $m = dn^{r/2}$ hyperedges contains an even cover of size $O(\log n)$.*

Proof. Construct the following auxiliary graph G. It has $\binom{n}{r/2} \leq n^{r/2}$ vertices, labelled by all possible sets of $r/2$ vertices of H. Every hyperedge e of H contributes one edge e' to G, connecting the vertex in G that is labelled by the set of $r/2$ vertices that make the prefix of the r-tuple that labels e and the vertex labelled by the set of $r/2$ vertices that make the suffix of the r-tuple that labels e. The average degree of G is (slightly larger than) $2d > 2$, and hence Lemma 2.1 implies that G has a cycle of length $O(\log n)$. The hyperedges of H that correspond to the edges of G along this cycle form an even cover in H. ■

We now return to the case that r is odd (in our case, $r = 3$), and consider an r uniform hypergraph with $m = dn^{r/2}$ edges.

For given n, r and d, let h be maximal such that h and s are positive integers satisfying $n^h < dn^{r/2}$ and $h + 2s = r$. For example, when $r = 3$ we have that $h = s = 1$, and for $r = 5$ we have that $h = 1$ when $d < \sqrt{n}$.

The following notion (used also in [11]) helps simplify later proofs.

Definition 2.3. For h and s as defined above, an r-uniform hypergraph satisfies the *small overlap* condition if no two hyperedges share $h+s$ vertices.

The following lemma (similar to [11]) shows that up to a negligible effect on d, we may assume that the small overlap condition holds.

Lemma 2.4. *Let H be an r-uniform hypergraph with $dn^{r/2}$ hyperedges, let h and s be as above, and let $\varepsilon > 0$ satisfy $\varepsilon dn^{r/2-s} > 1$. Then either*

- *H has a subhypergraph with $(d-2\varepsilon)n^{r/2}$ hyperedges that satisfies the small overlap condition,*

or

- *H has an even cover of size 4ℓ, where ℓ is the smallest integer satisfying $\left(2\varepsilon dn^{r/2-s}\right)^{\ell} > n^s$. For example, when $r = 3$ and $\varepsilon > 1/d$ this corresponds to an even cover of size 8.*

Proof. Given an r-uniform hypergraph H, consider an auxiliary graph G whose vertices are the hyperedges of H, and two vertices of G are connected by an edge if the respective hyperedges share at least $h + s$ vertices (in H). Consider an arbitrary maximal matching M in G. If the matching contains less than $\varepsilon dn^{r/2}$ edges, then remove the corresponding matched hyperedges from H. The number of hyperedges in H remains essentially unchanged, and H now satisfies the *small overlap* condition.

If the matching M contains $\varepsilon dn^{r/2}$ edges, then consider an auxiliary multigraph F (it may have parallel edges). The vertices of F are s-tuples of vertices of H. Every edge of the matching M contributes one edge to F as follows. Let the matching edge correspond to two hyperedges e_1 and e_2 in H, and without loss of generality, assume that e_1 and e_2 share their first $h+s$ vertices (in H). Then in F add an edge between the vertex that is labelled by the last s vertices of e_1 and the vertex that is labelled by the last s vertices of e_2. Observe that now any cycle in F corresponds in a natural way to an even cover in H (with twice as many hyperedges in H than edges if F). The average degree in F is at least $2\varepsilon dn^{r/2-s}$, which is greater than 2 by the conditions in the statement of the lemma. Hence Lemma 2.1 implies that H has an even cover with 4ℓ edges, where ℓ is the smallest value satisfying $\left(2\varepsilon dn^{r/2-s}\right)^{\ell} > n^s$. ■

We shall assume that the hypergraph H satisfies the small overlap condition. This assumption can be made essentially without loss of generality, because results proved under this assumption easily generalize to arbitrary hypergraphs with almost the same parameters, by Lemma 2.4.

Now is our main point of departure from [11]. We construct an auxiliary graph G that is different from the one constructed in [11], and this leads both to a considerable simplification in the proofs, and to a strengthening of the results. The graph that we construct is similar to the one constructed in [8] in their refutation algorithm for random 3SAT.

Each vertex of G corresponds to a set of $2s$ vertices of H. By our convention that vertices of H are sorted, a vertex of G will be labelled by a $2s$-tuple of vertices of H, for which the prefix of size s is sorted and the suffix of size s is sorted. The same vertex of H may appear both in the prefix and in the suffix. Hence G has $\binom{n}{s}^2$ vertices. The edges of G are derived from hyperedges of H as follows. Every hyperedge of H is an r-tuple. Every two hyperedges e_1 and e_2 of H whose r-tuples agree on the last h vertices contribute one edge to G. This edge connects the vertices v_1 and v_2 in G, if the labels of v_1 and v_2 satisfy the following conditions. The tuple labelling v_1 agrees on its first s coordinates with the first s coordinates of the tuple labelling e_1, and agrees in its last s coordinates with coordinates $s+1$ up to $2s$ of the tuple labelling e_2. The tuple labelling v_2 agrees on its first s coordinates with the first s coordinates of the tuple labelling e_2, and agrees in its last s coordinates with the coordinates $s+1$ up to $2s$ of the tuple labelling e_1. Moreover, we color this edge by the color c, where c is a tuple containing the last h vertices (the overlap vertices) in the tuples e_1

and e_2. Hence every edge of G (together with its color and the labels of its endpoints) uniquely determines which two hyperedges in H generated it.

Here are a few examples to illustrate the construction. When $r = 3$ we have that $h = s = 1$. Hence G contains n^2 vertices. Two hyperedges $e_1 = (1, 2, 5)$ and $e_2 = (3, 4, 5)$ in H would contribute the edge $\big((1, 4)(3, 2)\big)$ to G, and this edge would be colored (5). When $r = 5$ and d is small, then $h = 1$ and $s = 2$. Hence G contains $\binom{n}{2}^2 < n^4$ vertices. Two hyperedges $e_1 = (1, 3, 4, 6, 8)$ and $e_2 = (2, 4, 5, 7, 8)$ in H would contribute the edge $\big((1, 3, 5, 7)(2, 4, 4, 6)\big)$ to G, and this edge would be colored (8).

The following proposition is the key reason for introducing the *small overlap* property.

Proposition 2.5. *If H satisfies the small overlap condition, then the coloring of the edges of G is a legal coloring (no two edges of the same color are incident with the same vertex).*

Proof. Otherwise there would be two hyperedges in H whose overlap is at least $h + s$. ■

The key to finding even covers in H is by using cycles in G. Observe that for every cycle in G, every vertex of H appears an even number of times on this cycle (counting all its appearances in hyperedges of H that generated the cycle in G). Hence a cycle in G corresponds to an even cover in H. However, there is one potential problem in this correspondence. A hyperedge in H may generate several edges in G. Hence it might be the case that in a given cycle of G, some hyperedges of H appear more than once. Two appearances of the same hyperedge in an even cover can (and should) be removed – this still results in an even cover. Continuing in this fashion, if it happens that every hyperedge of H appears on the cycle an even number of times (say, either twice or not at all), all hyperedges are removed and one remains with the empty even cover. In this case we say that the cycle is *trivial*: it corresponds to the trivial even cover that contains no hyperedges. Hence a cycle in G corresponds to an even cover in H if and only if the cycle is not trivial. In this work, we shall consider one particular class of nontrivial cycles in G, namely, the class of distinguished cycles, as defined in Definition 1.5.

Lemma 2.6. *A distinguished cycle in G of length ℓ corresponds to an even cover in H with at most 2ℓ hyperedges.*

Proof. Every edge of G is generated by two hyperedges of H. Consider one appearance of the color c that appears either once or twice in the distinguished cycle, and the two hyperedges that generated this appearance. If c appears only once, then these two hyperedges each appear only once on the cycle (because any edge that any of them generates will be colored c). If c appears twice, then in can not be that both appearances were generated by the same pair of hyperedges, because then both appearances would correspond to the same edge, contradicting the requirement that the cycle is simple. Hence in a distinguished cycle there is at least one hyperedge that appears exactly once on the cycle, and hence the cycle cannot be trivial. A nontrivial cycle corresponds to an even cover, and the hyperedges of the even cover are those hyperedges that generated the edges of the cycle. Hence the corresponding even cover has at most 2ℓ hyperedges (and possible less, if some hyperedges generated more than one edge along the cycle). ∎

To show that G has short distinguished cycles, we first bound from below its average degree.

Lemma 2.7. *The graph G has average degree at least (roughly) d^2.*

Proof. Recall that G has $\binom{n}{s}^2$ vertices, that H has $dn^{r/2}$ hyperedges, and that $2s + h = r$. Group the hyperedges into $\binom{n}{h}$ disjoint groups, one for every possible h-suffix of a label of a hyperedge. A simple shifting argument implies that the number of edges in G is minimized when all groups are of the same size. Hence we assume that every group contains $dn^{r/2}/\binom{n}{h}$ hyperedges (ignoring rounding issues). Each group then generates $\binom{dn^{r/2}/\binom{n}{h}}{2}$ edges, and the total number of edges in G is roughly $d^2n^r/2\binom{n}{h}$. The average degree is at least as claimed because $\binom{n}{s}^2\binom{n}{h} \leq n^r$. ∎

The discussion so far leads to the problem of providing bounds for $L_{\text{avg}}(N, D)$ as defined below.

Definition 2.8. Let $L_{\text{avg}}(N, D)$ denote the minimum value of ℓ such that for every graph G with N vertices (in our case $N \simeq n^{r-h}$) and average degree at least D (in our case $D \geq d^2$), and for every legal coloring of its edges, G must contain a distinguished cycle of length at most 2ℓ.

In our application G need not be a simple graph. It may have parallel edges. But if it does, then it contains a distinguished cycle of length two.

When $D > c\frac{\log N}{\log \log N}$ (here c is some sufficiently large constant) then the existence of short distinguished cycles can be analyzed using the same approach as that used for the existence of short cycles in general. As we do not care for constant factors in the degree D, the analysis can be simplified by the following proposition (also used in [11] for the same purpose).

Proposition 2.9. *Every graph of average degree D has a subgraph of minimum degree $D/2$.*

Proof. Iteratively remove vertices of degree less than $D/2$ together with their incident edges. The total number of edges that can be removed in this process is strictly less than $nD/2$, and hence some subgraph remains. ∎

We define $L_{\min}(N, D)$ in a way similar to Definition 2.8, but with minimum degree replacing average degree. Proposition 2.9 implies that $L_{\text{avg}}(N, D) \leq \max_{N' \leq N} \left[L_{\min}(N', D/2)\right]$. (Our upper bounds on $L_{\min}(N, D)$ will be nondecreasing in N, and hence the value of N' to be used in the above inequality will be N.) The following lemma proves item 1 of Theorem 1.6.

Lemma 2.10. *For $L_{\min}(N, D)$ as defined above, $L_{\min}(N, D) \leq \ell$ where ℓ is the smallest integer for which $D!/(D-\ell)! \geq N$, if such an integer $\ell < D$ exists.*

Proof. Pick an arbitrary vertex r in G as the root, and develop a *colorful* version of a breadth first search tree from r. The root r is at level 0. All neighbors of r (there are at least D of them) are at level 1. Having developed level i, level $i+1$ is developed as follows. For every vertex v of level i, consider all edges incident with it that have colors different from the colors of the tree edges along the path from r to v. There are at least $D - i$ such edges. If any such edge connects to a different vertex v' at level i, then this closes a distinguished cycle (going through v, v' and their least common ancestor). Hence we may assume that all these edges go to level $i+1$. It follows that at level ℓ at the latest (with ℓ as in the lemma), some vertex has two different ancestors at one level below. This closes a cycle. As no color can appear more than twice on this cycle, it is a distinguished cycle. ∎

Remark 2.11. The proof of Lemma 2.10 shows the existence of cycles in which all colors on the cycle appear either once or twice, whereas for a cycle to be distinguished it suffices that one color appears either once or twice.

Corollary 2.12. *Given n, r, d and s as defined above, let $D = (d-1)^2/2$ and let $\ell < D$ be such that $D!/(D-\ell)! \geq n^{2s}$. Then every r-uniform hypergraph H on n vertices with $m = dn^{r/2}$ hyperedges has an even cover with no more than 4ℓ hyperedges. Moreover, such an even cover can be found in time polynomial in n and m.*

Proof. Given a hypergraph H, use Lemma 2.4 with $\varepsilon = 1$ to transform it to a hypergraph satisfying the small overlap condition, with the value of d replaced by $d-1$. (The other alternative in Lemma 2.4, if it holds, already implies an even cover of the desired size.) Construct from the resulting hypergraph the graph G. By Lemma 2.7 the average degree in G is at least $(d-1)^2$. By Proposition 2.9 G has a subgraph of minimum degree at least D. By Lemma 2.10, this subgraph has a distinguished cycle of length at most 2ℓ. By Lemma 2.6 this corresponds to an even cover of size at most 4ℓ in H. By inspection one can verify that all parts of the proof are algorithmic, leading to the desired polynomial time algorithm. ■

When D is much larger than ℓ, then the degree condition in Corollary 2.12 is essentially $d^{2\ell} \geq n^{2s}$ for the existence of an even cover of size $k = 4\ell$. Hence, to have an even cover of size k a value of $d = O(n^{4s/k})$ suffices. We remark that in [11] a somewhat different value is proved for d. Namely, for r divisible by 3, the bound in [11] is $d = O(n^{4r/3k})$ (and an error term is introduced in the exponent when k is not divisible by 3). For large r, we can choose $s \leq (r+3)/4$ (or smaller, if $r+1$ is divisible by 4) and we get a better bound of $d = O(n^{r/k+3/k})$.

Setting $r = 3$ and $d \simeq \sqrt{\log n/\log\log n}$ in Corollary 2.12 proves item 1 of Theorem 1.4.

3. Distinguished Cycles Longer than $\log n$

This section contains the proof of item (2) of Theorem 1.6. Only parts of this section (Proposition 3.3, Theorem 3.4 and Lemma 3.5) will be used in the proof of item 2 of Theorem 1.4, which will appear in Section 4.

The proof of Lemma 2.10 assumes the minimum degree to be $D = \Omega(\log n/\log\log n)$. The purpose of this section is to prove the existence of distinguished cycles when the minimum degree is much lower. We note that principles used in the analysis of [11] fail to work already when the minimum degree drops below $\log n$, because (translating their proof technique to our notation) they are using a stricter notion of distinguished cycle in which some color needs to appear exactly once. A hypercube in which edges are colored according to the coordinate of the bit that they flip is an example of a graph of degree $\log n$ that is legally colored and does not have any distinguished cycle in this stronger sense.

3.1. A digression

For the purpose of explaining our proof technique, let us temporarily change the setting in which we seek to find a distinguished cycle. Rather than having a legally-colored graph of minimum degree D, we shall assume that we have a graph in which edges are colored (not necessarily legally) by D colors, and every vertex is incident with at least one edge of every color.

Definition 3.1. Let $L_{\text{col}}(N, D)$ denote the minimum value of ℓ such that for every graph G with N vertices and any coloring of its edges by D colors, if every vertex is incident with edges of all colors, then G contains a distinguished cycle of length at most 2ℓ.

It may be useful to notice that $L_{\text{col}}(N, D)$, $L_{\text{avg}}(N, D)$ and $L_{\min}(N, D)$ have a common special case, namely D-regular graphs with a legal coloring by D colors. Such colorings exist for all bipartite D-regular graphs.

Observe that the proof of Lemma 2.10 applies to $L_{\text{col}}(N, D)$ as well. Hence for $D > \frac{2 \log n}{\log \log n}$ we have that $L_{\text{col}}(N, D) \leq \frac{2 \log n}{\log \log n}$ (the leading constant 2 is for illustrative purposes only and is not meant to be best possible). The following proposition improves the value of D.

Proposition 3.2. *For $D > \log \log N$, $L_{\text{col}}(N, D) \leq O(\log^3 N)$. Moreover, a distinguished cycle of this length can be found in polynomial time.*

Proof. Consider an arbitrary graph on N vertices, and an arbitrary coloring of its edges by D colors such that every vertex is incident with all colors. We will show that a distinguished cycle exists. Our proof also provides a polynomial time algorithm for finding such a cycle.

Remove all edges from G and put them back in, one color class at a time. We shall consider the minimum size of connected components that are formed at various steps of this process. We shall show that the assumption that there are no distinguished cycles implies the existence of a connected component of size larger than N, which is a contradiction. Moreover, throughout our proof we shall control the diameter of connected components, and this will lead to a proof that there is a distinguished cycle of length $O(\log^3 N)$.

Initially, all vertices are isolated and there are N components. After adding edges of the first color class, every vertex has degree at least one. We partition the graph into connected components as follows.

1. Iteratively, pick an arbitrary vertex that has not yet been marked. Mark it as a center vertex, and mark all its neighbors and all their neighbors as noncenter vertices.
2. Every center vertex will correspond to exactly one connected component. It will be connected to all its neighbors. All other noncenter vertices (those are at distant two from the set of center vertices) connect to the center vertex that originally marked them (other choices would work as well).

It is not hard to see that every connected component has size at least $s_1 = 2$ and diameter at most $d_1 = 5$, where for convenience diameter is measured here as the number of vertices (including endpoints) on the shortest path between the two most distant vertices in a connected component. (Hence for example, the diameter of C_4, the cycle on four vertices, is 3.)

Consider now what happens when edges of the second color class are added. If any such edge lies in an existing connected component, then this component must contain a distinguished cycle with this edge being the only edge of its color. Hence all second color edges join different components. Moreover, if there are two components that are joined by two edges of the second color, this leads to a distinguished cycle in which the second color appears twice. Hence we may assume that there is at most one edge of the second color joining any two components.

Consider now a graph for which the components after the first phase are the vertices, and edges of the second color are the edges. This must be a simple graph, and moreover, its minimum degree is 2 (because every vertex in every component is incident with at least one edge of the second color). Partitioning this new graph into connected components as described above we get components of size at least $s_2 = s_1(s_1 + 1) = 6$, and diameter at most $5d_1 = 25$.

Likewise, after adding edges of the third color, all components are of size at least $s_3 = s_2(s_2 + 1) = 42$ and the diameter is at most $5d_2 = 125$.

By induction, after the $i+1$th color is added, all components are of size larger than 2^{2^i}. Hence if there are more than $\log\log N$ colors there must be a distinguished cycle. The length of the distinguished cycle is at most twice the size of the maximum diameter reached (plus two, for the two edges of the last color connecting two components), and can readily be seen to be at most essentially $5^{\log\log N}$, and hence $O(\log^3 N)$. ■

3.2. Back to the main proof

We now return to the proof of item 2 in Theorem 1.6. The following technical proposition will be used in the proof of our next theorem.

Proposition 3.3. *For every integers $0 < b < a$:*

1. $\log a + \frac{b}{a} < \log(a+b)$.
2. $\frac{b}{a} < \frac{\log(b+1)}{\log(a+1)}$.
3. $\log\log a + \frac{b}{a} < \log\log\big(a(b+1)\big)$.

(All logarithms are in base 2.)

Proof.

1. For $b=0$ and for $b=a$, $\log a + \frac{b}{a} = \log(a+b)$. Hence the result for $0<b<a$ follows by concavity of the logarithm function.
2. For $b=0$ and for $b=a$, $\frac{b}{a} = \frac{\log(b+1)}{\log(a+1)}$. Again, the result for $0<b<a$ follows by concavity of the logarithm function.
3. In the derivation below, the first inequality follows from item (2) and the third inequality follows from item (1) (using $\log a$ as a).

$$\log\log a + \frac{b}{a} < \log\log a + \frac{\log(b+1)}{\log(a+1)} < \log\log a + \frac{\log(b+1)}{\log a}$$

$$< \log\big(\log a + \log(b+1)\big) = \log\log\big(a(b+1)\big). \qquad ■$$

We now want to prove a result similar to that of Proposition 3.2 also for $L_{\text{avg}}(N,D)$. We first do so without providing any bounds on the length of the distinguished cycle.

Theorem 3.4. *For every graph on n vertices and average degree $d > 4\log\log 2n$, every legal coloring of its edges creates at least one distinguished cycle.*

Proof. Given n and d, start with the empty graph on n vertices, and add in the edges one color class at a time, under the assumption that there is no distinguished cycle. We shall prove the following inductive hypothesis.

Inductive hypothesis. At no stage during the process there is a connected component with n' vertices and average degree larger than $d' = 4\log\log 2n'$.

Base case. All connected components are of size 1, with average degree $4\log\log 2 = 0$.

Inductive step. Assume that the theorem is true before adding color class c. When adding color class c, no new edge lies within an existing connected component, as this edge could be used to close a distinguished cycle with edges of previous colors. Likewise, no two previous components are connected by two new edges, as again these two new edges can be used to close a distinguished cycle with edges of previous colors. Hence any two previous components are connected by at most one new edge.

For every new edge connecting two components, charge both endpoints of the edge to the smaller of the two components (breaking ties arbitrarily), and there spread the charge evenly among all vertices of the smaller component. We show by induction that the total charge of a vertex v does not exceed $4\log\log 2n'$, where n' is the size of the component to which v belongs. Observe that for every connected component, the sum of the charges of all vertices is equal to the sum of the degrees. Hence the fact that in a component of size n' no charge exceeds $4\log\log 2n'$ implies the same with respect to average degree.

For a vertex v, let s be the size of its component before edges of color c are added. By the induction hypothesis, its charge at this point is at most $4\log\log 2s$. Assume that when adding edges of color c, the number of edges that are charged to the component of v is b. Hence the new charge for v is $4\log\log 2s + \frac{2b}{s}$. On the other hand, v must belong now to a component of size at least $s(b+1)$, because each one of the b edges must connect to a distinct component at least as large as s. Hence to establish the inductive step, it remains to see that

$$4\log\log 2s + \frac{2b}{s} \leq 4\left(\log\log 2s + \frac{b}{2s}\right) \leq 4\log\log\left(2s(b+1)\right)$$

where the last inequality follows from item (3) in Proposition 3.3 (replacing a by $2s$). ■

We shall now show that when the degree is $c \log\log 2n$ (for large enough c) there is a distinguished cycle which is not too long, and that such a cycle can be found efficiently. We shall use the following known lemma.

Lemma 3.5. *There is a polynomial time algorithm that given any graph on n vertices and m edges and a value $k > 1$, removes at most m/k edges and produces a graph in which every connected component has diameter $O(k \log n)$.*

Proof. Pick an arbitrary vertex and grow a ball of radius r around it, where r is the minimum value for which the number of boundary edges (that exit the ball) is smaller than $1/k$ times the number of internal edges (within the ball). Then discard the boundary edges (if any boundary edges exist). Repeat the process starting at an arbitrary vertex not already within a ball, as long as such vertices exist.

For every edge discarded, we keep at least k internal edges, and hence at most a $1/k$ fraction of the edges are discarded. The radius of a component cannot exceed $k \ln m$ because every new layer increases the number of edges by a factor of $(1 + 1/k)$, and we need to have $(1 + 1/k)^r \leq m$. ■

Our plan is to use Lemma 3.5, multiple times, once for each color class used in the legal coloring. For this reason, we first show that it suffices to consider legal colorings with only few colors.

Lemma 3.6. *Consider an arbitrary N-vertex graph of average degree D and an arbitrary legal coloring of it. Then there is a polynomial time procedure that generates a new graph G with at most $2N$ vertices and average degree at least $D/4$ together with a legal coloring of the edges of G using at most $4D$ colors, such that every distinguished cycle in G can be mapped back to a distinguished cycle of the same length in the original graph.*

Proof. Consider an arbitrary legally colored graph on N vertices with average degree D. Modify the input graph to be nearly regular, with all degrees between D and $D/2$. This can be done as follows. First, iteratively removing vertices of degree at most $D/2$, resulting (as in Proposition 2.9) in a graph of minimum degree at least $D/2$. Thereafter, iteratively split any vertex of degree $D' > D$ into two vertices, one of degree $\lceil D'/2 \rceil$ and the other of degree $\lfloor D'/2 \rfloor$. The splitting operation preserves the number of edges. Hence the resulting graph G' has minimum degree $D/2$, maximum degree D, and at most $2N$ vertices. It is legally colored and every distinguished cycle in G' corresponds to a distinguished cycle of at most the same length in the original graph.

The legal coloring of G' is arbitrary and there is no a-priori bound on the number of colors that it uses (other than not being larger than the number of edges). We now describe a procedure for replacing this legal coloring by a new legal coloring of the edges with only $4D$ new colors. This is done by going over the original colors one by one. For each of the original colors C_{old} pick one of the new colors C_{new} and recolor all edges of original color C_{old} by the color C_{new}. If this new coloring introduces conflicts (this can happen if an edge of C_{old} is incident with a vertex that already has some other incident edge colored C_{new}), then drop the edge of C_{old} that leads to the conflict. To avoid dropping too many edges, we use the following rule when mapping a color C_{new} to C_{old}: we choose the new color that will result in the smallest number of dropped edges from C_{old} (breaking ties arbitrarily). As there are $4D$ new colors and only at most $2(D-1)$ edges incident with the endpoints of any edge, there must be a choice of C_{new} that will result in dropping at most half the edges of C_{old}. Hence eventually the resulting graph is legally colored, and its average degree is at least $D/4$. Moreover, every distinguished cycle in the new graph (with respect to the new colors) is a distinguished cycle of the original graph (with respect to the original colors). ∎

We now reach the theorem that implies the proof of item (2) in Theorem 1.6.

Theorem 3.7. *Every legally colored graph on N vertices and average degree $D \geq 32 \log\log 4N$ has a distinguished cycle of length $O(\log N \log\log N)$. Moreover, such a cycle can be found in polynomial time.*

Proof. By Lemma 3.6, instead of the input graph we may consider a new graph G with at most $2N$ vertices, average degree at least $8 \log\log 4N$, and with a legal coloring that uses at most $4D = O(\log\log N)$ colors. Observe that the average degree is at least twice as large as that used in the proof of Theorem 3.4. This allows us to discard half the edges of the graph, and there still would be a distinguished cycle. We shall use this slackness so as to modify the proof of Theorem 3.4 so that after adding each color, no component will have diameter larger than $O(\log N \log\log N)$. This is done by applying Lemma 3.5 after each color class is added. The parameter k in the lemma is chosen to be $8D$, so that even after the lemma is applied $4D$ times (once for each color), at most half of the edges are discarded in total. The proof of Theorem 3.4 still works even though we discard at most half the edges, because there was a factor 2 slackness in the average degree that we started with. Hence eventually a distinguished cycle will be found (when an edge of a new color is placed inside an existing component, or when two edges of a new color join two existing components). The diameter of every component at the time that the distinguished cycle is found is at most $O(\log N \log\log N)$, and the length of the distinguished cycle need never be more than two plus sum of diameters of two components that are connected by two edges of the same color.

All steps of the proof are constructive and give a polynomial time algorithm for finding a distinguished cycle of the appropriate length. ∎

3.3. A negative example

A question that remains is whether for some constant average degree D and any legal edge coloring there must be a distinguished cycle. The only nontrivial negative result that the author is aware of is the following.

Proposition 3.8. *There are 3-regular graphs whose edges can be colored in such a way that no distinguished cycle exists.*

Proof. Consider an infinite tiling of the plain by hexagons (later we will modify the construction to be finite). This defines a 3-regular graph in a natural way. Legally color its edges by three colors so that every hexagon contains only two colors. This can be done by first coloring any two adjacent edges, and then this determines uniquely the color of every other edge (by alternating two colors along the edges of a hexagon, and using the third color for the other edges incident with the vertices of the hexagon).

We now show that there is no distinguished cycle. W.l.o.g., let red be the color that appears twice on a hypothetical distinguished cycle. (The case that some color appears only once is even simpler, and omitted.) Then the rest of the cycle is composed of paths that have only colors blue and green, and each such path must lie on a single hexagon. This requires two hexagons that are colored by the colors blue and green to be connected by two different edges of color red, but this never happens in the given 3-coloring.

Inspection shows that the coloring is periodic. Hence the infinite graph can be replaced by a finite graph as follows. Picking some orientation of edges as vertical, the hexagons are arranged in rows. Two even rows sufficiently far from each other can be identified to be one row. Each column makes a zigzag pattern. Two such columns sufficiently far from each other (at a distance that is a multiple of three) can be identified to be one column. This results in a finite graph. ∎

Proposition 3.8 refers to distinguished cycles, but does not imply anything for even covers. The graph there contains cycles of length 6, but it is possible to show that any hypergraph that satisfies the small overlap condition for which the corresponding graph has cycles of length 6 must have an even cover of size at most 12. This leads to the following question.

Question 3.9. Is there is a 3-uniform hypergraph H satisfying the small overlap condition that on the one hand does not contain any even cover, and on the other hand, the graph G associated with H has a cycle (in which case this cycle must be trivial)?

If the answer to Question 3.9 is negative, then item 2 of Theorem 1.4 can easily be improved: it would suffice to have $\beta = \Theta(\sqrt{n})$ in order for even covers of size $O(\log n)$ to exist.

3.4. Extensions

The bound on the degree stated in Theorem 3.7 is $32 \log\log 4N$. The leading constant of 32 is a consequence of our attempt to keep the proof simple, rather than optimize the leading constant. It can be drastically reduced, with only a modest loss in the cycle length (which will still asymptotically remain $O(\log N \log\log N)$). Let us mention a few places where there is slackness in our analysis.

The leading constant in the bound on the degree in Theorem 3.4 can be improved with more work. For example, the original reason for having $\log\log 2n$ in the theorem rather than $\log\log n$ is to handle cases that $n \leq 2$ in a unified way. However, this later costs a factor of 2 in the leading constant. Additional savings can be obtained by changing the charging mechanism. Rather than charging both end points of an edge to the smaller of the two components, one can allocate part of the charge to the larger component (not to mention the possibility of propagating the charge to other components).

In Lemma 3.6 we loose a factor of 4 in the average degree. However, there is no need to loose more than a factor of $(1 + \varepsilon)$ (for some small ε), at the cost of having the new legal coloring use more colors (which will eventually translate to longer distinguished cycles). This can be done by allowing G' to have maximum degree roughly D/ε, and thereafter using $O(D/\varepsilon^2)$ new colors. (Possibly, with tighter analysis, the number of colors would have a better dependency on ε than $1/\varepsilon^2$.)

In the proof of Theorem 3.7 we may take k to be much larger (say $1/\varepsilon$ times the number of colors), which again will reduce the degree requirement at a cost of increasing the diameter of connected components (and hence the length of the distinguished cycle).

Summarizing the above discussion, it should not be difficult to reduce the degree requirement in Theorem 3.7 to $c_1 \log\log 4N$ with c_1 being a constant much smaller than 32 (possibly, nearly 1), at the cost of finding distinguished cycles of length $c_2 \log N \log\log N$, with c_2 being a constant that depends on c_1.

In special families of graphs (that are probably not relevant to the application of refuting semirandom 3SAT instances), we can improve the bounds of Theorem 3.4. We first briefly recall a few well known facts. A graph H is a *minor* of graph G if it can be obtained from G by the operations of contracting edges, deleting edges, and removing isolated vertices. As shown by Robertson and Seymour, every minor closed family of graphs can be characterized by a finite list of forbidden minors. For example, the family of planar graphs is closed under minors, and the two forbidden minors are K_5 and $K_{3,3}$ (this was proved by Wagner, and is related to Kuratowski's theorem). For a minor closed family of graphs, if some graph F on f vertices is a forbidden minor, then so is K_f. Every graph of average degree d must have K_f as a minor, for some $f = \Omega(d/\log d)$ [10, 12]. Hence the average degree of any graph from a minor closed family is bounded by $O(f \log f)$, where f is the size of the smallest forbidden minor. Theorem 3.4 is not interesting (in an asymptotic sense) for minor closed families of graphs, because the degree bound $\log\log 2n$ in the theorem cannot be attained when n is large. For such graphs, the following corollary replaces the dependency on n by a similar dependency on f.

Corollary 3.10. *Let G be any graph with no K_f as a minor and of average degree d, with $d > c \log\log f + O(1)$, where $c \geq 1$ is some universal constant. Then for any legal coloring of the edges of G there is distinguished cycle.*

Proof. The proof follows that of Theorem 3.4, with the following change. When adding color c, let C be a new component formed by connecting some previous components $C_1, C_2, \ldots$, and let $|C_i|$ denote the number of vertices in component C_i. Redistribute the connecting edges so that every one of the original components C_i is incident with at most $\min\left[|C_i|, s\right]$ edges, where $s = f \log f$. This is possible, because otherwise G has K_f as a minor. Now apply the charging mechanism of Theorem 3.4.

We now compute how much a vertex v is charged overall. The total charge until its component has size s is $O(\log\log s)$, as in the proof of Theorem 3.4. Thereafter, for the component to grow from size S to fS, vertex v is charged at most s/S. Hence the total charge after size s is reached forms a decreasing geometric series that sums up to $O(1)$. ■

4. Nontrivial Cycles of Length $O(\log n)$

Here we prove item 2 of Theorem 1.4.

We recall some known facts about cycle bases of graphs. Given a connected graph with n vertices and m edges, order the edges in some arbitrary order, and with each set of edges associate an indicator vector in $\{0,1\}^m$ in a natural way. For the purpose of the discussion here, a cycle in a graph will be any collection of edges such that the degree induced on each vertex is even. (In particular, the union of two edge disjoint cycles is a cycle.) The vectors associated with all cycles in a graph form a vector space of dimension $m - n + 1$ (with vector addition modulo 2). A basis for this vector space can be obtained as follows. Consider an arbitrary spanning tree T of the graph. For each edge $e \notin T$, there is a unique cycle (called a *fundamental cycle*) that is a simple cycle containing (some of the) edges of the tree and the edge e. The $m - n + 1$ fundamental cycles form a basis for the cycle space.

The radius $R(G)$ of a graph G is defined to be the maximum distance between a center vertex u and any other vertex in the graph, where a center vertex u is any vertex that minimizes this maximum.

Proposition 4.1. *Every graph of radius R has a cycle basis in which every cycle has length at most $2R+1$.*

Proof. Let u be a center vertex for the graph, and consider the spanning tree corresponding to the breadth first search tree rooted at u. Then the fundamental cycles with respect to this tree each has length at most $2R+1$. ■

Recall that Lemma 3.5 shows that for every graph with m edges, we may discard half of its edges such that each connected component that remains has radius $O(\log m)$.

Corollary 4.2. *In every graph with m edges one may discard half the edges such that the remaining graph has a cycle basis in which each cycle is of length $O(\log m)$.*

Proof. Use Lemma 3.5 to choose which edges to discard so that each remaining component has radius $O(\log m)$. Thereafter, for each connected component separately, find a cycle basis as in Proposition 4.1. The union of these cycle bases is the desired cycle basis. ■

The following lemma motivates our degression to cycle bases of graphs.

Lemma 4.3. *Let G be a graph constructed from a hypergraph H as explained in Section 2. Let G' be an edge induced subgraph of G that has a cycle basis in which every cycle is of length at most ℓ. Then if G' has a distinguished cycle (of arbitrary length), then G' must also have a nontrivial cycle of length ℓ.*

Proof. Label every edge of G' by the two hyperedges of H that generate it. Then as we have shown in the proof of Lemma 2.6, there must be some hyperedge e of H that labels only one of the edges of the distinguished cycle. The distinguished cycle can be expressed as a sum (mod 2) of basis cycles. Then it must be the case that at least on one of these basis cycles (which has length at most ℓ), e labels an odd number of its edges. Hence this basis cycle must be nontrivial. ■

Note that the nontrivial cycle found in the proof of Lemma 4.3 need not be a distinguished cycle (because addition of basis cycles is done modulo 2 which may lead to cancellations of edges).

Theorem 4.4. *Let G be a graph constructed from a hypergraph H as explained in Section 2. If G has degree $8 \log \log 2n$, then it contains a nontrivial cycle of length $O(\log n)$.*

Proof. Use Corollary 4.2 to remove half the edges and remain with a graph G' of average degree at least $4 \log \log 2n$ and a cycle basis in which each cycle has length $O(\log n)$. Theorem 3.4 implies that G' has a distinguished cycle. Lemma 4.3 implies that G' has a nontrivial cycle of length $O(\log n)$. As G' is a subgraph of G, then also G has a nontrivial cycle of length $O(\log n)$. ∎

The leading constant of 8 in Theorem 4.4 was chosen for concreteness and simplicity. It can be reduced using arguments similar to those presented in Section 3.4.

Item 2 of Theorem 1.4 follows from Theorem 4.4 in a way similar to the proof of Corollary 2.12.

There is a straightforward correspondence between even covers in hypergraphs and linear dependency modulo 2 in vectors. We note that our proofs, going through the notion of nontrivial cycles, in fact correspond to linear dependencies of $\{0, 1\}$ vectors over any field (and this was also the case in [11]). The reason is as follows. Orient the edges of the nontrivial cycle so that it creates a directed cycle. An edge directed from v_1 to v_2 corresponds to two hyperedges. In the linear dependency we shall add one of them and subtract the other according to the following convention. The hyperedge whose prefix labels the prefix of v_1 is added, and the hyperedge whose prefix labels the prefix of v_2 is subtracted. It is not hard to see that going around the nontrivial cycle, all vertices cancel out.

Corollary 4.5. *For a sufficiently large constant c, in any set of $cn\sqrt{n \log \log n}$ vectors in $\{0, 1\}^n$ of hamming weight 3, there are two disjoint multisets of $O(\log n)$ vectors (the same vector may appear more than once in a multiset and then it is counted more than once) such that the two respective sums of all vectors in the multisets are identical.*

Acknowledgement. The author holds the Lawrence G. Horowitz Professorial Chair at the Weizmann Institute. Part of this work was done at Microsoft Research, Redmond, Washington. I thank Assaf Naor for bringing [11] to my attention, and suggesting that it is related to the work in [6].

References

[1] N. Alon and U. Feige, On the power of two, three and four probes, *manuscript* (2008).

[2] N. Alon, S. Hoory and N. Linial, The Moore bound for irregular graphs, *Graphs and Combinatorics,* **18** (2002), 53–57.

[3] A. Coja-Oghlan, A. Goerdt and A. Lanka, Strong Refutation Heuristics for Random k-SAT, in: *Combinatorics, Probability and Computing,* **16** (2007), 5–28.

[4] U. Feige, Relations between average case complexity and approximation complexity, in: *Proc. of the 34th Annual ACM Symposium on Theory of Computing* (2002), pp. 534–543.

[5] U. Feige, Refuting smoothed 3CNF formulas, in: *Proceedings of 48th Annual Symposium on Foundations of Computer Science* (2007), pp. 407–417.

[6] U. Feige, J. H. Kim and E. Ofek, Witnesses for non-satisfiability of dense random 3CNF formulas, in: *Proceedings of 47th Annual Symposium on Foundations of Computer Science* (2006), pp. 497–508.

[7] U. Feige and E. Ofek, Easily refutable subformulas of large random 3CNF formulas, *Theory of Computing*, Volume 3 (2007) Article 2, pp. 25–43. http://theoryofcomputing.org.

[8] J. Friedman, A. Goerdt and M. Krivelevich, Recognizing more unsatisfiable random k-SAT instances efficiently, *SIAM Journal on Computing,* **35(2)** (2005), 408–430.

[9] J. Hastad, Some optimal inapproximability results, *J. ACM,* **48(4)** (2001), 798–859.

[10] A. V. Kostochka, Lower bound of the Hadwiger number of graphs by their average degree. *Combinatorica,* **4 (4)** (1984), pp. 307–316.

[11] A. Naor and J. Verstraete, Parity check matrices and product representaions of squares. *Submitted for publication.* A preliminary version appeared in IEEE ISIT 2005.

[12] A. Thomason, An extremal function for contraction of graphs, *Math. Proc. Comb. Phil. Soc.,* **95** (1984), pp. 261–265.

Uriel Feige

Department of Computer Science
and Applied Mathematics
Weizmann Institute
Rehovot, Israel

e-mail: `uriel.feige@weizmann.ac.il`

BOLYAI SOCIETY
MATHEMATICAL STUDIES, 19

Building Bridges
pp. 309–320.

Plünnecke's Inequality for Different Summands

KATALIN GYARMATI*, MÁTÉ MATOLCSI† and IMRE Z. RUZSA‡

The aim of this paper is to prove a general version of Plünnecke's inequality. Namely, assume that for finite sets A, $B_1, \dots, B_k$ we have information on the size of the sumsets $A + B_{i_1} + \dots + B_{i_l}$ for all choices of indices $i_1, \dots, i_l$. Then we prove the existence of a non-empty subset X of A such that we have 'good control' over the size of the sumset $X + B_1 + \dots + B_k$. As an application of this result we generalize an inequality of [1] concerning the submultiplicativity of cardinalities of sumsets.

1. Introduction

Plünnecke [4] developed a graph-theoretic method to estimate the density of sumsets $A + B$, where A has a positive density and B is a basis. The third author published a simplified version of his proof [5, 6]. Accounts of this method can be found in Malouf [2], Nathanson [3], Tao and Vu [7].

The simplest instance of Plünnecke's inequality for finite sets goes as follows.

Theorem 1.1. *Let $l < k$ be integers, A, B sets in a commutative group and write $|A| = m$, $|A + lB| = \alpha m$. There exists an $X \subset A$, $X \neq \emptyset$ such*

*Supported by Hungarian National Foundation for Scientific Research (OTKA), Grants No. T 43631, T 43623, T 49693.

†Supported by Hungarian National Foundation for Scientific Research (OTKA), Grants No. PF-64061, T-049301, T-047276

‡Supported by Hungarian National Foundation for Scientific Research (OTKA), Grants No. T 43623, T 42750, K 61908.

that

$$|X + kB| \le \alpha^{k/l}|X|. \tag{1.1}$$

Plünnecke deduced his results from a property of the directed graph built on the sets A, $A + B$, …, $A + kB$ as vertices (in $k + 1$ different copies of the group), where from an $x \in A + iB$ edges go to each $x + b \in A + (i + 1)B$. This property (which he called 'commutativity') is the following. If $x, y, z_1, \dots, z_m$ are distinct vetices such that there is an edge from x to y, and from y to each z_i, then there are distinct vertices $y_1, \dots, y_m$ such that there are vertices from x to each y_i, and from y_i to z_i; also, the same property is required for the graph obtained by reversing the direction of all vertices. The fact that the addition graph has this property follows from the possibility of replacing a path from x to $x + b + b'$ through $x + b$ by a path through $x + b'$, so commutativity of addition and the assumption that we add the same set B repeatedly seemed to be central ingredients of this method. Still, it is possible to relax these assumptions. Here we concentrate on the second of them.

In [5] the case $l = 1$ of Theorem 1.1 is extended to the addition of different sets as follows.

Theorem 1.2. *Let $A, B_1, \dots, B_k$ be finite sets in a commutative group and write $|A| = m$, $|A + B_i| = \alpha_i m$, for $1 \le i \le k$. There exists an $X \subset A$, $X \ne \emptyset$ such that*

$$|X + B_1 + \dots + B_k| \le \alpha_1\alpha_2 \dots \alpha_k|X|. \tag{1.2}$$

The aim of this paper is to give a similar extension of the general case. This extension will then be applied in Section 6 to prove a conjecture from our paper [1].

Theorem 1.3. *Let $l < k$ be integers, and let $A, B_1, \dots, B_k$ be finite sets in a commutative group G. Let $K = \{1, 2, \dots, k\}$, and for any $I \subset K$ put*

$$B_I = \sum_{i \in I} B_i,$$

$$|A| = m, \qquad |A + B_I| = \alpha_I m.$$

(This is compatible with the previous notation if we identify a one-element subset of K with its element.) Write

$$\beta = \Big(\prod_{L \subset K,\ |L|=l} \alpha_L \Big)^{(l-1)!(k-l)!/(k-1)!} . \tag{1.3}$$

There exists an $X \subset A$, $X \neq \emptyset$ such that

$$|X + B_K| \le \beta |X|. \tag{1.4}$$

The problem of relaxing the commutativity assumption will be the subject of another paper. Here we just mention without proof the simplest case.

Theorem 1.4. *Let A, B_1, B_2 be sets in a (typically noncommutative group) G and write $|A| = m$, $|B_1 + A| = \alpha_1 m$, $|A + B_2| = \alpha_2 m$. There is an $X \subset A$, $X \neq \emptyset$ such that*

$$|B_1 + X + B_2| \le \alpha_1 \alpha_2 |X|. \tag{1.5}$$

The following result gives estimates for the size of this set X and a more general property than (1.4), but it is weaker by a constant. We do not make any effort to estimate this constant; an estimate could be derived from the proof, but we feel it is probably much weaker than the truth.

Theorem 1.5. *Let $l < k$ be positive integers, and let A, $B_1, \dots, B_k$ be finite sets in a commutative group G. Let K, B_I, α_I and β be as in Theorem 1.3. For any $J \subset K$ such that $l < j = |J| \le k$ define*

$$\beta_J = \Big(\prod_{L \subset J,\ |L|=l} \alpha_L \Big)^{(l-1)!(j-l)!/(j-1)!} . \tag{1.6}$$

(Observe that $\beta_K = \beta$ of (1.3).) Let furthermore a number ε be given, $0 < \varepsilon < 1$. There exists an $X \subset A$, $|X| > (1-\varepsilon)m$ such that

$$|X + B_J| \le c\beta_J |X| \tag{1.7}$$

for every $J \subset K$, $|J| \ge l$. Here c is a constant that depends on k, l and ε.

We return to the problem of finding large subsets in Section 5.

2. The Case $k = l + 1$

First we prove the case $k = l + 1$ of Theorem 1.3 in a form which is weaker by a constant.

Lemma 2.1. *Let l be a positive integer, $k = l+1$, and let $A, B_1, \dots, B_k$ be finite sets in a commutative group G. Let K, B_I, α_I be as in Theorem 1.3. Write*

$$\beta = \Bigl(\prod_{L \subset K,\ |L|=l} \alpha_L \Bigr)^{1/l}.$$

(Observe that this is the same as β of (1.3) in this particular case.) There exists an $X \subset A$, $X \neq \emptyset$ such that

$$|X + B_K| \le c_k \beta |X| \tag{2.1}$$

with a constant c_k depending on k.

Proof. Let $H_1, \dots, H_k$ be cyclic groups of order $n_1, \dots n_k$, respectively, let $H = H_1 \times H_2 \times \cdots \times H_k$, and consider the group $G' = G \times H = G \times H_1 \times \cdots \times H_k$. Introduce the notation $B_i' = B_i \times \{0\} \times \cdots \times \{0\} \times H_i \times \{0\} \times \cdots \times \{0\}$ which will be abbreviated as $B_i' = B_i \times H_i$, in the same manner as $A \times \{0\} \times \cdots \times \{0\}$ will still be denoted by A.

We introduce the notation $i^* = K \setminus \{i\} = \{1, \dots, i-1, i+1, \dots, k\}$ which gives naturally $B_{i^*} = \sum_{j \neq i} B_j$ and, correspondingly,

$$\alpha_{i^*} = \alpha_{\{1,2,\dots,i-1,i+1,\dots,k\}}.$$

Note that we have $\prod \alpha_{i^*} = \beta^l$.

Similarly, let $H_{i^*} = H_1 \times \cdots \times H_{i-1} \times \{0\} \times H_{i+1} \times \cdots \times H_k$, and $B'_{i^*} = \sum_{j \neq i} B_i' = B_{i^*} \times H_{i^*}$.

Let q be a positive integer (which should be thought of as a large number), and let $n_i = \alpha_{i^*} q$. We restrict q to values for which these are integers; such values exist, since the numbers α_L are rational. Then $|H| = n = \prod n_i = \beta^l q^k$ and $|H_{i^*}| = n/n_i = (\beta q)^l / \alpha_{i^*}$. Hence $|A + B'_{i^*}| = |A + B_{i^*}|\,|H_{i^*}| = m(\beta q)^l$ independently of i.

Now, let $B' = \bigcup_{i=1}^k B_i'$, and consider the cardinality of the set $A + (k-1)B'$. The point is that the main part of this cardinality comes from terms where the summands B'_i are all different, i.e. from terms of the form $A + B'_{i^*}$, $i = 1, 2, \dots, k$. There are k such terms, so their cardinality altogether is not greater than

$$km(\beta q)^l. \tag{2.2}$$

The rest of the terms all contain some equal summands, e.g. $A + B'_1 + B'_1 + B'_2 + B'_3 \cdots + B'_{k-2}$, containing two copies of B'_1, etc. The number of such terms is less than k^k, and each of them has 'small' cardinality for the simple reason that $H_i + H_i = H_i$. For instance, in the example above we have $|A + B'_1 + B'_1 + B'_2 + B'_3 \cdots + B'_{k-2}| \le m|B_1|\big(\prod_{j=1}^{k-2} |B_j|n_j\big) \le c(A, B_1, \dots B_k)q^{k-2}$ where $c(A, B_1, \dots B_k)$ is a constant depending on the sets $A, B_1, \dots, B_k$ but not on q. Therefore the cardinality of the terms containing some equal summands is not greater than

$$k^k c(A, B_1, \dots, B_k)q^{k-2} = c(k, A, B_1, \dots, B_k)q^{k-2} = o(q^l) \tag{2.3}$$

Therefore, combining (2.2) and (2.3) we conclude that

$$\big|A + (k-1)B'\big| \le 2km(\beta q)^l \tag{2.4}$$

if q is chosen large enough.

Finally, we apply Theorem 1.1 to the sets A and B' in G'. We conclude by (2.4) that there exists a subset $X \subset A$ such that

$$|X + kB'| \le |X|\big(2k(\beta q)^l\big)^{k/l} = c_k|X|(\beta q)^k. \tag{2.5}$$

Also, observe that $X + (B_K \times H) \subset X + kB'$, and $\big|X + (B_K \times H)\big| = n|X + B_K|$. From these facts and (2.5) we obtain

$$|X + B_K| \le c_k|X|(\beta q)^k/n = c_k\beta|X|$$

as desired. ∎

3. The General Case

In this section we prove Theorem 1.5.

As a first step we add a bound on $|X|$ to Lemma 2.1.

Lemma 3.1. *Let $k = l+1$, and let A, B_i, B_I, α_I and β be as in Lemma 2.1. Let a number ε be given, $0 < \varepsilon < 1$. There exists an $X \subset A$, $|X| > (1-\varepsilon)m$ such that*

$$|X + B_K| \le c(k,\varepsilon)\beta|X| \tag{3.1}$$

with a constant $c(k,\varepsilon) = c_k\varepsilon^{-\frac{k}{k-1}}$ depending on k and ε.

Proof. Take the largest $X \subset A$ for which (3.1) holds. If $|X| > (1-\varepsilon)m$, we are done. Assume this is not the case. Put $A' = A\setminus X$, and apply Lemma 2.1 with A' in the place of A. We know that $|A'| \ge \varepsilon m$. The assumptions will hold with

$$\alpha'_I = |A' + B_i|/|A'| \le |A + B_i|/|A'| \le \alpha_I/\varepsilon$$

in the place of α_I. We get a nonempty $X' \subset A'$ such that

$$|X' + B_K| \le c_k\beta'|X'|$$

with

$$\beta' = \Bigl(\prod_{L\subset K,\ |L|=l} \alpha'_L\Bigr)^{1/(k-1)} \le \beta\varepsilon^{-\frac{k}{k-1}}.$$

Then $X \cup X'$ would be a larger set, a contradiction. ■

Now we turn to the general case.

Lemma 3.2. *Let $J_1, \dots, J_n$ be a list of all subsets of K satisfying $l < |J| \le k$ arranged in an increasing order of cardinality (so that $J_n = K$); within a given cardinality the order of the sets may be arbitrary.*

Let A, B_i, B_I, α_I and β_I be as in Theorem 1.5, and let the numbers $0 < \varepsilon < 1$ and $1 \le r \le n$ be given. There exists an $X \subset A$, $|X| > (1-\varepsilon)m$ such that

$$|X + B_J| \le c(k,l,r,\varepsilon)\beta_J|X| \tag{3.2}$$

for every $J = J_1, \dots, J_r$ with a constant $c(k,l,r,\varepsilon)$ depending on k, l, r and ε.

Theorem 1.5 is the case $r = n$.

Proof. We shall prove the statement by induction on r. Since the sets are in increasing order of size, we have $|J_1| = l + 1$, and the claim for $r = 1$ follows from Lemma 3.1.

Now assume we know the statement for $r - 1$. We apply it with $\varepsilon/2$ in the place of ε, so we have a set $X \subset A$, $|X| > (1 - \varepsilon/2)m$ such that (3.2) holds for $J = J_1, \dots, J_{r-1}$ with $c(k, l, r - 1, \varepsilon/2)$. Write $A' = X$. This set satisfies the assumptions with

$$\alpha'_I = \alpha_I/(1 - \varepsilon/2).$$

We have $|J_r| = k'$ with some k', $l < k' \le k$. We are going to apply Lemma 3.1 with A', k' in the place of A, k and $\varepsilon/2$ in the place of ε. To this end we need bounds for $|A' + B_L|$ for every L such that $|L| = l' = k' - 1$. By the inductive assumption we know

$$|A' + B_L| \le c(k, l, r - 1, \varepsilon/2)\beta_L|A'|.$$

Lemma 3.1 gives us a set $X' \subset A'$ such that

$$|X'| > (1 - \varepsilon/2)|A'| > (1 - \varepsilon)m$$

and

$$|X' + B_{J_r}| \le c(l', \varepsilon/2)\beta'|X'|,$$

where

$$\beta' = \Bigg(\prod_{L \subset J_r,\ |L| = l'} c(k, l, r - 1, \varepsilon/2)\beta_L \Bigg)^{1/l} = c(k, l, r - 1, \varepsilon/2)\beta_{J_r}.$$

In the last step we used an identity among the quantities β_J which easily follows from their definition (1.6).

The desired set X will be this X', and the value of the constant is

$$c(k, l, r, \varepsilon) = c(l', \varepsilon/2)c(k, l, r - 1, \varepsilon/2). \qquad \blacksquare$$

4. Removing the Constant

In this section we prove Theorem 1.3. This is done with the help of Theorem 1.5 and the technique of taking direct powers of the appearing groups, sets, and corresponding graphs.

Proof of Theorem 1.3. Consider the following bipartite directed graph $\mathcal{G}^1$. The first collection of vertices V_1 are the elements of set A, and the second collection of vertices V_2 are the elements of set $A+B_K$ (taken in two different copies of the ambient group to make them disjoint). There is an edge in $\mathcal{G}^1$ from $v_1 = a_1 \in V_1$ to $v_2 = a_2 + b_{1,2} + \dots + b_{k,2} \in V_2$ if and only if there exist elements $b_{1,1}, \dots, b_{k,1}$ such that $a_1 + b_{1,1} + \dots + b_{k,1} = a_2 + b_{1,2} + \dots + b_{k,2}$. The image of a set $Z \subset V_1$ is the set $\operatorname{im} Z \subset V_2$ reachable from Z via edges. The *magnification ratio* γ of the the graph $\mathcal{G}^1$ is $\min\left\{\frac{|\operatorname{im} Z|}{|Z|},\ Z \subset V_1\right\}$. The statement of Theorem 1.3 in these terms is that $\gamma \le \beta$, with β as defined in the theorem.

Consider now the direct power $\mathcal{G}^r = \mathcal{G}^1 \times \mathcal{G}^1 \times \dots \times \mathcal{G}^1$ with collections of edges $V_1^r = V_1 \times \dots \times V_1$ and $V_2^r = V_2 \times \dots \times V_2$, and edges from $(v_1^1, v_2^1, \dots, v_r^1) \in V_1^r$ to $(v_1^2, v_2^2, \dots, v_r^2) \in V_2^r$ if and only if there exist $\mathcal{G}^1$-edges in each of the coordinates. Observe that the directed graph $\mathcal{G}^r$ corresponds exactly to the sets A^r and $A^r + \left(B_1^r + \dots + B_k^r\right)$ in the direct power group G^r. Applying Theorem 1.5 in the group G^r to the sets $A^r, B_1^r, \dots, B_k^r$ with any fixed ε, say $\varepsilon = 1/2$, we obtain that the magnification ratio γ_r of $\mathcal{G}^r$ is not greater than $c\beta^r$. On the other hand, the magnification ratio is multiplicative (see [5] or [3]), so that we have $\gamma_r = \gamma^r$. Therefore we conclude that $\gamma \le \sqrt[r]{c}\beta$ and, in the limit, $\gamma \le \beta$ as desired. ∎

5. Finding a Large Subset

We give an effective version of Theorem 1.5 in the original case, that is, when only $X + B_K$ needs to be small.

Theorem 5.1. *Let A, B_i, B_I, α_I and β be as in Theorem* 1.3.

(a) *Let an integer a be given, $1 \le a \le m$. There exists an $X \subset A$, $|X| \ge a$ such that*

$$(5.1)\qquad |X+B_K| \le$$
$$\le \beta m^{k/l}\big(m^{-k/l}+(m-1)^{-k/l}+\dots+$$
$$+(m-a+1)^{-k/l}+\big(|X|-a\big)(m-a+1)^{-k/l}\big).$$

(b) *Let a real number t be given, $0 \le t < m$. There exists an $X \subset A$, $|X| > t$ such that*

$$(5.2)\qquad |X+B_K| \le$$
$$\le \beta m^{k/l}\left(\frac{l}{k-l}\big((m-t)^{1-k/l}-m^{1-k/l}\big)+\big(|X|-t\big)(m-t)^{-k/l}\right).$$

Proof. To prove (a), we use induction on a. The case $a=1$ is Theorem 1.3. Now suppose we know it for a; we prove it for $a+1$. The assumption gives us a set X, $|X| \ge a$ with a bound on $|X+B_K|$ as given by (5.1). We want to find a set X' with $|X'| \ge a+1$ and

$$(5.3)\qquad |X'+B_K| \le$$
$$\le \beta m^{k/l}\big(m^{-k/l}+(m-1)^{-k/l}+\dots+$$
$$+(m-a)^{-k/l}+\big(|X|-a-1\big)(m-a)^{-k/l}\big).$$

If $|X| \ge a+1$, we can put $X'=X$. If $|X|=a$, we apply Theorem 1.3 to the sets $A'=A\setminus X, B_1,\dots,B_k$. In doing this the numbers α_I should be replaced by

$$\alpha'_I=\frac{|A'+B_I|}{|A'|}\le\frac{|A+B_I|}{|A'|}=\alpha_I\frac{m}{m-a}.$$

This yields a set $Y \subset A\setminus X$ such that

$$|Y+B_K| \le \beta'|Y|$$

with

$$\beta'=\Bigg(\prod_{L\subset K,\ |L|=l}\alpha'_L\Bigg)^{(l-1)!(k-l)!/(k-1)!}\le\beta\left(\frac{m}{m-a}\right)^{k/l}$$

and we put $X' = X \cup Y$.

To prove part (b) we apply (5.1) with $a = [t] + 1$. The right side of (5.2) can be written as $\beta m^{k/l} \int_0^{|X|} f(x)\, dx$, where $f(x) = (m - x)^{-k/l}$ for $0 \le x \le t$, and $f(x) = (m - t)^{-k/l}$ for $t < x \le |X|$. Since f is increasing, the integral is $\ge f(0) + f(1) + \dots + f\big(|X| - 1\big)$. This exceeds the right side of (5.1) by a termwise comparison. ■

6. An Application to Restricted Sums

We prove the following result, which was conjectured in [1].

Theorem 6.1. *Let $A, B_1, \dots, B_k$ be finite sets in a commutative group, and $S \subset B_1 + \dots + B_k$. We have*

$$|S + A|^k \le |S| \prod_{i=1}^{k} |A + B_1 + \dots + B_{i-1} + B_{i+1} + \dots + B_k|. \tag{6.1}$$

Two particular cases were established in [1]; the case when S is the complete sum $B_1 + \dots + B_k$, and the case $k = 2$. The proof in the sequel is similar to the proof of the case $k = 2$, the main difference being that we use the above generalized Plünnecke inequality, while for $k = 2$ the original was sufficient.

Proof. Let us use the notation $|A| = m$, $s = \prod_{i=1}^{k} |A + B_1 + \dots + B_{i-1} + B_{i+1} + \dots + B_k|$. Observe that if $|S| \le (s/m^k)^{\frac{1}{k-1}}$ then

$$|S + A| \le |S|\,|A| = |S|^{\frac{1}{k}} |S|^{\frac{k-1}{k}} m \le \big(|S| s\big)^{\frac{1}{k}} \tag{6.2}$$

and we are done.

If $|S| > (s/m^k)^{\frac{1}{k-1}}$, then we will use Theorem 5.1, part (b) with $l = k-1$. Note that the β of this theorem can be expressed by our s as

$$\beta = s^{1/(k-1)} m^{-k/(k-1)}.$$

We take $t = m - \left(\frac{s}{|S|^{k-1}}\right)^{1/k}$. Then there exists a set $X \subset A$ such that $|X| = r > t$ and (5.2) holds. For such an X we have

$$|S + X| \le |B_K + X| \le \tag{6.3}$$

$$\le (k-1)s^{\frac{1}{k-1}}\left((m-t)^{-\frac{1}{k-1}} - m^{-\frac{1}{k-1}}\right) + (r-t)\left(\frac{s}{(m-t)^k}\right)^{\frac{1}{k-1}}$$

and we add to this the trivial bound

$$\big|S + (A \setminus X)\big| \le |S|\,|A \setminus X| = |S|(m-r). \tag{6.4}$$

We conclude that

$$|S + A| \le |S + X| + \big|S + (A \setminus X)\big| \le \tag{6.5}$$

$$\le (k-1)s^{\frac{1}{k-1}}\left((m-t)^{-\frac{1}{k-1}} - m^{-\frac{1}{k-1}}\right) + (r-t)\left(\frac{s}{(m-t)^k}\right)^{\frac{1}{k-1}} +$$

$$+ |S|\big((m-t) - (r-t)\big) = ks^{1/k}|S|^{1/k} - (k-1)\left(\frac{s}{m}\right)^{\frac{1}{k-1}} \le k\big(s|S|\big)^{1/k}$$

This inequality is nearly the required one, except for the factor k on the right hand side. We can dispose of this factor as follows (once again, the method of direct powers). Consider the sets $A' = A^r$, $B'_j = B^r_j$ $(j = 1, \dots, k)$, and $S' = S^r$ in the r'th direct power of the original group. Applying equation (6.5) to A', etc., we obtain

$$|S' + A'| \le k\big(s'|S'|\big)^{1/k}. \tag{6.6}$$

Since $|S' + A'| = |S + A|^r$, $s' = s^r$ and $|S'| = |S|^r$, we get

$$|S + A| \le k^{1/r}\big(s|S|\big)^{1/k}. \tag{6.7}$$

Taking the limit as $r \to \infty$ we obtain the desired inequality

$$|S + A| \le \big(s|S|\big)^{1/k}. \quad \blacksquare \tag{6.8}$$

References

[1] K. Gyarmati, M. Matolcsi and I. Z. Ruzsa, A superadditivity and submultiplicativity property for cardinalities of sumsets, *Combinatorica,* to appear (also arXiv:0707.2707v1).

[2] J. L. Malouf, On a theorem of Plünnecke concerning the sum of a basis and a set of positive density, *J. Number Theory,* **54**.

[3] M. B. Nathanson, *Additive number theory: Inverse problems and the geometry of sumsets,* Springer, 1996.

[4] H. Plünnecke, Eine zahlentheoretische Anwendung der Graphtheorie, *J. Reine Angew. Math.,* **243** (1970), 171–183.

[5] I. Z. Ruzsa, An application of graph theory to additive number theory, *Scientia, Ser. A,* **3** (1989), 97–109.

[6] I. Z. Ruzsa, Addendum to: An application of graph theory to additive number theory, *Scientia, Ser. A,* **4** (1990/91), 93–94.

[7] T. Tao and V. H. Vu, *Additive combinatorics,* Cambridge University Press, Cambridge, 2006.

Katalin Gyarmati
Alfréd Rényi Institute of Mathematics
Budapest, Pf. 127
H-1364 Hungary
e-mail: gykati@cs.elte.hu

Máté Matolcsi
Alfréd Rényi Institute of Mathematics
Budapest, Pf. 127
H-1364 Hungary
e-mail: matomate@renyi.hu

Imre Z. Ruzsa
Alfréd Rényi Institute of Mathematics
Budapest, Pf. 127
H-1364 Hungary
e-mail: ruzsa@renyi.hu

To all authors:
e-mail: triola@renyi.hu

BOLYAI SOCIETY
MATHEMATICAL STUDIES, 19

Building Bridges
pp. 321–331.

Decoupling and Partial Independence

RAVI KANNAN

1. Introduction

The initial motivation of this note was the question: How many samples are needed to approximate the inertia matrix (variance-covariance matrix) of a density on $\mathbf{R}^n$? It first arose in a joint paper with L. Lovász and M. Simonovits on an algorithm for computing volumes of convex sets. Rudelson proved a very interesting result (answering the question) based on a classical theorem from Functional Analysis (see Square Form Theorem below) due to Lust-Piquard, which is proved using the beautiful technique of Decoupling. This note gives a self-contained proof of the theorem and its application to this problem as well as a different question dealing with extending the basic result of Random Matrix Theory to partially random matrices (see Theorem 3) below.

Suppose $y, y_1, y_2, \dots y_m$ are independent identically distributed (i.i.d.) samples, each drawn according to a log-concave probability density F with mean 0 on $\mathbf{R}^n$. The inertia matrix of F is the $n \times n$ matrix Eyy^T.[1] It is clear that as $m \to \infty$, the inertia matrix of the samples, namely $\frac{1}{m}\sum_{i=1}^m y_i y_i^T$ tends to the inertia matrix of F. The initial question raised in [4] was to show upper bounds on

$$\left\| \frac{1}{m} \sum_{i=1}^{m} \left(y_i y_i^T - Eyy^T \right) \right\|$$

[1]An element of $\mathbf{R}^n$ is to thought of as a column vector with n components. So, yy^T is a rank 1 matrix. We let E denote expectation of the entire expression which follows it; expectations of matrices are taken entry-wise.

in terms of m. ($\|\cdot\|$ denotes the spectral norm.) Such an upper bound can then be used (in a straightforward manner) to derive an upper bound on the number of samples needed to get a "relative error" approximation to the inertia matrix from the samples. [We do not discuss the details here.]

[4] proved that $m = O(n^2)$ samples suffice. Bourgain [1] using delicate geometric arguments showed that $O\big(n(\log n)^3\big)$ samples suffice for certain log-concave densities. Then Rudelson [9] showed that in fact $O(n \log n)$ samples suffice for these densitities; Rudelson and Vershynin [10] have since generalized these results. Lovász and Vempala [6] generalized Rudelson's results to all log-concave densities.

A different question was considered by Dasgupta, Hopcroft, Kannan and Mitra [2]. There is a classical theory of Random Matrices – symmetric matrices where the above-diagonal entries are mutually independent mean zero random variables. [We call these "Fully Random Matrices".] It is known that the largest eigenvalue of an $n \times n$ fully random matrix is at most

$$O\big(\nu\sqrt{n}\,\big),$$

where ν is the largest standard deviation of any entry. The theory started in Physics with Wigner; Füredi, Komlös [3] and Vu [11] prove the bound above. The upper bound on the largest eigenvalue has been widely used in Clustering and other problems, where given the data, one needs to infer the probabilistic model. More recently, similar probabilistic models have been proposed for several practical situations where Full Independence does not make sense. A typical example is when the columns of the matrix represent documents in a large collection and the rows represent terms (used in them) and the (i, j)-th entry is the number of occurrences of the term i in document j; this so-called "document-term" matrix is widely used. One often assumes that there is an underlying probabilistic model which "generates" the document-term matrix. In applying the above to such a situation, researchers had to assume full independence. But in this example, while it makes sense to assume that the documents are independent, one has to allow correlations between the terms occurring in each document. Thus a more natural assumption is:

The columns of the matrix are independent vector-valued random variables. [The rows may be dependent.]

[2] proved a Wigner-type theorem under this assumption of only column-independence, with almost the same bound as for the full-independence case with some extra log factors. Their proof follows closely Rudelson's. The

connection between the two problems is simple: denote by $y_1, y_2, \ldots y_m$ the (independently generated) columns of the matrix A, for which we assume without loss of generality that $Ey_i = 0$. (After one replaces y_i by $y_i - Ey_i$.) We wish to upper bound $\|A\|$. Since $AA^T = \sum_{i=1}^m y_i y_i^T$ and $\|AA^T\| = \|A\|^2$, it suffices to bound $\|\sum_{i=1}^m y_i y_i^T\|$ which is at most $\|\sum_{i=1}^m (y_i y_i^T - Eyy^T)\| + m\|Eyy^T\|$. The first term is precisely what Rudelson bounds and the second is a known constant. Rudelson's proof uses Theorem (1) below due to Lust-Piquard [7]. The proof of Theorem (1) is based on the beautiful classical technique of **Decoupling.** In a sense, the main purpose of this note is to present Decoupling in a manner accessible to Computer Scientists (so they need not spend as much time as the author did to understand it) and to present a proof of the theorem. [In any case, the Theorem as stated in [7] and used by [9] is only a special case of the statement here restricted to so-called Rademacher sums.] Rudelson uses some more recent developments (due to Lust-Piquard and Pisier [8]) to improve C_p in theorem to its best value – namely $O(\sqrt{p})$ which we will not present.

p will be a power of 2 throughout. We need the following:

Proposition 1. *For a matrix X, we let $\|X\|_p = \left(\sum_j (\sigma_j(X))^p\right)^{1/p}$ (where σ_j are the singular values). $\|X\|_p$ is a norm (called a Schatten $p-$norm). Hence it is a convex function of the entries of the matrix X.*

It is also known that $\|X\|_p \le \|X\|_q$ whenever $p \ge q$. $\|X\|_\infty$ (also denoted $\|X\|$) is the spectral norm. Note that $\|X\|_2^2 = \mathrm{Tr}\, XX^T$ equal to the sum of squares of all the entries and $\|X\|_\infty$ is the spectral norm of X.

Theorem 1 ([7], Square-Form Theorem). *Let $X_1, X_2, \ldots X_n$ be independent matrix-valued random variables with $EX_i = 0$. There is a constant C_p depending only on p such that*

$$E\left\|\sum_{i=1}^n X_i\right\|_p^p \le (C_p)^p \left(E\left\|\sum_{i=1}^n X_i X_i^T\right\|_{p/2}^{p/2} + E\left\|\sum_{i=1}^n X_i^T X_i\right\|_{p/2}^{p/2}\right).$$

From this, we will see that we can get a version of Rudelson's theorem (all c are constants):

Theorem 2 [9]. *Suppose $y_1, y_2, \ldots y_n$ are i.i.d. samples, each drawn according to a probability distribution F on $\mathbf{R}^m$. Suppose $|y_i| \le M$ and M*

satisfies $M \le c\sqrt{n}\|\,Eyy^T\|_p^{1/2}$. *Then, we have that with high probability,*

$$\left\| \sum_{i=1}^{n} \left(y_i y_i^T - Eyy^T\right) \right\| \le c_2 M \sqrt{n} (\ln n + \ln m)^{c'} \|\,Eyy^T\|^{1/2}.$$

We will also get a theorem on matrices whose columns are independent (vector-valued) random variables.

Theorem 3 [2]. *Suppose A is an $m \times n$ matrix with independent vector-valued random variables – $y_1, y_2, \dots y_n$ as its columns, with $Ey_i = 0$. Suppose the maximum variance of y_i in any direction is ν_i^2 and with probability at least $1 - \delta$, we have $|y_i| \le M$ for all i. Then for all $t > 0$,*

$$\Pr\left(\|A\| \ge \left(c(\log n + \log m)\right)^c t\left(M + \sqrt{n}\,\mathrm{MAX}_i\, \nu_i\right)\right) \le \delta + \frac{1}{n^{\log t/10}}.$$

In the Full Independent case, we can take $M = \nu\sqrt{n}\log n$ and δ very small. So, we can recover a Wigner-type result from Theorem above but with added log factors.

2. Proof of the Square-form Theorem

Proof. We will prove the Theorem with $C_p = 10p^7$ by induction on p. For $p = 2$, we have

$$E\left\| \sum_i X_i \right\|_2^2 = E\,\mathrm{Tr} \sum_{i,j} X_i X_j^T = \mathrm{Tr}\, E \sum_{i,j} X_i X_j^T = E\,\mathrm{Tr} \sum_i X_i X_i^T,$$

since $EX_iX_j^T = EX_iEX_j^T = 0$ for $i \ne j$. Now since $\sum_i X_i X_i^T$ is p.s.d., all its eigenvalues are non-negative and so, $\mathrm{Tr} \sum_i X_i X_i^T = \left\| \sum_i X_i X_i^T \right\|_1^1$ proving the case of $p = 2$.

We proceed to general p. We need the following (well-known) generalization of Hölder's inequality to matrices:

Proposition 2. *Suppose $A_1, A_2, \ldots A_m$ are matrices (of dimensions so that their product is defined). We have for any positive reals $r_1, r_2, \ldots r_m$ with $\sum_{i=1}^m \frac{1}{r_i} = 1$:*

$$\|A_1A_2 \ldots A_m\|_p \le \|A_1\|_{pr_1}\|A_2\|_{pr_2} \cdots \|A_m\|_{pr_m}.$$

We introduce an important ingredient in Decoupling – let $X_1', X_2', \ldots X_n'$ be independent copies of $X_1, X_2, \ldots X_n$ respectively (all $2n$ random variables are independent and X_i' and X_i have the same distribution).

$$E\Big\|\sum_{i=1}^n X_i\Big\|_p^p = E\Big\|\sum_i X_i \sum_j X_j^T\Big\|_{p/2}^{p/2}$$

$$\le 2^{p/2}E\Big\|\sum_i X_iX_i^T\Big\|_{p/2}^{p/2} + 8^{p/2}E\Big\|\sum_i X_iY^T\Big\|_{p/2}^{p/2}$$

where $Y = \sum_j X_j'$ and we have used "decoupling" – see Lemma below. Note that $EX_iY^T = EX_iEY^T = 0$ since X_i, Y are independent. We now use induction to get (with the notation that $[X_1 \mid X_2 \mid \ldots X_n]$ denotes the matrix with $X_1, X_2, \ldots X_n$ written in that order):

$$E\Big\|\sum_i X_iY^T\Big\|_{p/2}^{p/2} \le 10^{p/2}(p/2)^{3.5p}\Bigg(E\Big\|\sum_i\left(YX_i^TX_iY^T\right)\Big\|_{p/4}^{p/4}$$

$$+ E\Big\|\sum_i\left(X_iY^TYX_i^T\right)\Big\|_{p/4}^{p/4}\Bigg)$$

$$= 10^{p/2}(p/2)^{3.5p}\Bigg(E\Big\|Y\Big(\sum_i(X_i^TX_i)\Big)Y^T\Big\|_{p/4}^{p/4}$$

$$+ E\|[X_1 \mid X_2 \mid \ldots \mid X_n]Y^TY[X_1 \mid X_2 \mid \ldots X_n]^T)\|_{p/4}^{p/4}\Bigg)$$

$$\le 10^{p/2}(p/2)^{3.5p}\Bigg(E\|Y\|_p^{p/2}\Big\|\sum_i X_i^TX_i\Big\|_{p/2}^{p/4}$$

$$+ E\|Y\|_p^{p/2}\|[X_1 \mid X_2 \mid \ldots X_n]\|_p^{p/2}\Bigg) \text{using Prop. 2}$$

$$\leq 10^{p/2}(p/2)^{3.5p} E\|Y\|_p^{p/2} \left(\left\| \sum_i X_i^T X_i \right\|_{p/2}^{p/4} + \left\| \sum_i X_i X_i^T \right\|_{p/2}^{p/4} \right)$$

$$\leq 2 \cdot 10^{p/2}(p/2)^{3.5p} \left(E\left\| \sum_i X_i \right\|_p^p \right)^{1/2}$$

$$\times \left(E\left\| \sum_i X_i X_i^T \right\|_{p/2}^{p/2} + E\left\| \sum_i X_i^T X_i \right\|_{p/2}^{p/2} \right)^{1/2}$$

the last using Jensen and the fact that Y, $\sum_i X_i$ have same distribution. Letting $x = \sqrt{E\| \sum_i X_i \|_p^p}$ and $b = E\| \sum_i X_i X_i^T \|_{p/2}^{p/2} + E\| \sum_i X_i^T X_i \|_{p/2}^{p/2}$, this yields the following quadratic inequality for x:

$$x^2 \leq 2^{p/2} b + 2 \cdot 8^{p/2} 10^{p/2} (p/2)^{3.5p} \sqrt{b}\, x$$

It is easy to see that this implies that

$$x^2 \leq 10^p p^{7p} b,$$

completing the inductive proof. ■

2.1. Decoupling

We now introduce the beautiful technique developed by Probabilists and Functional Analysts called "decoupling" which helps get rid of some dependencies between random variables, making the analysis easier in many contexts. (See for example [5]). [Decoupling looks like sleight of hand, but it accomplishes a great deal. It has been extensively used precisely in contexts like the one we have here.]

Suppose f is any convex function from the set of matrices to non-negative reals with $f(A) = f(-A)$ and satisfying the condition that there is some $p > 0$ such that

$$f(A + B) \leq 2^p \big(f(A) + f(B) \big) .$$

[Typical examples of f will be p-th powers of norms.]

Lemma 1. *Suppose X_i, X_i' are as above and assume $EX_i = 0$. Then,*

$$Ef\Big(\sum_i X_i \sum_j X_j^T\Big) \le 8^p Ef\Big(\sum_i X_i \sum_j X_j'^T\Big) + 2^p Ef\Big(\sum_i X_i X_i^T\Big).$$

Remark 1. The point of the Lemma is that the first term on the r.h.s. is easier to handle than the l.h.s., since now X_i', X_i are independent.

Proof. We let $Y_i = \{X_i, X_i'\}$ (the set (without order) of the two elements X_i, X_i') and $Y = (Y_1, Y_2, \ldots Y_n)$. We define random variables $Z_1, Z_2, \ldots Z_n, Z_1', Z_2', \ldots Z_n'$ as follows: for each i, independently, with probability $1/2$ each, we let $(Z_i, Z_i') = (X_i, X_i')$ or $(Z_i, Z_i') = (X_i', X_i)$. Then, we clearly have

$$E\big(Z_i Z_j'^T \mid Y\big) = \frac{1}{4}\big(X_i X_j^T + X_i' X_j^T + X_i X_j'^T + X_i' X_j'^T\big) \text{ for } i \ne j$$

$$E\big(Z_i Z_i'^T \mid Y\big) = \frac{1}{2}\big(X_i X_i'^T + X_i' X_i^T\big).$$

$$Ef\Big(\sum_i X_i \sum_j X_j^T\Big) \le 2^p Ef\Big(\sum_i X_i X_i^T\Big) + 2^p Ef\Big(\sum_{i\ne j} X_i X_j^T\Big)$$

$$\le 2^p Ef\Big(\sum_i X_i X_i^T\Big)$$

$$+ 2^p Ef\Big(\sum_{i\ne j}\big(X_i X_j^T + EX_i X_j'^T + EX_i' X_j^T + EX_i' X_j'^T\big)$$

$$+ 2\sum_i \big(EX_i X_i'^T + EX_i' X_i^T\big)\Big)$$

$$\le 2^p Ef\Big(\sum_i X_i X_i^T\Big) + 2^p Ef\Big(\sum_{i\ne j}\big(X_i X_j^T + X_i X_j'^T + X_i' X_j^T + X_i' X_j'^T\big)$$

$$+ 2\sum_i \big(X_i X_i'^T + X_i' X_i^T\big)\Big)$$

using Jensen and convexity of f, so $f(EX) \le Ef(X)$

$$\le 2^p Ef\Big(\sum_i X_i X_i^T\Big) + 8^p Ef\bigg(\Big(\sum_i Z_i \sum_j Z_j'^T\Big)\,\Big|\, Y\bigg)$$

$$\le 2^p Ef\Big(\sum_i X_i X_i^T\Big) + 8^p Ef\Big(\sum_i Z_i \sum_j Z_j'^T\Big) \text{ Jensen again.}$$

Now, the Lemma follows noting that $\{(Z_i, Z_j') : i = 1, 2, \dots n\}$, and $\{(X_i, X_j') : i = 1, 2, \dots n\}$ have the same joint distributions. ■

3. Proof of Theorems 2 and 3

We will apply the Square Form theorem with

$$X_i = y_y y_i^T - E y_i y_i^T.$$

We first note

$$E\Big\|\sum_{i=1}^n X_i X_i^T\Big\|_{p/2}^{p/2} = E\Big\|\sum_i \big(y_i y_i^T - Eyy^T\big)^2\Big\|_{p/2}^{p/2}$$

$$\le E\Big\|\sum_i \big(y_i y_i^T\big)^2 + \big(Eyy^T\big)^2\Big\|_{p/2}^{p/2}$$

$$\le 2^{p/2} E\Big\|\sum_i \big(y_i y_i^T\big)^2\Big\|_{p/2}^{p/2} + (2n)^{p/2}\big\|\big(Eyy^T\big)^2\big\|_{p/2}^{p/2},$$

since for psd matrices A, B, $\|(A-B)^2\| \le \|A^2 + B^2\|$.

$$E\Big\|\sum_i \big(y_i y_i^T\big)^2\Big\|_{p/2}^{p/2} = E\Big\|\sum_i |y_i|^2 y_i y_i^T\Big\|_{p/2}^{p/2} \le E\Big\|\mathrm{MAX}_i\, |y_i|^2 \sum_i y_i y_i^T\Big\|_{p/2}^{p/2}$$

$$\le E\|\,\mathrm{MAX}_i\, |y_i|^2 I\|_p^{p/2}\Big\|\sum_i y_i y_i^T\Big\|_p^{p/2}$$

$$\leq m^{1/2}\big(E\,\mathrm{MAX}_i\,|y_i|^{2p}\big)^{1/2}\left(E\Big\|\sum_i y_iy_i^T\Big\|_p^p\right)^{1/2}$$

$$\leq \sqrt{m}\,M^p\left(2^{p/2}\left(E\Big\|\sum_i X_i\Big\|_p^p\right)^{1/2} + (2n)^{p/2}\|\,Eyy^T\|_p^{p/2}\right).$$

$$\|\,(Eyy^T)^2\|_{p/2}^{p/2} \leq \|\,Eyy^T\|_\infty^{p/2}\|\,Eyy^T\|_{p/2}^{p/2} \leq E|y|^p\|\,Eyy^T\|_{p/2}^{p/2}$$

$$\leq M^p\sqrt{m}\,\|\,Eyy^T\|_p^{p/2}.$$

So, we have

$$E\Big\|\sum_{i=1}^n X_iX_i^T\Big\|_{p/2}^{p/2} \leq 2^pM^p\sqrt{m}\left(E\Big\|\sum_{i=1}^n X_i\Big\|_p^p\right)^{1/2} + 2^{p+1}M^pn^{p/2}\|\,Eyy^T\|_p^{p/2}.$$

Plugging this into Theorem (1), we get the quadratic inequality for $x = \big(E\|\sum_{i=1}^n X_i\|_p^p\big)^{1/2}$:

$$x^2 \leq bx + a_0 \qquad \text{where,}$$

$$b = (C_p)^p2^{p+1}M^p\sqrt{m}\,; \quad a_0 = (C_p)^p2^{p+1}M^pn^{p/2}\sqrt{m}\;\|\,Eyy^T\|_p^{p/2}.$$

This implies that

$$(1) \qquad E\Big\|\sum_{i=1}^n X_i\Big\|_p^p \leq b^2 + 2a_0 \leq (c_1p)^{c_2p}n^{p/2}M^p\sqrt{m}\;\|\,Eyy^T\|_p^{p/2}$$

under the assumption that $M \leq c\sqrt{n}\|\,Eyy^T\|_p^{1/2}$. Now taking $p = c_3(\ln n + \ln m)$ and applying Markov inequality, we get Theorem (2).

To get Theorem (3), we first note that

$$\|A\|_p^p = \Big\|\sum_i y_iy_i^T\Big\|_{p/2}^{p/2} \leq 2^{p/2}\Big\|\sum_i X_i\Big\|_{p/2}^{p/2} + 2^{p/2}\Big\|\sum_i Ey_iy_i^T\Big\|_{p/2}^{p/2}.$$

Clearly,

$$\left\| \sum_i E y_i y_i^T \right\|_{p/2}^{p/2} \leq n^{(p/2)-1} \sum_i \| E y_i y_i^T \|_{p/2}^{p/2} \leq n^{(p/2)-1} m \sum_i \| E y_i y_i^T \|^{p/2}$$

$$\leq n^{(p/2)-1} m \sum_i \nu_i^p .$$

To bound $E\| \sum_i X_i \|_{p/2}^{p/2}$, we proceed similarly as above with a few modifications. [We do not supply all the details here.] Now the y_i are not i.i.d., but only independent. So, instead of (1), we get

$$E \left\| \sum_i X_i \right\|_{p/2}^{p/2} \leq (cp)^p M^p m + (cp)^p n^{(p/4)+1} M^{p/2} m \left(\sum_i \nu_i^p \right)^{1/2} .$$

From this, it is not difficult to get the Theorem.

References

[1] J. Bourgain, *Random points in isotropic convex sets,* Manuscript.

[2] A. Dasgupta, J. Hopcroft, R. Kannan and P. Mitra, Spectral Clustering with limited independance, *Symposium on Discrete Algorithms* (2007), 1036–1045.

[3] Z. Furedi and J. Komlos, The eigenvalues of random symmetric matrices, *Combinatorica 1,* **3** (1981), 233–241.

[4] R. Kannan, L. Lovász and M. Simonovits, Random Walks and an $O^*(n^5)$ volume algorithm for convex bodies, *Random Structures and Algorithms,* **11** (1997), 1–50.

[5] V. de la Pena and E. Gine, *Decoupling: From Dependence to Independence (Probability and its Applications),* Springer Verlag, New York (1999).

[6] L. Lovász and S. Vempala, The geometry of logconcave functions and sampling algorithms, *Random Structures and Algorithms,* Volume 30, Issue 3 (May 2007), 307–358

[7] F. Lust-Piquard, Inégalites de Khinchin dans C_p $(1 < p < \infty)$, C.R. Acad. Sci. Paris 303 (1986) 289-292.

[8] F. Lust-Piquard and G. Pisier, Non-commutative Khinchine and Payley inequalities, *Arkiv för Mat.,* **29** (1991), 241–260.

[9] M. Rudelson, Random vectors in isotropic positions, *Journal of Functional Analysis,* **164** (1999), 60–72.

[10] M. Rudelson and R. Vershynin, Sampling from large matrices: an approach through geometric functional analysis, *Journal of the Association for Computing Machinary* (2007).

[11] V. Vu, Spectral norm of random matrices, *Proceedings of the* 36th *annual ACM Symposium on Theory of computing* (2005), 619–626.

Ravi Kannan
Microsoft Research Labs.
Bangalore, India

e-mail: kannan@microsoft.com

BOLYAI SOCIETY
MATHEMATICAL STUDIES, 19

Building Bridges
pp. 333–368.

Combinatorial Problems in Chip Design

BERNHARD KORTE and JENS VYGEN

The design of very large scale integrated (VLSI) chips is an exciting area of applying mathematics, posing constantly new challenges.

We present some important and challenging open problems in various areas of chip design. Although the problems are motivated by chip design, they are formulated mathematically; understanding and solving them does not require any knowledge of chip design. We give some partial results and argue why a full resolution of one of the problems could result in an advance of the state of the art in algorithms for chip design.

Introduction

This paper is dedicated to Laci Lovász on the occasion of his 60th birthday. We have learned a lot from Laci about combinatorial problems and how to solve them. However, the paper presents mainly open problems and challenges the reader to solve them.

Chip design is one of the most interesting and important application areas of mathematics, in particular discrete mathematics. Many different mathematical techniques have been applied in chip design, and a lot of practical design problems have led to new interesting theoretical results. Still there are many challenging open problems, some of which known for decades, others raised only recently because of technological advances.

For many years we have been designing algorithms for chip design that are based on mathematics as much as possible. The resulting BonnTools (Korte, Rautenbach and Vygen [36]) are widely used in industry, but also competing industrial tools include more and more advanced algorithms. Nevertheless we are convinced that there is still a lot of room for improvement. We try to give several examples in this paper.

We also hope to stimulate more theoretical research that is relevant for practical problems. All examples in this paper are defined rigorously as abstract mathematical problems, which can be understood without any knowledge in chip design. Nevertheless we also mention the background briefly and argue why these problems are important.

Almost all problems arising in chip design are *NP*-hard. Sometimes it is not even clear whether exponential-time algorithms, or polynomial-time approximation algorithms exist. Nevertheless one can prove interesting positive results, for example by considering important special cases or by decomposing a problem into well-solved subproblems.

The design of a chip consists of many steps, which are of course not independent of each other. The main steps are logic synthesis, floorplanning, placement, timing optimization, clock tree design, and routing (roughly in this order). Each of these steps is decomposed further into sub-tasks. For example, routing is split into global and detailed routing; timing optimization comprises at least fanout tree design and gate sizing, etc. Most of these tasks will be explained briefly in later sections. For a detailed decscription we refer to [36].

Let us mention one of the few examples where an important problem can be solved optimally in polynomial time, and fast enough to be used in practice: clock skew scheduling. While traditionally designers were aiming for zero skew clock trees, i.e. simultaneously switching storage elements, this is far from optimal. By optimizing the arrival times one can obtain smaller cycle times and thus higher frequencies. Modelling the timing graph, where arcs correspond to signal propagations, as a digraph G with delays $d\colon E(G) \to \mathbb{R}$, and given a subset $F \subseteq E(G)$ and a threshold $\Theta \in \mathbb{R}$, the task is to find arrival times $\pi\colon V(G) \to \mathbb{R}$ with $\pi(x) + d(e) \le \pi(y)$ for $e = (x, y) \in E(G) \setminus F$ such that the vector of relevant *slacks*

$$\Big(\min \big\{ \Theta, \pi(y) - \pi(x) - d(e) \big\} \Big)_{e=(x,y)\in F},$$

after sorting entries in non-decreasing order, is lexicographically maximal. The edges in F are normally those incident to storage elements; their worst slack will eventually determine what frequency can be achieved.

This problem can be solved in $O(mn+n^2 \log n)$ time (Albrecht et al. [3]), where $n = \big|V(G)\big|$ and $m = \big|E(G)\big|$. For reasonable thresholds Θ it runs very fast in practice. See [28], [56], and [2] for generalizations and implementation aspects. See Figure 1 for an example.

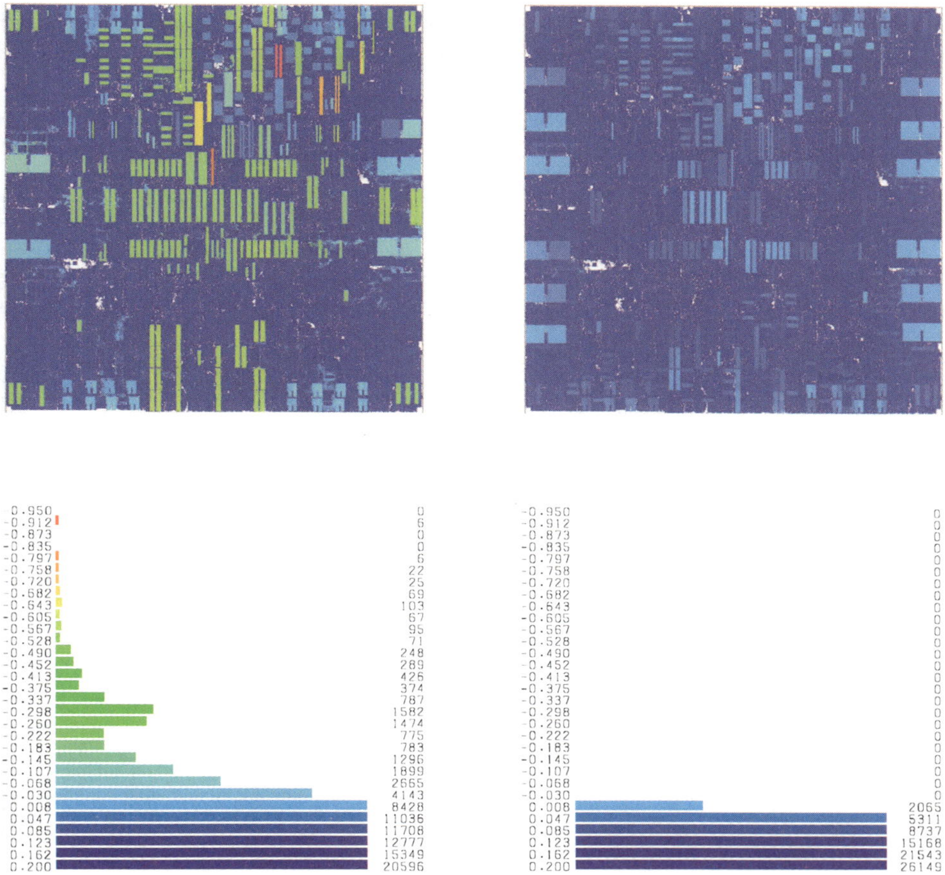

Fig. 1. The effect of clock skew scheduling on the Apple G5 system controller [29]. The left-hand side shows the timing result with zero skew, the right-hand side with optimal arrival times. This improved the speed (i.e., cycle time) by 27%. Placement (top) and slack histograms (showing the number of slacks in certain intervals) are colored in the same way: signals at red modules arrive much too late, while signals at blue modules have positive slack (i.e., arrive in time).

Held et al. [30] survey this and many other results that we achieved. The present paper complements this with a focus on unsolved problems.

There are well-known open problems that we do not list although they are very important for chip design (and many other applications). Examples are the questions what is the fastest running time of an algorithm for fundamental combinatorial optimization problems (such as the minimum cost flow problem), and of course whether $P = NP$. These problems have been studied for decades and seem to be very hard.

The rest of this paper consists of ten sections, each with a challenging open problem.

1. Floorplanning

Motivation

At an early design stage designers have a target chip area, i.e. a rectangle. They also have a list of rectangular objects (*modules* or *macros*) and their interconnection (the modules have *pins*, and sets of pins that must be connected are called *nets*). The task is to place these objects without overlaps such that an estimate of the interconnect length is minimized. This is called *floorplanning*. In *hierachical design* the number of objects is relatively small (about 20 to 200), as each object represents either a large memory array or a logic macro (which itself consists of many small objects). Another approach (*flat design*) is to deal with millions of small objects directly. Although the problem formulation is essentially the same, completely different approaches are needed due to the orders of magnitude (cf. Problem 2).

Instance

- a finite set M (of modules or macros)
- a rectangular chip area $A(\square) = [0, x_{\max}] \times [0, y_{\max}]$
- an outline $A(m) = \big[0, w(m)\big] \times \big[0, h(m)\big] \subseteq A(\square)$ for $m \in M$
- a finite set $\mathcal{N}$ of nets. Each net is a finite set with at least two elements; these elements are called pins
- an assignment $\gamma(p) \in M \mathbin{\dot\cup} \{\square\}$ and an offset $\big(x(p), y(p)\big)$ for each pin p (the pins p with $\gamma(p) = \square$ are fixed, e.g. I/O-ports). $(M, \mathcal{N}, \gamma)$ is called a *netlist*; it can be regarded as a hypergraph with a special vertex $\square$

Task

Compute

- a placement $(x, y)\colon M \rightarrow A(\Box)$ such that $\big(x(m) + w(m), y(m) + h(m)\big) \in A(\Box)$ for each $m \in M$, and for each $m, m' \in M$ at least one of the following conditions holds: $x(m) + w(m) \leq x(m')$ or $x(m') + w(m') \leq x(m)$ or $y(m) + h(m) \leq y(m')$ or $y(m') + h(m') \leq y(m)$
- such that $\sum_{N \in \mathcal{N}} \big(\max_{p \in N} \big(x\big(\gamma(p)\big) + x(p)\big) - \min_{p \in N} \big(x\big(\gamma(p)\big) + x(p)\big) + \max_{p \in N} \big(y\big(\gamma(p)\big) + y(p)\big) - \min_{p \in N} \big(y\big(\gamma(p)\big) + y(p)\big)\big)$ is minimum, where $x(\Box) := y(\Box) := 0$ (this is called the *bounding box netlength*)

or decide that no solution exists.

Challenge

Find an $O\big(k!4^k p\big(k, |\mathcal{N}|, |P|\big)\big)$-algorithm for this problem, where $k = |M|$ and p is a polynomial.

What is Known

The problem to decide whether a feasible placement exists is strongly *NP*-complete as it includes bin packing.

Suppose that we know which of the four conditions shall be satisfied for each pair of macros. Then one can formulate the problem as two separate linear programs (one for x- and one for y-coordinates), by adding variables representing the coordinates of the bounding box. These linear programs are duals of uncapacitated minimum cost flow instances and can thus be solved in $O\big(n \log n(m + n \log n)\big)$ time [43], where $n = |M| + |\mathcal{N}|$ and $m = |P|$. This was noted first by Cabot, Francis and Stary [13].

Of course one can enumerate all $4^{\binom{k}{2}}$ possibilities, but this is too slow and many of these contradict each other. Murata et al. [42] proposed a more efficient representation of the solution space by so-called *sequence pairs:* given any two permutations π, ρ of M, we require that m is to the west (south, north, east) of m' if m precedes m' in π and ρ (in π but not in ρ, in

ρ but not in π, neither in π nor in ρ). They showed that for every feasible placement there is a sequence pair implying constraints that are satisfied for this placement. Hence one needs to enumerate only $(k!)^2$ combinations. The resulting running time of $O\big((k!)^2 n \log n(m + n \log n)\big)$ seems to be the best known bound today.

Even with branch-and-bound techniques instances with $k \geq 15$ currently cannot be solved optimally. For larger instances local search heuristics (such as simulated annealing) based on these representations have been proposed and used in practice. There are hundreds of papers on floorplanning heuristics in the engineering literature, but there is no substantial progress in exact algorithms.

Other representations are more concise than sequence pairs but can only represent compactified placements, where no object can be moved downwards or to the left without destroying feasibility. The most efficient ones are O-trees [25, 53] and B*-trees [14], both with $k!C_k \approx \frac{(k-1)!4^k}{\sqrt{\pi k}}$ combinations, where $C_k = \frac{(2k)!}{k!(k+1)!}$ is the k-th Catalan number. However, it is not clear how to use them for finding a placement with minimum bounding box netlength.

If we just want to find a feasible solution or decide that none exists, the B*-representation together with the binary tree enumeration algorithm of Solomon and Finkel [51] does the job in $O\big((k-1)!4^k\sqrt{k}\,\big)$ time. Polynomial-time approximation algorithms in special cases where feasibility is trivial will be discussed in Problem 2.

Extensions

- Macros can often be flipped horizontally and/or vertically. In rare cases they can also be rotated, but normally not as this would require an adapation of the routing structure on top of the macro.
- Some areas of the chip may be forbidden (or, equivalently, some macros are pre-placed and must not be moved).
- Some of the objects may in fact be *soft macros* (also called random logic macros, RLMs). They come with an (estimated) area but no fixed aspect ratio. They have pins, but no pin locations. Soft macros correspond to units that are not designed yet and will be part of the chip. For each of them one needs to choose an aspect ratio and assign its pins to locations within its outline. Then they can be

placed just as hard macros. The aspect ratio can be subject to lower and upper bounds or part of the objective function (approximately squarish outlines are often preferable). Ibaraki and Nakamura [31] observed that there is a finite algorithm even in the case of soft macros by showing that this is a convex programming problem.

- Some soft macros may actually appear several times on a chip. In this case all realizations, i.e. aspect ratios and pin locations must be the same. In other words, for each group of identical soft macros one needs to choose a realization as hard macro first, and then place copies of this hard macro. The advantage is that this soft macro needs to be designed and checked only once and can be plugged in at several places on the chip.

Importance

Even an algorithm which can handle, say, twenty macros, would be useful. It could be used as part of a heuristic which decomposes larger instances and solves sub-instances optimally. Moreover, many practical instances are easy except for a specific local area where the problem may be very hard, but involves only relatively few macros.

Currently, due to the lack of good algorithms, floorplanning involves a lot of manual work.

2. Lower Bounds for Placement

Motivation

Placement is one of the most important tasks in chip design, but also (as shown in Problem 1) one of the hardest. Although studied for many decades, not much is known theoretically. In fact, even simplified formulations of the placement problem contain the notoriously hard quadratic assignment problem as special case. In practice the most successful approaches heuristically combine optimally solved subproblems; see e.g. Brenner, Struzyna and Vygen [9]. But almost nothing is known about how far from optimal they are.

Instance

- a graph G (whose vertices are called *modules* or *cells* and whose edges are called *nets*),
- an array of feasible locations $A = \{1, \ldots, x_{\max}\} \times \{1, \ldots, y_{\max}\}$

Task

Compute

- a placement $\pi\colon V(G) \to A$ satisfying $\pi(v) \neq \pi(w)$ for all $v, w \in V(G)$ with $v \neq w$
- such that
  (a) $\sum_{\{v,w\} \in E(G)} \left|\pi(v) - \pi(w)\right|$ is minimum.
  (b) $\sum_{\{v,w\} \in E(G)} \left(\pi(v) - \pi(w)\right)^2$ is minimum.

Challenge

Find an algorithm which computes the optimum value of (a) or (b) up to a constant factor, or even an $O\big(\log\left|V(G)\right|\big)$-factor, in polynomial time.

What is Known

The best known approximation algorithm for (a) has a performance guarantee of $O\big(\log\left|V(G)\right| \log\log\left|V(G)\right|\big)$ [19]. In the one-dimensional case ($y_{\max} = 1$) this is the famous optimum linear arrangement problem, for which an $O\big(\sqrt{\log\left|V(G)\right|} \log\log\left|V(G)\right|\big)$-factor approximation algorithm was found by Charikar et al. [15] and Feige and Lee [20]. However, the optimum linear arrangement problem is not known to be *MAXSNP*-hard (but see [5]). If some areas are forbidden, the problem has no constant-factor approximation algorithm [48].

Case (b) might be easier, but there is no positive result either. However, there are interesting connections to random walks and eigenvectors of the Laplacian of G [57, 52], which could potentially help.

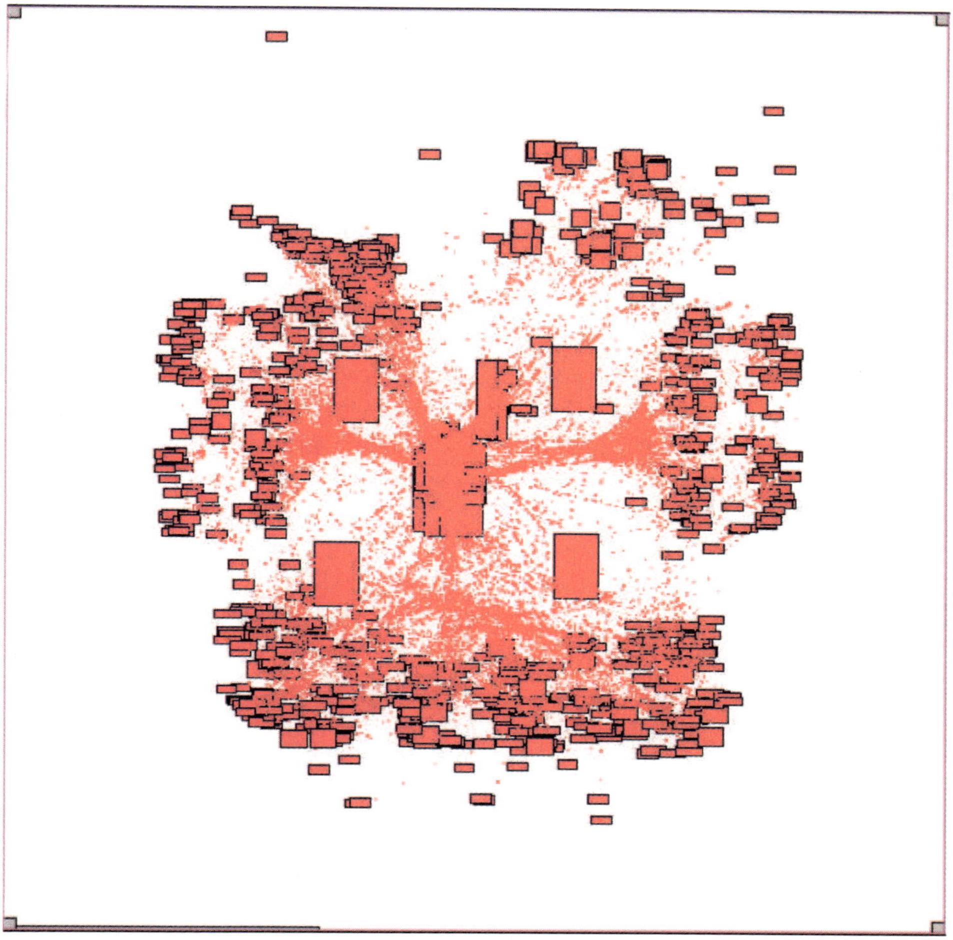

Fig. 2. Quadratic placement for the same chip as in Figure 1. The cells are colored red. Most cells are tiny and lie close to the center of the chip. Thus the placement is far from being legal.

In practice, there are only two techniques to compute lower bounds for large instances (which can be combined). The first one is to consider each net separately and assume the cells of this net to be placed as closely together as possible. The second one relaxes the problem by ignoring the constraint that no two cells can be placed at the same position. Assuming that some pins are fixed or some cells are preplaced, which is often the case in practice, this relaxation yields a positive lower bound, but a very weak one, in particular for (a). The relaxation for (b), called quadratic placement, is a basic ingredient of many placement tools [12]. See Figure 2 for an example.

Extensions

- To be really useful the algorithm should be fast enough to handle millions of cells.
- In practice the problem is more complex. Cell sizes vary (as in Problem 1), although most cells have the same height (standard cells) and are to be arranged in rows. Exceptions are macros where different problems arise (see Problem 1). Even the problem to decide whether a feasible solution exists is strongly *NP*-hard. But in practice it is very easy to find a feasible solution because a substantial portion of the available area is not used.

  Normally 99% of the cells are small (their height is uniform and their width varies by about a factor of ten), and different cell sizes typically mix in a good placement. Moreover, many cell sizes are actually determined only after placement (by a procedure called *gate sizing*, which chooses one out of several logically equivalent implementions for each cell). Hence it does not seem to be essential to work with non-uniform cell sizes.
- Pre-placed macros (as output of floorplanning; cf. Problem 1) or, equivalently, forbidden areas, also make the problem harder.
- In practice we have a hypergraph rather than a graph. Again, the vertices are cells that are to be placed and hyperedges are nets encoding the information that its pins need to be connected. More precisely, we have a netlist as in Problem 1. The pin offsets are rather small and can be ignored except at a very detailed level, but fixed pins really change the problem — just as pre-placed macros or forbidden areas do. But the number of fixed pins is typically quite small, and in some cases, where the I/O-ports are not yet placed, zero.

  It is common practice to represent hyperedges of cardinality more than two (*multi-terminal nets*) by a set of edges (two-terminal nets), e.g. a clique (cf. Brenner and Vygen [10]).
- Our formulation ignores important aspects like routing congestion and timing constraints. Nevertheless just minimizing netlength has proved a very useful model in practice.

Importance

Although our problem formulation is a radical simplification, it is the best approximation of the placement problem where we can hope for a positive solution even if $P \neq NP$.

It is not immediately clear how a positive answer would impact design practice. But the current status is that we have essentially no means to evaluate how close to optimum the placements that we compute are. Any progress here would be extremely interesting.

3. Legalizing a Placement

Motivation

Placement is usually divided into two subtasks: global placement and legalization. After global placement (cf. Problem 2), the size of many cells is changed and additional cells (in particular repeaters; cf. Problem 9) are added. Then the placement needs to be legalized without moving any cell too far. The (weighted) sum of quadratic movement is a good measure of legalization quality.

Instance

- an array of feasible locations $A = \{1, \ldots, x_{\max}\} \times \{1, \ldots, y_{\max}\}$
- a set C of cells to be placed, with $|C| \leq |A|$
- an initial placement $(x, y)\colon C \to A$

Task

Compute

- a legal placement $(\bar{x}, \bar{y})\colon C \to A$, i.e. one with $\bar{x}(c) \neq \bar{x}(c')$ or $\bar{y}(c) \neq \bar{y}(c')$ for $c, c' \in C$ with $c \neq c'$,

- such that

$$\sum_{c \in C} \left(\left| \bar{x}(c) - x(c) \right| + \left| \bar{y}(c) - y(c) \right| \right)^2$$

  is minimum.

Challenge

Find an $o\big(|C|^3\big)$ algorithm for this problem.

What is Known

This is a special case of the well-known linear assignment problem, which can be solved in $O\big(|A|^3\big)$ time even if the cost of assigning an element of C to an element of A is arbitrary (see, e.g., Korte and Vygen [37]).

It seems that no better algorithms are known for the special case. However, for linear movement costs (instead of quadratic) one can reduce the problem to an instance of the minimum cost flow problem, where the underlying graph is the two-dimensional grid graph given by A (with edges connecting nodes of Manhattan distance 1, oriented in both directions). Unit costs, infinite capacities, and balance values defined by $b(a) := \big|\{c \in C\colon (x,y)(c) = a\}\big| - 1$ for $a \in A$ complete the instance. An integral minimum cost flow exists and can be realized by moving cells. This yields an $O\big(|A|^2 \log^2 |A|\big)$-algorithm for the variant with linear movement costs. It seems plausible that $|A|$ can be replaced by $|C|$ in this time bound as locations that are never used do not require any computation.

This minimum cost flow approach has been used a lot in practice [54, 11]. Of course, when used as a heuristic for minimizing quadratic movement costs, the flow cannot be realized arbitrarily as it is better to move many cells a little rather than one cell far.

Extensions

- Typically some locations are pre-occupied by large macros and must not be used.

- Some cells are more important (because of timing constraints); so one has weights $w\colon C \to \mathbb{R}_+$ and should minimize the weighted sum of squared movements.
- Cells have different sizes, although in a typical legalization instance they all have the same height. But even with varying widths the problem to decide whether any feasible solution exists is strongly *NP*-hard. In practice this is not an issue, as rarely more than 90% of the chip area is used, and as the widths are typically small integer multiples of some minimum width.

  Nevertheless, with varying cell widths the problem becomes harder and can probably no longer be formulated as an assignment problem. Still [11] use the above minimum cost flow approach, construct an appropriate (more sophisticated) minimum cost flow instance, and solve knapsack problems to realize the flow. This paper also proposes an integer programming formulation whose LP relaxation provides reasonable lower bounds in practice.

Importance

Placement legalization is a key task in chip design and an increasing challenge with progressing technology. Poor legalization makes it hard to find a solution satisfying all timing constraints. A solution to the above problem formulation, although not capturing different cell widths, would be a significant step towards a practical and provably near-optimal solution.

4. Steiner Trees Minimizing Elmore Delay

Motivation

On a chip there are millions of signals that are generated at a certain place (the source) and must be transmitted to several destinations (sinks). Each net contains one source and at least one sink. The elements of a net need to be connected by a network of copper wires, which can be approximated well by a *rectilinear Steiner tree* (ignoring the third dimension). However, a shortest Steiner tree is often not best possible to meet tight constraints on the delay of the signal from the source to each sink.

To estimate the delay from the source to a sink we use the popular *Elmore delay* model [18] which is relatively simple and accurate (see below). To account for different timing constraints we have an extra additive term for each sink.

Instance

- a source $s \in \mathbb{R}^2$
- a nonempty finite set of sinks $T \subset \mathbb{R}^2$
- a delay adder $a_t \in \mathbb{R}$ for $t \in T$
- a source resistance $r_s > 0$
- a sink capacitance $c_t > 0$ for each $t \in T$
- the capacitance $c > 0$ and resistance $r > 0$ of a wire per unit length.

Task

Compute

- a rectilinear Steiner tree Y for $\{s\} \cup T$, oriented as an arborescence rooted at s whose vertices are elements of $\mathbb{R}^2$
- such that
$$\max_{t \in T} \left(a_t + \mathrm{ED}\,(s,t) \right)$$
is minimum,

where for $t \in T$ the Elmore delay from s to t is

$$\mathrm{ED}\,(s,t) := r_s C_s + \sum_{e=(v,w) \in E\left(Y_{[s,t]}\right)} r \|v-w\|_1 \left(\frac{c}{2} \|v-w\|_1 + C_w \right),$$

$Y_{[s,v]}$ denotes the path from s to $v \in V(Y)$ in Y, and

$$C_v := \sum_{e=(x,y)\colon\, v \in V\left(Y_{[s,x]}\right)} c\|x-y\|_1 + \sum_{t \in T\colon\, v \in V\left(Y_{[s,t]}\right)} c_t.$$

Challenge

Find a finite algorithm for this problem.

What is Known

The problem is known to be *NP*-hard [8]. For $|T| = 1$ it is trivial; a shortest path does the job. For $|T| \leq 3$ the problem can be solved in constant time [34, 44].

But even for $|T| = 2$ there are instances in which the only optimal solution is not part of the *Hanan grid*, the grid of horizontal and vertical lines induced by the source and the sinks [8]. This is not true for the variant in which $\sum_{t \in T} a_t \, \mathrm{ED}\,(s,t)$ is minimized for weights $a_t > 0$ $(t \in T)$; hence this variant can be solved in exponential time. See also [16] and [35] for other variants.

Extensions

- Actually we can choose the width of any wire (within a given range) and have several (currently up to 10) routing planes with different characteristics available. Routing planes normally have alternating preference directions (horizontal/vertical).

  Wider wires have smaller resistance but larger capacitance per unit length (and they consume more space). Wires on upper planes are often better, but the pins (source and sinks) are usually on the lower planes and the resistance of vias (metal contacts connecting adjacent planes) are not negligible.

  In practice, most of the nets are realized by a set of wires almost all of which (except short ones for pin access) have the same width and lie on two adjacent planes (one horizontal and one vertical) with similar electrical properties. Therefore the above planar simplification is meaningful.

- Our formulation ignores routing blockages. Sometimes there are significant blockages (in particular on top of macros) that must be avoided. But in most cases there are only small blockages (mostly due to the power supply) which can be neglected here.

- Pins are not single points but sets of metal shapes. Consequently we have a *group Steiner tree problem.* But since each pin spans only a small area, this is less important.
- The Elmore delay is the most accurate model with such a simple description. More accurate models depend on more parameters (in particular the so-called *slew rate:* the average rate of voltage changes); we do not go into details here.
- In some cases, in particular for clock networks, signals must not arrive too early either. In this case a reasonable objective function combines *latency* (maximum delay) and *skew* (maximum difference of delays), such as

$$(1+\alpha)\max_{t\in T}\left(a_t+\mathrm{ED}\left(s,t\right)\right)-\min_{t\in T}\left(a_t+\mathrm{ED}\left(s,t\right)\right)$$

  for some $\alpha > 0$. This variant is also unsolved.

Importance

The task to transmit a signal as fast as possible to a given set of destinations is a fundamental problem in chip design. Although shortest rectilinear Steiner trees are often good enough, one would of course like to solve such basic sub-tasks optimally.

5. Resource Sharing and Multiflows

Motivation

Routing is also a major task in chip design. Millions of nets have to be connected by wires, which must satisfy many rules. Routing is typically split into global routing and detailed routing. In global routing we determine a corridor for each net which restricts the solution space explored in detailed routing.

The simplest formulation of the global routing problem asks for packing Steiner trees for given terminal sets in a three-dimensional grid graph with edge capacities. In the case of two-terminal nets this is an integer multicommodity flow problem. But besides routing space (reflected by edge capacities) there are other resources that must be shared between nets, such as time (for signal delay), and space for buffers (needed for repowering signals; cf. Problem 9). A general resource sharing problem models all this and does not seem to be much harder than a standard multicommodity flow formulation. The common approach is to solve the fractional relaxation first and use randomized rounding for obtaining an integral solution.

Instance

- a finite set $\mathcal{R}$ of resources
- a finite set $\mathcal{C}$ of customers
- a number $u_r > 0$ specifying how many units of resource $r \in \mathcal{R}$ are available
- for each customer $c \in \mathcal{C}$ an implicitly given convex set

$$\mathcal{A}_c \subseteq \prod_{r \in \mathcal{R}} [0, u_r]$$

  of feasible resource allocation vectors (satisfying customer c). We assume an oracle for computing a function $f_c \colon \mathbb{R}_+^{\mathcal{R}} \to \mathcal{A}_c$ which approximately minimizes linear functions over $\mathcal{A}_c$: for a customer $c \in \mathcal{C}$ and a price vector $\omega \in \mathbb{R}_+^{\mathcal{R}}$ we require that $\omega^\top f_c(\omega) \leq (1+\varepsilon_0)\,\mathrm{opt}_c(\omega)$, where $\mathrm{opt}_c(\omega) := \min_{a \in \mathcal{A}_c} \omega^\top a$ and $\varepsilon_0 \geq 0$ is a constant

Task

Compute

- a feasible resource allocation vector for each customer
- such that the maximum relative utilization (or congestion) of all resources is minimized:

$$\min_{(a_c \in \mathcal{A}_c)_{c \in \mathcal{C}}} \max_{r \in \mathcal{R}} \frac{\sum_{c \in \mathcal{C}} (a_c)_r}{u_r}$$

Challenge

Find an algorithm which computes an $(1+\varepsilon_0+\varepsilon)$-approximate solution in $O^*\big(\frac{1}{\varepsilon}|\mathcal{C}|\gamma\big)$ time, for any $\varepsilon > 0$, where γ denotes the time for an oracle call and the O^*-notation suppresses logarithmic terms.

What is Known

If ε_0 can be chosen arbitrarily small, this is a special case of the equivalence of weak optimization and weak separation [24].

Combinatorial approximation algorithms (some of which assume $\varepsilon_0 = 0$) were proposed by Plotkin, Shmoys and Tardos [46], Grigoriadis and Khachiyan [23], Garg and Könemann [22], and Jansen and Zhang [32], based on earlier similar algorithms for the special case of multicommodity flows. All algorithms have a dependence on ε which is quadratic or worse. Müller and Vygen [41] (see [1] and [55] for special cases) found an $O\big(\frac{1}{\varepsilon^2}|\mathcal{C}|\gamma\ln^2|\mathcal{R}|\big)$-algorithm under some assumptions. Bienstock and Iyengar [7] showed how to reduce the term from $\frac{1}{\varepsilon^2}$ to $\frac{1}{\varepsilon}$, but by increasing the dependence on other parameters. It is not clear whether these techniques can be combined.

Extensions

- All known algorithms also compute an approximate dual solution. This can be useful in particular if the routing instance turns out to be infeasible.
- In practice one is of course interested in an integral solution. The corresponding problem includes the edge-disjoint paths problem and is of course *NP*-hard. Randomized rounding works quite well unless $\frac{a_r}{u_r}$ is large for some $r \in \mathcal{R}$, $c \in \mathcal{C}$ and $a \in \mathcal{A}_c$. However, this two-stage approach (first solve the fractional relaxation approximately then apply randomized rounding) may not be optimal.

Importance

Sharing resources optimally is a fundamental problem with numerous applications. In chip design it models global routing very well and can also handle various constraints and objectives that make global routing today much different from just packing Steiner trees subject to edge capacities. For example, Vygen [55] and Müller [40] showed how to model timing constraints and even manufacturing yield by resource constraints. The full value of this approach is yet to be explored.

6. Shortest Paths in Grids

Motivation

The main sub-task in detailed routing is to connect two sets of metal shapes by a set of wires of minimum total length. Each of these sets of shapes can be a pin or a previously routed set of wires connecting some pins. The problem can be modelled as a shortest path problem in a three-dimensional grid graph and solved by Dijkstra's algorithm [17].

The main problem is that we need to find more than 20 million paths, and the grid graph can have more than 100 billion vertices. Although the considered subgraphs for each net are restricted as a result of global routing, this remains an enormous computational challenge.

Routing layers normally have alternating preference directions (horizontal/vertical). Edges orthogonal to the preference directions, edges corresponding to vias (connecting adjacent planes), and edges on lower metal planes have higher cost to account for waste of space, effects on manufacturing yield, and electrical properties. See Figure 3 for an example.

Instance

- positive integers $x_{\max}$, $y_{\max}$, and $z_{\max}$
- for each odd $z \in \{1, \ldots, z_{\max}\}$ and each $y \in \{1, \ldots, y_{\max}\}$ an integer $i \geq 0$ and a list of coordinates $1 \leq x_1 < x_2 < \cdots < x_{2i-1} < x_{2i} \leq x_{\max}$ defining i intervals $\big\{(x_1, y, z), (x_1 + 1, y, z), \ldots, (x_2, y, z)\big\}, \ldots, \big\{(x_{2i-1}, y, z), (x_{2i-1} + 1, y, z), \ldots, (x_{2i}, y, z)\big\}$
- for each even $z \in \{1, \ldots, z_{\max}\}$ and each $x \in \{1, \ldots, x_{\max}\}$ an integer $i \geq 0$ and a list of coordinates $1 \leq y_1 < y_2 < \cdots < y_{2i-1} < y_{2i} \leq y_{\max}$ defining i intervals $\big\{(x, y_1, z), (x, y_1 + 1, z), \ldots, (x, y_2, z)\big\}, \ldots, \big\{(x, y_{2i-1}, z), (x, y_{2i-1} + 1, z), \ldots, (x, y_{2i}, z)\big\}$

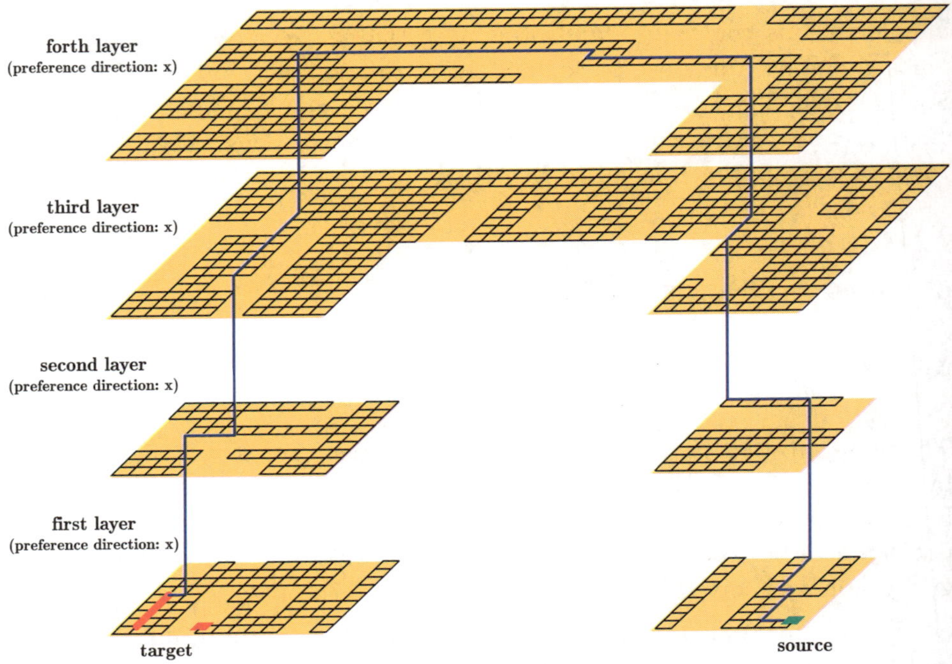

Fig. 3. This example shows a shortest path between source (green) and target (red) in a subgrid with four layers determined by global routing (yellow). The cost of an edge running in and orthogonal to the preference direction is 1 and 4, respectively, and the cost of a via is 13. With these costs the blue path has length 153, which is shortest possible. Note that the graph can be represented by relatively few intervals.

- G is defined as follows: $V(G)$ is the union of all these intervals, and $E(G) := \big\{\{(x,y,z),(x',y',z')\} : (x,y,z),(x',y',z') \in V(G), |x-x'| + |y-y'| + |z-z'| = 1\big\}$
- numbers $c_{z,i} \in \mathbb{N}$ for $z = \{1,\ldots,z_{\max}\}$ and $i \in \{1,2,3\}$, defining edge weights $c\colon E(G) \to \mathbb{N}$ by $c(e) := c_{z,i}$ for $e = \big\{(x,y,z),(x',y',z')\big\} \in E(G)$ with $z \le z'$, where $i = 1$ if $x \ne x'$, $i = 2$ if $y \ne y'$, and $i = 3$ if $z + 1 = z'$
- vertex sets $S, T \subseteq V(G)$

Task

Compute

- a shortest S-T-path in (G, c), i.e. a path P in G from a vertex $s \in S$ to a vertex $t \in T$ such that $\sum_{e \in E(P)} c(e)$ is minimum.

Challenge

Find an algorithm with running time $O(K \log K)$, where K is the number of intervals I for which there is an S-I-path of length L, where L is the length of a shortest S-T-path.

What is Known

If all intervals are singletons, Dijkstra's algorithm with a heap implementation does the job. For the general case there is an $O(LK \log K)$-algorithm by Peyer, Rautenbach and Vygen [45].

Extensions

- A well-known idea for speeding up Dijkstra's algorithm in practice is to replace the cost of an edge (v, w), originally $c(\{v, w\})$, by $c'(v, w) := c(\{v, w\}) - \pi(v) + \pi(w)$, where $\pi\colon V(G) \to \mathbb{R}$ satisfies $\pi(t) = 0$ for $t \in T$ and $\pi(v) \leq c(\{v, w\}) + \pi(w)$ for $\{v, w\} \in E(G)$ (and hence $\pi(v)$ is a lower bound on the length of a path from v to T) [47, 50]. The better this lower bound is, the shorter is the distance from S to T with respect to c', and the fewer vertices are labelled by Dijkstra's algorithm.

  Of course, this technique (often called *goal-oriented search*) changes the instance: it makes the graph directed and changes some of the costs. By splitting intervals when necessary the edges within an interval can be assumed to have constant cost in each direction. The result of [45] extends to this more general situation.

- G may actually not be an induced subgraph of the complete three-dimensional grid graph. Some edges, in particular some of those representing vias, are often missing (e.g. in order to model via distance rules).

- Many nets have more than two terminals; we need a Steiner tree connecting their pins. It may be possible to extend the algorithm to k-terminal nets for any fixed constant k.
- Not all rules which must be followed in routing can be encoded by the incomplete grid graph as above. An examples of such a rule is that there must be a certain minimum amount of metal (i.e., horizontal or vertical wiring) between two adjacent vias. As an approximate formulation in graph-theoretic terms, each connected component of the subgraph of the path that is induced by one routing layer must have minimum size.
- The rules for pin access are quite involved and cannot be discussed here, but the good news is that they require only local techniques and can be ignored when searching for global (long) connections.

Importance

Many hours of computing time are spent with routing on a chip in leading-edge technology. In particular long-distance paths in regions with many obstacles are difficult to find. Computing such a path can take several seconds. Speeding this up would have a great benefit.

7. Sink Clustering

Motivation

Some signals on a chip are distributed to thousands, and sometimes even millions of receivers (sinks). An example is given by clock networks distributing a clock signal to all storage elements (registers, flip-flops, latches) in a clock domain. Such networks consume a lot of resources, and in particular a lot of electric power.

Often the network has two levels. A set of special drivers receives the signal from a top-level network, and each of these drivers distributes it further to a set of sinks. The placement of drivers and assignment of sinks to drivers is a kind of facility location problem that we call sink clustering (cf. Figure 4).

The electrical capacitance of a net depends on the total length of the wires and the input capacitance of its sinks. There is a limit on the electrical capacitance that a driver can cope with. The overall power consumption also depends on the total capacitance and on the number of drivers. This yields the following problem formulation.

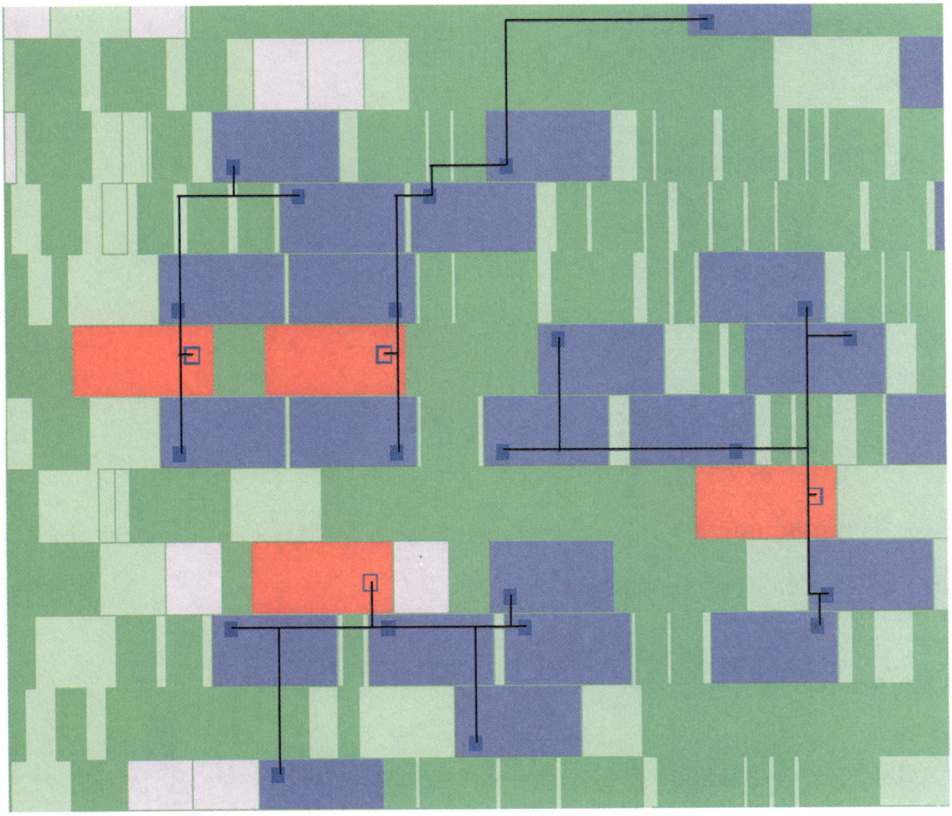

Fig. 4. Example for sink clustering (small part of a chip). The blue squares are the sinks (here: input pins of storage elements which need to receive a periodic clock signal). The red objects are the drivers that are to be placed. We may assume that each driver can be placed on a shortest Steiner tree connecting the sinks that it drives (or very close). The green objects are irrelevant here.

Instance

- a finite set $D \subseteq \mathbb{R}^2$ of sinks
- demands (sink capacitances) $d\colon D \to \mathbb{R}_+$
- a facility opening cost $f \in \mathbb{R}_+$
- a capacity $u \in \mathbb{R}_+$

Task

Compute

- a partition $D = D_1 \dot\cup \cdots \dot\cup D_k$,
- a rectilinear Steiner tree T_i for each D_i $(i = 1, \ldots, k)$
- such that $c(T_i) + \sum_{s \in D_i} d(s) \leq u$ for $i = 1, \ldots, k$
- and

$$\sum_{i=1}^{k} c(T_i) + kf$$

  is minimum.

Here $c(T_i)$ denotes the length of T_i.

Challenge

Find a polynomial-time 2-approximation algorithm.

What is Known

Maßberg and Vygen [39] found a 4-approximation algorithm which runs in $O\big(|D| \log |D|\big)$ time. Unless $P = NP$ there is no approximation algorithm with performance guarantee better than 2, which can easily be seen by a reduction from the *NP*-hard rectilinear Steiner tree problem [39]: it is *NP*-complete to decide whether there is a feasible solution with $k = 1$.

Extensions

- The above description uses just one type of drivers. In practice one often has several types, cheaper ones that can drive less and stronger ones that can drive more capacitance. This corresponds to several pairs (u_i, f_i), $i = 1, \ldots, t$. Then $u := \max_{i=1}^{t} u_i$, and the objective function becomes

$$\sum_{i=1}^{k} \left(c(T_i) + \min_{i:\, u_i \geq c(T_i)+d(D_i)} f_i \right)$$

- Our description assumes that drivers can be placed anywhere. This is a good approximation except for macros which block a larger part of a chip. A proper formulation would take a list of rectangular blockages into account, although we do not see the need for this in practice.
- The drivers must be connected by a top-level network, often a tree. Therefore one could think of including the length of a Steiner tree connecting the drivers in the objective function.
- The problem can also be considered in general metric spaces. Here the best known approximation algorithm has a performance guarantee of 4.099 [39].

Importance

On many chips the clock networks consume a substantial portion of the power. Reducing power consumption becomes more and more important. The objective function of the sink clustering problem models power consumption (of the lowest stage, which typically makes up more than 80% of the clock tree's overall power consumption) quite accurately.

8. Octagon Representation

Motivation

An *octilinear octagon* is a nonempty set of the form $\{ (x,y) \in \mathbb{R}^2 : ix + jy \geq c_{ij} \text{ for } (i,j) \in \{-1,0,1\}^2 \}$ for some constants c_{ij}. In the special case when $c_{ij} = -\infty$ for $|i| + |j| \neq 1$ they are called *rectilinear rectangles.*

Octilinear octagons arise from ℓ_1-discs by subtracting blockages. More precisely, consider all sets that can be generated from points by the following operations: (a) replace $X \subseteq \mathbb{R}^2$ by $\{ y \in \mathbb{R}^2 : \|y - x\|_1 \leq d \text{ for some } x \in X \}$ for some $d > 0$; (b) replace $X \subseteq \mathbb{R}^2$ by the closure of $X \setminus R$, where R is a rectilinear rectangle. All such sets can be written as union of octilinear octagons. They arise for example in clock tree design [29].

Instance

- a set S of n octilinear octagons

Task

Compute

- a set S' of octilinear octagons such that the union of S' equals the union of S and the interiors of any two elements of S' are disjoint.

Challenge

Find an $O(n \log n + k)$-algorithm, where k is the cardinality of a smallest set S' which is a feasible output.

What is Known

This problem can be reduced to the so-called contour problem, asking for the contour of the union of S: given the contour with k corners one can generate a solution S' with $|S'| \leq k$. As every corner of the contour must be a corner of an element of S', this is optimum up to a factor of 8.

There are simple examples for which $k = \Theta(n^2)$, even when we restrict to rectilinear rectangles. But such instances will probably not arise in practice.

For rectilinear rectangles (i.e. $c_{ij} = -\infty$ for $|i| + |j| = 2$) the contour problem can be solved by a sweepline algorithm [26] or by divide-and-conquer [27].

Extensions

- Ideally the algorithm would compute a set S' of minimum possible cardinality. Then there is no longer a simple reduction to the contour problem.
- The most interesting instances are those that arise from a set S of n octagons for which the interiors of any two elements are disjoint by an operation (a) followed by a sequence of n' operations (b). Assuming that S and the sequence of these operations is given explicitly, Then an $O\big((n + n')\log(n + n')\big)$-time algorithm could be possible.

Importance

This is the task of a basic subroutine in bottom-up algorithms for clock tree design and could also be used by fanout tree algorithms (cf. Section 9). Currently we are using an algorithm with a quadratic worst-case running time and apply heuristics to bound the number of octagons. This is not a satisfactory solution.

9. Short and Fast Fanout Trees

Motivation

If a signal is sent over a large distance or distributed to many sinks (i.e., has a large fanout), it has to be re-powered. This is done by buffers or pairs of inverters (implementing the identity function). Buffers and inverters are also called *repeaters*. Optimal buffering makes the delay per unit distance grow almost linearly. However, the delay along a path also depends on the number of bifurcations (more precisely, on the electrical capacitance driven by the repeaters).

Designing trees with repeaters in an optimal way, such that signals arrive in time but not too many resources are consumed, is an important task. Traditional approaches started with a shortest Steiner tree and inserted buffers and inverters, but this is far from optimal. Better trees are needed to meet tight timing constraints.

Instance

- a source $r \in \mathbb{R}^2$
- a finite set $S \subset \mathbb{R}^2$ of sinks
- required arrival times a_s $(s \in S)$, assuming that the signals starts at r at time 0
- a delay $d > 0$ per unit distance
- a bifurcation cost $c > 0$

Task

Compute

- an arborescence A with $\{r\} \dot\cup S \subseteq V(A) \subset \mathbb{R}^2$ and $\left|\delta^+(r)\right| = 1$ and $\left|\delta^+(v)\right| = 2$ for $v \in V(A) \setminus \big(\{r\} \cup S\big)$
- such that the worst slack

$$\min_{s \in S}\left(a_s - \sum_{e=(v,w)\in E\left(A_{[r,s]}\right)} d\|v-w\|_1 - c\Big(\big|E\big(A_{[r,s]}\big)\big| - 1\Big)\right)$$

is as large as possible up to an additive constant of c,

- and among all such solutions $\sum_{e=(v,w)\in E(A)} \|v-w\|_1$ is minimum.

Here $A_{[r,s]}$ denotes the unique r-s-path in A.

Challenge

Find a polynomial-time $\frac{3}{2}$-approximation algorithm.

What is Known

Just maximizing the worst slack is trivial by *Huffman coding* as long as all numbers $a_s - d\|r - s\|_1$ $(s \in S)$ are integer multiples of c. Here all $|S| - 1$ Steiner points can be placed at the position of r. Defining the criticality of a sink s by $a_s - d\|r - s\|_1$, the arborescence is constructed in a bottom-up fashion by iteratively replacing the two sinks with minimum criticality by a new sink whose criticality is by c less than the smaller of the two.

It is easy to see that this yields optimum worst slack (up to a rounding error of at most c), but it will generally lead to intolerably large length. Bartoschek et al. [6] showed how the length can be improved by successively inserting the sinks at an optimal place, in an order of nondecreasing criticality. However, this guarantees neither optimum worst slack nor short length. If we always insert a sink that is closest to the current tree in a shortest possible way, we get a well-known $\frac{3}{2}$-approximation algorithm (sometimes called Prim heuristic) for the rectilinear Steiner tree problem.

The question is whether the two results can be combined somehow. However, this will not be straightforward, as Alon and Azar [4] gave an example showing that for the online rectilinear Steiner tree problem the best appproximation ratio we can hope for is $\Theta(\log n / \log\log n)$, where n is the number of terminals.

Extensions

- Ideally the algorithm should have at most quadratic runnning time.
- Some areas of the chip are blocked and must not be used. All vertices and edges must be embedded in the remaining area. Other areas are blocked by macros and cannot be used for inverters and buffers but can be used for wiring. Maximal connected subtrees of the arborescence that run across such blockages must not exceed a certain length (corresponding to the capacitance that a repeater can drive).
- There are several routing planes available. Upper planes are faster but also more expensive in terms of resource consumption. Moreover, the planes can be used either only for horizontal or only for vertical wires. The extended task includes the assignment of each edge to a pair of orthogonal planes.

- Of course it would be even better if one could directly find approximately optimal repeater trees rather than first constructing a topology and then inserting inverters and buffers. However, this seems to be even harder.

Importance

A state-of-the-art chip has millions of buffers and inverters, and timing constraints are very hard to meet. Buffering, in particular designing fanout trees, is a major part of timing optimization and a key problem in chip design today.

10. Logic Synthesis

Motivation

Logic synthesis is the problem of implementing a given Boolean function $f\colon \{0,1\}^m \to \{0,1\}^n$ by a *Boolean circuit*. A Boolean circuit can be represented by an acyclic digraph with n sources and m sinks. The other vertices are *gates*, each of which has an elementary Boolean function, such as an inverter, AND, OR, NAND, NOR, XOR, etc. A Boolean circuit computes a Boolean function in the natural way. It is essentially equivalent to a netlist. f is typically given either in a certain hardware description language (similar to a programming language), or by a Boolean circuit.

The main criteria for the quality of a solution are area (which is estimated by the number of gates) and timing (see below).

In spite of its tremendous relevance, not much is known about algorithms for logic synthesis. Quite simple heuristics are used in practice. Of course it is *coNP*-complete to decide whether a given Boolean circuit implements a constant function, but this does not mean that all problems are hopeless.

Instance

- a Boolean function $f\colon \{0,1\}^m \to \{0,1\}^n$ by a list of all 2^m function values (each consisting of n bits)

Task

Compute

- a Boolean circuit (using some fixed library) with minimum number of gates implementing f.

Challenge

Find a polynomial-time algorithm, or at least an approximation algorithm.

What is Known

Essentially nothing. Even in the case $n = 1$ we know no approximation algorithm. The library is probably less important, for example we could restrict to NANDs only. To the best of our knowledge the only known theoretical results consider variants of this problem (see below).

Extensions

- As a variation, f might be given by some Boolean circuit using k gates; now we ask for an algorithm whose running time is $2^{O(k)}$.
- Often we also want to bound the depth, i.e. the length of a longest path. This roughly corresponds to delay. The variation where the Boolean circuit must have depth 2 was shown to be *NP*-complete by Masek [38] and inapproximable by Feldman [21]. See also the work of Kabanets and Cai [33]. The problem seems to be open even for depth 3.
- The inputs $1, \ldots, m$ may be associated with arrival times $a_1, \ldots, a_m$, and the outputs have required times $r_1, \ldots, r_n$. Then we want to maximize $\min_{i,j}(r_j - a_i - d_{ij})$, where d_{ij} denotes the length of a longest path from i to j. Rautenbach, Szegedy and Werber [49] suggested an approximation scheme for a special case of the case $n = 1$.
- In fact we often have an incompletely specified Boolean function $f\colon \{0,1\}^m \to \{0,1,d\}^n$, where d stands for *don't care.* We look for a *Boolean circuit* computing some concretion of f, i.e. some

$g \colon \{0,1\}^m \to \{0,1\}^n$ with $f_i(x) \in \{g_i(x), d\}$ for all $x \in \{0,1\}^m$ and $i = 1, \ldots, n$.

Importance

Logic synthesis is key for the quality of a netlist. The state of the art is rather poor, compared to other areas of chip design. Therefore we see much room for improvement here. While physical chip layout has improved substantially in the last two decades due to better algorithms, logic synthesis has not been approached seriously by mathematicians. Consequently an improvement of logic synthesis by mathematical ideas is badly needed.

Conclusion

Although a lot of progress has been made, and many theoretical results and new algorithms made their way to real design tools, chip design practice is also a constant source of interesting problems, many of which are of combinatorial nature. On the one hand, new problems come up or become important due to technological advances. On the other hand, new ideas lead to new formulations in classical problem areas. We hope that we could demonstrate the variety of interesting problems and could give a fresh view on some of the most important combinatorial problems in chip design.

References

[1] C. Albrecht, Global routing by new approximation algorithms for multicommodity flow, *IEEE Transactions on Computer Aided Design of Integrated Circuits and Systems,* **20** (2001), 622–632.

[2] C. Albrecht, Efficient incremental clock latency scheduling for large circuits, *Design, Automation and Test in Europe, Proceedings, IEEE* (2006), 1091–1096.

[3] C. Albrecht, B. Korte, J. Schietke and J. Vygen, Maximum mean weight cycle in a digraph and minimizing cycle time of a logic chip, *Discrete Applied Mathematics,* **123** (2002), 103–127.

[4] N. Alon and Y. Azar, On-line Steiner trees in the Euclidean plane, *Discrete and Computational Geometry,* **10** (1993), 113–121.

[5] C. Ambühl, M. Mastrolilli and O. Svensson, Inapproximability results for sparsest cut, optimal linear arrangement, and precedence constrained scheduling, *Proceedings of the 48th Annual IEEE Symposium on Foundations of Computer Science* (2007), 329–337.

[6] C. Bartoschek, S. Held, D. Rautenbach and J. Vygen, Efficient generation of short and fast repeater tree topologies, *Proceedings of the International Symposium on Physical Design* (2006), 120–127.

[7] D. Bienstock and G. Iyengar, Solving fractional packing problems in $O^*\left(\frac{1}{\varepsilon}\right)$ iterations, *SIAM Journal on Computing,* **35** (2006), 825–854.

[8] K. D. Boese, A. B. Kahng, B. A. McCoy and G. Robins, Rectilinear Steiner trees with minimum Elmore delay, *Proceedings of the 31st IEEE/ACM Design Automation Conference* (1994), 381–386.

[9] U. Brenner, M. Struzyna and J. Vygen, BonnPlace: placement of leading-edge chips by advanced combinatorial algorithms, *IEEE Transactions on Computer Aided Design of Integrated Circuits and Systems,* to appear

[10] U. Brenner and J. Vygen, Worst-case ratios of networks in the rectilinear plane, *Networks,* **38** (2001), 126–139.

[11] U. Brenner and J. Vygen, Legalizing a placement with minimum total movement, *IEEE Transactions on Computer Aided Design of Integrated Circuits and Systems,* **23** (2004), 1597–1613.

[12] U. Brenner and J. Vygen, Analytical methods in VLSI placement, in: *Handbook of Algorithms for VLSI Physical Design Automation* (C. J. Alpert, D. P. Mehta, S. S. Sapatnekar, eds.), Taylor and Francis (2008).

[13] A. V. Cabot, R. L. Francis and A. M. Stary, A network flow solution to a rectilinear distance facility location problem, *AIIE Transactions,* **2** (1970), 132–141.

[14] Y.-C. Chang, Y.-W. Chang, G.-M. Wu and S.-W. Wu, B*-trees: a new representation for non-slicing floorplans, *Proceedings of the 37th ACM/IEEE Design Automation Conference* (2000), 458–463.

[15] M. Charikar, M. Hajiaghayi, H. Karloff and S. Rao, ℓ_2^2 spreading metrics for vertex ordering problems, *Proceedings of the 17th ACM-SIAM Symposium on Discrete Algorithms* (2006), 1018–1027.

[16] J. Cong, L. He, C.-K. Koh and P. H. Madden, Performance optimization of VLSI interconnect layout, *Integration, the VLSI Journal,* **21** (1996), 1–94.

[17] E. W. Dijkstra, A note on two problems in connexion with graphs, *Numerische Mathematik,* **1** (1959), 269–271.

[18] W. C. Elmore, The transient response of damped linear networks with particular regard to wide-band amplifiers, *Journal of Applied Physics,* **19** (1948), 55–63.

[19] G. Even, J. Naor, S. Rao and B. Schieber, Divide-and-conquer approximation algorithms via spreading metrics, *Journal of the ACM,* **47** (2000), 585–616.

[20] U. Feige and J. R. Lee, An improved approximation ratio for the minimum linear arrangement problem, *Information Processing Letters,* **101** (2007), 26–29.

[21] V. Feldman, Hardness of approximate two-level logic minimization and PAC learning with membership queries, *Proceedings of the 38th Annual ACM Symposium on the Theory of Computing* (2006), 363–372.

[22] N. Garg and J. Könemann, Faster and simpler algorithms for multicommodity flow and other fractional packing problems, *SIAM Journal on Computing,* **37** (2007), 630–652.

[23] M. D. Grigoriadis and L. D. Khachiyan, Coordination complexity of parallel price-directive decomposition, *Mathematics of Operations Research,* **21** (1996), 321–340.

[24] M. Grötschel, L. Lovász and A. Schrijver, *Geometric Algorithms and Combinatorial Optimization,* Springer, Berlin (1988).

[25] P. N. Guo, C.-K. Cheng and T. Yoshimura, An O-tree representation of non-slicing floorplan and its applications, *Proceedings of the 36th ACM/IEEE Design Automation Conference* (1999), 268–273.

[26] R. H. Güting, An optimal contour algorithm for iso-oriented rectangles, *Journal of Algorithms,* **5** (1984), 303–326.

[27] R. H. Güting, Optimal divide-and-conquer to compute measure and contour for a set of iso-oriented rectangles, *Acta Informatica,* **21** (1984), 271–291.

[28] S. Held, *Algorithmen für Potential-Balancierungs-Probleme und Anwendungen im VLSI-Design,* Diploma thesis, University of Bonn (2001).

[29] S. Held, B. Korte, J. Maßberg, M. Ringe and J. Vygen, Clock scheduling and clocktree construction for high performance ASICs, *Proceedings of the IEEE International Conference on Computer-Aided Design* (2003), 232–239.

[30] S. Held, B. Korte, D. Rautenbach and J. Vygen, Combinatorial optimization in VLSI design, in: *Combinatorial Optimization: Methods and Applications* (V. Chvátal, N. Sbihi, eds.), IOS Press, to appear.

[31] T. Ibaraki and K. Nakamura, Packing problems with soft rectangles, in: *Hybrid Metaheuristics;* Proceedings of the 3rd International Workshop on Hybrid Metaheuristics (HM 2006); LNCS 4030 (F. Almeida et al., eds.), Springer, Berlin (2006), pp. 13–27.

[32] K. Jansen and H. Zhang, Approximation algorithms for general packing problems and their application to the multicast congestion problem, *Mathematical Programming,* **114** (2008), 183–206.

[33] V. Kabanets and J.-Y. Cai, Circuit minimization problem, *Proceedings of the 32nd Annual ACM Symposium on the Theory of Computing,* (2000), 73–79.

[34] T. Kadodi, *Steiner routing based on Elmore delay model for minimizing maximum propagation delay,* Master's Thesis, Japan Advanced Institute of Science and Technology (1999).

[35] A. B. Kahng and G. Robins, *On Optimal Interconnections for VLSI,* Kluwer Academic Publishers, Boston (1995).

[36] B. Korte, D. Rautenbach and J. Vygen, BonnTools: mathematical innovation for layout and timing closure of systems on a chip, *Proceedings of the IEEE,* **95** (2007) 555–572.

[37] B. Korte and J. Vygen, *Combinatorial Optimization: Theory and Algorithms.* Fourth Edition, Springer, Berlin (2008).

[38] W. J. Masek, *Some NP-complete set covering problems,* Unpublished manuscript, M.I.T., Cambridge (1978).

[39] J. Maßberg and J. Vygen, Approximation algorithms for a facility location problem with service capacities, *ACM Transactions on Algorithms,* to appear. A preliminary extended abstract appeared in: *Approximation, Randomization and Combinatorial Optimization;* Proceedings of the 8th International Workshop on Approximation Algorithms for Combinatorial Optimization Problems (APPROX 2005); LNCS 3624 (C. Chekuri, K. Jansen, J. D. P. Rolim, L. Trevisan, eds). Springer, Berlin (2005), pp. 158–169.

[40] D. Müller, Optimizing yield in global routing, *Proceedings of the IEEE International Conference on Computer-Aided Design* (2006), 480–486.

[41] D. Müller and J. Vygen, *Faster min-max resource sharing and applications,* manuscript, in preparation.

[42] H. Murata, K. Fujiyoshi, S. Nakatake and Y. Kajitani, Rectangle-packing-based module placement, *Proceedings of the IEEE International Conference on Computer-Aided Design* (1995), 472–479.

[43] J. B. Orlin, A faster strongly polynomial minimum cost flow algorithm, *Operations Research,* **41** (1993), 338–350.

[44] S. Peyer, Elmore–Delay-optimale Steinerbäume im VLSI-Design. Diploma thesis, University of Bonn (2000).

[45] S. Peyer, D. Rautenbach and J. Vygen, A generalization of Dijkstra's shortest path algorithm with applications to VLSI routing. Report No. 06964-OR, Research Institute for Discrete Mathematics, University of Bonn (2006).

[46] S. A. Plotkin, D. B. Shmoys and É. Tardos, Fast approximation algorithms for fractional packing and covering problems, *Mathematics of Operations Research,* **2** (1995), 257–301.

[47] I. Pohl, Bi-directional search, *Machine Intelligence,* **6** (1971), 124–140.

[48] M. Queyranne, Performance ratio of polynomial heuristics for triangle inequality quadratic assignment problems, *Operations Research Letters,* **4** (1986), 231–234.

[49] D. Rautenbach, C. Szegedy and J. Werber, Asymptotically optimal Boolean circuits for functions of the form $g_{n-1}\left(g_{n-2}\left(\dots g_3(g_2(g_1(x_1,x_2),x_3),x_4)\dots,x_{n-1}\right),x_n\right)$. Report No. 03931, Research Institute for Discrete Mathematics, University of Bonn (2003).

[50] F. Rubin, The Lee path connection algorithm, *IEEE Transactions on Computers,* **23** (1974), 907–914.

[51] M. Solomon and R. A. Finkel, A note on enumerating binary trees, *Journal of the ACM,* **27** (1980), 3–5.

[52] D. A. Spielman, Spectral graph theory and its applications, *Proceedings of the 48th Annual IEEE Symposium on Foundations of Computer Science* (2007), 29–38.

[53] T. Takahashi, A new encoding scheme for rectangle packing problem, *Proceedings of the Asia and South Pacific Design Automation Conference* (2000), 175–178.

[54] J. Vygen, Algorithms for detailed placement of standard cells, *Design, Automation and Test in Europe, Proceedings, IEEE* (1998), 321–324.

[55] J. Vygen, Near-optimum global routing with coupling, delay bounds, and power consumption, in: *Integer Programming and Combinatorial Optimization; Proceedings of the 10th International IPCO Conference; LNCS 3064* (G. Nemhauser, D. Bienstock, eds.), Springer, Berlin (2004), pp. 308–324.

[56] J. Vygen, Slack in static timing analysis, *IEEE Transactions on Computer-Aided Design of Integrated Circuits and Systems,* **25** (2006), 1876–1885.

[57] J. Vygen, New theoretical results on quadratic placement, *Integration, the VLSI Journal,* **40** (2007), 305–314.

[58] N. Young, Randomized rounding without solving the linear program, *Proceedings of the 6th Annual ACM-SIAM Symposium on Discrete Algorithms* (1995), 170–178.

Bernhard Korte and Jens Vygen
Research Institute for Discrete Mathematics
University of Bonn
Lennéstr. 2
53113 Bonn
Germany

BOLYAI SOCIETY
MATHEMATICAL STUDIES, 19

Building Bridges
pp. 369–426.

STRUCTURAL PROPERTIES OF SPARSE GRAPHS

JAROSLAV NEŠETŘIL* and PATRICE OSSONA DE MENDEZ*

CONTENTS

---

*Supported by grant 1M0021620808 of the Czech Ministry of Education and AEOLUS.

1. Introduction

In this chapter we briefly outline the main motivation of our work and we relate it to other research. We do not include any definition here.

1.1. Dense graphs

Dense graphs have been extensively studied in the context of Extremal Graph Theory. The outstanding Szemerédi Regularity Lemma [106] states that any dense network has properties which are close to the ones of a random graph. In particular, a large dense network cannot be too irregular. This structural result is one of the cornerstones of contemporary combinatorics (and one would like to say mathematics in general). It also led to manifold applications and generalizations, see e.g. [63, 62, 69, 108, 38]. The closest to our topic covered in this paper is the recent development which is based on the study of *homomorphisms* of graphs (and structures). (It is perhaps of interest note in how many different areas and a variety contexts the notion of a homomorphism recently appeared, see [57]). Regularity is viewed here as a structural approximation in a proper metrics and also as a convergence. For a survey of this development see [13]. The main idea here is to study the local structure of a large graph G by counting the homomorphisms from various small graphs F into G (this relates to the area called *property testing*), and to study the global structure of G by *counting* its homomorphisms into various small graphs H (sometimes interpreted as *templates*). Very schematically this may be outlined by the schema:

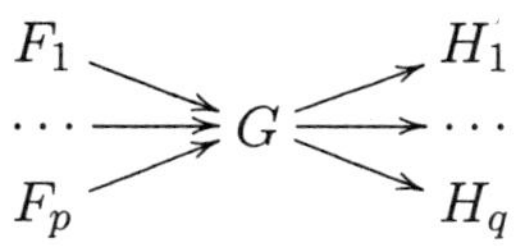

This approach proved to be very fruitful and relates (among others) to the notion of *quasi-random* graph, see e.g. [21], and to the full characterizations of testable graph properties, see e.g. [5, 13]. Nevertheless, such an approach fails when the considered structures become too sparse. In particular, Szemerédi's regularity lemma concerns graphs which have (at least locally) a number m of edges which is quadratic with respect to the number n of vertices, or at least as large as $n^{1+\varepsilon}$ if one consider extensions and generalizations of this lemma to the sparse context, see e.g. [62]. It is our ambition to deal exactly with sparse graphs which are not covered by this spectrum of results. Yet our goals are similar: we are aiming for regular and highly regular partitions.

1.2. Sparse graphs

We aim (as in the Szemerédi regularity lemma) for structural theorems for *all* graphs. The dense graphs display a remarkable stability (and many of their properties do not change by deletions and additions of a small proportion of all edges, see e.g. [112]) and, as has been discovered recently, they may be studied by number of homomorphisms and by limit objects of geometrical nature, [13, 38]. But our graphs have typically linearly many edges, large independent sets and exponentially many endomorphism. As a consequence we do not consider statistical properties but rather existential properties, i.e. properties defined by the existence (and non-existence) of mappings. In other words we deal with the simplification of category of graphs (or the *homomorphism order*), see e.g. [57]. But the first difficulty we shall meet is the definition of what a "sparse graph" is. Let us consider various approaches to this problem.

Of course, if we consider any dense graph and break every link by inserting a new vertex, the obtained graph has a number of links less than twice its number of vertices and nevertheless inherit most of the structure of the original graph. So the degeneracy (or maximal average degree of a subgraph) of our graphs is not sufficient. This also indicates that tests which contain a bounded number vertices are not sufficient to our purposes.

Another possible (and finer) restriction is to consider graphs with no minor belonging to some fixed family. In this way we get for example the class of all planar graphs. The interest of such a restriction is twofold: first it ensures a number of efficient algorithms, and also a large scientific literature. One way of describing such a family is the following: if you consider disjoint connected parts, you will never be able to find more than (fixed) p parts which are pairwise adjacent. In the other words the complete graph K_{p+1} is a forbidden minor. Classes like these are called *proper minor closed*. Such restrictions are natural for geometric networks, but for our purposes do not seem to be general enough. For instance, a very simple operation which is to clone every vertex (with its incident edges) does not preserve such properties. Another feature is the lack of parametrization: one graph is "forbidden" at all levels.

Another interesting restriction is to consider bounded degree graphs. Such graphs almost surely have nice properties when large (they are almost surely expanders). Nevertheless, important real networks like the WEB surely does not fit this restriction. And this class does not include even the class of all trees (which should be considered as sparse graphs).

A more general framework (a framework which include the above examples) concerns *proper topologically closed classes* of graphs. These classes are characterized as follows: whenever a subdivision of a graph G belongs to the class then G belongs to the class; moreover, not every graph belong to the class. Such classes are obviously defined by a (maybe infinite) set of forbidden configurations. These classes naturally catch the classes from geometrical origin, and also appear as a good approximation base for real-world networks. Notice that such graphs still have a number of edges which is bounded by a linear function of their orders. But still this lacks a parametrization and our classes will strictly include these classes.

Our principal notion for sparse graphs is the notion of *bounded expansion class* of graphs. These classes are characterized by the fact that the average degree of minors obtained by contracting disjoint subgraphs each of radius at most r is bounded by a function of r only. This means that local contractions cannot make the graphs too dense. These classes will be introduced in detail in the next chapter and we shall also indicate the various equivalent definition and regularity properties of graphs belonging to these classes. The characterization theorems are then summarized in the last chapter of this article.

1.3. Nowhere dense graphs

For any class with bounded expansion all graphs in the class have linear number of edges. There is numerous evidence that graphs with $n^{1+\varepsilon}$ edges share many properties of random graphs (for example such graphs include graphs with large girth and high chromatic number, a seminal result of Erdős). Thus $n^{1+\varepsilon}$ edges of a graph with n vertices seems to be a natural bound for our investigations of sparse graphs. This bound is natural. As we will show (and motivated by problems from model theory) a new type of graph classes arises here: *classes of nowhere dense graphs.* These classes are characterized by the fact that the number m of edges of a graph in the class is bounded by $n^{1+o(1)}$, where n is the order of the graph, and that such a statement holds for the class of the minors obtained by contracting disjoint balls of radius at most r for each fixed r. Again, this definition should be compared with the fact that every sufficiently big graph G having at least $n^{1+\varepsilon}$ edges has a big dense minor obtained by contracting balls of radius at most $r(\varepsilon)$ (by dense we mean: having a quadratic number of edges). But not only that; the classes of nowhere dense graphs have a characterization which combines virtually all concepts which were developed for the study of bounded expansion classes and expose them in the new light. To demonstrate this explicitly we included all characterization theorem in the final section of this article.

2. Measuring Sparsity

The *distance* in a graph G between two vertices x and y is the minimum length of a path linking x and y (or ∞ if x and y do not belong to the same connected component of G) and is denoted by $\mathrm{dist}_G(x, y)$. Let $G = (V, E)$ be a graph and let d be an integer. The *d-neighborhood* $N_d^G(u)$ of a vertex $u \in V$ is the subset of vertices of G at distance at most d from u in G: $N_d^G(u) = \{v \in V\colon\ \mathrm{dist}_G(u, v) \le d\}$.

We use standard graph theory terminology however we find it useful to introduce the following: for a graph $G=(V,E)$, we denote by $|G|$ the *order* of G (that is: $|V|$) and by $\|G\|$ the *size* of G (that is: $|E|$).

2.1. Shallow minors and grads

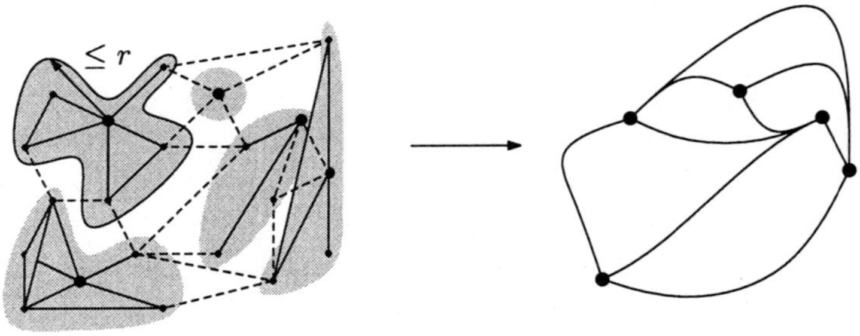

Fig. 2.1. A shallow minor of depth r of a graph G is a simple subgraph of a minor of G obtained by contracting vertex disjoints subgraphs with radius at most r

For any graphs H and G and any integer d, the graph H is said to be a *shallow minor* of G at *depth* d ([95] attributes this notion, then called *low depth minor,* to Ch. Leiserson and S. Toledo) if there exists a subset $\{x_1,\dots,x_p\}$ of G and a collection of disjoint subsets $V_1\subseteq N_d^G(x_1),\dots,V_p\subseteq N_d^G(x_p)$ such that H is a subgraph of the graph obtained from G by contracting each V_i into x_i and removing loops and multiple edges (see Fig. 2.1). The set of all shallow minors of G at depth d is denoted by $G\,\triangledown\,d$. In particular, $G\,\triangledown\,0$ is the set of all subgraphs of G.

The *greatest reduced average density* (shortly *grad*) with rank r of a graph G [84] is defined by formula

$$\nabla_r(G)=\max\left\{\frac{\|H\|}{|H|}:H\in G\,\triangledown\,r\right\}. \tag{1}$$

Also we denote by $\nabla(G)=\nabla_\infty(G)$ the maximum edge-density of a minor of G. Notice that this last invariant is related to the order of the largest complete graph which is a minor of G, that is: the so-called *Hadwiger number* $\eta(G)$ of G. It follows from the definition that

$$\eta(G)\le 2\nabla(G)+1.$$

By extension, for a class of graphs $\mathcal{C}$, we denote by $\mathcal{C} \triangledown i$ the set of all shallow minors at depth i of graphs of $\mathcal{C}$, that is:

$$\mathcal{C} \triangledown i = \bigcup_{G \in \mathcal{C}} (G \triangledown i).$$

Hence we have

$$\mathcal{C} \subseteq \mathcal{C} \triangledown 0 \subseteq \mathcal{C} \triangledown 1 \subseteq \cdots \subseteq \mathcal{C} \triangledown i \subseteq \cdots \subseteq \mathcal{C} \triangledown \infty.$$

Here we denoted by $C \triangledown \infty$ the class of all minors of graphs from $\mathcal{C}$. This is of course a minor closed class of graphs (which may coincide with the class of all finite simple graphs; think e.g. of the class of all cubic graphs).

Also, for a class $\mathcal{C}$ of graphs we define the *expansion* of the class $\mathcal{C}$ as:

$$\nabla_i(\mathcal{C}) = \sup_{G \in \mathcal{C}} \nabla_i(G)$$

$$\nabla(\mathcal{C}) = \sup_{G \in \mathcal{C}} \nabla(G)$$

Notice that $\nabla_r(G) = \nabla_0(G \triangledown r)$.

A *proper minor closed class* of graphs $\mathcal{C}$ is a minor closed class of graphs excluding at least one minor, i.e. such that $\mathcal{C}$ is not the class of all finite simple graphs. Every proper minor closed class of graphs $\mathcal{C}$ is such that $\nabla(\mathcal{C}) < \infty$. Conversely, if $\mathcal{C}$ is a class of graphs such that $\nabla(\mathcal{C}) < \infty$ then $\mathcal{C}$ is a subclass of a proper minor closed class of graphs (the smallest being $\mathcal{C} \triangledown \infty$).

Also, a grad of particular importance is ∇_0. It is related to the *maximum average degree* (mad) of a graph by $\mathrm{mad}\,(G) = 2\nabla_0(G)$. A class $\mathcal{C}$ of graphs such that $\nabla_0(G) < (k+1)/2$ (where k is an integer) is called k*-degenerate.* The equivalent defining property of a k-degenerate class of graphs is that every non-empty subgraph contains at least a vertex of degree at most k. Thus there is also an easy (greedy) algorithm to determine $\nabla_0(G)$.

It has to be noticed [35] that the determination of $\nabla_r(G)$ is a difficult problem whenever $r \geq 1$.

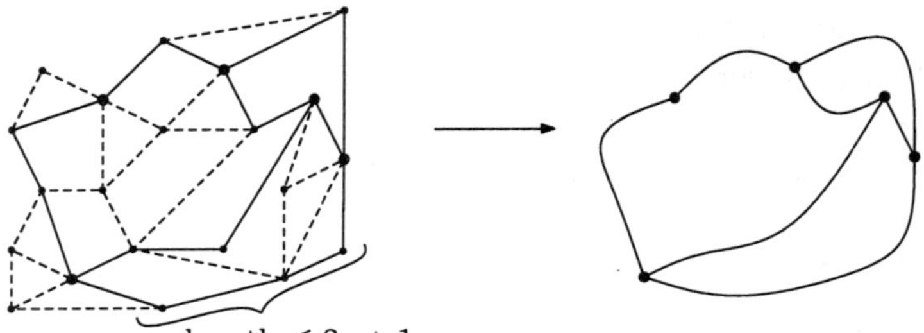

Fig. 2.2. A shallow topological minor of depth r of a graph G is a simple subgraph of a minor of G obtained by contracting internally vertex disjoint paths of length at most $2r+1$

2.2. Shallow topological minors and top-grads

Our approach makes it possible to treat minors and topological subgraphs similarly. For any (simple) graphs H and G and any integer d, the graph H is said to be a *shallow topological minor* of G at *depth* d if there exists a subset $\{x_1,\ldots,x_p\}$ of G and a collection of internally vertex disjoint paths $P_1 \ldots P_q$ each of length at most $d+1$ of G with endpoints in $\{x_1,\ldots,x_p\}$ whose contraction into single edges define on $\{x_1,\ldots,x_p\}$ a graph isomorphic to H (see Fig. 2.2).

The set of all the shallow topological minors of G at depth d is denoted by $G \mathbin{\widetilde{\triangledown}} d$. In particular, $G \mathbin{\widetilde{\triangledown}} 0$ is the set of all the subgraphs of G. Notice that for every graph G and every integer i we clearly have $(G \mathbin{\widetilde{\triangledown}} i) \subseteq (G \mathbin{\triangledown} i)$.

The *topological greatest reduced average density* (*top-grad*) with rank r of a graph G is:

$$\widetilde{\nabla}_r(G) = \max\left\{\frac{\|H\|}{|H|} : H \in G \mathbin{\widetilde{\triangledown}} r\right\}. \tag{2}$$

Also, we denote by $\widetilde{\nabla}(G)$ the limit value $\widetilde{\nabla}_\infty(G)$.

By extension, for a class of graphs $\mathcal{C}$, we denote by $\mathcal{C} \mathbin{\widetilde{\triangledown}} i$ the set of all shallow topological minors at depth i of graphs of $\mathcal{C}$, that is:

$$\mathcal{C} \mathbin{\widetilde{\triangledown}} i = \bigcup_{G\in\mathcal{C}} (G \mathbin{\widetilde{\triangledown}} i).$$

Hence we have

$$\mathcal{C} \subseteq \mathcal{C} \mathbin{\widetilde{\triangledown}} 0 \subseteq \mathcal{C} \mathbin{\widetilde{\triangledown}} 1 \subseteq \cdots \subseteq \mathcal{C} \mathbin{\widetilde{\triangledown}} i \subseteq \cdots \subseteq \mathcal{C} \mathbin{\widetilde{\triangledown}} \infty$$

For a class $\mathcal{C}$ of graphs we define the *topological expansion* of $\mathcal{C}$ as:

$$\widetilde{\nabla}_i(\mathcal{C}) = \sup_{G\in\mathcal{C}} \widetilde{\nabla}_i(G);$$

$$\widetilde{\nabla}(\mathcal{C}) = \sup_{G\in\mathcal{C}} \widetilde{\nabla}(G).$$

Notice that $\widetilde{\nabla}_i(\mathcal{C}) = \widetilde{\nabla}_0(\mathcal{C} \mathbin{\widetilde{\triangledown}} i)$.

Also, a class $\mathcal{C}$ is topologically closed if $\mathcal{C} = \mathcal{C} \mathbin{\widetilde{\triangledown}} \infty$. A topologically closed class $\mathcal{C}$ is *proper* if it is different from the class of all simple finite graphs. Notice that a class $\mathcal{C}$ is a subclass of a proper topologically closed class of graphs if and only if $\widetilde{\nabla}(\mathcal{C}) < \infty$.

2.3. Hajós or Hadwiger?

Although any proper minor closed class of graphs is also a proper topologically closed class, the converse is not true. Also, some important properties which holds from the former do not hold for the latter. A striking example stands in the fundamental difference Hadwiger conjecture (which is at least satisfied by almost every graphs) and Hajós conjecture (which is satisfied by almost no graphs).

Hence it seems to be of great importance to decide whether we will choose to define the sparsity of a class of graphs using the grad or the top-grad. However, a bit surprisingly, this does not make a difference at all. This is expressed by the following result of Zdeněk Dvořák, [35]:

Theorem 2.1. *For every integer r, the invariants ∇_r and $\widetilde{\nabla}_r$ are polynomially equivalent. Precisely, for every graph G:*

$$\frac{1}{4}\left(\frac{\nabla_r(G)}{4}\right)^{\frac{1}{(r+1)^2}} \leq \widetilde{\nabla}_r(G) \leq \nabla_r(G).$$

Similar correspondence as for edge density (expressed in terms of grads and top-grads) but also for clique number $\omega(G)$:

Lemma 2.2 [87]. *Let $r \in \mathbb{N}$. For any graph G:*

$$\omega(G \mathbin{\widetilde{\triangledown}} r) \leq \omega(G \triangledown r) \leq 2^{2^r-1}\left(\omega\left(G \mathbin{\widetilde{\triangledown}} \frac{9^{r+1}-5}{2}\right)\right)^{2^{r+1}}. \qquad \blacksquare$$

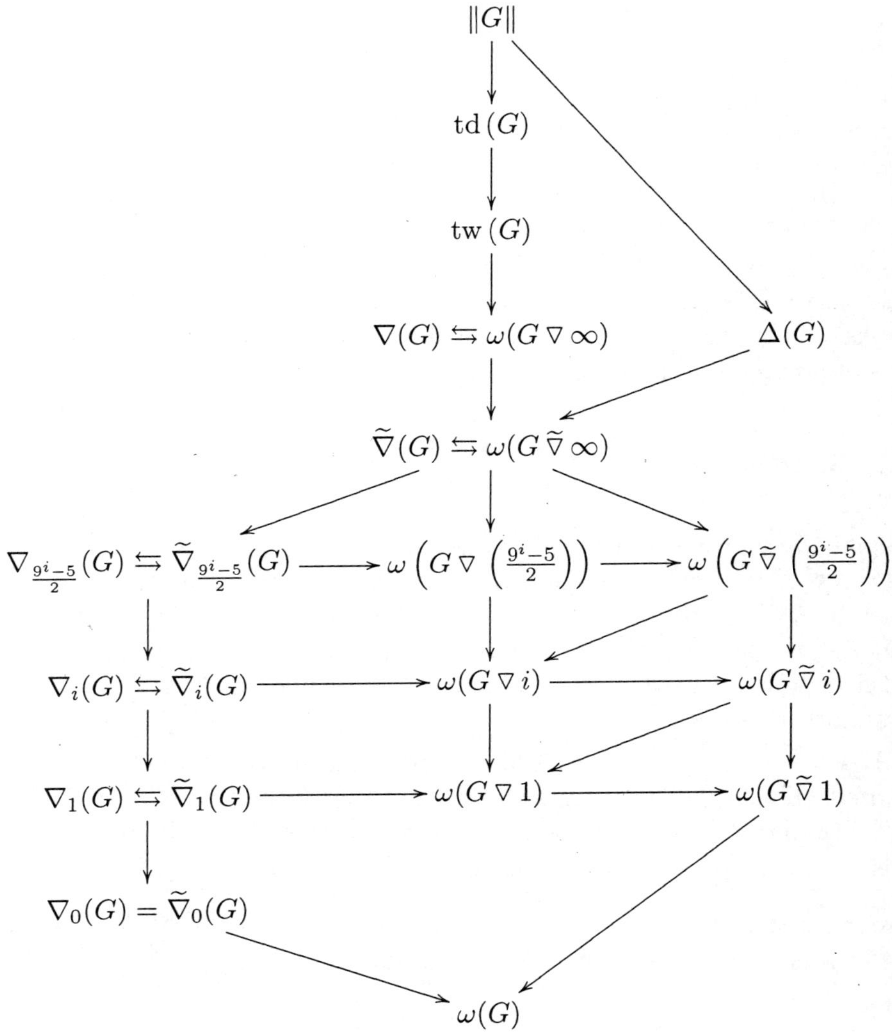

Fig. 2.3. Dominance of Invariants (up to polynomial transformation)

These two results are related by the following theorem, which has been proved by Z. Dvořák in his thesis [35]:

Theorem 2.3. *For each ε $(0 < \varepsilon \leq 1)$ there exist integers n_0 and c_0 and a real number $\mu > 0$ such that every graph G with $n \geq n_0$ vertices* *and minimum degree at least n^ε contains the c-subdivision of K_{n^μ} as a subgraph, for some $c \leq c_0$.*

2.4. Stability with respect to lexicographic product

Let G, H be graphs. The *lexicographic product* $G \bullet H$ is defined by

$$V(G \bullet H) = V(G) \times V(H)$$

$$E(G \bullet H) = \big\{\{(x,y),(x',y')\colon \{x,x'\} \in E(G) \text{ or } x = x' \text{ and } \{y,y'\} \in E(H)\big\}.$$

Note that the lexicographic product (or blowing up of vertices) is incompatible with minors, since it is easily seen that every graph is a minor of $G \bullet K_2$ for some planar graph G. However the lexicographic product and blow-up are natural constructions in the context of homomorphisms and quasi-randomness.

The long and difficult proof of the following Lemma is omitted here.

Lemma 2.4 [84]. *For every integer r there exists a polynomial P_r of degree $O\big(\frac{(2r+1)!}{2^r r!}\big)$ such that for every graphs G and H:*

$$\nabla_r(G \bullet H) \leq P_r\big(|H|\nabla_r(G)\big).$$

We may notice a slight difference between the treatment of dense and sparse graphs: In the case of dense graphs, it is usual to consider that any blow-up of the vertices of a graph G produce a graph which is intrinsically equivalent to G (hence the definition of the distance in [13]). However, in the sparse case, we only allow to blow the vertices of the graphs a bounded number of times, and the obtained graphs although not "equivalent" have characteristic which are polynomially equivalent to the ones of the original graph.

Also, Lemma 2.4 is the core of the proof of the existence of bounded transitive fraternal augmentations for graphs with bounded grads, the heart of our decomposition result for sparse graphs (see Section 3.6).

Notice also that we have an easy inequality the other way:

Lemma 2.5. *For every integer r and for every graphs G and H:*

$$\nabla_r(G \bullet H) \geq \nabla_r(G)|H|.$$

Hence, for fixed r, $\nabla_r(G)|H|$ and $\nabla_r(G \bullet H)$ are polynomially equivalent.

Proof. Consider a shallow minor G' of G of depth r such that $\nabla_r(G) = \frac{\|G'\|}{|G'|}$. Then $G' \bullet H$ is obviously a minor of $G \bullet H$ and $\frac{\|G' \bullet H\|}{|G' \bullet H|} \geq \frac{|H|^2\|G'\|}{|H||G'|} = |H|\frac{\|G'\|}{|G|}$. ■

3. Sparse Classes of Graphs

3.1. Basic definitions

A class $\mathcal{C}$ of graphs is *hereditary* if every induced subgraph of a graph in $\mathcal{C}$ to $\mathcal{C}$, and it is *monotone* of every subgraph of a graph in $\mathcal{C}$ belongs to $\mathcal{C}$. For a class of graphs $\mathcal{C}$, we denote by $H(\mathcal{C})$ the class containing all the induced subgraphs of graphs in $\mathcal{C}$, that is the inclusion-minimal hereditary class of graphs containing $\mathcal{C}$.

3.1.1. Limits. Let $\mathcal{C}$ be an infinite class of graphs and let $f\colon \mathcal{C} \to \mathbb{R}$ be a graph invariant. Let $\mathrm{Inj}\,(\mathbb{N}, \mathcal{C})$ be the set of all injective mappings from $\mathbb{N}$ to $\mathcal{C}$. Then we define:

$$\limsup_{G \in \mathcal{C}} f(G) = \sup_{\phi \in \mathrm{Inj}\,(\mathbb{N},\mathcal{C})} \limsup_{i \to \infty} f\big(\phi(i)\big).$$

Notice that $\limsup_{G \in \mathcal{C}} f(G)$ always exist and is either a real number or $\pm\infty$.

If $\limsup_{G \in \mathcal{C}} f(G) = \alpha \in \overline{\mathbb{R}} = \mathbb{R} \cup \{-\infty, \infty\}$ we have the following two properties:

- for every $\phi \in \mathrm{Inj}\,(\mathbb{N}, \mathcal{C})$, $\limsup_{i \to \infty} f\big(\phi(i)\big) \leq \alpha$;
- there exists $\phi \in \mathrm{Inj}\,(\mathbb{N}, \mathcal{C})$, $\limsup_{i \to \infty} f\big(\phi(i)\big) = \alpha$.

The second property is easy to prove: consider a sequence $\phi_1, \dots, \phi_i, \dots$ such that $\lim_{i \to \infty} \limsup_{j \to \infty} f\big(\phi_i(j)\big) = \alpha$. For each i, let $s_i(1) < \dots < s_i(j) < \dots$ be such that $\limsup_{j \to \infty} f\big(\phi_i(j)\big) = \lim_{j \to \infty} f\big(\phi_i\big(s_i(j)\big)\big)$. Then iteratively define $\phi \in \mathrm{Inj}$ by $\phi(1) = \phi_1\big(s_1(1)\big)$ and $\phi(i) = \phi_i\big(s_i(j)\big)$, where j is the minimal integer greater or equal to i such that $\phi_i\big(s_i(j)\big)$ will be different from $\phi(1), \ldots, \phi(i-1)$. Then $\limsup_{j\to\infty} f\big(\phi(j)\big) = \alpha$.

3.1.2. Derived classes. Graph operations naturally define operations on graph classes: for a class $\mathcal{C}$, an integer r and a graph H, we define:

$$\begin{aligned} \mathcal{C} \mathbin{\triangledown} r &= \bigcup_{G\in\mathcal{C}} G \mathbin{\triangledown} r;\\ \mathfrak{B}_r(\mathcal{C}) &= \big\{G \in \mathcal{C} \mathbin{\triangledown} 0\colon \rho(G) \le r\big\};\\ \mathcal{C} \bullet H &= \big\{G \bullet H\colon G \in \mathcal{C}\big\};\\ \mathcal{C} + H &= \{G + H\colon G \in \mathcal{C}\}. \end{aligned}$$

(Here $G + H$ of course means the disjoint union of graphs G and H.)

3.2. When is a class sparse or dense?

Defining the boundary between sparse and dense classes is not an easy task. Several definitions have been given for "sparse graphs", which do not allow a dense/sparse dichotomy (for instance: a graph is sparse if it has a size which is linear with respect to its order, dense if it is quadratic). Instead of defining what is a "sparse graph" or a "dense graph", we define "sparse classes of graphs" and "dense classes of graphs" by the limit behavior of the "biggest" graphs in the class when their order tends to infinity. Moreover, we will demand that our definition stays invariant in the context of derived classes, i.e. when we perform lexicographic products with small graphs, contractions of small balls, etc. It appears that the right measure of the growth of edge densities is the fraction of logarithms. This leads to the following *trichotomy* which is the starting point of our classification:

Lemma 3.1 [87]. *Let $\mathcal{C}$ be an infinite class of graphs. Then*

$$\lim_{r\to\infty} \limsup_{G\in\mathcal{C}\mathbin{\triangledown} r} \frac{\log \|G\|}{\log |G|} \in \{0, 1, 2\} \qquad \blacksquare$$

The first case of Lemma 3.1, that is: $\lim_{r\to\infty} \limsup_{G\in\mathcal{C}\mathbin{\triangledown} r} \frac{\log \|G\|}{\log |G|} = 0$ corresponds to a class of graphs $\mathcal{C}$ such that the number of edges of the graphs in $\mathcal{C}$ is bounded (for otherwise $\limsup_{G\in\mathcal{C}\,\triangledown\,0} \frac{\log\|G\|}{\log|G|} > 0$). Then the graphs in $\mathcal{C}$ only contain isolated vertices with the exception of a bounded number of vertices. We say that such a class is a class of *bounded size* graphs.

The third case of Lemma 3.1, that is: $\lim_{r\to\infty} \limsup_{G\in\mathcal{C}\,\triangledown\,r} \frac{\log\|G\|}{\log|G|} = 2$ corresponds to a class of graphs $\mathcal{C}$ such that by considering shallow minors at some "reasonable" depth, one will find infinitely many dense graphs. Actually (as shown in [87]) the property of such classes is even stronger: there exists some threshold integer $r_\mathcal{C}$ such that $\mathcal{C}\,\triangledown\,r_\mathcal{C}$ contains all finite graphs! Such classes we call *classes of somewhere dense graphs.*

Between these two extreme cases which seem to be well characterized lie the classes $\mathcal{C}$ such that:

$$\lim_{r\to\infty} \limsup_{G\in\mathcal{C}\,\triangledown\,r} \frac{\log\|G\|}{\log|G|} \leq 1.$$

Such classes we call *classes of nowhere dense graphs.* They are alternatively defined by the fact that there exists no integer r such that $\mathcal{C}\,\triangledown\,r$ contains all finite graphs (i.e. such that $\omega(\mathcal{C}\,\triangledown\,r) = \infty$). The intrinsic structure of this class and of its subclasses is the main subject of this paper. The situation is summarized in the following diagram:

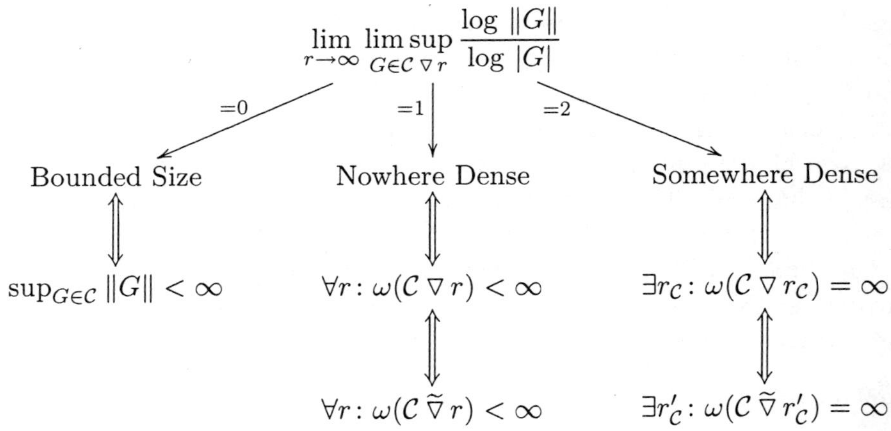

3.3. Within the nowhere dense world

Why do we have to consider shallow minors (i.e. classes $\mathcal{C} \triangledown r$)? Could a possible way to classify classes of nowhere dense graphs be to look precisely at the behavior of $\frac{\log \|G\|}{\log |G|} - 1$ for $G \in \mathcal{C}$ and $|G| \to \infty$? Alas, it happens that this value can be equivalent to any function of $|G|$ which tends to zero:

Lemma 3.2. *Let $\varepsilon\colon \mathbb{N} \to \mathbb{R}$ be a function such that $\varepsilon(n) > 0$ and $\lim_{n\to\infty} \varepsilon(n) = 0$. Then there exists an infinite hereditary class of nowhere dense graphs $\mathcal{C}_\varepsilon$ such that*

$$\limsup_{G\in\mathcal{C}} \left(\frac{\log \|G\|}{\log |G|} - 1\right) \sim \varepsilon(|G|)\,.$$

Proof. We can use well known constructions of expanders and even weaker construction of [105], where a deterministic algorithm is given that constructs a graph of girth $log_k(n) + O(1)$ and minimum degree $k-1$, n is the number of vertices and number of edges is $e = \lfloor nk/2 \rfloor$ (where $k < \frac{n}{3}$). The degree of each vertex is guaranteed to be $k-1$, k, or $k+1$, where k is the average degree.

As $\lim_{n\to\infty} \varepsilon(n) = 0$, there exists $N \in \mathbb{N}$ such that $\varepsilon(N) < 1$ and $N^{\varepsilon(N)} < N/3$. For $n \geq N$, let G_n be a graph of order n, average degree $n^{\varepsilon(n)}$ and girth $g_n = \frac{1}{\varepsilon(n)} + O(1)$. Let $\mathcal{C} = \{G_n\}_{n\in\mathbb{N}} \triangledown 0$.

For $n, p, r \in \mathbb{N}$, assume $K_p \in G_n \,\widetilde{\triangledown}\, r$. Then the girth of G_n is at most $3(2r+1)$ hence $\frac{1}{\varepsilon(n)} + O(1) \geq 6r$ thus $n \leq h(r)$ for some function $h\colon \mathbb{N} \to \mathbb{N}$. As obviously $p \leq n$ we deduce $p \leq h(r)$. It follows that $\omega(\mathcal{C} \,\widetilde{\triangledown}\, r) \leq h(r)$ hence $\mathcal{C}$ is a class of nowhere dense graphs. ■

Hence we will be more modest in our tentative classification: we will base the classification on the rough behavior of $\mathcal{C}$ with respect to bounded depth contractions. From the fact that $\lim_{r\to\infty} \limsup_{G\in\mathcal{C}\triangledown r} \frac{\log \|G\|}{\log |G|} = 1$, we can prove that the grads are "almost bounded" in the sense that $\nabla_r(G) = |G|^{o(1)}$ for $G \in \mathcal{C}$ and $|G| \to \infty$. This property suggest to consider the particular case where the function $\nabla_r(G)$ is actually bounded for every integer r. The classes for which $\nabla_r(G)$ is bounded by some value $f(r)$ independent of G are called *classes with bounded expansion.* That is:

$$\mathcal{C} \text{ has bounded expansion} \iff \forall i \geq 0\colon \sup_{G\in\mathcal{C}} \nabla_i(G) < \infty.$$

The *expansion* of a class $\mathcal{C}$ with bounded expansion is the function f defined by:

$$f(r) = \nabla_r(\mathcal{C}) = \sup_{G \in \mathcal{C}} \nabla_r(G).$$

Let us remark here that we are explaining our definitions in the reverse chronological order. Classes with bounded expansions were defined in 2005 (see e.g. [79, 80] while the importance of nowhere dense classes was realized recently (see e.g. [83]). An intermediate level between classes with bounded expansion and general classes of nowhere dense graphs are classes with *bounded local expansion*, defined by the fact that for every ρ, the class $\mathfrak{B}_\rho(\mathcal{C})$ of all balls of radius ρ in graphs in $\mathcal{C}$ has bounded expansion. Alternatively, this may be expressed as follows:

$$\mathcal{C} \text{ has bounded local expansion } \iff$$

$$\iff \forall \rho, i \geq 0\colon \sup_{v \in G \in \mathcal{C}} \nabla_i\big(G\big[N_\rho^G(v)\big]\big) < \infty.$$

The interest in these classes is limited by the fact that adding an apex (that is: a new vertex linked to all the original vertices) to the graphs in the class destroys the property that the class has bounded local expansion if it does not actually have a bounded expansion. However classes with bounded local expansion strictly contain classes with locally forbidden minors and they in turn minor closed classes. They were studied extensively, see e.g. [27]. A standard example of a class with bounded local expansion is the class $\mathcal{G}$ of graphs G such that $\operatorname{girth}(G) \geq \Delta(G)$: consider any fixed integer r and the subgraph G_v of $G \in \mathcal{G}$ induced by the r-neighborhood of v. Either $\Delta(G) < 2r$ and thus $|G_v| \leq (2r)^r$ or $\Delta(G) > 2r$ thus $\operatorname{girth}(G) > 2r$ hence G_v is a tree. Thus, except for a bounded number of graphs, the class $\mathfrak{B}_r(\mathcal{G})$ only includes forests.

Another approach to sparsity is to look for subsets of vertices which are far away from each other. Intuitively, for any integer d, if a graph is sparse and sufficiently large it will be sufficient to delete few vertices to find a big subset of vertices, any two of which are at distance at least d. Such a deletion is necessary (as we shall see) if we do not want to restrict "sparsity" to "bounded degree".

Let $r \geq 1$ be an integer. A subset A of vertices of a graph G is r-*independent* if the distance between any two distinct elements of A is strictly greater than r. We denote by $\alpha_r(G)$ the maximum size of an r-independent set of G. Thus $\alpha_1(G)$ is the usual independence number $\alpha(G)$ of G. A subset A of vertices of G is *d-scattered* if $N_d^G(u) \cap N_d^G(v) = \emptyset$ for every two distinct vertices $u, v \in A$. Thus A is d-scattered if and only if it is $2r$-independent.

A class of graphs is *wide* if every sufficiently large graph in the class contains an arbitrarily big d-scattered set. Following Dawar [26], a class if *almost wide* if deleting at most some number of vertices (bounded independently to d) makes it possible to find an arbitrarily big d-scattered set in a sufficiently large graph in the class. The class is *quasi-wide* when the number of vertices to delete may depend on d. Precisely:

$$\mathcal{C} \text{ is wide} \iff \qquad \forall d \in \mathbb{N},\ \liminf_{G\in\mathcal{C}} \alpha_d(G) = \infty$$

$$\mathcal{C} \text{ is almost wide} \iff \exists s \in \mathbb{N},\quad \forall d \in \mathbb{N},\ \liminf_{G\in\mathcal{C}} \max_{|S|\le s} \alpha_d(G-S) = \infty$$

$$\mathcal{C} \text{ is quasi wide} \iff \exists s\colon \mathbb{N} \to \mathbb{N},\ \forall d \in \mathbb{N},\ \liminf_{G\in\mathcal{C}} \max_{|S|\le s(d)} \alpha_d(G-S) = \infty.$$

It has been proved in [9] that classes with bounded degree are wide, and in [25] that proper minor closed classes of graphs are almost wide. In [83] we characterized these classes and showed how they relate to the classes of nowhere dense graphs. In particular, we prove that a hereditary class of graph is quasi-wide if and only if it is a class of nowhere dense graphs, see Section 5.2.

3.4. Classes with bounded expansion

For an extensive study of bounded expansion classes we refer the reader to [84, 85, 86, 35, 36].

Let us list some examples of classes with bounded expansion. Some inclusions of these classes are schematically depicted on Fig 3.1. However, we should remark that these classes may correspond to different expansion functions.

- **d-dimensional meshes with bounded aspect ratio.** Classes of graphs which occur naturally in finite-element and finite-difference problems are introduced in [72]. Theses classes, the classes of *d-dimensional meshes with bounded aspect ratio*, are formed by the interior skeletons of a family of d-dimensional simplicial complexes with bounded aspect ratio. As such graphs exclude K_h as a depth L minor

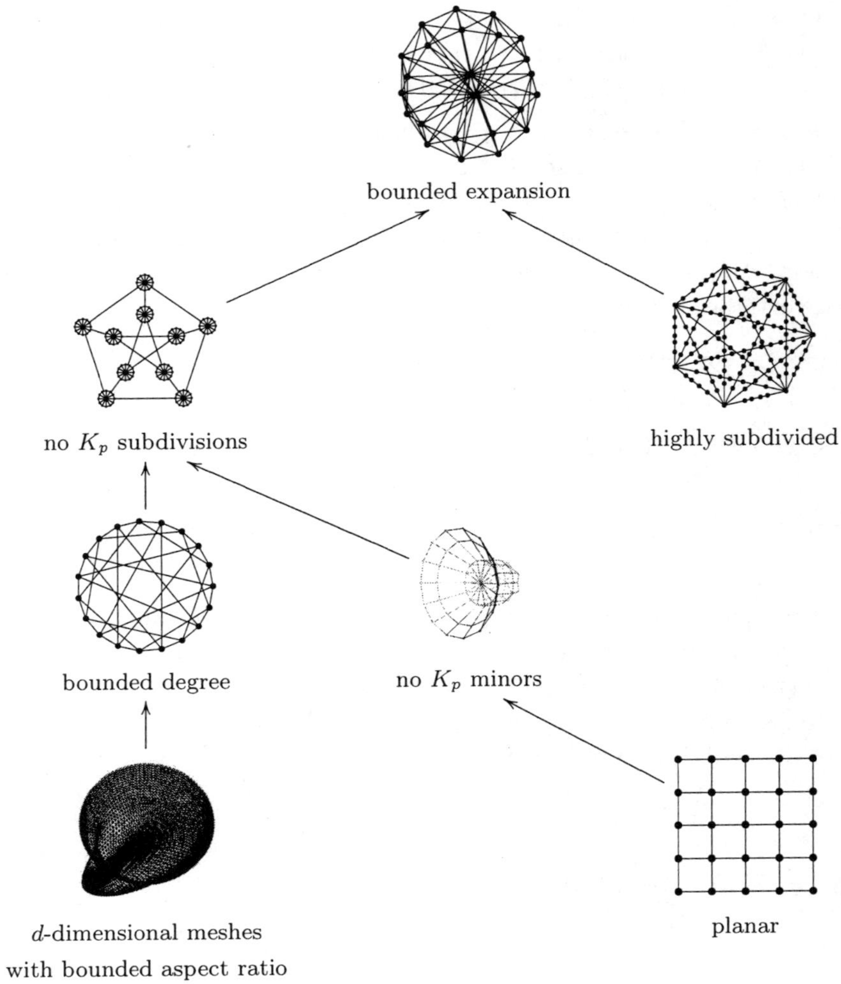

Fig. 3.1. Classes with Bounded Expansion

if $h = \Omega(L^d)$ [109] they form (for each d) a class with polynomially bounded expansion. Our results (and particularly linear algorithm for low tree depth decompositions, see Sections 3.5 and 3.6) present a natural link of applicable results [72].

- **bounded degree classes.** Let Δ be an integer. Then the class of graphs with maximum degree at most Δ has expansion bounded by the exponential function $f(r) = \Delta^{r+1}$.

- **planar graphs.** Any planar graph graph G of order n has size at most $3n-6$, hence $\nabla_0(G) < 3$ for every planar graph. As any minor of a planar graph is also planar, $\nabla_r(G) < 3$ for every integer $r \geq 0$ and any planar graph G. Hence the class of planar graphs has bounded expansion.

- **proper minor closed classes.** More generally, any proper minor closed class of graphs has expansion bounded by a constant function. Conversely, any class of graphs with expansion bounded by a constant is included in some proper minor closed class of graphs.

- **proper topologically closed classes.** These classes are defined by a (possibly infinite) set $\mathcal{S}$ of forbidden configurations, in the sense of Kuratowski's configurations: a graph G belongs to the class if no subdivision of a graph in $\mathcal{S}$ is isomorphic to a subgraph of G. Such classes have expansion bounded by a double exponential function $f(r) = 2^{r-1}\big(\min_{H\in\mathcal{S}}|V(H)|\big)^{2^{r+1}}$ (see [79]).

- **highly subdivided cliques.** For any non-decreasing function $f\colon \mathbb{N} \to \mathbb{N}\setminus\{0,1,2\}$ we may construct a class $\mathcal{C}_f$ of graphs with expansion f by including (for each integer r) the complete graph of $2f(r)+1$ vertices whose edges are subdivided 3^r-1 times.

- **union of bounded expansion classes.** Union of finitely many classes each with bounded expansion is itself a class with bounded expansion.

3.5. Proper minor closed classes

Minor closed classes have been extensively studied by Robertson and Seymour (see [97] for instance). From our point of view, proper minor closed classes of graphs (that is: minor closed classes excluding at least one minor) form the very extreme case where the expansion of the class is uniformly bounded by a constant.

Important results have been obtained concerning proper minor closed graphs, such as the celebrated proof of Wagner's conjecture (the minor relation is a well quasi-order) and the Structure Theorem. This field is also strongly connected to the study of another fundamental conjecture, namely Hadwiger's conjecture.

In their study of classes of graphs excluding a minor, Robertson and Seymour have shown the particular importance of the *tree-width* $\mathrm{tw}(G)$ and of classes with *bounded tree-width.* Structural and algorithmic importance of tree-width [98] also appeared in the context of Monadic Second-order Logic (MSL) through the results of Courcelle [23, 24].

In [80], we introduced yet a more restrictive type of classes of graphs, related to a new invariant: the *tree-depth* $\mathrm{td}(G)$. Although a class of graphs has bounded tree-width if and only if it excludes some grid as a minor, it has bounded tree-depth if and only if it excludes some path as a minor. Classes with bounded tree-depth appear to behave like classes of "almost finite" graphs. For instance, only a bounded number of graphs with tree-depth at most fixed k have no non-trivial involutive automorphism (see Section 3.2).

3.6. The full picture

The hierarchy of some important properties of hereditary sparse classes of graphs is depicted Fig. 3.2. It is interesting to note that all the properties shown in Fig. 3.2 are preserved when considering depth 1 shallow minors. This means that the considered properties are "weakly minor closed". For instance, $\mathcal{C}$ has bounded degree if and only if $\mathcal{C} \triangledown 1$ has bounded degree. We give a short proof for the case of bounded local tree-width for completeness (a similar proof applies for locally excluded minors):

Lemma 3.3. *Let $\mathcal{C}$ be a class of graphs. Then $\mathcal{C}$ has bounded local tree-width if and only if $\mathcal{C} \triangledown 1$ has bounded local tree-width.*

Proof. It is sufficient to prove that if $\mathcal{C}$ has bounded local tree-width, so has $\mathcal{C} \triangledown 1$. Let $f\colon \mathbb{N} \to \mathbb{N}$ be such that for every connected $H \subseteq G \in \mathcal{C}$ and every $t \in \mathbb{N}$ we have $\rho(H) \leq t \Longrightarrow \mathrm{tw}(H) \leq f(t)$ (where $\rho(H)$ is the radius of H). Let $G \in \mathcal{C}$ and let $H \in G \triangledown 1$. Then there is $G' \subseteq G$ such that $H \in G' \triangledown 1$ and $\rho(G') \leq 3\rho(H)$. As tw is minor-monotone, we deduce $\mathrm{tw}(H) \leq \mathrm{tw}(G') \leq f\big(\rho(G')\big) \leq f\big(3\rho(H)\big)$. It follows that $\mathcal{C} \triangledown 1$ has bounded local tree-width. ■

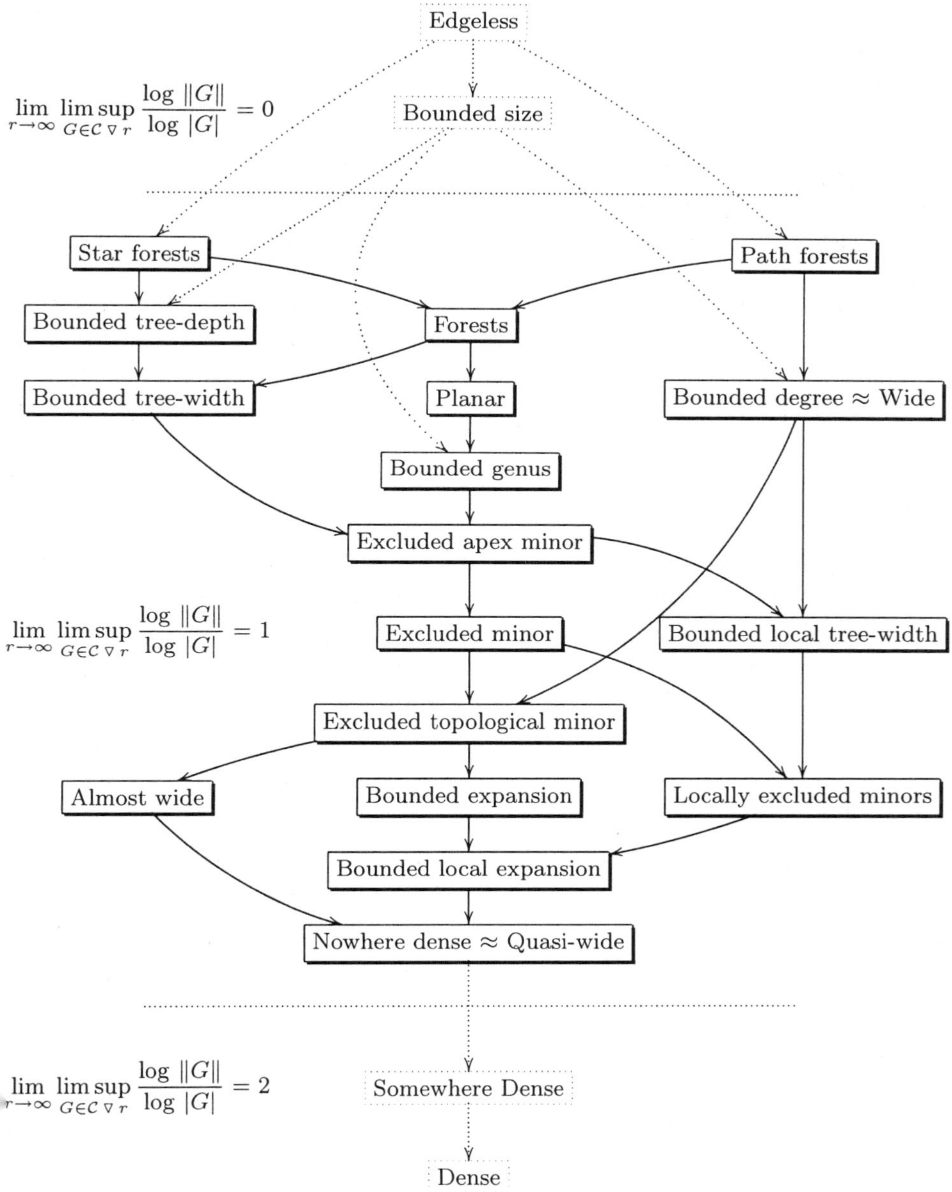

Fig. 3.2. The Nowhere Dense World: inclusion map of some important properties of **hereditary** classes of graphs

4. Regular Partitions of Sparse Graphs

4.1. Tree-width

The concept of tree-width [56, 96, 111] is central to the analysis of graphs with forbidden minors of Robertson and Seymour. This concept gained much algorithmic attention thanks to the general complexity result of Courcelle about monadic second-order logic graph properties decidability for graphs with bounded tree-width [23, 24]. It appeared that many NP-complete problems may be solved in polynomial time when restricted to a class with bounded tree-width. However, bounded tree-width is quite a strong restriction, as planar graphs for instance do not have bounded tree-width.

Let k be an integer. A *k-tree* is a graph which is either a clique of size at most k or a graph G inductively constructed from a k-tree G' with order one less by adding a vertex adjacent to a clique of size at most k in G'. A *partial k-tree* is a subgraph of a k-tree. Although it is not the usual definition of tree-width (but is equivalent to it) we define the tree-width of a graph in terms of partial k-trees: The tree-width $\mathrm{tw}(G)$ of a graph G is the minimum k such that G is a partial k-tree. Notice that a graph G with tree-width k is *k-degenerate* is the sense that every non-empty subgraph of G has at least one vertex of degree at most k (this of course does not hold conversely).

It is NP-complete to determine whether a given graph G has tree-width at most a given variable k [6]. However, when k is any fixed constant, the graphs with tree-width k can be recognized in linear time [11].

The notion of tree-width is closely related to that of vertex-separator. An *α-vertex separator* of a graph G of order n is a subset S of vertices such that every connected component of $G - S$ contains at most αn vertices. It is proved in [98] that any graph of tree-width at most k has a $\frac{1}{2}$-vertex separator of size at most $k + 1$.

4.2. Tree-depth

The concept of tree-depth has been introduced in [76, 80] to study generalized chromatic numbers of graphs (which will be introduced in Section 3.5).

A *rooted forest* is a disjoint union of rooted trees. The *height* of a vertex x in a rooted forest F is the number of vertices of a path from the root (of the tree to which x belongs to) to x and is noted $\text{height}\,(x, F)$. The *height* of F is the maximum height of the vertices of F. Let x, y be vertices of F. The vertex x is an *ancestor* of y in F if x belongs to the path linking y and the root of the tree of F to which y belongs to. The *closure* $\text{clos}\,(F)$ of a rooted forest F is the graph with vertex set $V(F)$ and edge set $\big\{\{x, y\}\colon x$ is an ancestor of y in F, $x \neq y\big\}$. A rooted forest F defines a partial order on its set of vertices: $x \leq_F y$ if x is an ancestor of y in F. The comparability graph of this partial order is obviously $\text{clos}\,(F)$.

The *tree-depth* $\text{td}\,(G)$ of a graph G is the minimum height of a rooted forest F such that $G \subseteq \text{clos}\,(F)$. This definition is analogous to the definition of *rank function* of a graph which has been used for analysis of countable graphs, see e.g. [91]. The concept also plays a key role in the recent beautiful proof of Rossmann [100].

The tree-depth of a graph may alternatively be defined inductively as follows: Let G be a graph and let $G_1, \dots, G_p$ be its connected components. Then

$$\text{td}\,(G) = \begin{cases} 1, & \text{if } \big|V(G)\big| = 1; \\ 1 + \min\limits_{v \in V(G)} \text{td}\,(G - v), & \text{if } p = 1 \text{ and } \big|V(G)\big| > 1; \\ \max\limits_{i=1}^{p} \text{td}\,(G_i), & \text{otherwise.} \end{cases}$$

The tree-depth is minor monotone: if H is a minor of G then $\text{td}\,(H) \leq \text{td}\,(G)$. The tree-depth $\text{td}\,(G)$ of a graph G is related to the order $l(G)$ of a longest path of G by:

$$l(G) \leq \text{td}\,(G) \leq 2^{l(G)}$$

and to its tree-width (see [80, 12]) by:

$$\text{tw}\,(G) + 1 \leq \text{td}\,(G) \leq \big(\text{tw}(G) + 1\big) \log_2 n.$$

The upper bound is, for instance, attained for paths (see Fig 4.1).

The tree-depth is also related to vertex-separators: for a graph G of order n and an integer $i \leq n$, let $s_G(i)$ be the maximum size of a $\frac{1}{2}$-vertex

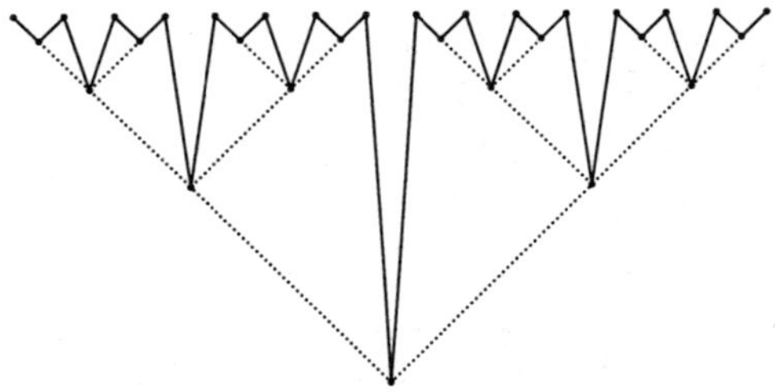

Fig. 4.1. The tree-depth of a path is logarithmic in the order of the path

separator of a subgraph of G of order at most i. Then:

$$\operatorname{td}(G) \leq \sum_{i=1}^{\log_2 n} s_G\left(\frac{n}{2^i}\right)$$

what implies that every graph G of order n with no minor isomorphic to K_h has tree-depth at most $\left(2+\sqrt{2}\right)\sqrt{h^3 n}$ (as a graph of order i with no K_h minor has a $\frac{1}{2}$-vertex separator of size at most $\sqrt{h^3 i}$ [4]).

Although there is an (easy) polynomial algorithm to decide whether $td(G) \leq k$ for any fixed k, if P $\neq$ NP then no polynomial time approximation algorithm for the tree-depth can guarantee an error bounded by n^ε, where ε is a constant with $0 < \varepsilon < 1$ and n is the order of the graph [12].

One of the strongest properties of tree-depth is "finiteness" of graphs with bounded tree-depth. Precisely, there exists a function $F\colon \mathbb{N} \times \mathbb{N} \to \mathbb{N}$ with the following property: For any integer N, any graph G of order $n > F\left(N, \operatorname{td}(G)\right)$ and any coloring $g\colon V(G) \to \{1, \ldots, N\}$, there exists a non trivial involuting g-preserving automorphism $\mu\colon G \to G$. As a consequence, any asymmetric (or rigid) graph of tree-depth t has order at most $F(1,t)$. Also, any graph G is hom-equivalent to one of its induced subgraph of order at most $F\left(1, \operatorname{td}(G)\right)$. Hence the class $\mathcal{D}_k$ of all graphs G with $td(G) \leq k$ includes a finite subset $\hat{\mathcal{D}}_k$ such that, for every graph $G \in \mathcal{D}_k$, there exists $\hat{G} \in \hat{\mathcal{D}}_k$ which is hom-equivalent to G and isomorphic to an induced subgraph of G.

The finiteness is a deep property of finite structures which are "spanned" by a branching and it has many forms. For example we can consider the category of all pairs (G,T) where G is a graph (or a structure), T a rooted tree (or branching) and $G \subseteq \operatorname{clos}(F)$. Such objects can be called *graphtr.* The morphisms between graphtrs are mappings which preserve both edges of the graph and arcs of the branchings. The above results about involutive automorphisms and finitely many hom-equivalent objects hold also in this category. Many variations are possible, see also [100] where tree depth corresponds to the quantifier rank. At its place we want to mention that the above function F grows as quickly as the Ackerman function.

The tree-depth is intimately related to special types of colorings:

A *centered coloring* of a graph G is a vertex coloring such that, for any (induced) connected subgraph H, some color $c(H)$ appears exactly once in H. Note that a centered coloring is necessarily proper. Actually, the minimum number of colors in a centered coloring of a graph G is exactly $\mathrm{td}(G)$ [80].

We can also relate the minimum number of colors in a centered coloring to the notion of vertex ranking number which has been investigated in [28, 101]: The *vertex ranking* (or *ordered coloring*) of a graph is a vertex coloring by a linearly ordered set of colors such that for every path in the graph with end vertices of the same color there is a vertex on this path with a higher color. A vertex-coloring $c\colon V(G) \to \{1, \dots, t\}$ with this property is a *vertex t-ranking* of G. The minimum t such that G has a vertex t-ranking is the *vertex ranking number* of G (see [28, 101]). This parameter also equals $\mathrm{td}(G)$ [80].

4.3. Generalized coloring numbers

Consider the following ordering game played by Alice and Bob with Alice playing first. The players take turns choosing vertices from the set of unchosen vertices. This creates a linear order L of the vertices of G with $x < y$ if and only if x is chosen before y. Given a linear order L on V, the *back degree* of a vertex x *relative to* L is the number of neighbors of x which precedes x in L. The *back degree of* L is the maximum back degree of a vertex relative to L. Alice's goal is to minimize the back degree of L, while Bob's goal is to maximize the back degree of L. This is a zero-sum two person game. Therefore each player has an optimal strategy. The *game coloring number* $\mathrm{col}_g(G)$ is the smallest (largest) integer t for which Alice (Bob) has a strategy to ensure that the linear order produced by playing the game has back degree at most (at least) $t-1$.

For instance, the complete bipartite graph $K_{n,n}$ has game coloring number $n+1$. It was proved by Faigle et al. [44] that the game coloring number of a forest is at most 4, and that the game coloring number of an interval graph G is at most $3\omega-2$. It was proved by Zhu [113] that the game coloring number of the planar graphs is at most 19 and this bound has been further reduced by Kierstead to 18 [59] and by Zhu to 17 [115]. It has also been shown by Guan and Zhu [52] that the outerplanar graphs have game coloring number at most 7. The game coloring number of graphs with bounded ∇_1 is bounded (see [81, 115, 35]).

As a generalization of both arrangeability and coloring number Kierstead and Yang introduced in [60] two new series of invariants col_k and wcol_k, that is: the *coloring number* of rank k and the *weak coloring number* of rank k.

Let L be a linear order on the vertex set of a graph G, and let x, y be vertices of G. We say y is *weakly k-accessible from* x if $y <_L x$ and there exists an x–y-path P of length at most k (i.e. with at most k edges) with minimum vertex y with respect to $<_L$ (see Fig. 4.2). The vertex is *k-accessible* from x if $y <_L x$ and there exists an x–y-path P of length at most k with minimum vertex y and second minimum vertex x with respect to $<_L$.

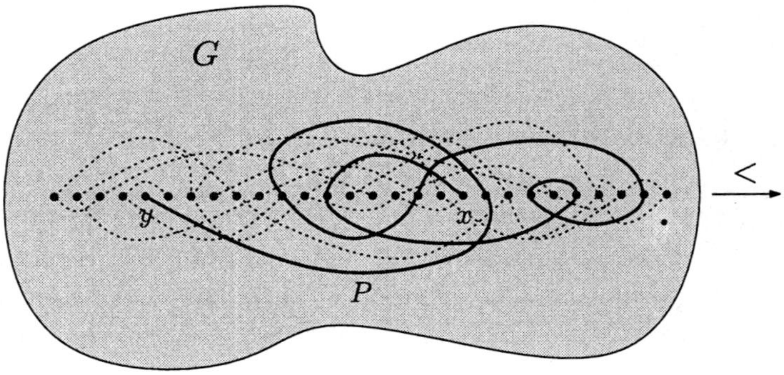

Fig. 4.2. The vertex y is weakly 8-accessible from x

Let $Q_k(G, L, x)$ and $R_k(G, L, x)$ be the sets of vertices that are respectively weakly k-accessible and k-accessible from x:

$$Q_k(G, L, x) = \{y \colon \exists\ x\text{–}y \text{ path } P \text{ such that } \min P = y\}$$

$$R_k(G, L, x) = \Big\{y \colon \exists\ x\text{–}y \text{ path } P \text{ such that } \min P = y \\ \text{and } \min(P - y) = x\Big\}$$

The *weak k-coloring number* $\mathrm{wcol}_k(G)$ and the *k-coloring number* $\mathrm{col}_k(G)$ of G are defined by:

$$\mathrm{wcol}_k(G) = 1 + \min_L \max_{v \in V(G)} \left| Q_k(G, L, v) \right|,$$

$$\mathrm{col}_k(G) = 1 + \min_L \max_{v \in V(G)} \left| R_k(G, L, v) \right|.$$

These two graph invariants are polynomially dependent, as shown in [60]:

$$\mathrm{col}_k(G) \le \mathrm{wcol}_k(G) \le \left(\mathrm{col}_k(G) \right)^k$$

They form two non-decreasing sequences, the sequence of weak-coloring numbers having the tree-depth as its maximum:

$$\begin{aligned} \mathrm{col}\,(G) &= \mathrm{wcol}_1(G) \le \mathrm{wcol}_2(G) \le \cdots \le \mathrm{wcol}_k(G) \le \cdots \le \mathrm{wcol}_\infty(G) \\ &= \mathrm{td}\,(G) \end{aligned}$$

Generalized coloring numbers are strongly related to grads: it has been proved by X. Zhu that there exists polynomials F_k such that the following holds:

Theorem 4.1 [116]. *For every integer k and every graph G:*

$$\nabla_{\frac{k-1}{2}}(G) \le \mathrm{wcol}_k(G) \le F_k\big(\nabla_{\frac{k-1}{2}}(G)\big)$$

4.4. Low tree-width coloring

A class $\mathcal{C}$ has a *low tree-width coloring* if, for any integer $p \ge 1$, there exists an integer $N(p)$ such that any graph $G \in \mathcal{C}$ may be vertex-colored using $N(p)$ colors so that each of the connected components of the subgraph induced by any $i \le p$ parts has tree-width at most $(i-1)$. According to this definition, the result of DeVos et al. may be expressed as

Theorem 4.2 [29]. *Any minor closed class of graphs excluding at least one graph has a low tree-width coloring.*

4.5. Low tree-depth coloring and p-centered colorings

As we introduced low tree-width coloring, we say that a class $\mathcal{C}$ has a *low tree-depth coloring* if, for any integer $p \geq 1$, there exists an integer $N(p)$ such that any graph $G \in \mathcal{C}$ may be vertex-colored using $N(p)$ colors so that each of the connected components of the subgraph induced by any $i \leq p$ parts has tree-depth at most i. As $\operatorname{td}(G) \geq \operatorname{tw}(G) - 1$, a class having a low-tree depth coloring has a low tree-width coloring.

Vertex Partitions		
Parameter	*Tree-Width*	*Tree-Depth*
1	proper coloring	
2	acyclic coloring [14]	star coloring [51]
p	low tree-width decomposition [29]	low tree-depth decomposition [80]

Following [80], we will make use of the notation $\chi_p(G)$ for the minimum number of colors need for a vertex coloring of G such that $i < p$ parts induce a subgraph of tree-depth at most i. These graph invariants ("generalized chromatic numbers") form a non-decreasing sequence:

$$\chi(G) = \chi_1(G) \leq \chi_2(G) \leq \cdots \leq \chi_p(G) \leq \cdots \leq \chi_\infty(G) = \operatorname{td}(G).$$

Also, we say that a vertex coloring of a graph G is a *p-centered coloring* if, for any (induced) connected subgraph H, either some color $c(H)$ appears exactly once in H, or H gets at least p colors.

The main result of [84] is the proof that these notions are related and that asking for the χ_p's to be bounded on a class of graph (with bounds depending on p) is the same as requiring that the class has bounded expansion:

Theorem 4.3 [84]. *Let $\mathcal{C}$ be a class of graphs. The following conditions are equivalent:*

(1) *$\mathcal{C}$ has low tree-width colorings,*

(2) *$\mathcal{C}$ has low tree-depth colorings,*

(3) *for every integer p, $\{\chi_p(G) : G \in \mathcal{C}\}$ is bounded,*

(4) *for every integer p, there exists an integer $X(p)$ such that every graph $G \in \mathcal{C}$ has a p-centered colorings using at most $X(p)$ colors,*

(5) *$\mathcal{C}$ has bounded expansion.* ■

More precisely, the properties of having bounded χ_p and bounded ∇_r are related in [84] as shown Fig. 4.3:

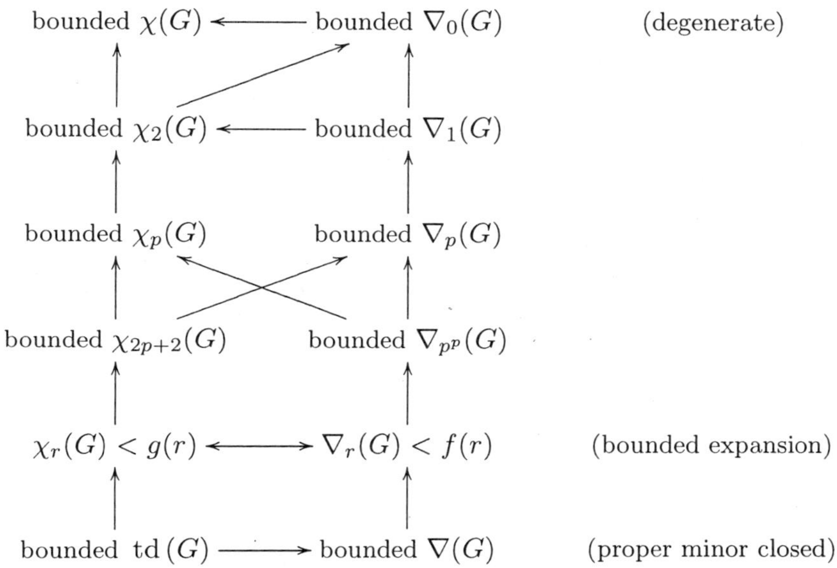

Fig. 4.3. Invariant dependence

Further improvements have been obtained in bounding $\chi_p(G)$ in terms of the grads of G [116, 35]. The best bound up to now is:

Theorem 4.4 [35]. *For each $p > 0$, there exists a polynomial P of degree $O(8^p)$ such that for each graph G,*

$$\chi_p(G) \leq P\big(\nabla_{2^{p-1}-1}(G)\big)$$

As a consequence of these results (and of the above dependency schema), we also have the following equivalence:

Theorem 4.5 [87]. *Let $\mathcal{C}$ be an infinite (and not size bounded) class of graphs. The following conditions are equivalent:*

- $\lim_{p\to\infty} \limsup_{G\in\mathcal{C}} \dfrac{\log \chi_p(G)}{\log |G|} = 0,$

- $\lim_{r\to\infty} \limsup_{G\in\mathcal{C}} \dfrac{\log \nabla_r(G)}{\log |G|} = 0,$

- *$\mathcal{C}$ is a class of nowhere dense graphs.*

Further characterizations are stated in the final chapter.

4.6. Algorithmic considerations

The decomposition algorithms we present here are those described in [79, 85]. They are based on indegree bounded orientations and transitive fraternal augmentations of these (see [84] for the relation between transitive fraternal augmentations and low tree-depth decompositions; see also [75, 78]).

Let $\vec{G}$ be a directed graph. A 1-*transitive fraternal augmentation* of $\vec{G}$ is a directed graph $\vec{H}$ with the same vertex set, including all the arcs of $\vec{G}$ and such that, for every vertices x, y, z,

- if (x,z) and (z,y) are arcs of $\vec{G}$ then (x,y) is an arc of $\vec{H}$ (*transitivity*),
- if (x,z) and (y,z) are arcs of $\vec{G}$ then (x,y) or (y,x) is an arc of $\vec{H}$ (*fraternity*).

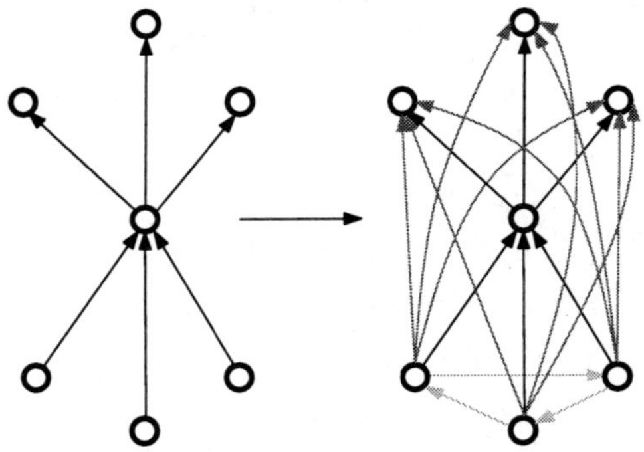

Fig. 4.4. The Transitive Fraternal Augmentation of a Graph

A 1-transitive fraternal augmentation $\vec{H}$ of $\vec{G}$ is *tight* if for each arc (x,y) in $\vec{H}$ which is not in $\vec{G}$ there exists a vertex z so that (x,z) and at least one of $(z,y),(y,z)$ are arcs of $\vec{G}$.

A *transitive fraternal augmentation* of a directed graph $\vec{G}$ is a sequence $\vec{G} = \vec{G}_1 \subseteq \vec{G}_2 \subseteq \cdots \subseteq \vec{G}_i \subseteq \vec{G}_{i+1} \subseteq \cdots$, such that $\vec{G}_{i+1}$ is a 1-transitive fraternal augmentation of $\vec{G}_i$ for every $i \geq 1$. The transitive fraternal augmentation is *tight* if all the 1-transitive fraternal augmentations of the sequence are tight.

4.6.1. Computing a Transitive Fraternal Augmentation. We describe an algorithm which computes, given any directed graph $\vec{G}$ of order n, a 1-tight transitive fraternal augmentation $\vec{H}$ in $O\big(\Delta^-(\vec{G})^2 n\big)$-time, such that:

$$\Delta^-(\vec{H}) \leq f\big(\Delta^-(\vec{G}, \nabla_1(\vec{G})\big)$$

$$\nabla_r(\vec{H}) \leq g_r\big(\Delta^-(\vec{G}), \nabla_{2r+1}(\vec{G})\big)$$

for some polynomials f and g_r ($r \in \mathbb{N}$).

In the augmentation process, we add two kind of arcs: transitivity arcs and fraternity arcs. Let us start with transitivity ones:

```
Require: D represents the directed graph to be augmented.
Ensure: D' represents the array of the added arcs.
  Initialize D'.
  for all v ∈ {1, ..., n} do
    for all (u, e) ∈ D[v] do
      for all (x, f) ∈ D[u] do
        m ← m + 1; append (x, m) to D'[v].
      end for
    end for
  end for
```

This algorithm runs in $O\big(\Delta^-(\vec{G})^2 n\big)$ time, where $\Delta^-(\vec{G})$ is the maximum indegree of the graph to be augmented. It computes the list array D' of the transitivity arcs which are missing in $\vec{G}$, missing arcs may appear more than once in the list, but the number of added edges cannot exceed $\Delta^-(\vec{G})^2 n$.

Now, we shall consider the fraternity edges.

```
Require: D represents the directed graph to be augmented.
Ensure: L represents the list of edges to be added.
  L = ().
  for all v ∈ {1, ..., n} do
    for all (x, e) ∈ D[v] do
      for all (y, f) ∈ D[v] do
        if x < y then
          append (x, y) to L.
        end if
```

end for
end for
end for

This algorithm runs in $O\big(\Delta^-(\vec{G})^2 n\big)$-time and computes the list of the fraternity edges. Edges may appear in this list more than once but the length of the list L cannot exceed $\Delta^-(\vec{G})^2 n/2$.

The simplification of L (that is the removal of multiple instances of a same edge in the list), the computation of an acyclic orientation of the graph with edge set L having minimum possible maximum indegree and the merge/simplification with the arcs in D and D' may be achieved in $O\big(\Delta^-(\vec{G})^2 n\big)$-time.

Let G be a graph. Define $f(r) = \nabla_r(G)$ and $F(x, y) = x^2 + 2y$ and let $R(p) = 1 + (p-1)\big(2 + \lceil \log_2 p \rceil\big)$. The tight fraternal augmentation $\vec{G} = \vec{G}_1 \subseteq \vec{G}_2 \subseteq \cdots \subseteq \vec{G}_{R(p)}$ of G computed by iterating $R(p)$ times the tight 1-transitive fraternal augmentation algorithm is such that any proper coloring of $G_{R(p)}$ defines a p-centered coloring of G. Using the fact that a proper coloring of $G_{R(p)}$ using at most $\big\lfloor 2\nabla_0\big(G_{R(p)}\big)\big\rfloor + 1$ colors may easily been computed in $O(n)$-time, we get an algorithm which computes a p-centered coloring of G using at most $C_p\big(\nabla_{p^p}(G)\big)$ colors in time $C'_p\big(\nabla_{p^p}(G)\big)\, n$ where C_p and C'_p are polynomials.

From this follows, in particular that for every fixed p, our p-centered coloring algorithm has the following properties:

p-centered coloring characteristics		
Class type	*Max number of color*	*Max running time*
Classes with bounded expansion	$O(1)$	$O(n)$
Classes of nowhere dense graphs	$n^{o(1)}$	$\leq n^{1+o(1)}$

5. Algorithmic Applications

In this Chapter we give a sample of algorithmic applications. Such applications seem to be typified by this situation: Results which were proved earlier for planar graphs, and later sometimes generalized for proper minor closed classes can be sometimes proved for general classes with bounded expansion. And for bounded degree graphs we can sometimes proceed similarly. We review sample of such instances, for other applications see [79].

5.1. Subgraph isomorphism problem

Eppstein [41] gives a linear time algorithm to solve the subgraph isomorphism problem for a fixed pattern in a planar graph. He also gives a linear time bound for a fixed pattern and an input graph with bounded tree width decomposition. From this lemma and using our p-centered coloring algorithm, we deduce an extension of Eppstein's result of [41, 42] to classes with bounded expansion:

Theorem 5.1. *Let $\mathcal{C}$ be a class with bounded expansion and let H be a fixed graph. Then there exists a linear time algorithm which computes, from a pair (G, S) formed by a graph $G \in \mathcal{C}$ and a subset S of vertices of G, the number of isomorphs of H in G that include some vertex in S. There also exists an algorithm running in time $O(n) + O(k)$ listing all such isomorphism where k denotes the number of isomorphs (thus represents the output size).*

It is also possible to extend this result to classes of nowhere dense graphs, with a complexity increasing from $O(n)$ to $n^{1+o(1)}$. All of these results are summarized in the following table:

Subgraph isomorphism problem		
Context	*Complexity*	*Reference(s)*
General	$O\left(n^{0.792\,\lvert V(H)\rvert}\right)$	[88] using [22]
Bounded tree-width	$O(n)$	[41] (also [23, 24])
Planar	$O(n)$	[40] [41]
Bounded genus	$O(n)$	[42]
Bounded expansion (includes the three previous classes)	$O(n)$	[79, 85]
Nowhere dense	$\leq n^{1+o(1)}$	[87]

5.2. Small distance checking

The following result is a weighted extension of the basic observation that bounded orientations allows $O(1)$-time checking of adjacency [20].

Theorem 5.2. *For any class $\mathcal{C}$ with bounded expansion and for any integer k, there exists a linear time preprocessing algorithm so that for any preprocessed $G \in \mathcal{C}$ and any pair $\{x, y\}$ of vertices of G the value $\min\big(k, \operatorname{dist}(x, y)\big)$ may be computed in $O(1)$-time.* ∎

Also, this result may be extended to classes of nowhere dense graphs, using a preprocessing algorithm in $n^{1+o(1)}$-time allowing $\min\big(k, \operatorname{dist}(x, y)\big)$ to be computed in $n^{o(1)}$-time.

5.3. Existential first-order properties

Monadic second-order logic (MSOL) is an extension of first-order logic (FOL) that includes subsets of vertex sets (i.e. we expand our language by monadic predicates). The following theorem of Courcelle has been applied to solve many optimization problems.

Theorem 5.3 Courcelle [23, 24]. *Let $\mathcal{K}$ be class of finite graphs $G = \langle V, E, R\rangle$ represented as τ_2-structures, that is: by two sorts of elements (vertices V and edges E) and an incidence relation R. Let ϕ be a $MSOL(\tau_2)$ sentence. If $\mathcal{K}$ has bounded tree width and $G \in \mathcal{K}$, then checking whether $G \models \phi$ can be done in linear time.*

From this theorem and our results (especially low tree depth decomposition) we deduce (see e.g. [79]):

Theorem 5.4. *Let $\mathcal{C}$ be a class with bounded expansion and let p be a fixed integer. Let ϕ be an existential $FOL(\tau_2)$ sentence. Then there exists a linear time algorithms to check whether an input graph $G \in \mathcal{C}$ satisfies ϕ or not.*

Thus for instance we have([79]):

Theorem 5.5. *Let $\mathcal{K}$ be a class with bounded expansion and let H be a fixed graph. Then, for each of the next properties there exists a linear time algorithm to decide whether a graph $G \in \mathcal{K}$ satisfies them:*

- *H has a homomorphism to G,*
- *H is a subgraph of G,*
- *H is an induced subgraph of G.*

5.4. Dominating sets

The DOMINATING SET problem (DSP) is defined as follows.

Input: A graph $G = (V, E)$ and an integer parameter k.

Question: Does there exist a dominating set of size k or less for G, i.e., a set $V' \subseteq V$ with $|V'| \leq k$ and such that for all $u \in V - V'$ there is a v in V' for which $uv \in E$?

This is a classic NP-complete problem [47] which is also apparently not fixed parameter tractable (with respect to the parameter k) because it is known to be W[2]-complete in the W-hierarchy of fixed parameter complexity theory [32]. In this theory, any graph problem for which there is an algorithm with time complexity $O(f(k)n^{\alpha})$, for some problem parameter k, where n is the number of vertices in the graph and where α is a constant independent of k and n, is said to be *fixed parameter tractable* (fpt).

DSP is fixed parameter tractable with respect to, for example, tree-width [6] and tree decompositions are computable in linear time, for fixed tree-width [11]. DSP is similar in definition to the vertex cover problem (VCP), but they seem to differ considerably in their fixed parameter tractability properties. The Robertson-Seymour theory of graph minors [99] can be used to show that VCP is a fixed parameter tractable problem because vertex cover is closed with respect to taking minors, and fpt algorithms have been described [32] for VCP. But DSP is not closed with respect to taking minors.

DSP remains NP-complete when restricted to planar graphs [47]. Fellows and Downey [31, 32] gave a search tree algorithm for this problem which has time complexity $O(11^k n)$, when the input is restricted to planar graphs.

In [39] it is shown, using the search tree approach, that the dominating set problem is fixed parameter tractable for graphs of bounded genus, with time complexity of $O\big((4g+40)^k n^2\big)$ for graphs of genus $g \geq 1$.

The idea to make this problem tractable, is to consider the strong properties of small dominating sets on classes with bounded tree-depth.

Let $G = (V, E)$ be a graph. A subset $X \subseteq V$ of G is a *dominating set* of G if every vertex of G not in X is adjacent to some vertex in X. We note $\mathfrak{D}(G)$ the family of all dominating sets of G and by $\mathfrak{D}_k(G)$ the family of the dominating sets of G having cardinality at most k.

For subsets $X, W \subseteq V$, we say that X *dominates* W if every vertex in $W \setminus X$ has a neighbor in X. We denote $\mathfrak{D}_k(G, W)$ the family of the subsets dominating W and having cardinality at most k.

Lemma 5.6. *For every integers $k, l \geq 1$ for every graph $G = (V, E)$ with tree-depth at most l and for every subset $W \subseteq V$ of vertices, there exists a (blocker) subset $A = A(G, W) \subseteq V$ of at most kl vertices meeting every $X \in \mathfrak{D}_k(G, W)$. Moreover, if a rooted forest Y of height l is given such that $G \subseteq \operatorname{clos}(Y)$ then we can find the blocker set A in $O(kl)$-time.*

From this Lemma, using a low tree-depth decomposition, we deduce:

Lemma 5.7. *Let $\mathcal{C}$ be a class with bounded expansion. Then there exists a function $f\colon \mathbb{N} \to \mathbb{N}$ such that for every integer k, for every $G = (V, E) \in \mathcal{C}$ and for every $W \subseteq V$ a set $A(G, W)$ of cardinality at most $f(k)$ may be computed in $O(n)$-time (where n is the order of G) which meets every set in $\mathfrak{D}_k(G, W)$.*

Hence, by an easy induction on k:

Theorem 5.8. *Let $\mathcal{C}$ be a class with bounded expansion. Then there exists a function $g\colon \mathbb{N} \to \mathbb{N}$ such that for every integer k, every $G = (V, E) \in \mathcal{C}$ and every $W \subseteq V$ one may compute in time $O(g(k)n)$ a set X which is either minimal set cardinality at most k dominating W or the empty set if G has no dominating set of cardinality at most k.*

Actually, we also deduce that any graph G has at most $F\big(k, \nabla_{k^k}(G)\big)$ dominating sets of size at most k and that they may be all enumerated in time $O\big(\phi\big(k, \nabla_{k^k}(G)\big)\, n\big)$. Notice that the result does not extend to the problem of finding a set X of cardinality at most k such that every vertex not in X is at distance at most 2 from X (consider k disjoint stars of order n/k, giving $(n/k)^k$ possible solutions to the problem.

5.5. Induced matchings

A *matching* in a graph G is a subset of pairwise non-adjacent edges. An *induced matching* in a graph G is a matching of G which is an induced subgraph of G, that is a matching with the property that no endpoint of an edge in the matching is adjacent to an endpoint of another edge in the matching.

The problem of finding a *maximum induced matching* (that is: an induced matching with maximum cardinality) has been introduced by Stockmeyer and Vazirani [103] as the "risk-free marriage problem" and it was studied extensively [34, 43, 45, 50, 102]. For a graph G we denote by $\beta^*(G)$ the size of a maximum induced matching.

It is known that the problem of deciding whether a given graph has an induced matching of size at least k (for given k) is NP-complete [103], even for bipartite graphs of maximum degree 4. However, this problem has been shown to be solvable in polynomial time for several graph classes [15, 16, 17, 18, 19, 49, 50, 61, 70, 71] and even in linear time for trees [46, 50, 117].

A vertex v of a graph G is a *clone* if G has a vertex $u \neq v$ with the same neighborhood as v. In that say we say that v is a *clone* of u. We denoted by $\sim$ be the equivalence relation defined by $x \sim y$ if x and y have the same neighbors (i.e. are clones). The authors proved in [82] that clone-free graphs with bounded $\nabla_1(G)$ have an induced matching of linear size (that is: of size $\varepsilon\big(\nabla_1(G)\big)\,|G|$).

Theorem 5.9. *Let G be a clone-free connected graph. Then*

$$\beta^*(G) \geq \frac{|G|}{f\big(\nabla_0(G), \nabla_1(G)\big)}$$

where

$$f(x,y) = 4x(2^{2y}+y+1)\big(2x(2^{2y}+y+1)+1\big)^2 \qquad \blacksquare$$

Actually, a more general result is proved in [82]:

Theorem 5.10. *For every integer $k > 2$ and every $C > 0$ there exists > 0 such that every connected graph G of order n with no involutive automorphism φ exchanging two connected P_k-free subgraphs and such that*

$\nabla_{\lfloor k/2 \rfloor}(G) < C$ has a subset of $k\varepsilon n$ vertices inducing εn disjoint paths of order k. ■

5.6. Vertex separators

A celebrated theorem of Lipton and Tarjan [68] states that any planar graph has a separator of size $O(\sqrt{n})$. Alon, Seymour and Thomas [4, 3] showed that excluding K_h as a minor ensures the existence of a separator of size at most $O(h^{3/2}\sqrt{n})$. Gilbert, Hutchinson, and Tarjan [48] further proved that graphs with genus g have a separator of size $O(\sqrt{gn})$ (this result is optimal).

Plotkin et al. [95] introduced the concept of *limited-depth minor* exclusion and have shown that exclusion of small limited-depth minors implies the existence of a small separator. Precisely, they prove that any graph excluding K_h as a depth l minor has a separator of size $O(lh^2 \log n + n/l)$ hence proving that excluding a K_h minor ensures the existence of a separator of size $O(h\sqrt{n}\log n)$.

Plotkin et al. [95] proved that for graphs with m edges and n vertices, and integers l and h, there is an $O(mn/l)$ time algorithm that will either produce a K_h-minor of depth at most $l \log n$ or will find a separator of size at most $O(n/l + 4lh^2 \log n)$. We deduce that classes with sub-exponential expansions have separators of sub-linear size. Random cubic graphs having expansion bounded by $f(x) = 2^x$ and almost surely $\Omega(n)$ bisection width [64] (thus $\Omega(n)$ separators) show that this result is optimal.

Theorem 5.11. *Let $\mathcal{C}$ be a class of graphs with expansion bounded by a function f such that $\log f(x) = o(x)$.*

Then the graphs of order n in $\mathcal{C}$ have separators of size $s(n) = o(n)$ which may be computed in time $O(ns(n)) = o(n^2)$. ■

As random cubic graphs almost surely have bisection width at least $0.101n$, they have almost surely no separator of size smaller than $n/20$ It follows that if $\log f(x) = (\log 2)x$, the graphs have no sublinear separators any more. This shows the optimality of Theorem 5.11. More: as proved by Dvořák, the absence of small vertex separators implies that the expansion of a class of graphs has to be sub-exponential. Precisely:

Theorem 5.12 [35]. *If $\mathcal{C}$ is a monotone class of graphs such that each graph in $\mathcal{C}$ of order n has a vertex separator of size $o\big(\frac{n}{\log n}\big)$, then $\mathcal{C}$ has subexponential expansion.*

6. Homomorphisms and Logic

In this chapter we relate our theory to some problems treated in the context of model theory and mathematical logic: In Section 6.1 we deal with dualities and in Section 6.2 we deal with homomorphisms preservation theorems. Both these questions were intensively studied in the unrestricted cases [93, 100, 7] as well as under various restrictions (to minor closed classes and classes of bounded degree graphs; see e.g. [8, 10, 25, 27, 33, 55, 80]. This research continued by considering classes of bounded expansion and, more recently, classes of nowhere dense graphs. A bit surprisingly, all the main results may be generalized by restriction to these classes.

This is proper place to say that the results of this section hold for more general structures than undirected graphs. They hold for oriented graphs, for colored graphs, hypergraphs and finite relational structures. It is easy to transform the results for graphs to results for hypergraphs and relational systems. This can be done using *incidence graphs* and, in most cases alternatively, using 2-*sections* (known in model theory as *Gaifman graphs*), see [87]. However to keep the style of this paper uniform we state most of the results for graphs only.

6.1. Restricted dualities

Recall that a *homomorphism* from a graph G to a graph H is a mapping $f\colon V \to V(H)$ which preserves adjacency: $\big\{f(x), f(y)\big\} \in E(H)$ whenever $\{x,y\} \in E(G)$. We denote by $G \xrightarrow{f} H$ or $f\colon G \longrightarrow H$ that f is a homomorphism from G to H. The existence of a homomorphism from G to H is denoted by $G \longrightarrow H$, while the non-existence of such a homomorphism is denoted by $G \not\longrightarrow H$. Graphs G, G' are said to be *homomorphism equivalent* if we have both $G \longrightarrow G'$ and $G' \longrightarrow G$. It is also clear that the relation $G \leq H$ defined as $G \longrightarrow H$ is a quasiorder on the class of all finite graphs. This quasiorder becomes a partial order if we restrict it to the class of all non-isomorphic minimal retracts (i.e. *cores*). This partial order is called the *homomorphism order.* All graphs considered in this paper are finite. A *class* is a (possibly infinite) class of finite graphs. See [57] for a recent introduction to graphs and homomorphisms.

The following definition is the central definition of this section:

Definition 6.1. A class of graphs $\mathcal{K}$ has *all restricted dualities* if, for any finite set of connected graphs $\mathcal{F} = \{F_1, F_2, \ldots, F_t\}$, there exists a set of finite graphs $\mathcal{D}_{\mathcal{F}}^{\mathcal{K}}$ such that $F_i \not\longrightarrow D$ for $i = 1, \ldots, t$ and every $D \in \mathcal{D}_{\mathcal{F}}^{\mathcal{K}}$, and such that for all $G \in \mathcal{K}$,

$$(3)\ (F_i \not\longrightarrow G \text{ for all } i = 1, 2, \ldots, t) \iff \left(G \longrightarrow D \text{ for some } D \in D_{\mathcal{F}}^{\mathcal{K}}\right).$$

Any instance of (3) is called a *restricted finite duality* (for the class $\mathcal{K}$), or $\mathcal{K}$-restricted duality.

In the extremal case that $\mathcal{F}$ and $\mathcal{D}_{\mathcal{F}}^{\mathcal{K}}$ consists from single element sets we speak about *restricted singleton duality* (this case is however the key to the general case). Also note that if all graphs are connected then the set $\mathcal{D}_{\mathcal{F}}^{\mathcal{K}}$ can be chosen with one element.

We now justify this general definition by the following two examples and by the context in which this definition crystallized.

Example 6.2. Grötzsch's celebrated theorem (see e.g. [110]) says that every triangle-free planar graph is 3-colorable. In the language of homomorphisms this says that for every triangle-free planar graph G there is a homomorphism of G into K_3. Using the partial order terminology, Grötzsch's theorem says that K_3 is an upper bound (in the homomorphism order) for the class $\mathcal{P}_3$ of all planar triangle-free graphs. It is $K_3 \notin \mathcal{P}_3$ and this suggests a natural question (first formulated in [74]): Is there yet a smaller bound? The answer, which may be viewed as a strengthening of Grötzsch's theorem, is positive: there exists a triangle free 3-colorable graph H such that $G \longrightarrow H$ for every graph $G \in \mathcal{P}_3$. This has been proved in [80, 78] in a stronger form for all proper minor-closed classes. (The case of triangle-free planar graphs is interesting in its own and it has been related in [73] to a conjecture by Seymour and to Guenin's theorem [53] and seems to find a proper setting in the context of TT-continuous mappings [90].) One can view these results as restricted dualities (which hold in the class of planar graphs). Restricted duality results have since been generalized not only to proper minor closed classes of graphs and but also to other forbidden subgraphs, in fact to any finite set of connected graphs, [80]. This then implies that Grötzsch's theorem can be strengthened by a sequence of even stronger bounds and that the supremum (in the homomorphism order) of the class of all triangle free planar graphs does not exist, [77].

Example 6.3. A graph is *sub-cubic* if the degrees of all its vertices are ≤ 3. By Brooks theorem (see e.g. [30]) every sub-cubic connected graph is 3-colorable with the single exception of K_4. What about the class of all sub-cubic *triangle-free* graphs? Does there exists a triangle free 3-colorable bound? The positive answer to this question is given in [33, 55]. In fact for every finite set $\mathcal{F} = \{F_1, F_2, \ldots, F_t\}$ of connected graphs there exists a graph H with the following property:

$$G \longrightarrow H \text{ for every sub-cubic graph } G \in \mathrm{Forb_h}(\mathcal{F}).$$

(Here $\mathrm{Forb_h}(\mathcal{F})$ is the class of all graphs G which satisfy $F_i \not\longrightarrow G$ for every $i = 1, 2, \ldots, t$. Thus $\mathrm{Forb_h}(K_3)$ is the class of all triangle free graphs.) It follows that the class of all sub-cubic graphs has all restricted dualities.

Note that while sub-cubic graphs, and more generally graphs with bounded degrees, have all restricted dualities, this is not true for classes of degenerate graphs [74, 77].

Where lies the boundary for validity of restricted dualities? This is the central question of this area. We give a very general sufficient condition for a class to have all restricted dualities. To motivate these results we first introduce the original context of (unrestricted) dualities.

The following is a partial order formulation of an important homomorphism (or coloring; or Constraint Satisfaction) problem (this time we formulate the definition for finite structures):

Definition 6.4. A pair F, D of structures is called a *dual pair* if for every structure G,

$$F \not\longrightarrow G \iff G \longrightarrow D. \tag{4}$$

We also say that F and D form a duality, D is called *(singleton) dual of* F. Dual pairs of graphs and of finite relational structures were characterized in [93], the notion itself goes back to [89]. Equivalently, one can describe a dual pair F, D by saying that the structure D is the maximum graph in the class $\mathrm{Forb_h}(F)$ (maximum in the homomorphism order).

It appears (and this is the main result of [93]) that (up to homomorphism equivalence) all the dualities are of the form (T, D_T) where T is a finite (relational) tree. Every dual D_T is uniquely determined (up to homomorphism equivalence) by the tree T (but its structure is far more difficult to describe, see e.g. [94, 92, 66]). These results imply infinitely many examples of dualities. But a much richer spectrum (and in fact a surprising richness of results) is obtained by restricting the validity of (4) to a particular class of graphs $\mathcal{K}$. This then is expressed by the notion of a restricted duality.

It is easy to see that using the homomorphism order we can reformulate the restricted duality as follows: A class $\mathcal{K}$ has all restricted dualities if for any finite set of connected graphs $\mathcal{F} = \{F_1, F_2, \dots, F_t\}$ the class $\mathrm{Forb}_{\mathrm{h}}(\mathcal{F}) \cap \mathcal{K}$ has an upper bound in the homomorphism order (namely $D^{\mathcal{K}}_{\mathcal{F}}$) which belongs to the class $\mathrm{Forb}_{\mathrm{h}}(\mathcal{F})$.

Bounded expansion classes of graphs and structures provide a rich spectrum of restricted dualities. This has been shown in [84, 85, 86, 79]. The following may be see as one of the main results:

Theorem 6.5. *Any class of graphs (and more generally structures) with bounded expansion has all restricted dualities.*

As both proper minor closed classes and bounded degree graphs form classes of bounded expansion this result generalizes both Examples 1 and 2. In fact the seeming incomparability of bounded degree graphs and minor closed classes led us to the definition of bounded expansion classes.

6.2. Homomorphism preservation

Homomorphisms are one of the key concept of model theory as they are naturally related to the satisfiability of formulas. This we shall illustrate on *homomorphism preservation theorems.* This application provided the motivation for the concept of nowhere dense graphs and structures.

Classical model theory studies properties of abstract mathematical structures (finite or not) expressible in first-order logic, see e.g. [58]. In this context, three classical fundamental preservation theorems have been proved, which connect syntactic and semantic properties of first-order formulas:

- the Łoś–Tarski theorem, which asserts that a first-order formula is preserved under extensions on all structures if, and only if, it is logically equivalent to an existential formula;

- Lyndon's theorem, which asserts that a first-order formula is preserved under surjective homomorphisms on all structures if, and only if, it is logically equivalent to a positive formula;
- the homomorphism preservation theorem which asserts that a first-order formula is preserved under homomorphisms on all structures if, and only if, it is logically equivalent to an existential-positive formula.

The terms "all structures", which means finite and infinite structures, is crucial in the statement of these theorems.

Finite model theory is the study of the first-order logic (and its various extensions) on finite structures [37, 67]. In this context, it has been proved that the two first theorems fail when relativized to the finite, that is: there exists a first-order formula that is preserved under extensions on finite structures, but is not equivalent in the finite to an existential formula [107, 54, 2] and there exists a first-order formula that is preserved under surjective homomorphisms on finite structures, but is not equivalent in the finite to a positive formula [1, 104]. However, a bit surprisingly, the relativized version of the homomorphism preservation theorem to the finite has been recently proved by B. Rossman [100].

Relativizations of homomorphism preservation theorem to specific classes of structures have been studied and in this context e.g. in [9, 8, 10] and in this context A. Dawar defined classes of graphs called *wide, almost wide and quasi-wide* (see e.g. [25]) and they were introduced in Section 2.3. Here we treat these interesting classes in a greater detail.

For instance, it has been proved in [8] that the extension preservation theorem holds in any class $\mathcal{C}$ that is wide, hereditary (i.e. closed under taking substructures) and closed under disjoint unions. Wide classes includes classes with bounded maximum degree. We prove here (and see [83] that an hereditary class of graphs is actually wide if and only if it has a bounded degree (Theorem 6.6).

Also, it has been proved in [9, 10] that the homomorphism preservation theorem holds in any class $\mathcal{C}$ that is almost wide, hereditary and closed under disjoint unions. Almost wide classes of graphs include classes of graphs which exclude a minor [65].

Dawar [26] recently announced that the homomorphism preservation theorem holds in any hereditary quasi-wide class that is closed under disjoint unions. This is a strengthening of the result proved in [9]. Clearly, quasi-wide quasi-wide classes of graphs include classes of graphs locally excluding a minor [27]. Using the theory developed for classes of sparse graphs we shall give a complete characterization of hereditary classes of graphs which are wide, almost wide and quasi-wide. In fact this led us to the definition of classes of *nowhere dense structures.*

We find it useful to study wide (and almost wide and quasi-wide) classes (defined already in Section 2.3) by means of the following functions $\Phi_{\mathcal{C}}$ and $\overline{\Phi}_{\mathcal{C}}$ defined for classes of graphs. It is essential for our approach that we also define the uniform version of these concepts.

Function $\Phi_{\mathcal{C}}$. This function has domain $\mathbb{N}$ and range $\mathbb{N} \cup \{\infty\}$ and $\Phi_{\mathcal{C}}(d)$ is defined for $d \geq 1$ as the minimum s such that the class $\mathcal{C}$ satisfies the following property:

"There exists a function $F\colon \mathbb{N} \to \mathbb{N}$ such that for every integer m, every graph $G \in \mathcal{C}$ with order at least $F(m)$ contains a subset S of size at most s so that $G - S$ has a d-independent set of size m."

We put $\Phi_{\mathcal{C}}(d) = \infty$ if $\mathcal{C}$ does not satisfy the above property for any value of s. Moreover, we define $\Phi_{\mathcal{C}}(0) = 0$.

Function $\overline{\Phi}_{\mathcal{C}}$. This function has domain $\mathbb{N}$ and range $\mathbb{N}\cup\{\infty\}$ and $\overline{\Phi}_{\mathcal{C}}(d)$ is defined for $d \geq 1$ as the minimum s such that $\mathcal{C}$ satisfies the following property:

"There exists a function $F\colon \mathbb{N} \to \mathbb{N}$ such that for every integer m, every graph $G \in \mathcal{C}$ and every subset A of vertices of G of size at least $F(m)$, the graph G contains a subset S of size at most s so that A includes a d-independent set of size m of $G - S$."

We put $\overline{\Phi}_{\mathcal{C}}(d) = \infty$ if $\mathcal{C}$ does not satisfy the above property for any value of s. Moreover, we define $\overline{\Phi}_{\mathcal{C}}(0) = 0$.

Notice that obviously $\overline{\Phi}_{\mathcal{C}} \geq \Phi_{\mathcal{C}}$ for every class $\mathcal{C}$ and for every integer d.

Using these functions we can formulate notions of wide, almost wide and quasi-wide classes (which were defined in Section 3.3) as follows:

A class of graphs $\mathcal{C}$ is *wide* (resp. *almost wide*, resp. *quasi-wide*) if $\Phi_{\mathcal{C}}$ is identically 0 (resp. bounded, resp. finite) [25]:

$$\begin{aligned}
\mathcal{C} \text{ is wide} &\iff \forall d \in \mathbb{N}\colon \Phi_{\mathcal{C}}(d) = 0\\
\mathcal{C} \text{ is almost wide} &\iff \sup_{d\in\mathbb{N}} \Phi_{\mathcal{C}}(d) < \infty\\
\mathcal{C} \text{ is quasi-wide} &\iff \forall d \in \mathbb{N}\colon \Phi_{\mathcal{C}}(d) < \infty
\end{aligned}$$

Notice that a hereditary class $\mathcal{C}$ is wide (resp. almost wide, resp. quasi-wide) if and only if $\mathcal{C} \mathbin{\triangledown} 0$ is wide (resp. almost wide, resp. quasi-wide) as deleting edges cannot make it more difficult to find independent sets.

We introduce the following variation of the above definitions: A class of graphs $\mathcal{C}$ is *uniformly wide* (resp. *uniformly almost wide*, resp. *uniformly quasi-wide*) if $\overline{\Phi}_{\mathcal{C}}$ is identically 0 (resp. bounded, resp. finite):

$$\begin{aligned} \mathcal{C} \text{ is uniformly wide} \quad &\iff \quad \forall d \in \mathbb{N}\colon \overline{\Phi}_{\mathcal{C}}(d) = 0 \\ \mathcal{C} \text{ is uniformly almost wide} \quad &\iff \quad \sup_{d\in\mathbb{N}} \overline{\Phi}_{\mathcal{C}}(d) < \infty \\ \mathcal{C} \text{ is uniformly quasi-wide} \quad &\iff \quad \forall d \in \mathbb{N}\colon \overline{\Phi}_{\mathcal{C}}(d) < \infty \end{aligned}$$

Notice that a class $\mathcal{C}$ is uniformly wide (resp. uniformly almost wide, resp. uniformly quasi-wide) if and only if $\mathcal{C} \mathbin{\triangledown} 0$ is uniformly wide (resp. uniformly almost wide, resp. uniformly quasi-wide) as the property is hereditary in nature and deleting edges cannot make it more difficult to find independent sets.

Theorem 6.6 [83]. *Let $\mathcal{C}$ be a hereditary class of graphs. Then the following are equivalent:*

- $\Delta(\mathcal{C}) < \infty$,
- *$\mathcal{C}$ is wide,*
- *$\mathcal{C}$ is uniformly wide.*

Before characterizing almost wide classes we state a quantitative result relating these classes to bounded expansion.

Theorem 6.7 [83]. *Let $\mathcal{C}$ be a class with bounded expansion. Then $\overline{\Phi}_d(\mathcal{C}) \leq \nabla_{\lfloor d/2 \rfloor - 1}(\mathcal{C})$.*

As a consequence we have the following characterization of hereditary almost wide classes of graphs, which gives a positive answer to a question of Dawar whether classes of graphs more general than those excluding a minor could be proved to be almost wide [25].

Theorem 6.8 [83]. *Let $\mathcal{C}$ be a hereditary class of graphs. Then the following are equivalent:*

- *$\mathcal{C}$ is almost wide;*
- *$\mathcal{C}$ is uniformly almost wide;*
- *There are $s \in \mathbb{N}$ and $t\colon \mathbb{N} \to \mathbb{N}$ such that $K_{s,t(r)} \notin \mathcal{C} \mathbin{\triangledown} r$ (for all $r \in \mathbb{N}$).*

If $\mathcal{C}$ is actually a minor closed class the we can be more precise:

Theorem 6.9 [83]. *Let $\mathcal{C}$ be a minor closed class of graphs and let s be an integer. Then the following are equivalent:*

- *$\mathcal{C}$ is almost wide and $\Phi_{\mathcal{C}}(d) < s$ for every integer $d \geq 2$;*
- *$\mathcal{C}$ is uniformly almost wide and $\overline{\Phi}_{\mathcal{C}}(d) < s$ for every integer $d \geq 2$;*
- *$\mathcal{C}$ excludes some graph $K_{s,t}$.*

For instance, consider a surface Σ and let $\mathcal{C}_\Sigma$ be the class of the graphs which embed on Σ. It has been proved in [10] that $\mathcal{C}_\Sigma$ is almost wide (for every surface Σ) and that $\Phi_{\mathcal{C}_\Sigma}(d)$ is at most equal to the order of the smallest clique which does not embed on Σ. Actually, $\Phi_{\mathcal{C}_\Sigma}(d) = \overline{\Phi}_{\mathcal{C}_\Sigma}(d) = 2$ for every integer d, as every $K_{2,n}$ embed on any surface but not every $K_{3,n}$ does.

Finally, we have the following characterization of quasi-wide classes:

Theorem 6.10 [83]. *Let $\mathcal{C}$ be a hereditary class of graphs. The following conditions are equivalent:*

- *$\mathcal{C}$ is quasi-wide;*
- *$\mathcal{C}$ is uniformly quasi-wide;*
- *for every integer d there is an integer N such that $K_N \notin \mathcal{C} \triangledown d$;*
- *$\mathcal{C}$ is a class of nowhere dense graphs.*

This then implies (using the above mentioned result of Dawar and Molod [26]) that the relativized homomorphism preservation theorem holds for all classes of nowhere dense graphs. Perhaps these result indicate that classes with bounded expansion and classes of nowhere dense graphs provide a proper setting for this type of questions (about wide, semi-wide and quasi-wide classes) and we obtain characterization theorems (which are reviewed in the last chapter).

6.3. Richness of first order

A class $\mathcal{K}$ is said to be *first order definable* if there exists a first order formula Φ such that $\mathcal{K}$ is the class of all structures which are models of Φ. This can be obviously relativized: A subclass $\mathcal{L}$ of $\mathcal{K}$ is said to be *first order definable in the class* $\mathcal{K}$ if $\mathcal{L}$ is just the class of all structures in $\mathcal{K}$ which model Φ. However, if a class $\mathcal{L}$ is defined by an existentially positive first order formula then $\mathcal{L}$ is defined as the class of all structures in $\mathcal{K}$ for which there exist a homomorphism from a finite set $\mathcal{F}$ of structures. This in turn means that the complementary class $\mathcal{L}'$ of $\mathcal{L}$ is the class of all structures $\mathbf{A}$ for which there is no homomorphism $\mathbf{F} \longrightarrow \mathbf{A}$ for any $\mathbf{F} \in \mathcal{F}$. In other words the complementary class is the class $\mathrm{Forb_h}(\mathcal{F})$. This setting is close to (homomorphism) dualities and to homomorphism preservation theorems.

Combining the above Theorems 6.7, 6.5 we obtain the following:

Theorem 6.11. *Let $\mathcal{K}$ be a bounded expansion class of structures. For a homomorphism closed subclass $\mathcal{L}$ of $\mathcal{K}$ are the following statements equivalent:*

- *$\mathcal{L}$ is first order definable in $\mathcal{K}$;*
- *$\mathcal{L}' = \mathrm{Forb_h}(\mathcal{F})$ for a finite set $\mathcal{F}$ of structures;*
- *$\mathcal{L}$ is defined by a (finite) $\mathcal{K}$-restricted duality.*

Combining with the results of [86] we prove perhaps surprising fact that any homomorphism closed first order property when restricted to a class with bounded expansion is a restricted finite duality. It follows that any First Order definable subset of a class with bounded expansion is a restriction of a constrained satisfaction problem. This should be compared with the following characterization of constrained satisfaction classes by means of (unrestricted) dualities:

Theorem 6.12. *For a subclass* CSP (H) *of graphs (structures) the following statements are equivalent:*

- CSP (H) *is first order definable;*
- CSP $(H) = \mathrm{Forb_h}(\mathcal{F})$ *for a finite set $\mathcal{F}$ of trees;*
- CSP (H) *is defined by a (finite) duality.*

This is a combination of [93] and [7] (and also [100]).

7. Summary (Characterization Theorems)

7.1. Polynomial dependence

Two graph invariants f and g are *polynomially dependent*, and we denote this by $f \asymp g$, if there exists polynomials P, Q such that for every graph G:

$$f(G) \le P\big(g(G)\big) \qquad \text{and} \qquad g(G) \le Q\big(f(G)\big).$$

Notice that $f \asymp g$ is equivalent to $\log f = \Theta(\log g)$.

For instance, according to Section 1.3, Section 2.3 and Theorem 4.1:

$$\widetilde{\nabla}_r \asymp \nabla_r \asymp \mathrm{wcol}_{2r+1} \asymp \mathrm{col}_{2r+1}$$

Also, we may extend this property to functions of more than one graph and express concisely the main result of Section 1.4:

$$\nabla_r(G \bullet H) \asymp \nabla_r(G)|H|.$$

(This is a direct consequence of Lemma 2.4 and Lemma 2.5.)

We also consider a weaker form of dependence for invariant sequences: $(f_i)_{i\in\mathbb{N}}$ and $(g_i)_{i\in\mathbb{N}}$ which are said to be *weakly polynomially dependent*, and we denote this by $(f_i)_{i\in\mathbb{N}} \overset{\star}{\asymp} (g_i)_{i\in\mathbb{N}}$ if there exists $\alpha, \beta \colon \mathbb{N} \to \mathbb{N}$ and polynomials $(P_i)_{i\in\mathbb{N}}, (Q_i)_{i\in\mathbb{N}}$ such that for every integer i and every graph G:

$$f_i(G) \le P_i\big(g_{\alpha(i)}(G)\big) \qquad \text{and} \qquad g_i(G) \le Q_i\big(f_{\beta(i)}(G)\big).$$

In this notation we have for instance:

$$(\chi_i)_{i\in\mathbb{N}} \overset{\star}{\asymp} (\nabla_i)_{i\in\mathbb{N}}$$

and

$$(\omega_i)_{i\in\mathbb{N}} \overset{\star}{\asymp} (\widetilde{\omega}_i)_{i\in\mathbb{N}},$$

where $\omega_i(G) = \omega(G \,\triangledown\, i)$ and $\widetilde{\omega}_i(G) = \omega(G \,\widetilde{\triangledown}\, i)$.

7.2. Characterizations

In this section, we state some characterizations of sparse classes, which are mainly consequences of two aspects:

- the polynomial dependence (and weak polynomial dependence) of certain graph invariants, like ∇_r, $\widetilde{\nabla}_r$, χ_p, col_p, wcol_p, etc.
- the characterization of uniformly quasi-wide classes.

These three characterization theorems perhaps present a fitting conclusion for this survey.

7.2.1. Classes of nowhere dense graphs.

Theorem 7.1 [87, 83]. *Let $\mathcal{C}$ be an unbounded size infinite class of graphs. Then the following conditions are equivalent:*

(1) *$\mathcal{C}$ is a class of nowhere dense graphs,*

(2) *for every integer r, $\mathcal{C} \triangledown r$ is not the class of all finite graphs,*

(3) *for every integer r, $\mathcal{C} \,\widetilde{\triangledown}\, r$ is not the class of all finite graphs,*

(4) *$\mathcal{C}$ is a uniformly quasi-wide class,*

(5) *$H(\mathcal{C})$ is a quasi-wide class,*

(6) $$\lim_{r\to\infty} \limsup_{G\in\mathcal{C}\triangledown r} \frac{\log \|G\|}{\log |G|} = 1,$$

(7) $$\lim_{r\to\infty} \limsup_{G\in\mathcal{C}\,\widetilde{\triangledown}\, r} \frac{\log \|G\|}{\log |G|} = 1,$$

(8) $$\lim_{r\to\infty} \limsup_{G\in\mathcal{C}} \frac{\log \nabla_r(G)}{\log |G|} = 0,$$

(9) $$\lim_{r\to\infty} \limsup_{G\in\mathcal{C}} \frac{\log \widetilde{\nabla}_r(G)}{\log |G|} = 0,$$

(10) $$\lim_{p\to\infty} \limsup_{G\in\mathcal{C}} \frac{\log \chi_p(G)}{\log |G|} = 0,$$

(11) $$\lim_{p\to\infty} \limsup_{G\in\mathcal{C}} \frac{\log \mathrm{col}_p(G)}{\log |G|} = 0,$$

(12) $\lim_{p\to\infty} \limsup_{G\in\mathcal{C}} \frac{\log \mathrm{wcol}_p(G)}{\log |G|} = 0$,

(13) *for every integer* c*, the class* $\mathcal{C} \bullet K_c = \{G \bullet K_c \colon G \in \mathcal{C}\}$ *is a class of nowhere dense graphs,*

(14) *for every integer* p*, every graph* $G \in \mathcal{C}$ *has a* p*-centered colorings using at most* $|G|^{o(1)}$ *colors,*

(15) *for every polynomial* P*, the class* $\mathcal{C}'$ *of the* 1*-transitive fraternal augmentations of directed graphs* $\vec{G}$ *with* $\Delta^-(\vec{G}) \le P\big(\nabla_0(G)\big)$ *and* $G \in \mathcal{C}$ *form a class of nowhere dense graphs,*

7.2.2. Bounded expansion classes.

Theorem 7.2 [84, 116]. *Let* $\mathcal{C}$ *be a class of graphs. Then the following conditions are equivalent:*

(1) $\mathcal{C}$ *has bounded expansion,*

(2) *for every integer* r*,* $\sup_{G\in\mathcal{C}} \nabla_r(G) < \infty$,

(3) *for every integer* r*,* $\sup_{G\in\mathcal{C}} \widetilde{\nabla}_r(G) < \infty$,

(4) *for every integer* p*,* $\sup_{G\in\mathcal{C}} \chi_p(G) < \infty$,

(5) *for every integer* p*,* $\sup_{G\in\mathcal{C}} \mathrm{col}_p(G) < \infty$,

(6) *for every integer* p*,* $\sup_{G\in\mathcal{C}} \mathrm{wcol}_p(G) < \infty$,

(7) *for every integer* c*, the class* $\mathcal{C} \bullet K_c = \{G \bullet K_c \colon G \in \mathcal{C}\}$ *has bounded expansion,*

(8) $\mathcal{C}$ *has low tree-width colorings,*

(9) $\mathcal{C}$ *has low tree-depth colorings,*

(10) *for every integer* p*, there exists an integer* $X(p)$ *such that every graph* $G \in \mathcal{C}$ *has a* p*-centered colorings using at most* $X(p)$ *colors,*

(11) *for every integer* k*, the class* $\mathcal{C}'$ *of the* 1*-transitive fraternal augmentations of directed graphs* $\vec{G}$ *with* $\Delta^-(\vec{G}) \le k$ *and* $G \in \mathcal{C}$ *form a class with bounded expansion,*

(12) *the class* $\mathcal{C}$ *is a degenerate class of graphs (that is:* $\nabla_0(G)$ *is bounded on* $\mathcal{C}$*) and there exists a function* F *such that every orientation* $\vec{G}$ *of a*

graph $G \in \mathcal{C}$ has a transitive fraternal augmentation $\vec{G} = \vec{G}_1 \subseteq \vec{G}_2 \subseteq \cdots \subseteq \vec{G}_i \subseteq \cdots$ where $\Delta^-(\vec{G}_i) \leq Q\big(\Delta^-(\vec{G}), i\big)$,

(13) *there exists a function f such that every graph $G \in \mathcal{C}$ has a transitive fraternal augmentation $\vec{G} = \vec{G}_1 \subseteq \vec{G}_2 \subseteq \cdots \subseteq \vec{G}_i \subseteq \cdots$ where $\Delta^-(\vec{G}_i) \leq f(i)$.*

7.2.3. Bounded tree-depth classes.

Theorem 7.3. *Let $\mathcal{C}$ be a class of graphs. The following conditions are equivalent:*

(1) *$\mathcal{C}$ has bounded tree-depth,*

(2) *there exists an integer $l(\mathcal{C})$ such that no graph $G \in \mathcal{C}$ includes a path of length greater than $l(\mathcal{C})$,*

(3) *$\mathcal{C}$ is degenerate (i.e. $\nabla_0(\mathcal{C}) < \infty$) and there exists an integer $L(\mathcal{C})$ such that no graph $G \in \mathcal{C}$ includes an induced path of length greater than $L(\mathcal{C})$,*

(4) $\lim_{p\to\infty} \chi_p(\mathcal{C}) < \infty$,

(5) $\lim_{p\to\infty} \mathrm{col}_p(\mathcal{C}) < \infty$,

(6) $\lim_{p\to\infty} \mathrm{wcol}_p(\mathcal{C}) < \infty$.

References

[1] M. Ajtai and Y. Gurevich, Monotone versus positive, *Journal of the ACM,* **34** (1987), 1004–1015.

[2] N. Alechina and Y. Gurevich, *Syntax vs. semantics on infinite structures,* Structures in Logic and Computer Science (1997), 14–33.

[3] N. Alon, P. D. Seymour and R. Thomas, A separator theorem for graphs with excluded minor and its applications, *Proceedings of the 22nd Annual ACM Symposium on Theory of Computing* (1990), pp. 293–299.

[4] N. Alon, P. D. Seymour and R. Thomas, A separator theorem for nonplanar graphs, *Journal of the American Mathematical Society,* **3** (1990), 801–808.

[5] N. Alon and A. Shapira, A characterization of the (natural) graph properties testable with one-sided error, *Proc. 46th IEEE FOCS* (2005), pp. 429–438.

[6] S. Arnborg and A. Proskurowski, Linear time algorithms for NP-hard problems restricted to partial k-trees, *Discrete Applied Mathematics,* **23** (1989), no. 1, 11–24.

[7] A. Atserias, On digraph coloring problems and treewidth duality, *20th IEEE Symposium on Logic in Computer Science (LICS)* (2005), pp. 106–115.

[8] A. Atserias, A. Dawar and M. Grohe, *Preservation under extensions on well-behaved finite structures,* 32nd International Colloquium on Automata, Languages and Programming (ICALP) (Springer-Verlag, ed.), Lecture Notes in Computer Science, vol. 3580, 2005, pp. 1437–1449.

[9] A. Atserias, A. Dawar and P. G. Kolaitis, *On preservation under homomorphisms and unions of conjunctive queries,* Proceedings of the twenty-third ACM SIGMOD-SIGACT-SIGART symposium on Principles of database systems, ACM Press, 2004, pp. 319–329.

[10] A. Atserias, A. Dawar and P. G. Kolaitis, On preservation under homomorphisms and unions of conjunctive queries, *Journal of the ACM,* **53** (2006), 208–237.

[11] H. L. Bodlaender, A linear-time algorithm for finding tree-decompositions of small treewidth, *SIAM Journal of Computing,* **25** (1996), no. 6, 1305–1317.

[12] H. L. Bodlaender, J. R. Gilbert, H. Hafsteinsson and T. Kloks, Approximating tree-width, pathwidth, frontsize and shortest elimination tree, *Journal of Algorithms,* **18** (1995), 238–255.

[13] C. Borgs, J. Chayes, L. Lovász, V. T. Sós and K. Vesztergombi, Counting graph homomorphisms, in: *Topics in Discrete Mathematics* (M. Klazar, J. Kratochvíl, M. Loebl, J. Matoušek, R. Thomas and P. Valtr, eds.), Algorithms and Combinatorics, vol. 26, Springer Verlag, 2006, (dedicated to Jarik Nešetřil on the Occasion of his 60th birthday), pp. 315–371.

[14] O. V. Borodin, On acyclic colorings of planar graphs, *Discrete Mathematics,* **25** (1979), no. 3, 211–236.

[15] A. Brandstädt, H.-O. Le and J.-M. Vanherpe, Structure and stability number of chair-, co-P- and gem-free graphs revisited, *Information Processing Letters,* **86** (2003), 161–167.

[16] K. Cameron, Induced matchings, *Discrete Applied Mathematics,* **24** (1989), 97–102.

[17] K. Cameron, Induced matchings in intersection graphs, *Discrete Mathematics,* **278** (2004), 1–9.

[18] K. Cameron, R. Sritharan and Y. Tang, Finding a maximum induced matching in weakly chordal graphs, *Discrete Mathematics,* **266** (2003), 133–142.

[19] J.-M. Chang, Induced matchings in asteroidal triple-free graphs, *Discrete Applied Mathematics,* **132** (2004), 67–78.

[20] M. Chrobak and D. Eppstein, Planar orientations with low out-degree and compaction of adjacency matrices, *Theoretical Computer Science,* **86** (1991), 243–266.

[21] F. R. K. Chung, R. L. Graham and R. M. Wilson, Quasi-random graphs, *Combinatorica,* **9** (1989), no. 4, 345–362.

[22] D. Coppersmith and S. Winograd, Matrix multiplication via arithmetic progressions, *J. Symbolic Comput.*, **9** (1990), 251–280.

[23] B. Courcelle, Graph rewriting: an algebraic and logic approach, in: *Handbook of Theoretical Computer Science* (J. van Leeuwen, ed.), vol. 2, Elsevier, Amsterdam (1990), pp. 142–193.

[24] B. Courcelle, The monadic second-order logic of graphs I: recognizable sets of finite graphs, *Information Computation,* **85** (1990), 12–75.

[25] A. Dawar, Finite model theory on tame classes of structures, in: *Mathematical Foundations of Computer Science 2007* (L. Kučera and A. Kučera, eds.), Lecture Notes in Computer Science, vol. 4708, Springer (2007), pp. 2–12.

[26] A. Dawar, *On preservation theorems in finite model theory,* Invited talk at the 6th Panhellenic Logic Symposium – Volos, Greece, July 2007.

[27] A. Dawar, M. Grohe and S. Kreutzer, *Locally excluding a minor,* Proc. 22nd IEEE Symp. on Logic in Computer Science (2007).

[28] J.S. Deogun, T. Kloks, D. Kratsch and H. Muller, On vertex ranking for permutation and other graphs, in: *Proceedings of the 11th Annual Symposium on Theoretical Aspects of Computer Science* (P. Enjalbert, E .W. Mayr and K. W. Wagner, eds.), Lecture Notes in Computer Science, vol. 775, Springer, 1994, pp. 747–758.

[29] M. DeVos, G. Ding, B. Oporowski, D. P. Sanders, B. Reed, P. D. Seymour and D. Vertigan, Excluding any graph as a minor allows a low tree-width 2-coloring, *Journal of Combinatorial Theory, Series B,* **91** (2004), 25–41.

[30] R. Diestel, *Graph theory,* Springer Verlag (1997).

[31] R. G. Downey and M. R. Fellows, Parameterized computational feasibility, in: P. Clote, J. Remmel (eds.): *Feasible Mathematics II,* Boston: Birkhäuser (1995), pp. 219–244.

[32] R. G. Downey and M. R. Fellows, *Parameterized complexity,* Springer (1999).

[33] P. Dreyer, Ch. Malon and J. Nešetřil, Universal h-colorable graphs without a given configuration, *Discrete Math.,* **250** (2002), 245–25.

[34] W. Duckworth, N. C. Wormald and M. Zito, Maximum induced matchings of random cubic graphs, *Journal of Computational and Applied Mathematics,* **142** (2002), no. 1, 39–50.

[35] Z. Dvořák, *Asymptotical structure of combinatorial objects,* Ph.D. thesis, Charles University, Faculty of Mathematics and Physics (2007).

[36] Z. Dvořák, On forbidden subdivision characterization of graph classes, *European Journal of Combinatorics* (2007), (in press).

[37] H.-D. Ebbinghaus and J. Flum, *Finite model theory,* Springer-Verlag (1996).

[38] G. Elek and B. Szegedy, *Limits of hypergraphs, removal and regularity lemmas, a non-standard approach,* preprint.

[39] J. Ellis, H. Fan and M. Fellows, The dominating set problem is fixed parameter tractable for graphs of bounded genus, *J. Algorithms,* **52** (2004), no. 2, 152–168.

[40] D. Eppstein, Subgraph isomorphism in planar graphs and related problems, in: *Proc. 6th Symp. Discrete Algorithms,* ACM and SIAM (January 1995), pp. 632–640.

[41] D. Eppstein, Subgraph isomorphism in planar graphs and related problems, *Journal of Graph Algorithms & Applications,* **3** (1999), no. 3, 1–27.

[42] D. Eppstein, Diameter and treewidth in minor-closed graph families, *Algorithmica,* **27** (2000), 275–291, Special issue on treewidth, graph minors and algorithms.

[43] P. Erdős, Problems and results in combinatorial analysis and graph theory, *Discrete Mathematics,* **72** (1988), no. 1-3, 81–92.

[44] U. Faigle, U. Kern, H. A. Kierstead and W. T. Trotter, On the game chromatic number of some classes of graphs, *Ars Combinatoria,* **35** (1993), 143–150.

[45] R. J. Faudree, A. Gyárfás, R. H. Schelp and Zs. Tuza, Induced matchings in bipartite graphs, *Discrete Mathematics,* **78** (1989), no. 1-2, 83–87.

[46] G. Fricke and R. Laskar, Strong matchings in trees, *Congressus Numerantium,* **89** (1992), 239–244.

[47] M. R. Garey and D. S. Johnson, *Computers and intractability; a guide to the theory of np-completeness,* W. H. Freeman & Co., New York, NY, USA (1990).

[48] J. R. Gilbert, J. P. Hutchinson and R. E. Tarjan, A separator theorem for graphs of bounded genus, *J. Algorithms* (1984), no. 5, 375–390.

[49] M. C. Golumbic and M. C. Laskar, Irredundancy in circular arc graphs, *Discrete Applied Mathematics,* **44** (1993), 79–89.

[50] M. C. Golumbic and M. Lewenstein, New results on induced matchings, *Discrete Applied Mathematics,* **101** (2000), no. 1–3, 157–165.

[51] B. Grünbaum, Acyclic colorings of planar graphs, *Israel Journal of Mathematics,* **14** (1973), 390–408.

[52] D. Guan and X. Zhu, The game chromatic number of outerplanar graphs, *Journal of Graph Theory,* **30** (1999), 67–70.

[53] B. Guenin, *Edge coloring plane regular multigraphs,* manuscript.

[54] Y. Gurevich, *Toward logic tailored for computational complexity,* Computation and Proof Theory (M. M. Richter et al., ed.), Lecture Notes in Mathematics, Springer-Verlag, 1984.

[55] R. Häggkvist and P. Hell, Universality of A-mote graphs, *Europ. J. Combinatorics,* **14** (1993), 23–27.

[56] R. Halin, S-functions for graphs, *Journal of Geometry,* **8** (1976), 171–176.

[57] P. Hell and J. Nešetřil, *Graphs and homomorphisms,* Oxford Lecture Series in Mathematics and its Applications, vol. 28, Oxford University Press (2004).

[58] W. Hodges, *Model theory,* Cambridge University Press, 1993.

[59] H. A. Kierstead, *personal communication cited in* [114].

[60] H. A. Kierstead and D. Yang, Orderings on graphs and game coloring number, *Order,* **20** (2003), 255–264.

[61] D. Kobler and U. Rotics, Finding maximum induced matchings in subclasses of claw-free and P_5-free graphs and in graphs with matching and induced matching of equal maximum size, *Algorithmica,* **37** (2003), 327–346.

[62] Y. Kohayakawa and V. Rödl, Szemerédi's regularity lemma and quasi-randomness, in: *Recent Advances in Algorithmic Combinatorics* (B. Reed and C. Linhares-Sales, eds.), CMS Books Math./Ouvrages Math. SMC, vol. 11, Springer, New-York (2003), pp. 289–347.

[63] J. Komlós and M. Simonovits, Szemerédi's regularity lemma and its applications in graph theory, in: *Combinatorics, Paul Erdős is Eighty,* vol. 2, János Bolyai Math. Soc. (1993), pp. 295–352.

[64] A. V. Kostochka and L. S. Melnikov, On bounds of the bisection width of cubic graphs, in: *Fourth Czechoslovakian Symposium on Combinatorics, Graphs and Complexity* (J. Nesetril and M. Fiedler, eds.), Elsevier (1992), pp. 151–154.

[65] M. Kreidler and D. Seese, Monadic NP and graph minors, in: *Computer Science Logic,* Lecture Notes in Computer Science, vol. 1584, Springer (1999), pp. 126–141.

[66] B. Larose, C. Loten and C. Tardif, A characterisation of first-order constraint satisfaction problems, *Logical Methods in Computer Science,* **3** (2007), no. 4, 6, 22pp, (electronic).

[67] L. Libkin, *Elements of finite model theory,* Springer-Verlag (2004).

[68] R. Lipton and R. E. Tarjan, A separator theorem for planar graphs, *SIAM Journal on Applied Mathematics,* **36** (1979), no. 2, 177–189.

[69] L. Lovász and B. Szegedy, Szemerédi lemma for the analyst, *Geom. Func. Anal.,* **17** (2007), 252–270.

[70] V. V. Lozin, On maximum induced matchings in bipartite graphs, *Information Processing Letters,* **81** (2002), 7–11.

[71] V. V. Lozin and D. Rautenbach, Some results on graphs without long induced paths, *Information Processing Letters,* **88** (2003), 167–171.

[72] G. L. Miller, S.-H. Teng, W. Thurston and S. A. Vavasis, Geometric separators for finite-element meshes, *SIAM J. on Scientific Computing,* **19** (1998), no. 2, 364–386.

[73] R. Naserasr, Homomorphisms and edge-coloring of planar graphs, *Journal of Combinatorial Theory, Series B* (2005), to appear.

[74] J. Nešetřil, Aspects of structural combinatorics, *Taiwanese J. Math.,* **3** (1999), no. 4, 381–424.

[75] J. Nešetřil and P. Ossona de Mendez, Colorings and homomorphisms of minor closed classes, in: *The Goodman-Pollack Festschrift* (B. Aronov, S. Basu, J. Pach and M. Sharir, eds.), Algorithms and Combinatorics, vol. 25, Discrete & Computational Geometry (2003), pp. 651–664.

[76] J. Nešetřil and P. Ossona de Mendez, How many colors may we require?, in: *Prague Midsummer Combinatorial Workshop IX* (M. Mareš, ed.), KAM-DIMATIA Series, no. 2004-686 (2004), abstract, pp. 27–30.

[77] J. Nešetřil and P. Ossona de Mendez, Cuts and bounds, *Discrete Mathematics, Structural Combinatorics – Combinatorial and Computational Aspects of Optimization, Topology and Algebra,* **302** (2005), no. 1–3, 211–224.

[78] J. Nešetřil and P. Ossona de Mendez, Folding, *Journal of Combinatorial Theory, Series B,* **96** (2006), no. 5, 730–739.

[79] J. Nešetřil and P. Ossona de Mendez, Linear time low tree-width partitions and algorithmic consequences, in: *STOC'06. Proceedings of the 38th Annual ACM Symposium on Theory of Computing,* ACM Press (2006), pp. 391–400.

[80] J. Nešetřil and P. Ossona de Mendez, Tree depth, subgraph coloring and homomorphism bounds, *European Journal of Combinatorics,* **27** (2006), no. 6, 1022–1041.

[81] J. Nešetřil and P. Ossona de Mendez, Fraternal augmentations, arrangeability and linearly bounded Ramsey numbers, *SIAM Journal on Discrete Mathematics* (2007), submitted.

[82] J. Nešetřil and P. Ossona de Mendez, *Induced matchings and induced paths in graphs,* Tech. Report 2007-810, KAM-DIMATIA Series, 2007.

[83] J. Nešetřil and P. Ossona de Mendez, First order properties on nowhere dense structures, *Journal of Symbolic Logic* (2008), submitted.

[84] J. Nešetřil and P. Ossona de Mendez, Grad and classes with bounded expansion I. decompositions, *European Journal of Combinatorics,* **29** (2008), no. 3, 760–776.

[85] J. Nešetřil and P. Ossona de Mendez, Grad and classes with bounded expansion II. algorithmic aspects, *European Journal of Combinatorics,* **29** (2008), no. 3, 777–791.

[86] J. Nešetřil and P. Ossona de Mendez, Grad and classes with bounded expansion III. restricted graph homomorphism dualities, *European Journal of Combinatorics,* **29** (2008), no. 4, 1012–1024.

[87] J. Nešetřil and P. Ossona de Mendez, On nowhere dense graphs, *European Journal of Combinatorics* (2008), submitted.

[88] J. Nešetřil and S. Poljak, Complexity of the subgraph problem, *Comment. Math. Univ. Carol.,* **26.2** (1985), 415–420.

[89] J. Nešetřil and A. Pultr, On classes of relations and graphs determined by subobjects and factorobjects, *Discrete Mathematics,* **22** (1978), 287–300.

[90] J. Nešetřil and R. Šámal, *Tension continuous maps-their structure and applications, European J. Comb.,* **29,4** (2008), 1025–1054.

[91] J. Nešetřil and S. Shelah, Order of countable graphs, *European J. Comb.,* **24** (2003), 649–663.

[92] J. Nešetřil and I. Švejdarová, Diameter of duals are linear, *SIAM J. Discrete Math.,* **21,2** (2007), 374–384.

[93] J. Nešetřil and C. Tardif, Duality theorems for finite structures (characterizing gaps and good characterizations), *Journal of Combinatorial Theory, Series B,* **80** (2000), 80–97.

[94] J. Nešetřil and C. Tardif, Short answers to exponentially long questions: Extremal aspects of homomorphism duality, *SIAM Journal of Discrete Mathematics,* **19** (2005), no. 4, 914–920.

[95] S. Plotkin, S. Rao and W. D. Smith, *Shallow excluded minors and improved graph decomposition,* 5th Symp. Discrete Algorithms, SIAM (1994), 462–470.

[96] N. Robertson and P. D. Seymour, Graph minors. I. Excluding a forest, *Journal of Combinatorial Theory, Series B,* **35** (1983), 39–61.

[97] N. Robertson and P. D. Seymour, Graph minors – a survey, in: *Surveys in Combinatorics* (I. Anderson, ed.), Cambridge University Press (1985), pp. 153–171.

[98] N. Robertson and P. D. Seymour, Graph Minors. II. Algorithmic aspects of tree-width, *Journal of Algorithms,* **7** (1986), 309–322.

[99] N. Robertson and P. D. Seymour, Graph Minors. VIII., *Journal of Combinatorial Theory, Series B,* **48** (1990), no. 2, 227–254.

[100] B. Rossman, Homomorphisms and first-order logic, *Journal of the ACM* (2007), submitted.

[101] P. Schaffer, Optimal node ranking of trees in linear time, *Information Processing Letters,* **33** (1989/90), 91–96.

[102] A. Steger and M.-L. Yu, On induced matchings, *Discrete Mathematics,* **120** (1993), no. 1–3, 291–295.

[103] L. J. Stockmeyer and V. V. Vazirani, NP-completeness of some generalizations of the maximum matching problem, *Information Processing Letters,* **15** (1982), no. 1, 14–19.

[104] A. Stolboushkin, Finite monotone properties, in: *Proc. 10th IEEE Symp. on Logic in Computer Science* (1995), pp. 324–330.

[105] L. Sunil Chandran, A high girth graph construction, *SIAM J. Discret. Math.,* **16** (2003), no. 3, 366–370.

[106] E. Szemerédi, Regular partitions of graphs, *Problémes combinatoires et théorie des graphes, CNRS,* 1976, pp. 399–401.

[107] W. Tait, A counterexample to a conjecture of Scott and Suppes, *Journal of Symbolic Logic,* **24** (1959), 15–16.

[108] T. Tao, The dichotomy between structure and randomness, arithmetic progression and the primes, in: *Proceedings of the International Congress of Mathematicians* (Madrid 2006) (European Math. Society, ed.), vol. 1 (2007), pp. 581–608.

[109] S.-H. Teng, Combinatorial aspects of geometric graphs, *Computational Geometry* (1998), no. 9, 277–287.

[110] C. Thomassen, A short list color proof of Grötzsch theorem, *Journal of Combinatorial Theory, Series B,* **88** (2003), 189–192.

[111] K. Wagner, Über eine Eigenschaft der Ebenen Komplexe, *Mathematische Annalen,* **114** (1937), 570–590.

[112] R. Yuster, Combinatorial and computational aspects of graph packing and graph decompositions, *Computer Science Review,* **1** (2007), no. 1, 12–26.

[113] X. Zhu, The game coloring number of planar graphs, *Journal of Combinatorial Theory, Series B,* **75** (1999), no. 2, 245–258.

[114] X. Zhu, The game coloring number of pseudo partial k-trees, *Discrete Mathematics,* **215** (2000), 245–262.

[115] X. Zhu, Refined activation strategy for the marking game, *Journal of Combinatorial Theory, Series B* (2007), in press.

[116] X. Zhu, Colouring graphs with bounded generalized colouring number, *Discrete Mathematics* (2008), in press.

[117] M. Zito, Maximum induced matchings in regular graphs and trees, in: *Proceedings of WG '99: the 25th International Workshop on Graph-Theoretic Concepts in Computer Science,* Lecture Notes in Computer Science, vol. 1665 (1999), pp. 89–100.

Jaroslav Nešetřil
Department of Applied Mathematics and Institute of Theoretical Computer Science (ITI)
Charles University
Malostranské nám.25
11800 Praha 1
Czech Republic

e-mail:
nesetril@kam.ms.mff.cuni.cz

Patrice Ossona de Mendez
Centre d'Analyse et de Mathématiques Sociales
CNRS, UMR 8557
54 Bd Raspail, 75006 Paris
France

e-mail: pom@ehess.fr

BOLYAI SOCIETY
MATHEMATICAL STUDIES, 19

Building Bridges
pp. 427–454.

Recent Progress in Matching Extension

MICHAEL D. PLUMMER

Dedicated to László Lovász on the occasion of his 60th birthday

Let G be a graph with at least $2n+2$ vertices, where n is a non-negative integer. The graph G is said to be *n-extendable* if every matching of size n in G extends to (i.e., is a subset of) a perfect matching. The study of this concept began in earnest in the 1980's, although it was born out of the study of canonical matching decompositions carried out in the 1970's and before. As is often the case, in retrospect it is apparent that there are roots of this topic to be found even earlier.

In the present paper, we will begin with a brief history of the subject and then concentrate on reviewing results on n-extendability and closely related areas obtained in the last ten-fifteen years, as there already exist two surveys of the subject in 1994 and 1996, respectively.

1. Introduction and a Brief History

Let G denote a finite simple undirected graph with vertex set $V(G)$ and edge set $E(G)$. A set $M \subseteq E(G)$ is a *matching* if no two edges share a vertex. A matching M is *perfect* if it covers all of $V(G)$. A perfect matching is also often called a $1-$*factor*. Historically speaking, perfect matchings were studied as early as the latter part of the nineteenth and early part of the twentieth century with such names as Frobenius [34, 35], Petersen [87] and König [51, 52], and later, in the 1960's and 1970's, Kotzig [53, 54, 55], Edmonds [27], Gallai [36, 37], and Lovász [68, 69] figuring prominently.

A graph G is *factor-critical* if $G-v$ has a perfect matching for every vertex $v \in V(G)$. A graph G is *bicritical* if $G-u-v$ has a perfect matching for every pair of distinct vertices $u, v \in V(G)$. Factor-critical and bicritical graphs play important roles in what we know today as a canonical decomposition theory for arbitrary graphs in terms of their matchings. We will explore this in brief below. The interested reader is referred to [76] for much more on this subject.

Strongly motivated by the importance of factor-critical and bicritical graphs mentioned above, the present author was led to make the following definition. Let n be a non-negative integer and let G be an even graph with at least $2n+2$ vertices. Then G is said to be *n-extendable* if every matching of size n in G extends to (i.e., is a subset of) a perfect matching in G. (We will take "0-extendable" to mean simply that G contains a perfect matching.) In particular, a 1-extendable graph denotes the same property as the older term *matching-covered.* 1-extendable graphs are the graphs which are natural candidates for the "brick-brace" decomposition which is the main subject of Section 6 of this survey.

In [113, 114] we already find a matching extension property of sorts. There the author defines a graph to have the *uniform matching property* (or UMXP) if every matching of i edges extends to a matching of j edges in the same number of ways.

The subject of n-extendability has been surveyed twice before [93, 95], the most recent of these two papers appearing in 1996. For this reason, in the present paper we will concentrate mostly on results which have been obtained within the past ten-fifteen years.

In Section 2 we will discuss recent results dealing with n-extendability in general. In Section 3 we will consider matching extension for graphs embedded in surfaces. In Section 4 we will review recent work in the area of restricted matchings. Here "restricted" means that not only do we want to extend partial matchings to perfect matchings, but in addition, we want to avoid other matchings. Restricted matchings will be discussed both in general and in the framework of embedded graphs.

Section 5 deals with generalizations and variations which have grown out of the original concept of n-extendability. In Section 6 we will review briefly the deep results on brick and brace decomposition obtained in the last ten years or so. This work then leads naturally to some powerful new results on Pfaffian graphs and their connection with the motivating problem of counting the number of perfect matchings in a graph. This will be surveyed in our final Section 7.

2. n-EXTENDABILITY IN GENERAL

Without doubt, the cornerstone of the theory of matchings in graphs is the classical theorem of Tutte [106]. Let G be a graph and $S \subseteq V(G)$. Denote by $c_o(G - S)$ the number of components of $G - S$ having odd order.

Theorem 2.1 [106]. *A graph G has a perfect matching if and only if for every set $S \subseteq V(G)$, $c_o(G - S) \leq |S|$.*

Yu [111] characterized n-extendable graphs in a similar manner.

Theorem 2.2 [111]. *Let $n \geq 1$. A graph G is n-extendable if and only if for all $S \subseteq V(G)$,*

(1) *$c_o(G - S) \leq |S|$ and*

(2) *$c_o(G - S) = |S| - 2k$, $(0 \leq k \leq n - 1)$ implies that $F(S) \leq k$, where $F(S)$ is the size of a maximum matching in $G[S]$.*

Let us first focus on several recent results involving n-extendability and vertex degree. Perhaps the earliest theorem of this kind is the following which gives a sufficient degree condition to guarantee n-extendability.

Theorem 2.3 [88]. *Let G be a graph on $2p$ vertices and let n be an integer such that $1 \leq n \leq p - 1$. If $\delta(G) \geq p + n$, then G is n-extendable.*

For the special case of bipartite graphs, Ananchuen and Caccetta proved that a much lower minimum degree suffices for n-extendability.

Theorem 2.4 [16]. *If G is a balanced bipartite graph on $2p$ vertices and $1 \leq n \leq p - 1$, then if $\delta(G) \geq (1/2)(p + n)$, G is n-extendable.*

On the other hand, the same authors discovered the following necessary degree condition for n-extendability which says essentially that for a certain range of the value of n, a certain interval of values for the minimum degree is forbidden.

Theorem 2.5 [16]. *If G is n-extendable on $2p$ vertices, where $1 \leq n \leq p-1$, then either $n + 1 \leq \delta(G) \leq p$ or $\delta(G) \geq 2n + 1$. (In other words, if $n > \big(|V(G)| - 2\big)/4$, then $\delta(G)$ must lie in the union of two disjoint intervals $[n + 1, |V(G)|/2] \cup [2n + 1, |V(G)| - 1]$.)*

Moreover, it is shown that for every value j in this union of intervals, there is an n-extendable graph.

A graph G is said to be *n-minimal* if it is n-extendable, but $G-e$ is not n-extendable, for every edge $e \in E(G)$. The same two authors have shown the following.

Theorem 2.6 [12, 16]. *Let G be an n-minimal graph on $2p$ vertices. Then:*

(a) *if $G \neq K_{2p}$, then $\delta(G) \leq p+n-1$, while*

(b) *if G is bipartite, $G \neq K_{p,p}$, and $n \leq p-3$, then $\delta(G) < (1/2)(p+n)$.*

Next we turn to connectivity issues. Let $\kappa(G)$ denote the vertex connectivity of graph G. An early theorem involving the vertex connectivity of n-extendable graphs was the following.

Theorem 2.7 [88]. *If $n \geq 1$ and G is an n-extendable graph on at least $2n+2$ vertices, then $\kappa(G) \geq n+1$.*

Lou and Yu [67] have shown that if n is large enough with respect to $|V(G)|$, the above bound on n can be significantly improved.

Theorem 2.8 [67]. *If G is an n-extendable graph on $2p$ vertices and $n \geq p/2$, then either G is bipartite or $\kappa(G) \geq 2n$.*

The same authors also obtain the following corollary.

Corollary 2.9 [67]. *If G is an n-extendable graph on $2p$ vertices and $n \geq p/2$, then G is Hamiltonian.*

They conjecture that the bound $p/2$ in the above corollary can be replaced by $(p-1)/3$, but this remains open.

Another connectivity parameter which has been studied in connection with n-extendability is cyclic connectivity. The *cyclic connectivity* of G, denoted by $c_\lambda(G)$, is the size of a smallest edge cutset the deletion of which leaves two components each containing at least one cycle. (If no such cutset exists, we define c_λ to be $+\infty$.) In [43] it was proved that regular bipartite graphs with cyclic connectivity sufficiently large are n-extendable. In contrast to this result, it was shown in [65] that for every $k \geq 2$ and $m \geq 1$, there exists a $(k+1)$-regular $(k+1)$-connected graph G having $c_\lambda(G) \geq m$, which is not n-extendable. So large cyclic connectivity by itself is not enough to guarantee n-extendability, even if the graph is regular. So what other hypotheses taken together with cyclic connectivity might be enough to guarantee n-extendability in regular graphs? One answer is provided by the next theorem.

Theorem 2.10 [1]. *Suppose $k \geq 3$ and $\lceil (k+1)/2 \rceil \leq n \leq k-1$. Then every cyclically $(k-1)(2n-1)$-edge-connected transitive k-regular even graph is n-extendable.*

Degree sum conditions have been much studied in various areas of graph theory such as the theory of Hamiltonian cycles. Xu and Yu [110] have sharpened a sufficient degree sum condition for n-extendability from [91] by excluding a certain family of graphs. On the other hand, Kawarabayashi, Ota and Saito have obtained a result which says that an n-extendable graph under certain conditions on its degree sums, will either be Hamiltonian or be one of two small exceptional graphs. Let $\sigma_2(G)$ denote the minimum taken over all non-adjacent pairs of vertices u and v in G of the quantity $d(u) + d(v)$.

Theorem 2.11 [47]. *Let n be a non-negative integer and let G be a connected n-extendable graph. If $\sigma_2(G) \geq p - n - 1$, then either*

(1) *G is Hamiltonian, or*

(2) *$n = 1$ and $2K_1 + 3K_2 \subset G \subset K_2 + 3K_2$.*

Independence number too has been studied in relation to n-extendability. In the first result below, Maschlanka and Volkmann obtain an upper bound on the independence number $\alpha(G)$ of any n-extendable graph.

Theorem 2.12 [77]. *Suppose $n \geq 0$ and let G be an n-extendable non-bipartite graph on $2p$ vertices. Then $\alpha(G) \leq p - n$. Moreover, this bound is sharp for all n and p.*

On the other hand, in the next result it is shown that for a certain range of possible values for n, the independence number can be used to give a necessary *and* sufficient condition for the graph to be n-extendable.

Theorem 2.13 [14]. *Let G be a graph of $2p$ vertices and n any positive integer such that $p/2 \leq n \leq p - 2$ and $p - n$ is even. Suppose further that $\delta(G) \geq p + n - 1$. Then G is n-extendable if and only if $\alpha(G) \leq p - n$.*

Lou has investigated local independence versus n-extendability and has proved the following theorems. Here $N_2(v)$ denotes the set of vertices at distance two from vertex v.

Theorem 2.14 [62, 63]. *Let G be a connected graph of even order. If for each vertex $v \in V(G)$ and for each independent set $S \subseteq N_2(v)$, $|N(v) \cap N(S)| \geq |S| + 2n$, then G is n-extendable.*

Theorem 2.15 [64]. *Let G be a triangle-free graph of finite girth and even order. Then, if for each vertex $v \in V(G)$, $\alpha\big(G\big[N_2(v)\big]\big) \leq d(v) - 1$, G is regular and $\lceil d(v)/4 \rceil$-extendable.*

Let us now turn to some algorithmic complexity questions. For a fixed value of n, it is, of course, easy to see that testing whether or not a given graph is n-extendable can be done in time polynomial in the size of the input graph. But in 1989, Frank, Györi and Sebő [33] posed a more challenging question: Suppose n_0 denotes the maximum value of n for which a given graph G is n-extendable. Is there a polynomial algorithm to determine n_0?

In 1998 Lakhal and Litzler [56] found the first polynomial algorithm to determine n_0 in the case when G is bipartite. Their algorithm runs in $O\big(m \times \min\{n_0^3 + p, n_0 p\}\big)$ time, where $\big|E(G)\big|$ and $p = \big|V(G)\big|$. Zhang and Zhang [116] subsequently gave a faster algorithm which is $O(mn)$. However, the complexity of determining n_0 for non-bipartite graphs remains unsettled.

In [12] it was shown that a graph G on $2p$ vertices is $(p-1)$-extendable if and only if it is K_{2p} or $K_{p,p}$. The class of $(p-2)$-extendable graphs turns out to have a much more complicated structure, but nevertheless these graphs too have been characterized. (Cf. [13, 15].) However, for $n < p - 2$, no similar characterization is known.

Two properties closely related to that of n-extendability are random matchability and equimatchability. Sumner [102] defined a graph G to be *randomly matchable* if every matching in G extends to a perfect matching. He then showed that such graphs are very special indeed; namely, they must be either a K_{2n} or a $K_{n,n}$.

Generalizing this property, Lesk, Plummer and Pulleyblank [58] defined a graph G to be *equimatchable* if every matching in G extends to a maximum matching. Although the structure of such graphs is not so apparent, the three authors showed that the recognition problem is polynomial. (See also [29].) In [50] all equimatchable cubic and all equimatchable 3-connected planar (i.e., polytopal) graphs were determined, there being exactly two of the former type and twenty-three of the latter type. A further result showing that equimatchable graphs of any fixed genus are bounded in size appears in Section 3 below.

3. Matching Extension in Embedded Graphs

Matching extension for graphs embedded in surfaces has various historical roots. As early as 1973, Lovász and the author [73, 74, 75] discovered that a family of planar graphs used by Halin in studying minimally 3-connected graphs also provided an infinite family of examples of *minimal* bicritical graphs. Let T be a tree in which every non-leaf has degree at least three. Embed T in the plane and pass a cycle C through all leaves of T in such a way that the resulting graph $T \cup C$ is plane. (Not surprisingly, such graphs have come to be called *Halin* graphs.) A bicritical graph G is said to be *minimal* if $G - e$ is not bicritical, for every $e \in E(G)$. It turns out that every even Halin graph H is minimal bicritical and, moreover, with a single exception, that $\Phi(H) \geq (2/3)\big(|V(H)| - 1\big)$ and this bound is sharp. (Here and henceforth, $\Phi(G)$ denotes the number of perfect matchings in G.)

The subject of extending matchings in embedded graphs seems to have then remained dormant for some time, until the following result appeared in 1989.

Theorem 3.1 [90]. *No planar graph is 3-extendable.*

There soon followed a series of papers dealing with 1- and 2-extendability in the plane. (Cf. [42, 41, 92, 94, 61].) For example, we have the next result.

Theorem 3.2 [92, 61]. *Every 5-connected even planar graph is 2-extendable.*

This result was only recently generalized by Lou and Yu as follows.

Theorem 3.3 [66]. *Any graph of even order having a 5-connected spanning planar subgraph is 2-extendable.*

Somewhat earlier, the author [89] obtained an upper bound for the extendability of a graph in terms of its (orientable) genus. Let $\mu(\Sigma)$ denote the smallest integer such that no graph embeddable in the surface Σ is $\mu(\Sigma)$-extendable. (For example, $\mu = 3$ for the sphere by Theorem 3.1 above.) Dean [18] extended this work as follows.

Theorem 3.4 [18]. *If Σ is any surface (orientable or non-orientable) other than the sphere, then $\mu(\Sigma) = 2 + \left\lfloor \sqrt{4 - 2\chi} \right\rfloor$, where χ is the Euler characteristic of the surface.*

Thus in particular, $\mu(\Sigma) = 3, 4$ and 4 for the projective plane, the torus and the Klein bottle respectively.

Although there are clearly graphs of arbitrarily high extendability (K_{2n} and $K_{n,n}$, for example), the bound given by the above theorem is, in some sense, rather misleading. For example, see part (1) of the following result obtained quite recently by Aldred, Kawarabayashi and the author [4]. If G is embedded in a surface Σ, the *representativity* (or *face-width*) of the embedding is the minimum number of times any non-contractible curve in the surface intersects the graph.

Theorem 3.5 [4].

(1) *If G is embedded in a surface of Euler characteristic χ and the number of vertices of G is large enough, then G is not 4-extendable;*

(2) *given $g > 0$ there are infinitely many graphs of orientable genus g which are 3-extendable, and given $\overline{g} \geq 2$ there are infinitely many graphs of non-orientable genus $\overline{g}$ which are 3-extendable, and*

(3) *if G is a 5-connected even triangulation with genus $g > 0$ and sufficiently large representativity, then G is 2-extendable.*

In some sense, it seems more likely that a set of independent edges in a graph will extend to a perfect matching if they are mutually far enough apart. Aldred and the author obtained the following theorem in this direction for graphs in the plane which should be contrasted with Theorem 3.1. A matching M in graph G is said to be *induced* if no two edges in M are joined by a third edge of G.

Theorem 3.6 [7]. *Let G be a 5-connected even planar triangulation and let M be an induced matching of size three. Then M extends to a perfect matching.*

Aldred and Jackson have studied similar edge proximity conditions for extendability in the family of cubic bipartite graphs.

Theorem 3.7 [3]. *Let G be a cubic bipartite graph containing a perfect matching, M, a set of m edges in G and suppose $c_\lambda(G) \geq 3m - 2$. Then M extends to a perfect matching whenever the edges of M are pairwise at* *distance at least $f(m)$ apart, where*

$$f(m) = \begin{cases} 1, & m = 2 \\ 3, & 3 \le m \le 4 \\ 4, & 5 \le m \le 8 \\ 5, & m \ge 9. \end{cases}$$

Equimatchable graphs embedded in surfaces have also been investigated. In [50] the precisely twenty-three 3-connected planar equimatchable graphs were determined. Very recently, an extension [49] of this result shows that equimatchable graphs on any fixed surface must have bounded order.

Theorem 3.8. *Let G be a 3-connected equimatchable graph of genus g (respectively, non-orientable genus $\overline{g}$). Then if G is non-bipartite or if G is bipartite and the representativity of the embedding is at least three, $\left|V(G)\right| \le \max\left\{ f_1(g), f_2(g), f_3(\overline{g}), f_4(\overline{g})\right\}$, where*

$$f_1(g) = \left(\frac{7+\sqrt{1+48g}}{2}\right)\binom{8}{3}(4g+3)+9,$$

$$f_2(g) = 4\left(1+\sqrt{g}\right)\binom{8}{3}(4g+3)+9,$$

$$f_3(\overline{g}) = \left(\frac{7+\sqrt{1+24\overline{g}}}{2}\right)\binom{8}{3}(2\overline{g}+3)+9,$$

and

$$f_4(\overline{g}) = \left(4+2\sqrt{2\overline{g}}\right)\binom{8}{3}(2\overline{g}+3)+9.$$

We will present some additional results for embedded graphs in the next section on so-called "restricted" matching.

4. Restricted Matching Extension: the $E(m, n)$ Properties

Motivated by previous results on paths and cycles restricted to include, or not include, sets of vertices and edges of a certain size, Porteous and Aldred [97] extended the concept of n-extendability by means of the following definition. Let m and n be non-negative integers and let G be a graph on at least $2(m+n+1)$ vertices which contains a perfect matching. Then G is said to have the property $E(m,n)$ (or simply, "G is $E(m,n)$") if, for every pair of disjoint matchings M and N with $|M| = m$ and $|N| = n$, there is a perfect matching F in G such that $M \subseteq F$, but $F \cap N = \emptyset$.

Clearly, this concept generalizes that of n-extendability since a graph is $E(m,0)$ if and only if it is m-extendable. There are a number of implications involving $E(m,n)$ for various values of m and n. We list only a few; for others see [97, 96, 81].

Theorem 4.1.

(1) *For all $m, n \geq 0$, if G is $E(m,n)$, it is $E(m,0)$.*

(2) *If $m \geq 1, n \geq 0$ and G is $E(m,n)$, then it is $(m+1)$-connected. Moreover, this bound is sharp.*

(3) *For all $m \geq 1, n \geq 0$, if G is $E(m,n)$, then it is $E(m-1,n+1)$.*

More surprising, perhaps, is the following *non*-implication result.

Theorem 4.2 [97]. *For $m \geq 1$ and $n \geq 6m+1$, there exists a graph G having property $E(m,n)$, but not property $E(m,n-1)$.*

However, such "bad" graphs *must* necessarily have m bounded by a function of n as in the next result.

Theorem 4.3 [97]. *If $n \leq 2m+2$ and graph G is $E(m,n)$, then G is $E(m,n-1)$.*

In addition, these "bad" graphs are of bounded size as is seen by the next theorem.

Theorem 4.4 [97]. *Suppose $m \geq 1$. If graph G has property $E(m,n)$, but not $E(m,n-1)$, then $\left|V(G)\right| \leq 4(n-1)-2m$.*

The one missing case ($m = 0$) in the preceding theorem was settled in [81].

Theorem 4.5 [81]. *Let $n > n' \geq 0$. Then:*

(a) *if $n' < 3$, $E(0,n)$ implies $E(0,n')$, while*

(b) *if $n' \geq 3$, there exists a graph which is $E(0,n)$, but not $E(0,n')$.*

The following diagram illustrates a portion of the resulting lattice of implications (and non-implications) involving $E(m,n)$.

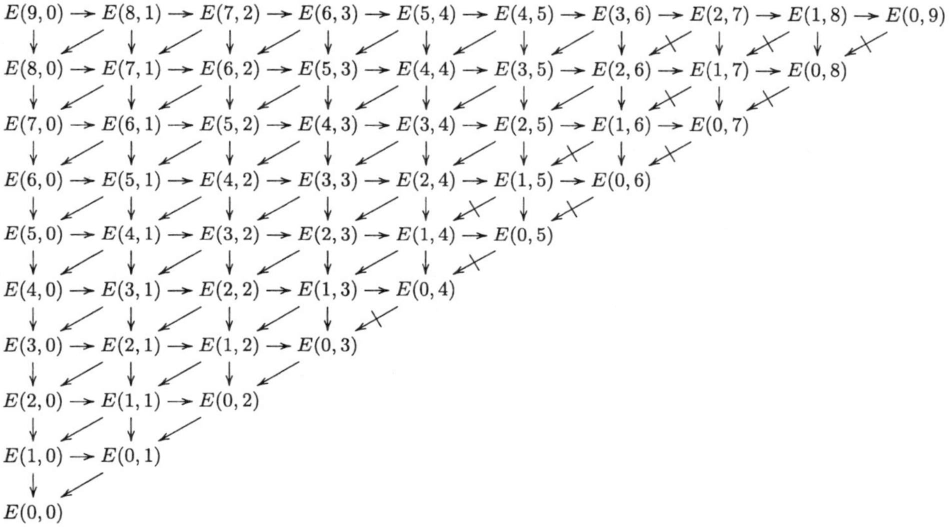

Fig. 4.1. The lattice of implications

Two recent results in the direction of sufficient conditions guaranteeing $E(m,n)$ are the following. The first deals with bipartite graphs and is an extension of an earlier result (cf. [2]) of the same type.

Theorem 4.6 [5]. *Let m, n and r be non-negative integers with $r \geq \max\{2n+1, m+2\}$. Let G be an r-regular bipartite graph with $|V(G)| \geq 2m+2n+2$ and*

$$c_\lambda(G) \geq \begin{cases} 0, & \text{when } m = 0; \\ (m-1)r + 2n + 1, & \text{for all } m \geq 1. \end{cases}$$

then G is $E(m,n)$.

The second deals with so-called "star-free" graphs, a generalization of the well-known family of claw-free graphs. As usual, let $K_{1,n}$ denote the complete bipartite graph with one black vertex and n white vertices. (Graph

$K_{1,n}$ is often called a *star* and $K_{1,3}$, a *claw*.) A graph is said to be $K_{1,n}$*-free* if it does not contain $K_{1,n}$ as an induced subgraph.

A classical result of Sumner and Las Vergnas [100, 57] on matchings in claw-free graphs is the following.

Theorem 4.7 [100, 57]. *Every connected claw-free graph of even order has a perfect matching.*

The following result represents a generalization of Sumner and Las Vergnas's archetypal result to $E(m, n)$. It generalizes two previous results (cf. [17] and [2]).

Theorem 4.8 [5]. *Let m, n and r be non-negative integers with $m \geq 1$, $r \geq 3$ and let G be a $(2m + n + r - 2)$-connected $K_{1,r}$-free graph of even order at least $2m + 2n + 2$. Then G is $E(m, n)$.*

We turn now to the properties $E(m, n)$ as applied to graphs embedded in surfaces. Theorem 3.1 above clearly translates into the language of this section as "no planar graph is $E(3, 0)$". We also see from Theorem 4.1 of the present section that property $E(3, 0)$ implies property $E(2, 1)$. The following result [6], therefore, represents a strengthening of Theorem 3.1.

Theorem 4.9 [6]. *No planar graph is $E(2, 1)$.*

The proof of this result appeals to the so-called *theory of Euler contributions* (indeed, as did its predecessor Theorem 3.1).

If one consults the implications involving $E(m, n)$ for various values of m and n listed in Theorem 4.1 above, it is immediately apparent that the possible cases of $E(m, n)$ which can possibly hold for a planar graph are few in number. In fact, we need only consider properties of the form $E(2, 0)$, $E(1, n)$ and $E(0, n)$, for $n \geq 0$.

We note that a 3-connected even planar graph need not even be $E(0, 0)$. The well-known *Kleetopes* provide an infinite family of such graphs. Let $\{w, x, y, z\}$ be the vertices of a plane K_4. Subdivide the edge wx in inserting any even number (> 0) of new vertices and join each new vertex to both y and z. Consider the vertices of the resulting triangulation T to be white. Now insert a new black vertex inside each of the faces of T and join it to each of the three white vertices bounding the face of T to which it belongs. Clearly, there can be no perfect matching in the resulting graph since it contains more black vertices than white and the black vertices are independent. Hence it is of interest only to consider 4- and 5-connected planar even graphs. For these classes, we have the following result.

Theorem 4.10 [6].

(a) *If G is a 4-connected even planar graph, then G is $E(1,1)$, $E(1,0)$, $E(0,0)$, $E(0,1)$, $E(0,2)$ and $E(0,3)$.*

(b) *If G is a 5-connected even planar graph, then G is, in addition, $E(2,0)$, $E(1,2)$ and $E(0,4)$.*

These results are best possible.

On the other hand, if the planar graphs involved are triangulations, one can do better in some cases.

Theorem 4.11 [7]. *If G is a 5-connected even planar triangulation, then G is $E(1,3)$ and $E(0,7)$.*

It should be pointed out that if one assumes only 4-connectivity, then there are even planar triangulations that are not $E(1,2)$.

More recently, similar restricted matching results have been obtained for graphs of "small" genus; namely those which minimally embed on the projective plane, torus or Klein bottle. The following is a summary of what is known. (Cf. [8].)

Theorem 4.12.

(a) *No projective planar even graph is $E(2,1)$ (and therefore, none is $E(3,0)$).*

(b) *If G is a k-connected even graph minimally embedded in the projective plane, then G is $E(m,n)$, where k, m and n are related as follows:*

(b1) *if $k = 4$ and $m = 1$, then $n = 0$;*

(b2) *if $k = 4$ and $m = 0$, then $n = 0, 1, 2$;*

(b3) *if $k = 5$ and $m = 2$, then $n = 0$;*

(b4) *if $k = 5$ and $m = 1$, then $n = 0, 1, 2$;*

(b5) *if $k = 5$ and $m = 0$, then $n = 0, 1, 2, 3, 4$.*

In each of these cases, the results are best possible.

Theorem 4.13.

(a) *No toroidal even graph is $E(3,1)$ (and therefore, none is $E(4,0)$).*

(b) *If a toroidal even graph is $E(3,0)$, then it is a 4-regular quadrangulation.*

(c) *If a 5-connected even toroidal graph is* $E(2,1)$, *then it is a 6-regular triangulation.*

(d) *If* G *is a* k*-connected even graph minimally embedded in the torus, then* G *is* $E(m,n)$, *where* k, m *and* n *are related as follows:*

(d1) *if* $k = 4$ *and* $m = 0$, *then* $n = 0, 1$;

(d2) *if* $k = 5$ *and* $m = 1$, *then* $n = 0, 1, 2$;

(d3) *if* $k = 5$ *and* $m = 0$, *then* $n = 0, 1, 2, 3, 4$;

(d4) *if* $k = 6$ *and* $m = 2$, *then* $n = 0, 1$;

(d5) *if* $k = 6$ *and* $m = 1$, *then* $n = 0, 1, 2, 3$.

(d6) *if* $k = 6$ *and* $m = 0$, *then* $0, 1, 2, 3, 4, 5$.

In cases (d1)–(d4), the results are best possible.

Theorem 4.14.

(a) *No even graph on the Klein bottle is* $E(3,1)$ *(and therefore, none is* $E(4,0)$*).*

(b) *If a 5-connected even graph on the Klein bottle is* $E(2,1)$, *then it is a 6-regular triangulation.*

(c) *If* G *is a* k*-connected even graph minimally embedded in the torus, then* G *is* $E(m,n)$, *where* k, m *and* n *are related as follows:*

(c1) *if* $k = 4$ *and* $m = 0$, *then* $n = 0, 1$;

(c2) *if* $k = 5$ *and* $m = 1$, *then* $n = 0, 1, 2$;

(c3) *if* $k = 5$ *and* $m = 0$, *then* $n = 0, 1, 2, 3, 4$;

(c4) *if* $k = 6$ *and* $m = 2$, *then* $n = 0, 1$;

(c5) *if* $k = 6$ *and* $m = 1$, *then* $n = 0, 1, 2, 3$.

(c6) *if* $k = 6$ *and* $m = 0$, *then* $0, 1, 2, 3, 4, 5$.

In cases (c1)–(c4), the results are best possible.

5. Variations on a Theme

The concept of n-extendability has recently been generalized in several different directions. We briefly discuss some of these variations.

Yu [111] and Favaron [30] independently formulated the definition of a k-factor-critical graph. For a given integer k, $0 \le k \le p$, a graph of order p is *k-factor-critical* if $G - S$ has a perfect matching for every $S \subseteq V(G)$ with $|S| = k$. Each also independently produced the following Tutte-like characterization of k-factor-critical graphs.

Theorem 5.1 [111, 30]. *A graph G is k-factor-critical if and only if $\big|V(G)\big| \equiv k \pmod 2$ and for all $S \subseteq V(G)$ with $|S| \ge k$, $c_o(G-S) \le |S|-k$.*

Clearly, for p and k even, every k-factor-critical graph is $k/2$-extendable. However, the converse is clearly false; simply consider bipartite graphs. For non-bipartite graphs, the quest for a converse of sorts becomes more complicated. However, building on a number of previous results (cf. [30, 31, 32, 60, 66, 112]), Zhang, Wang and Lou obtained the following characterization, when the extendability is sufficiently large in terms of the order of the graph.

Theorem 5.2 [115]. *If $n \ge \big(\big|V(G)\big| + 2\big)/4$, then a non-bipartite graph G is n-extendable if and only if it is $2n$-factor-critical.*

Lou and Yu also generalized Theorem 3.2 above by obtaining the next result for planar graphs.

Theorem 5.3 [66]. *Let G be a 5-connected planar graph of order p. Then G is $(4-\varepsilon)$-factor-critical, where $\varepsilon = 0$ or 1 and $\varepsilon \equiv p \pmod 2$.*

In the same vein as their earlier paper on n-extendable graphs, Kawarabayashi, Ota and Saito [48] determined a degree sum criterion guaranteeing that a k-factor-critical graph have Hamiltonian cycle.

Theorem 5.4 [48]. *Let n be a non-negative integer and let G be a 2-connected n-factor-critical graph of order p. If $\sigma_3(G) \ge (3/2)(p-n-1)$, then G is Hamiltonian.*

Also motivated by the property of bicriticality, Ananchuen defined a yet another new variation called *strong extendability*. (This is not to be confused with another concept called "strong matching", a term often used to mean induced matching.) Given integers p and n with $0 \le n \le p-2$, a graph G on $2p$ vertices is *strongly n-extendable* (written "n^*-extendable") if $G-u-v$ is n-extendable. The property of n^*-extendability, at least for non-bipartite graphs, turns out to be intermediate between $(n+2)$-extendability and $(n+1)$-extendability.

Theorem 5.5 [9]. *Let G a graph on $2p$ vertices.*

(a) *Suppose $0 \le n \le p-3$ and G is non-bipartite. Then if G is $(n+2)$-extendable, G is n^*-extendable.*

(b) *Suppose $0 \le n \le p-2$. Then if G is n^*-extendable, G is t-extendable for all t, $0 \le t \le n+1$.*

Other properties of n^*-extendable graphs are explored in [10, 11], where many theorems paralleling similar results for n-extendability are derived.

6. Bricks and Braces: a Brief Outline

The problem of computing $\Phi(G)$, that is, determining the number of perfect matchings in a graph G, was raised by Kasteleyn [44, 45, 46] who wanted to count the number of matchings in the graphs of certain hydrocarbon molecules. The number of such matchings is called by chemists the *resonance energy* of the molecule.

In 1979, Valiant [107] proved that this counting problem is NP-hard in general. Therefore, attention became increasingly focused on two main related problems: (a) determining which graphs G admitted a polynomial algorithm for determining $\Phi(G)$ *exactly* and (b) *bounding* $\Phi(G)$ for all graphs.

Work on both of these problems has led to some of the deepest results known in matching theory: the decomposition of matching covered graphs into so-called "bricks" and "braces". A *brick* is a 3-connected bicritical graph and a *brace* is a 2-extendable (and therefore 3-connected) bipartite graph. It is known [88] that every 2-extendable graph is either a brick or a brace.

Work in both the areas of Pfaffians and the brick-brace decomposition theory have been very recently, and very thoroughly, surveyed in several papers. For Pfaffians, we refer the reader to Thomas's survey presented to the International Congress in 2006 [104] and to McCuaig's 2004 paper [79]. For the theory of bricks and braces, the reader should consult the survey paper of de Carvalho, Lucchesi and Murty [24] on the subject of the matching lattice, to McCuaig [78] for building braces, and to de Carvalho, Lucchesi and Murty [26] and Norine and Thomas [86] for building bricks. Because these excellent and recent surveys exist, we will confine ourselves to a skeletal discussion of the subject.

Let us begin with bricks and braces. By the 1970's, a beautiful decomposition theory for graphs had developed which is based upon the maximum matchings in the graph. (See [76, 71] for more detailed historical background.) Let $D(G) \subseteq V(G)$ denote the set of vertices v such that G has a maximum matching missing v. Let $A(G)$ denote the set of neighbors of vertices in $D(G)$ which are not themselves in $D(G)$ Finally, let $C(G) = V(G) \backslash \big(D(G) \cup A(G) \big)$. The partition $\big\{ D(G), A(G), C(G) \big\}$ is called the *Gallai-Edmonds decomposition* of the graph G and has a number of very useful properties. For example, each component of $D(G)$ is factor-critical (and therefore of odd order) and every maximum matching M of G decomposes into a near-perfect matching of each component of $D(G)$, a matching covering each vertex of $A(G)$ and a perfect matching of each component of $C(G)$.

Three important extreme situations with respect to this decomposition occur when (a) G is factor-critical in which case $D(G) = V(G)$ (and hence $A(G) = C(G) = \emptyset$), (b) G has a perfect matching in which case $C(G) = V(G)$ (and hence $A(G) = D(G) = \emptyset$), and (c) the bipartite graph obtained from G by deleting $C(G)$ and the edges spanned by $A(G)$, and by contracting each component of $D(G)$ to a separate single vertex has "positive surplus" when viewed from $A(G)$. (Surplus is defined as follows. Let X be any subset of $A(G)$. Then the *surplus of* X $= \sigma_G(X) = \big| N(X) \big| - |X|$ and the *surplus of* G (viewed from $A(G)$) $= \min \big\{ \sigma(X) \mid X \subseteq A(G),\ X \neq \emptyset \big\}$.)

Factor-critical graphs are, in some sense, the easiest class of "atoms" in this decomposition to understand. For example, they have a nice "ear structure". (Cf. [70].) This ear structure leads to the following result, first proved by Pulleyblank [98] in 1973.

Theorem 6.1 [98]. *Every 2-connected factor-critical graph G contains at least $\big| E(G) \big|$ near-perfect matchings.*

The class of graphs with perfect matchings presents a more difficult challenge. From this point on, remembering that we are motivated by trying to find a lower bound on $\Phi(G)$, we will discard all edges in the graphs in this class that belong to no perfect matching. The resulting graphs were originally called *matching-covered,* but of course in light of this survey, this is the same thing as 1-extendable. These 1-extendable graphs may be further decomposed via a so-called "tight cut" procedure to arrive at graphs having no "non-trivial tight cut". (A cutset L of edges in G is said to be *tight* (also sometimes called *strict* in the literature) if every perfect matching in G uses exactly one edge of L. A cutset is *trivial* if it is the star at a vertex.) Lovász proved a key result about these graphs.

Theorem 6.2 [72]. *A 1-extendable graph has no non-trivial tight cut if and only if it is a brick or a brace.*

Using polyhedral theory, Edmonds, Lovász and Pulleyblank [28] were able to show that if G is a 1-extendable graph, then

$$r(G) = \big|E(G)\big| - \big|V(G)\big| + 2 - b(G),$$

where $r(G)$ denotes the maximum number of perfect matchings in G the incidence vectors of which are linearly independent over $\Re$, and $b(G)$ denotes the number of bricks resulting from any tight set (or "brick") decomposition of G. Thus a knowledge of the number of bricks in such a decomposition yields a lower bound on $\Phi(G)$ for any 1-extendable graph. In particular, if G is bipartite, we obtain $r(G) = \big|E(G)\big| - \big|V(G)\big| + 2$, whereas, if G is a brick, $r(G) = \big|E(G)\big| - \big|V(G)\big| + 1$. It should be noted also that implicit in the above result is the non-trivial fact that the number of bricks in any brick decomposition of a 1-extendable graph G is an invariant of G. Later, in a deep paper on the matching lattice [72], Lovász proved more, namely that any two applications of the tight cut decomposition procedure on a 1-extendable graph results in the same list of bricks and braces (except for possibly the multiplicities of edges). It is also important to note that the tight cut decomposition procedure can be carried out in polynomial time.

An *ear decomposition* of a 1-extendable graph G is a sequence

$$K_2 = G_1 \subset G_2 \subset \cdots \subset G_r = G$$

where in this sequence one adds either a single ear (path of odd length) or a double ear (simultaneous addition of two single ears) to obtain the next larger graph in the sequence. It is proved in [76] that every 1-extendable graph has an ear decomposition. Moreover, if the 1-extendable graph G is bipartite, it has an ear decomposition consisting of all single ear additions. (Cf. [40].)

A 1-extendable graph may have many different ear decompositions. It turns out that we are most interested in those decompositions which have the smallest possible number of 2-ear additions. Such decompositions are called *optimal.* The reason for this interest is clear from the next result.

Theorem 6.3 [21]. *If a 1-extendable graph G has an ear decomposition containing d double ears, then there exist $|E(G)| - |V(G)| + 2 - d$ perfect matchings the incidence vectors of which are linearly independent over $\Re$.*

It turns out that a certain special brick–namely the Petersen graph–plays a special role in improving the above lower bound on the number of perfect matchings in a 1-extendable graph. The three authors proceed to show that every 1-extendable graph has an optimal ear decomposition which contains exactly $b+p$ double ears, where b is the number of all bricks and p is the number of Petersen bricks in the brick decomposition of G.

The story does not stop here. There are a number of papers dealing with further decomposition and generation of bricks. In addition to the references already mentioned above, the interested reader is referred to [19, 20, 22, 23, 25, 85]. In particular, we draw the reader's attention to [86, 26] on brick generation and to [78] for brace generation.

7. Pfaffian Graphs

Finally, we turn to the subject of Pfaffian graphs. The motivation again is to try to decide for what graphs G can one determine $\Phi(G)$ in polynomial time. Let G be any graph on n vertices. Let $\overrightarrow{G}$ be the oriented graph obtained by orienting the edges of G in some arbitrary manner. Define next a skew-symmetric $n \times n$ matrix $Sk(G) = (a_{ij})_{n \times n}$ as follows:

$$a_{ij} = \begin{cases} +1, & \text{if } (u_i, u_j) \in E(\overrightarrow{G}); \\ -1, & \text{if } (u_j, u_i) \in E(\overrightarrow{G}). \end{cases}$$

One then forms a certain algebraic quantity depending upon $Sk(G)$ called the *Pfaffian* of G and denoted by $pf\big(Sk(G)\big)$, the formulation of which may be found in [76, Section 8.3]. The important fact about this Pfaffian function is that it can be computed in polynomial time; more precisely, we have the next result, due to Muir [82, 83].

Theorem 7.1 [82, 83]. *If M is a skew-symmetric matrix, then $pf(M) = \sqrt{\det M}$.*

Returning to our oriented graph $\overrightarrow{G}$, it is easy to see that $|pf(Sk(\overrightarrow{G})| \leq \Phi(G)$. It is then a natural question to ask when in fact equality holds, for if equality holds, we can then evaluate $\Phi(G)$ exactly in polynomial time by evaluating a determinant.

Let us call any orientation of G which leads to the above equality a *Pfaffian orientation.* Then call G *Pfaffian* if it admits a Pfaffian orientation. (There are a number of equivalent ways of defining Pfaffian graphs; we will be satisfied with one of these.)

Let us call a cycle C in G *nice* if $G\backslash V(C)$ contains a perfect matching. Now let $\overrightarrow{G}$ denote any orientation of G. If C is any undirected cycle in G, we shall say C is *evenly oriented* if it has an even number of edges oriented in the direction of the orientation of $\overrightarrow{G}$. Otherwise, C is *oddly oriented.*

The next result asserts that three different decision problems involving Pfaffians are polynomial-time equivalent. (Cf. [76, 108, 109].)

Theorem 7.2. *Let G be any graph of even order and $\overrightarrow{G}$, an orientation of G. Then the following decision problems are reducible to each other in polynomial time:*

(a) *$\overrightarrow{G}$ is a Pfaffian orientation of G.*

(b) *Every nice cycle in G is oddly oriented relative to $\overrightarrow{G}$.*

(c) *There exists a Pfaffian orientation for G.*

Perhaps the most surprising aspect of this result is the equivalence of (a) and (c). That they are polynomially reducible to each other was shown in [108, 109]. The same authors also show that the number of Pfaffian orientations of any graph is either zero or a power of two.

Unfortunately, no polynomial algorithm is known for checking whether or not an arbitrary graph is Pfaffian. It is known that deciding if a graph G has a Pfaffian orientation is in co-NP. (Cf. [76, 108, 109].) Whether or not the problem is in NP, however, remains unsettled.

Kasteleyn showed that if the graph G is planar, it always has a Pfaffian orientation and showed how to obtain one. (Cf. [76].) For the study of Pfaffians of graphs embedded on surfaces of higher genus, see [38, 39, 103].

There has been remarkable success recently in the study of Pfaffian bipartite graphs and this work involves bricks and braces.

Little obtained the first characterization of this class.

Theorem 7.3 [59]. *A bipartite graph is Pfaffian if and only if it contains no even subdivision of* $K_{3,3}$.

Unfortunately, there is no known efficient way to algorithmically test for such subdivisions of $K_{3,3}$ and hence Little's characterization does not seem to lead to a polynomial algorithm for recognizing bipartite Pfaffian graphs. This difficulty was recently overcome, however, by two teams of researchers (McCuaig and Robertson-Seymour-Thomas) working independently, but at nearly the same time.

Let us define a graph operation known as the 4-*sum* operation. Let G_0 be any graph and let C be a nice cycle in G_0 of length four. Let G_1 and G_2 be two subgraphs of G_0 such that $G_1 \cup G_2 = G_0$, $G_1 \cap G_2 = C$, $V(G_1)\backslash V(G_2) \neq \emptyset$, and $V(G_2)\backslash V(G_1) \neq \emptyset$. Now let G be obtained from G_0 by deleting some (possibly none) of the edges of cycle C. We then say that G is a 4-sum of G_1 and G_2. The main theorem then can be stated as follows.

Theorem 7.4 [79, 99, 80]. *A brace has a Pfaffian orientation if and only if either it is isomorphic to the Heawood graph (see Figure 7.1) or it can be obtained from planar braces by repeated application of the 4-sum operation.*

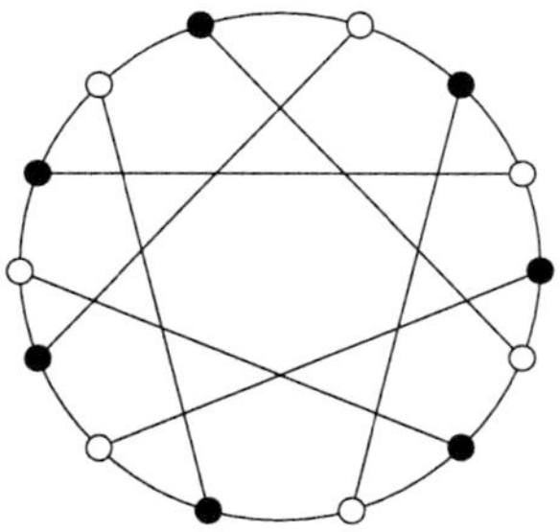

Fig. 7.1. The Heawood graph

The authors then design a polynomial algorithm for testing a bipartite graph for the Pfaffian property using this decomposition theorem.

How does all this relate to bricks and braces? Very nicely, as is illustrated by the next theorem. (See also [105].)

Theorem 7.5 [108, 109]. *A graph G has a Pfaffian orientation if and only if each of its bricks and braces has a Pfaffian orientation.*

The study of bricks, braces and Pfaffian graphs continues.

References

[1] R. E. L. Aldred, D. A. Holton and D. Lou, N-extendability of symmetric graphs, *J. Graph Theory,* **17** (1993), 253–261.

[2] R. E. L. Aldred, D. A. Holton, M. D. Plummer and M. I. Porteous, Two results on matching extensions with prescribed and proscribed edge sets, *Discrete Math.,* **206** (1999), 35–43.

[3] R. E. L. Aldred and B. Jackson, Edge proximity conditions for extendability in cubic bipartite graphs, *J. Graph Theory,* **55** (2007), 112–120.

[4] R. E. L. Aldred, K. Kawarabayashi and M. D. Plummer, On the matching extendability of graphs in surfaces, *J. Combin. Theory Ser. B,* **98** (2008), 105–115.

[5] R. E. L. Aldred and M. D. Plummer, On matching extensions with prescribed and proscribed edge sets II, *Discrete Math.,* **197/198** (1999), 29–40.

[6] R. E. L. Aldred and M. D. Plummer, On restricted matching extension in planar graphs, *Discrete Math.,* **231** (2001), 73–79.

[7] R. E. L. Aldred and M. D. Plummer, Edge proximity and matching extension in planar triangulations, *Australas. J. Combin.,* **29** (2004), 215–224.

[8] R. E. L. Aldred and M. D. Plummer, Restricted matching extension in graphs of small genus, *Discrete Math.* 2008, (to appear).

[9] N. Ananchuen, On strongly k-extendable graphs, *J. Combin. Math. Combin. Comput.,* **38** (2001), 3–19.

[10] N. Ananchuen, On minimum degree of strongly k-extendable graphs, *J. Combin. Math. Combin. Comput.,* **38** (2001), 149–159.

[11] N. Ananchuen, On a minimum cutset of strongly k-extendable graphs, *J. Combin. Math. Combin. Comput.,* **45** (2003), 63–78.

[12] N. Ananchuen and L. Caccetta, On minimally k-extendable graphs, *Australas. J. Combin.,* **9** (1994), 153–168.

[13] N. Ananchuen and L. Caccetta, On $(n-2)$-extendable graphs, *J. Combin. Math. Combin. Comput.,* **16** (1994), 115–128.

[14] N. Ananchuen and L. Caccetta, A note of k-extendable graphs and independence number, *Australas. J. Combin.,* **12** (1995), 59–65.

[15] N. Ananchuen and L. Caccetta, On $(n-2)$-extendable graphs – II, *J. Combin. Math. Combin. Comput.,* **20** (1996), 65–80.

[16] N. Ananchuen and L. Caccetta, Matching extension and minimum degree, *Discrete Math.,* **170** (1997), 1–13.

[17] C. Chen, Matchings and matching extensions in graphs, *Discrete Math.,* **186** (1998), 95–103.

[18] N. Dean, The matching extendability of surfaces, *J. Combin. Theory Ser. B,* **54** (1992), 133–141.

[19] M. de Carvalho and C. Lucchesi, Matching covered graphs and subdivisions of K_4 and $\overline{C_6}$, *J. Combin. Theory Ser. B,* **66** (1996), 263–268.

[20] M. de Carvalho, C. Lucchesi and U. S. R. Murty, Ear decompositions of matching covered graphs, *Combinatorica,* **19** (1999), 151–174.

[21] M. de Carvalho, C. Lucchesi and U. S. R. Murty, Optimal ear decompositions of matching covered graphs and bases for the matching lattice, *J. Combin. Theory Ser. B,* **85** (2002), 59–93.

[22] M. de Carvalho, C. Lucchesi and U. S. R. Murty, On a conjecture of Lovász concerning bricks I. the characteristic of a matching covered graph, *J. Combin. Theory Ser. B,* **85** (2002), 94–136.

[23] M. de Carvalho, C. Lucchesi and U. S. R. Murty, On a conjecture of Lovász concerning bricks II. Bricks of finite characteristic, *J. Combin. Theory Ser. B,* **85** (2002), 137–180.

[24] M. de Carvalho, C. Lucchesi and U. S. R. Murty, The matching lattice, *Recent advances in algorithms and combinatorics, CMS Books Math./Ouvrages Math. SMC,* **11** 2003, 1–25.

[25] M. de Carvalho, C. Lucchesi and U. S. R. Murty, Graphs with independent perfect matchings, *J. Graph Theory,* **48** (2005), 19–50.

[26] M. de Carvalho, C. Lucchesi and U. S. R. Murty, How to build a brick, *Discrete Math.,* **306** (2006), 2383–2410.

[27] J. Edmonds, Paths, trees and flowers, *Canad. J. Math.,* **17** (1965), 449–467.

[28] J. Edmonds, L. Lovász and W. Pulleyblank, Brick decompositions and the matching rank of graphs, *Combinatorica,* **2** (1982), 247–274.

[29] O. Favaron, Equimatchable factor-critical graphs, *J. Graph Theory,* **10** (1986), 439–448.

[30] O. Favaron, On k-factor-critical graphs, *Discuss. Math. Graph Theory,* **16** (1996), 41–51.

[31] O. Favaron, Extendability and factor-criticality, *Discrete Math.,* **213** (2000), 115–122.

[32] O. Favaron and M. Shi, Minimally k-factor-critical graphs, *Australas. J. Combin.,* **17** (1998), 89–97.

[33] A. Frank, E. Györi and A. Sebő, personal communication, 1989.

[34] G. Frobenius, Über Matrizen aus nicht negativen elementen, *Sitzungsber. König. Preuss. Akad. Wiss.,* **26** (1912), 456–477.

[35] G. Frobenius, Über zerlegbare Determinanten, *Sitzungsber. König. Preuss. Akad. Wiss.,* **XVIII** (1917), 274–277.

[36] T. Gallai, Kritische Graphen II, *Magyar Tud. Akad. Mat. Kutató Int. Közl.,* **8** (1963), 373–395.

[37] T. Gallai, Maximale Systeme unabhängiger Kanten, *Magyar Tud. Akad. Mat. Kutató Int. Közl.,* **9** (1964), 401–413.

[38] A. Galluccio and M. Loebl, On the theory of Pfaffian orientations. I. Perfect matchings and permanents, *Electron. J. Combin.,* **6** (1999), Research Paper 6, 18pp.

[39] A.Galluccio and M. Loebl, On the theory of Pfaffian orientations. II. T-joins, k-cuts, and duality of enumeration, *Electron. J. Combin.,* **6** (1999), Research Paper 7, 10pp.

[40] G. Hetyei, Rectangular configurations which can be covered by 2×1 rectangles, *Pécsi Tan. Főisk. Közl.,* **8** (1964), 351–367. (in Hungarian).

[41] D. A. Holton, D-J. Lou and M. D. Plummer, On the 2-extendability of planar graphs *Discrete Math.,* **96** (1991), no. 2, 81–99. (Corrigendum: *Discrete Math.,* **118** (1993), 295–297.)

[42] D.A . Holton and M. D. Plummer, 2-extendability in 3-polytopes, Combinatorics (Eger, 1987), Colloq. Math. Soc. János Bolyai, **52**, North-Holland, Amsterdam, 1988, 281–300.

[43] D. A. Holton and M. D. Plummer, Matching extension and connectivity in graphs. II. Graph theory, combinatorics, and applications. Vol. 2 (Kalamazoo, MI, 1988), Wiley-Intersci. Publ., Wiley, New York, 1991, 651–665.

[44] P. Kasteleyn, The statistics of dimers on a lattice, *Physica,* **27** (1961), 1209–1225.

[45] P. Kasteleyn, Dimer statistics and phase transitions, *J. Math. Phys.,* **4** (1963), 287–293.

[46] P. Kasteleyn, Graph theory and crystal physics, *Graph Theory and Theoretical Physics*, Academic Press, London, 1967, 43–110.

[47] K. Kawarabayashi, K. Ota and A. Saito, Hamiltonian cycles in n-extendable graphs, *J. Graph Theory,* **40** (2002), 75–82.

[48] K. Kawarabayashi, K. Ota and A. Saito, Hamiltonian cycles in n-factor-critical graphs, *Discrete Math.,* **240** (2001), 71–82.

[49] K. Kawarabayashi and M. D. Plummer, Bounding the size of equimatchable graphs of fixed genus, 2008, (submitted).

[50] K. Kawarabayashi, M. D. Plummer and A. Saito, On two equimatchable graph classes, *Discrete Math.,* **266** (2003), 263–274.

[51] D. König, Über Graphen und ihre Andwendung auf Determinantentheorie und Mengenlehre, *Math. Ann.,* **77** (1916), 453–465.

[52] D. König, Graphok és alkalmazásuk a determinánsok és a halmazok elméletére, *Math. Termész. Ért.,* **34** (1916), 104–119.

[53] Kotzig, On the theory of finite graphs with a linear factor I, *Mat.-Fyz.Časopis Slovensk. Akad. Vied,* **9** (1959), 73–91. (Slovak).

[54] Kotzig, On the theory of finite graphs with a linear factor II, *Mat.-Fyz.Časopis Slovensk. Akad. Vied,* **9** (1959), 136–159. (Slovak).

[55] Kotzig, On the theory of finite graphs with a linear factor III, *Mat.-Fyz.Časopis Slovensk. Akad. Vied,* **10** (1960), 205–215. (Slovak).

[56] J. Lakhal and L. Litzler, A polynomial algorithm for the extendability problem in bipartite graphs, *Inform. Process. Lett.*, **65** (1998), 11–16.

[57] M. Las Vergnas, A note on matchings in graphs, *Colloque sur la Théorie des Graphes (Paris, 1974)*, Cahiers Centre Études Rech. Opér. **17** (1975), 257–260.

[58] M. Lesk, M. D. Plummer and W. R. Pulleyblank, Equi-matchable graphs, *Graph theory and combinatorics (Cambridge,* 1983), Academic Press, London, 1984, 239–254.

[59] C. H. C. Little, A characterization of convertible $(0, 1)$-matrices, *J. Combin. Theory Ser. B,* **18** (1975), 187–208.

[60] G. Liu and Q. Yu, on N-edge-deletable and N-critical graphs, *Bull. Inst. Combin. Appl.,* **24** (1998), 65–72.

[61] D. J. Lou, On the 2-extendability of planar graphs, *Acta Sci. Natur. Univ. Sunyatseni,* **29** (1990), 124–126.

[62] D. Lou, A local neighbourhood condition for n-extendable graphs, *Australas. J. Combin.,* **14** (1996), 229–233.

[63] D. Lou, Local neighborhood conditions for n-extendability of a graph, *Kexu Tongbao,* **41** (1996), 1899–1901 (Chinese).

[64] D. Lou, A local independence number condition for n-extendable graphs, *Discrete Math.,* **195**, 263–268.

[65] D. Lou and D. A. Holton, Lower bound of cyclic edge connectivity for n-extendability of regular graphs, *Discrete Math.,* **112** (1993), 139–150.

[66] D. Lou and Q. Yu, Sufficient conditions for n-matchable graphs, *Australas. J. Combin.,* **29** (2004), 127–133.

[67] D. Lou and Q. Yu, Connectivity of k-extendable graphs with large k, *Discrete Appl. Math.,* **136** (2004), 55–61.

[68] L. Lovász, On the structure of factorizable graphs I, *Acta Math. Acad. Sci. Hungar.,* **23** (1972), 179–195.

[69] L. Lovász, On the structure of factorizable graphs II, *Acta Math. Acad. Sci. Hungar.,* **23** (1972), 465–478.

[70] L. Lovász, A note on factor-critical graphs, *Studia Sci. Math. Hungar.,* **7** (1972), 279–280.

[71] L. Lovász, Factors of graphs, *Proceedings of the Fourth Southeastern Conference on Combinatorics, Graph Theory, and Computing (Florida Atlantic Univ., Boca Raton, Fla.,* 1973), Utilitas Math., Winnipeg, Man., 1973, 13–22.

[72] L. Lovász, Matching structure and the matching lattice, *J. Combin. Theory Ser. B,* **43** (1987), 187–222.

[73] L. Lovász and M. D. Plummer, On a family of planar bicritical graphs, *Combinatorics (Proc. British Combinatorial Conf., Univ. Coll. Wales, Aberystwyth, 1973),* London Math. Soc. Lecture Note Ser., No. 13, Cambridge Univ. Press, London, 1974, 103–107.

[74] L. Lovász and M. D. Plummer, On a family of planar bicritical graphs, *Proc. London Math. Soc.,* **30** (1975), 160–176.

[75] L. Lovász and M. D. Plummer, On bicritical graphs, *Infinite and finite sets (Colloq., Keszthely, 1973; dedicated to P. Erdős on his 60th birthday)*, Vol. II, Colloq. Math. Soc. János Bolyai, Vol. 10, North-Holland, Amsterdam, 1975, 1051–1079.

[76] L. Lovász and M. D. Plummer, *Matching Theory*, Ann. Discrete Math., **29**, North-Holland, Amsterdam (1986).

[77] P. Maschlanka and L. Volkmann, Independence number for n-extendable graphs, *Discrete Math.,* **154** (1996), 167–178.

[78] W. McCuaig, Brace generation, *J. Graph Theory,* **38** (2001), 124–169.

[79] W. McCuaig, Pólya's permanent problem, *Electron. J. Combin.,* **11** (2004), Research Paper 79, 83pp.

[80] W. McCuaig, N. Robertson, P. Seymour and R. Thomas, Permanents, Pfaffian orientations, and even directed circuits (extended abstract), *Proc. Twenty-ninth Annual ACM Symp. Theory Comput.* May, 1997, 402–405.

[81] A. McGregor-Macdonald, The $E(m,n)$ property, M.S. thesis, Univ. of Otago, New Zealand, 2000.

[82] T. Muir, *A treatise on the theorem of determinants*, MacMillan and Co., London, 1882.

[83] T. Muir, *The theory of determinants*, MacMillan and Co., London, 1906.

[84] D. Naddef, Rank of maximum matchings in a graph, *Math. Programming,* **22** (1982), 52–70.

[85] S. Norine and R. Thomas, Minimal bricks, *J. Combin. Theory Ser. B,* **96** (2006), 505–513.

[86] S. Norine and R. Thomas, Generating bricks, *J. Combin. Theory Ser. B,* **97** (2007), 769–817.

[87] J. Petersen, Die Theorie der regulären Graphen, *Acta Math.,* **15** (1891), 193–220.

[88] M. D. Plummer, On n-extendable graphs, *Discrete Math.,* **31** (1980), 201–210.

[89] M. D. Plummer, Matching extension and the genus of a graph, *J. Combin. Theory Ser. B,* **44** (1988), 329–337.

[90] M. D. Plummer, A theorem on matchings in the plane, *Graph theory in memory of G. A. Dirac* (*Sandbjerg*, 1985), Ann. Discrete Math., 41, North-Holland, Amsterdam, 1989, 347–354.

[91] M. D. Plummer, Degree sums, neighborhood unions and matching extension in graphs, *Contemporary methods in graph theory, bibliographisches Inst., Mannheim*, 1990, 489–502.

[92] M. D. Plummer, Extending matchings in planar graphs IV, *Discrete Math.,* **109** (1992), 207–219.

[93] M. D. Plummer, Extending matchings in graphs: a survey, *Discrete Math.,* **127** (1994), 277–292.

[94] M. D. Plummer, Extending matchings in planar graphs V., *Discrete Math.,* **150** (1996), 315–324.

[95] M. D. Plummer, Extending matchings in graphs: an update, Surveys in graph theory (San Francisco, CA, 1995), *Congr. Numer.,* **116** (1996), 3–32.

[96] M. I. Porteous, Generalizing matching extensions, M.S. Thesis, Univ. of Otago, New Zealand, 1995.

[97] M. I. Porteous and R. E. L. Aldred, Matching extensions with prescribed and forbidden edges, *Australas. J. Combin.,* **13** (1996), 163–174.

[98] W. R. Pulleyblank, Faces of matching polyhedra, Ph.D. thesis, Dept. of Combinatorics and Optimization, Univ. of Waterloo, 1973.

[99] N. Robertson, P. Seymour and R. Thomas, Permanents, Pfaffian orientations, and even directed circuits, *Ann. Math.,* **150** (1999), 929–975.

[100] D. Sumner, Graphs with 1-factors, *Proc. Amer. Math. Soc.,* **42** (1974), 8–12.

[101] D. Sumner, 1-factors and antifactor sets, *J. London Math. Soc.,* **13** (1976), 351–359.

[102] D. Sumner, Randomly matchable graphs, *J. Graph Theory,* **3** (1979), 183–186.

[103] G. Tesler, Matchings in graphs on non-orientable surfaces, *J. Combin. Theory Ser. B,* **78** (2000), 198–231.

[104] R. Thomas, A survey of Pfaffian orientations of graphs, *International Congress of Mathematicians, Vol. III, Eur. Math. Soc., Zürich*, 2006, 963–984.

[105] C. Thomassen, Sign-nonsingular matrices and even cycles in directed graphs, *Linear Algebra Appl.,* **75** (1986), 27–41; Erratum: *Linear Algebra Appl.,* **240** (1996), 238.

[106] W. T. Tutte, The factorization of linear graphs, *J. London Math. Soc.,* **22** (1947), 107–111.

[107] L. Valiant, The complexity of computing the permanent, *Theoret. Comput. Sci.,* **8** (1979), 189–201.

[108] V. V. Vazirani and M. Yannakakis, Pfaffian orientations, $0-1$ permanents, and even cycles in direct graphs, *Discrete Appl. Math.,* **25** (1989), 179–190.

[109] V. V. Vazirani and M. Yannakakis, Pfaffian orientations, $0-1$ permanents, and even cycles in direct graphs, *Automata, languages and programming* (*Tampere*, 1988, *Lecture Notes in Comput. Sci.,* **317**, Springer, Berlin, 1988, 667–681.

[110] R. Xu and Q. Yu, Degree-sum conditions for k-extendable graphs, *Congr. Numer.,* **163** (2003), 189–195.

[111] Q. Yu, characterizations of various matching extensions in graphs, *Australas. J. Combin.,* **7** (1993), 55–64.

[112] Q. Yu, A note on extendability and factor-criticality, *Ann. Comb.,* **6** (2002), 479–481.

[113] T. Zaslavsky, Complementary matching vectors and the uniform matching extension property, *European J. Combin.,* **2** (1981), 91–103.

[114] T. Zaslavsky, Correction to: "Complementary matching vectors and the uniform matching extension property", *European J. Combin.,* **2** (1981), 305.

[115] Z-B. Zhang, T. Wang and D. Lou, Equivalence between extendibility and factor-criticality, *Ars Combin.*, **85** (2007), 279–285.

[116] F. Zhang and H. Zhang, Construction for bicritical graphs and k-extendable bipartite graphs, *Discrete Math.*, **306** (2006), 1415–1423.

Michael D. Plummer
Department of Mathematics
Vanderbilt University
Nashville, Tennessee 37240, USA

e-mail: `michael.d.plummer@vanderbilt.edu`

BOLYAI SOCIETY
MATHEMATICAL STUDIES, 19

Building Bridges
pp. 455–486.

The Structure of the Complex of Maximal Lattice Free Bodies for a Matrix of Size $(n+1) \times n$

HERBERT E. SCARF*

To Laci Lovász on his 60th Birthday

The complex of maximal lattice free bodies associated with a well behaved matrix A of size $(n+1) \times n$ is generated by a finite set of simplicies, $K_0(A)$, of the form $\{0, h^1, \dots, h^k\}$, with $k \le n$, and their lattice translates. The simplicies in $K_0(A)$ are selected so that the plane $a_0 x = 0$, with a_0 the first row of A, passes through the vertex 0. The collection of simplicies $\{h^1, \dots, h^k\}$ is denoted by Top. Various properties of Top are demonstrated, including the fact that no two interior faces of Top are lattice translates of each other. Moreover, if g is a generator of the cone generated by the set of neighbors $\{h\}$ with $a_0 h > 0$, then the set of simplicies of Top which contain g is the union of linear intervals of simplicies with special features. These features lead to an algorithm for calculating the simplicies in $K_0(A)$ as a_0 varies and the plane $a_0 x = 0$ passes through the generator g.

1. Introduction

The purpose of this paper is to demonstrate some elementary structural properties of the simplicial complex of maximal lattice free bodies associated with a *"generic"* matrix A with $n+1$ rows and n columns. The properties are quite easy to establish, but they are not properties that I would have expected when I began studying this subject many years ago.

*I am grateful to Sasha Barvinok, Dave Bayer, Anders Björner, Raymond Hemmecke, Roger Howe, Bjarke Roune, David Shallcross, Bernd Sturmfels, and Kevin Woods for their intellectual company over the course of many years. I would also like to acknowledge my great debt to Imre Bárány, Ravi Kannan, Niels Lauritzen, Rekha Thomas and Zaifu Yang for their thoughtful, patient and constructive reading of the material in this paper.

The structure that is revealed permits an easy calculation of the simplicial complex associated with matrices obtained by perturbing A if the perturbed matrix is also generic. The calculation involves finding the generators of a cone, and comparing simplicies to see whether they are lattice translates of each other; no integer programs need be solved. The simplicity of the calculation suggests that homotopy methods may be useful in computing this simplicial complex and the test set for integer programming that is given by its edges. But it is imperative that both the initial matrix and the perturbed matrix be generic. The paper is silent if this is not correct.

Let A be a real matrix with $n+1$ rows and n columns satisfying the assumption:

*A*1. There exists a unique (aside from scale) strictly positive vector π such that $\pi A = 0$.

This assumption implies that the $n \times n$ minors of A are non-singular and that the bodies $K_b = \{x\colon Ax \geq b\}$ are bounded for any b. A *maximal lattice free body* is such a body K_b containing no lattice points in its interior, and such that any convex body that properly contains K_b does have a lattice point in its interior.

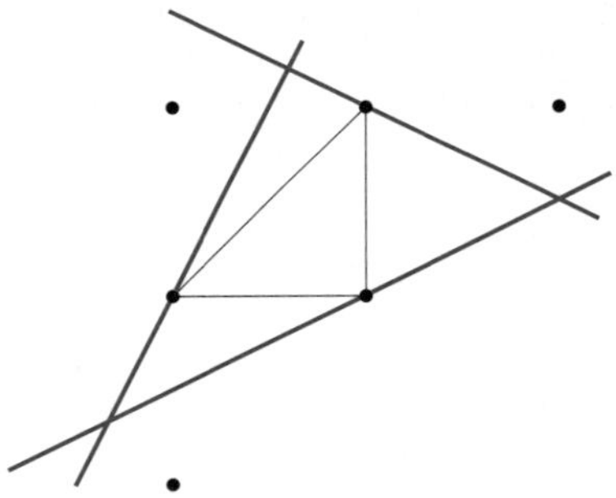

Fig. 1. A Maximal Lattice Free Body

Each of the $n+1$ faces of a maximal lattice free body will contain at least one lattice point. The possibility of several lattice points lying on a single face is a nuisance that can be avoided by perturbing the matrix A slightly. Let us assume that this has been done, and that the following assumption of *genericity* holds:

*A*2. Each face of a maximal lattice free body contains precisely one lattice point.

According to this assumption, each maximal lattice free body will be associated with $n+1$ lattice points $\{h^0, h^1, \ldots, h^n\}$ with, say, $a_i h^i = b_i$ and $a_i h^j > b_i$ for all i and all $j \neq i$. The simplicial complex $K(A)$, occasionally called the Scarf Complex, defined by A consists of these simplicies, and all of their proper subsimplicies.

The Scarf Complex was introduced, of course under a different name, as an essential ingredient in a constructive proof that a balanced n person game has a non empty core [12]. In this treatment, the set of vertices Y was an arbitrary finite set of points in R^{n+1} rather than the lattice $Y = \{Ah\colon h \in \mathbb{Z}^n\}$. The rule for finding an adjacent simplex, given in Section 3 of the current paper, was fully described. The term "Scarf Complex" was first used in [3], and [2].

The complex can be illustrated by a figure which is quite familiar to algebraic geometers. Let Y be the lattice generated by the columns of A. Append a negative orthant to each point $y \in Y$.

The union of these translated negative orthants has a "staircase" structure as in Figure 2 in which $n = 2$. The maximal points on the upper surface are the vectors in the lattice Y.

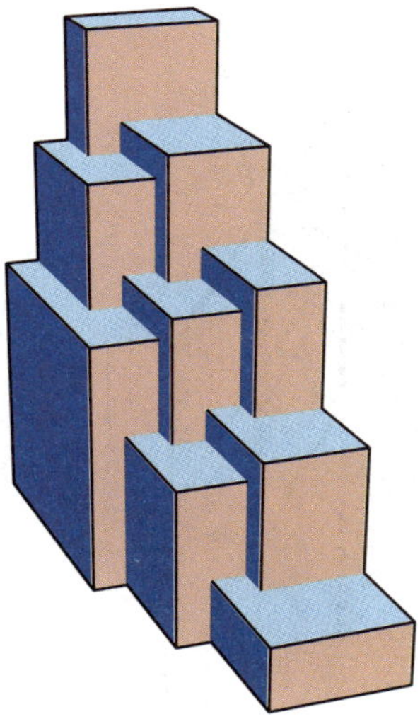

Fig. 2. The Staircase

In this example, each minimal point on the upper surface lies on 3 negative orthants associated with 3 different vectors, say $y^{j_1}, y^{j_2}, y^{j_3} \in Y$. Figure 3 shows a triangle whose vertices are one of these triples.

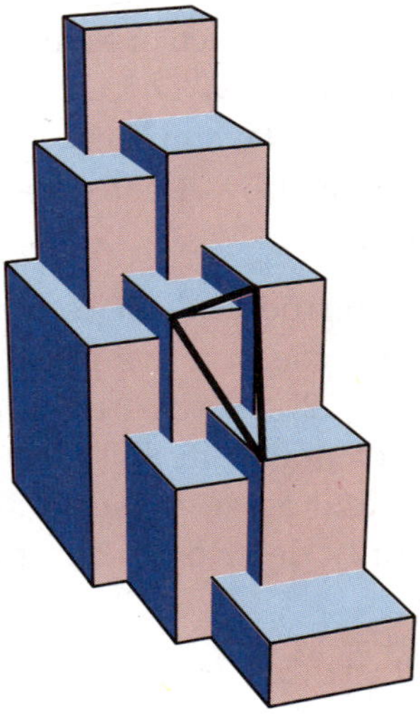

Fig. 3. A Simplex

If we construct the triangle associated with every minimal point on this surface we get a collection of triangles (**and edges and vertices**) that forms the simplicial complex associated with this set Y.

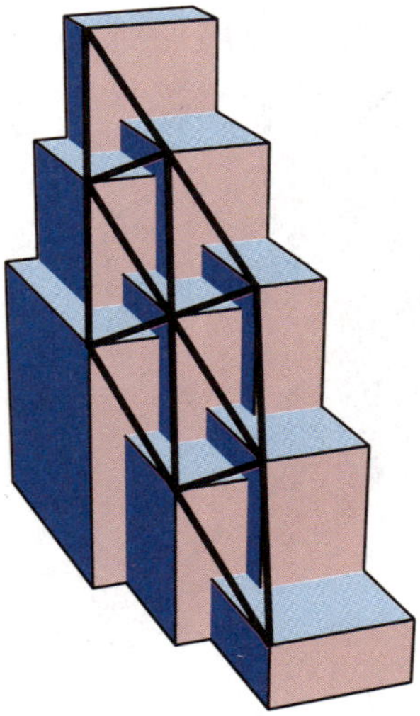

Fig. 4. The Complex

If A is an $(n+1) \times n$ matrix, then the simplicial complex will typically be of dimension n. It will consist of n dimensional simplicies and all of their faces of arbitrary dimensions. As in the example with $n = 2$, we append a negative orthant to each of the points

$$y \in Y$$

and take the union of these translated negative orthants. If A is generic, each minimal point on the upper surface will be associated with a set of $n + 1$ points

$$y^{j_0}, \ldots, y^{j_n} \in Y.$$

The simplicies S of maximal dimension in $K(A)$ are these collections of $n+1$ vectors.

The edges of this complex provide the unique, minimal *test set* for the family of integer programs

$$\begin{aligned} \max\ & a_0 h \text{ subject to} \\ & a_i h \geq b_i, \text{ for } i = 1, \ldots, n \text{ and } h \text{ integral,} \end{aligned}$$

obtained by selecting a single row of A, say row 0, as the objective function and imposing constraints derived from the remaining rows [13, 14]. More specifically, let $N(A)$ be the finite, symmetric set of those non-zero lattice points k that are contained in a maximal lattice free body, one of whose other vertices is the origin. The lattice points in $N(A)$ are called *neighbors of the origin.* (It is a consequence of Assumption $A2$ that $a_j h \neq 0$ for any neighbor h.) It can be shown that a lattice point h satisfying the constraints of this integer program is the optimal solution to the integer program if for every neighbor of the origin, k, with $a_0 k > 0$, the lattice point $h + k$ violates at least one of the constraints. Moreover, under the assumption of genericity this test set is minimal: If an arbitrary element, and its negative, of $N(A)$ are deleted, then some right hand side b, and some feasible solution h can be found, such that h is not optimal, but its lack of optimality cannot be detected using this smaller test set.

The minimal test set $N(A)$ for an alternative formulation

$$\begin{aligned} & \min a_0 h \text{ subject to} \\ & a_i h = b_i \ \text{ for } \ i = 1, \ldots, n, \ h \in \mathbb{N}^n, \end{aligned}$$

is given in [19]. This test set is shown to be a particular Gröbner basis for a binomial ideal associated with the columns of A. Sturmfels and Thomas [18] study the relationship between the set of optimal solutions of two families of integer programming problems, with identical constraint sets, but with different objective functions a_0 and $\widehat{a}_0$. They show that the optimal solutions to the two problems will be the same for each right hand side b, if, and only if, the two associated matrices have indentical sets of neighbors.

Let $c = (c_1, \ldots, c_n)$ be a vector of positive integers with no common factor. The Frobenius problem is to find the largest integer b, termed the Frobenius number, which **cannot** be written as ch with $h \in \mathbb{Z}^n_+$. In [16] a direct relationship is exhibited between the Frobenius problem and the simplicial complex $K(A)$ where A is a matrix whose columns generate the lattice $L = \{h \in \mathbb{Z}^n : ch = 0\}$. There has recently been a considerable renewal of interest in the Frobenius problem. A variety of techniques have been developed to calculate the Frobenius number using Gröbner Bases [9], [11] and the entirely novel approach taken in [8].

The simplicial complex $K(A)$ has been carefully studied and its topological structure is quite well known [5], [7] and [4]. The complex (more precisely, its realization in R^n) is homeomorphic to R^n. The immediate presentation of the complex obtained by drawing all of its simplicies in R^n is quite intricate. Aside from a few elementary cases, many of the simplicies overlap with each other, so that the embedding into R^n is elaborately folded.

If B is another matrix with $n+1$ rows and n columns such that

$$\operatorname{sign}(a_i h) = \operatorname{sign}(b_i h) \text{ for all } i \text{ and } h \in N(A),$$

then B has the same set of neighbors and the same simplicial complex as A [6]. The result is also correct when the two matrices have m rows and n columns with $m > n$.

An arbitrary lattice translate of a simplex in $K(A)$ is also in the complex, and it is frequently convenient to choose one simplex from each class of lattice translates. We shall be concerned with the behavior of the collection of simplicies, of arbitrary dimension, as the first row of the matrix a_0 varies. For this purpose it will be useful to select a specific representative from each set of lattice translates by requiring the origin to lie on the plane $a_0x = b_0 = 0$. In our notation, we consider the collection of simplicies in $K(A)$ of the form

$$\{0, h^1, \ldots, h^k\},$$

with $a_0 h^j > 0$ for $j = 1, \ldots, k$. We give the name $K_0(A)$ to this special subset of $K(A)$.

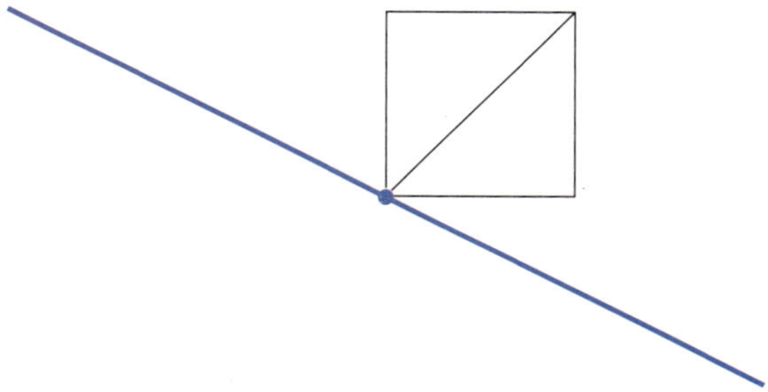

Fig. 5. $K_0(A)$

It will be useful for us to have a simple example of an $n+1 \times n$ matrix A for which $K_0(A)$ is easy to describe. Let A be an $(n+1) \times n$ matrix, with rows $0, 1, \ldots, n$, columns $1, \ldots, n$, with the sign pattern

$$A = \begin{bmatrix} + & + & \cdots & + \\ - & + & \cdots & + \\ + & - & \cdots & + \\ \vdots & \vdots & \ddots & \vdots \\ + & + & \cdots & - \end{bmatrix}$$

which satisfies

$$\sum_j a_{ij} < 0 \quad \text{for} \quad i = 1, \ldots, n.$$

Then the full dimensional simplicies in $K_0(A)$ are given by

$$S = \left(0, e_1, e_1 + e_2, \ldots, \sum_{j=1,n} e_j\right)$$

for $e_1, e_2, \ldots, e_n$ an arbitrary permutation of the n unit vectors in R^n. The neighbors of the origin are the non-zero lattice points $h = \{h_1, \ldots, h_n\}$ with $h_j = 0, 1$, and their negatives [15].

2. Top

Let us define Top to be the collection of simplicies $\{h^1, \ldots, h^k\}$ such that $\{0, h^1, \ldots, h^k\}$ is a simplex in $K_0(A)$, with the plane $a_0 x = b_0$ passing through the origin.

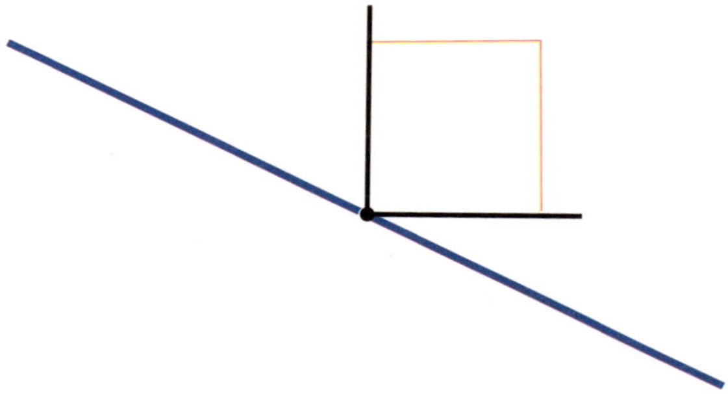

Fig. 6. Top

The vertices of Top consist of the neighbors h with $a_0h > 0$. Let C be the cone generated by this half-set of neighbors. For each generator g of C, we define Top $[g]$ to be the collection of simplicies in Top, one of whose vertices is that generator. We use the notation Top / Top $[g]$ for the collection of $n-1$ simplicies in Top but not in Top $[g]$.

The number of generators of C may be shown to be polynomial in the bit size of A for fixed n [10].

Let us consider an example of Top based on the matrix

$$A = \begin{bmatrix} 1 & 1 & 1 \\ -101 & 31 & 43 \\ 29 & -301 & 173 \\ 41 & 131 & -203 \end{bmatrix}$$

For this example, Top is the figure consisting of the following 6 triangles

$$\{(1,0,0),(1,1,0),(1,1,1)\}$$
$$\{(1,0,0),(1,0,1),(1,1,1)\}$$
$$\{(0,1,0),(1,1,0),(1,1,1)\}$$
$$\{(0,1,0),(0,1,1),(1,1,1)\}$$
$$\{(0,0,1),(1,0,1),(1,1,1)\}$$
$$\{(0,0,1),(0,1,1),(1,1,1)\}$$

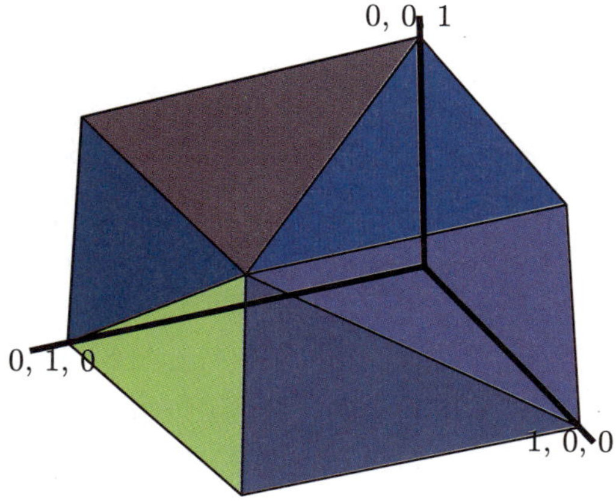

Fig. 7. Top

The generators of Top are given by the following three neighbors

$$(1, 0, 0)$$
$$(0, 1, 0)$$
$$(0, 0, 1)$$

Let $g = (1, 0, 0)$. Then Top $[g]$, the set of simplicies containing g, is shown in Figure 8.

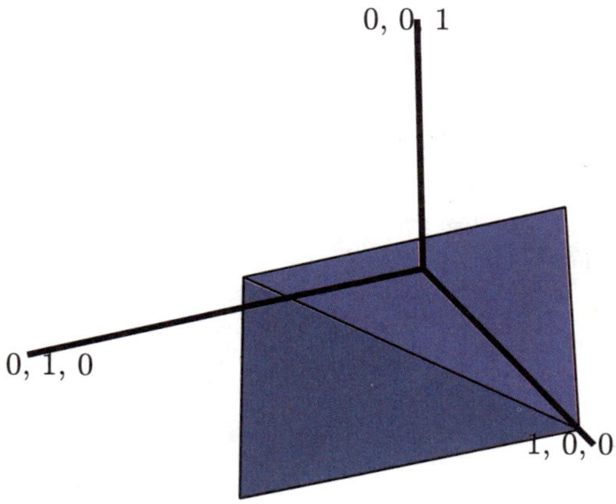

Fig. 8. Top $[(1, 0, 0)]$

The plane $a_0 x = 0$ passes through the origin. The complex $K(A)$ is unchanged by a perturbation of this plane as long as the plane does not touch one of the generators of Top. But as soon as the plane passes through one of the generators the entire complex is transformed into a new simplicial complex. For example, if the plane $a_0 x = 0$ passes through the generator $g = (1, 0, 0)$ the new version of Top, say Top*, can be shown to consist of the 12 triangles

$$\begin{array}{l}
\{\{1,1,0\},\{0,1,0\},\{1,1,1\}\}\\
\{\{2,1,1\},\{1,1,0\},\{1,1,1\}\}\\
\{\{4,0,1\},\{4,1,1\},\{3,0,1\}\}\\
\{\{5,1,1\},\{4,1,1\},\{4,0,1\}\}\\
\{\{1,1,1\},\{0,1,0\},\{-1,0,0\}\}\\
\{\{4,1,1\},\{3,1,1\},\{-1,0,0\}\}\\
\{\{3,1,1\},\{2,1,1\},\{-1,0,0\}\}\\
\{\{2,1,1\},\{1,1,1\},\{-1,0,0\}\}\\
\{\{4,1,1\},\{-1,0,0\},\{3,0,1\}\}\\
\{\{1,0,1\},\{-1,0,0\},\{0,0,1\}\}\\
\{\{2,0,1\},\{-1,0,0\},\{1,0,1\}\}\\
\{\{3,0,1\},\{-1,0,0\},\{2,0,1\}\}
\end{array}$$

shown in Figure 9.

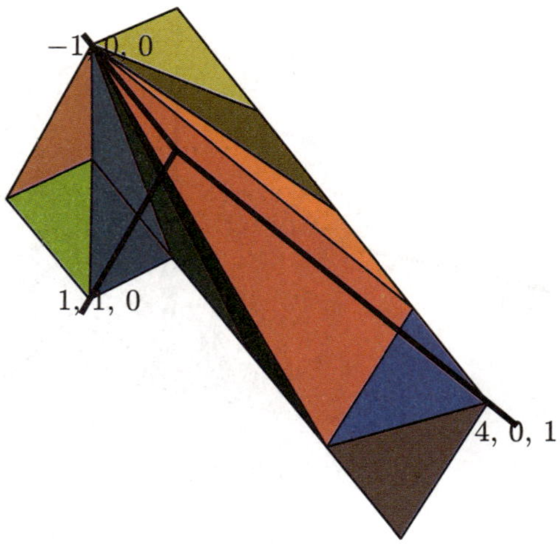

Fig. 9. Top*

The new generators are

$$\begin{array}{c}(1,1,0)\\(4,0,1)\\(-1,0,0)\end{array}$$

$\mathrm{Top}^*[-g]$ is the collection of simplicies in Top^* containing the new generator $-g = (-1,0,0)$. It is displayed in Figure 10.

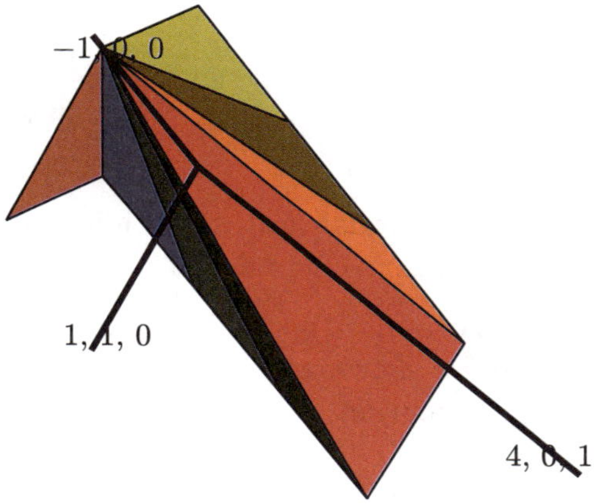

Fig. 10. $\mathrm{Top}^*[-g]$

As we see $\mathrm{Top}^*[-g]$ contains 8 distinct simplicies.

2.1. The Structure of Top. An Informal Presentation

I would like to present an informal summary of the basic structure of Top, reserving formal mathematical proofs for later sections. The two examples that we have constructed, Top and Top^*, are quite different and the properties we are about to discuss may be seen more readily in the more complex example.

Top is a connected $n-1$ dimensional piece-wise linear manifold: every $n-2$ dimensional face of Top is incident to either one or two $n-1$ dimensional simplicies. Top is homeomorphic to the $n-1$ ball [5].

An $n-2$ facet of Top incident to a single $n-1$ dimensional simplex is on the *boundary* of Top, which we denote by $\partial\,\mathrm{Top}$. An $n-2$ facet incident to two $n-1$ dimensional simplicies is *interior* to Top. The simplicies of Top have faces of dimension $0, 1, \ldots, n-2$. A $k-1$ dimensional face of Top is defined by a set $F = \{h^1, \ldots, h^k\}$ with $a_0 h^j > 0$ such that the $n-2$ facet

$$S = \{h^1, \ldots, h^k, \alpha^{k+1}, \ldots, \alpha^{n-1}\} \in \text{Top}$$

for some $\alpha^{k+1}, \ldots, \alpha^{n-1}$. There are, of course, many such $\alpha^{k+1}, \ldots, \alpha^{n-1}$ that can be used to complete F. (Section 4, and, in particular, Lemma 1)

- **The face F is in the boundary of** Top **if, and only if, there is some completion to a facet which is itself on the boundary of** Top**.**

The next set of features are unexpected (to me) properties of lattice translates of *interior* faces of Top. Two faces E and F are lattice translates if $F = E + h$ for some lattice point h.

- **No two interior k dimensional simplicies of** Top **are lattice translates of each other** (Lemma 3 in Section 5).

As a special case of this result, Top has a single interior vertex [4]. The single interior vertex may be shown to be the lattice point h^* where h^* is the solution to the integer program

$$\min a_0 h \quad \text{subject to}$$

$$a_i h < 0, \quad \text{for } i = 1, \ldots, n, \quad \text{and } h \text{ integral.}$$

- **Every k dimensional simplex on the boundary of** Top **has a lattice translate that is interior to** Top (Lemma 2 in Section 4).

Each $n-2$ simplex on the boundary of Top is a lattice translate of an interior simplex, but many of them may also be lattice translates of other boundary simplicies. Of particular interest are those $n-2$ simplicies that are translates by a generator of the cone C. Let g be such a generator. Two $n-2$ simplicies, E and F, of Top will be said to be congruent mod (g), if $F = E + jg$, for some integral j.

- **Lattice Translates of $n-2$ simplicies by a generator. The following properties are demonstrated in Theorem 4 of Section 6.**

Let g be a generator of the cone C. The set of $n-2$ simplicies of Top $[g]$ that are congruent mod (g) to a specific $n-2$ simplex consists of an interval

$$E + g, \ldots, E + ug,$$

with $E = \{h^1, \ldots, \text{not } i, \ldots, h^n\}$ a simplex in Top / Top $[g]$. ($u = 0$, if there are no simplicies of this form.) Moreover

1. All of the $n-2$ simplicies $E, E+g, \ldots, E+(u-1)g$ are in the boundary of Top.
2. All of the $n-1$ simplicies $\{g, E+g\}, \ldots, \{g, E+ug\}$ are in Top $[g]$.
3. The first $n-2$ simplex E is contained in an n simplex $S = \{0, h^1, \ldots, \alpha, \ldots, h^n\}$ with $\alpha \neq g$.

The last simplex, $E+ug$, in such an interval may, or may not be *interior* to Top. It is important to differentiate between these two cases; let us call such a sequence of $n-2$ simplicies *interior* if the last simplex in the interval is interior to Top, and call it a *boundary* sequence if the last simplex is on the boundary. If $u < t$ the interval is definitely a boundary interval. If $u = t$ the interval may be of either type.

Figure 11 illustrates a boundary sequence of Top$^*[-g]$, with $g = (1, 0, 0)$, derived from our basic example using the matrix A. As we see, the first such simplex is not in Top$^*[-g]$ and the last simplex is on the boundary of Top*. I have taken the liberty, in this and in subsequent drawings, of including the $n-1$ simplicies containing the members of the sequence and also the initial $n-1$ simplex which does not belong to Top$^*[-g]$.

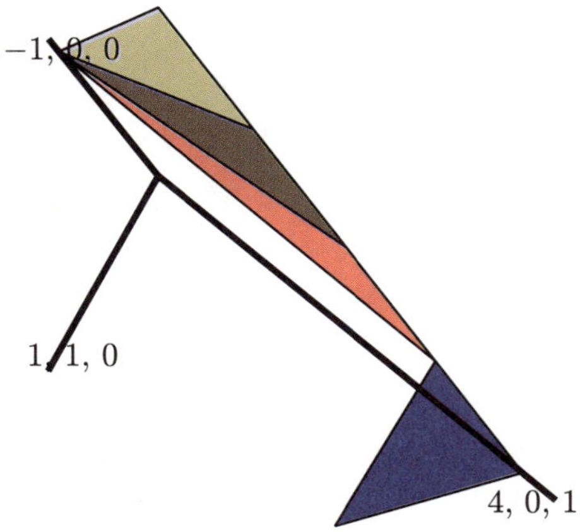

Fig. 11. A Boundary Sequence

In this example, Top$^*[-g]$ has three interior sequences which are shown in the next three figures. Again I have included the $n-1$ simplicies containing the edges of the sequence and the pair of $n-1$ simplicies at the two ends of the interior sequence.

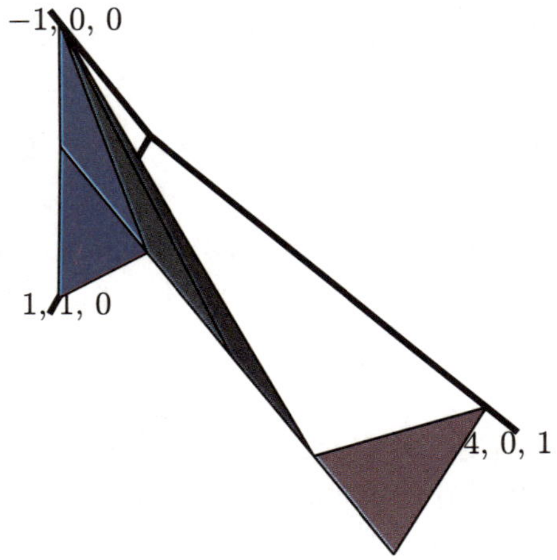

Fig. 12. Interior Sequence 1

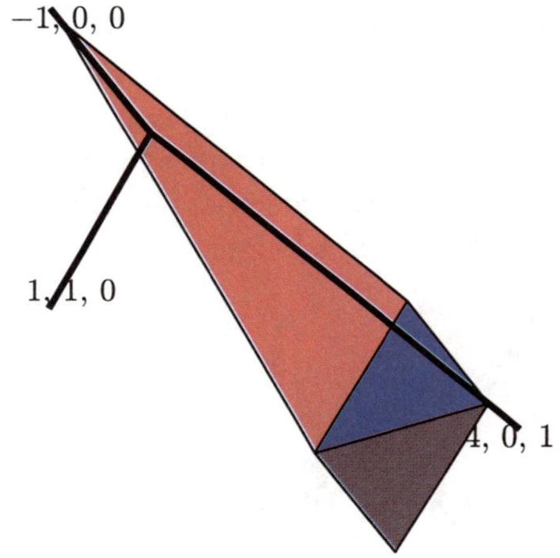

Fig. 13. Interior Sequence 2

If $\{g, E\}$ is an $n - 1$ simplex in Top $[g]$, then E belongs either to an interior or a boundary sequence. **It would be extremely interesting if the number of sequences in** Top $[g]$ **could be shown to be small in fixed dimension.**

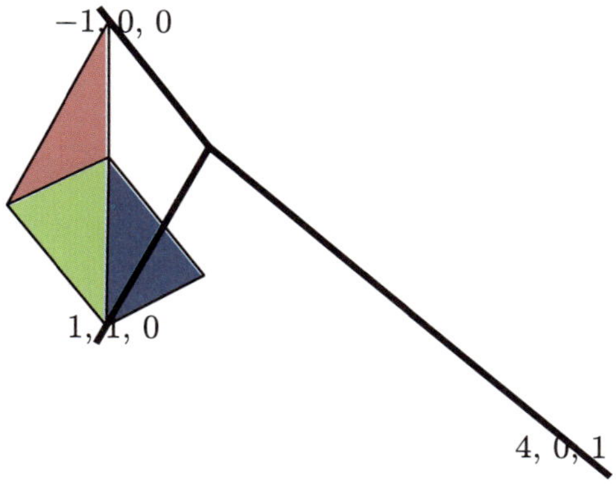

Fig. 14. Interior Sequence 3

3. Adjacent n Simplicies

In order to differentiate the faces of Top which are interior to Top from those which are on the boundary we begin by examining adjacent n simplicies in $K(A)$.

Let $S = \{h^0, h^1, \ldots, h^n\}$ be an n simplex in $K(A)$. For each such S we introduce the $(n+1)$ by $(n+1)$ matrix

$$AH = \left[a_i h^j\right], i, j = 0, \ldots, n.$$

If the matrix A is *generic,* then the row minima of AH lie in different columns. We have agreed to name the vectors $\{h^j\}$ so that the row minima lie on the main diagonal of AH. If this is done then the smallest body of the form $Ax \geq b$, containing S has

$$b_i = a_i h^i.$$

Let

$$\{h^0, h^1, \ldots \text{ not } h^i, \ldots, h^n\}$$

be a particular $n-1$ face of S. We remind the reader of the simple rule for finding the unique, different maximal lattice free body which shares this $n-1$ face with S. Define $j = j(i)$ to be the index j which gives the **second**

smallest value of $a_i h^j$. Then the replacement for h^i is that unique lattice point h which solves

$$\begin{aligned}&\max a_j h \ \text{ subject to}\\&a_k h > a_k h^k \ \text{ for } \ k \neq i, j \ \text{ and}\\&a_i h > a_i h^j.\end{aligned}$$

4. Which $k-1$ Faces are Interior to Top?

A $k-1$ dimensional face of Top is defined by a set $F = \{h^1, \ldots, h^k\}$ with $a_0 h^j > 0$, for $j = 1, \ldots, k$, such that

$$S = \{h^1, \ldots, h^k, \alpha^{k+1}, \ldots, \alpha^n\} \in \text{Top}$$

for some $\alpha^{k+1}, \ldots, \alpha^n$ with $a_0 \alpha^l > 0$, for $l = k+1, \ldots, n$. There are, of course, many such $\alpha^{k+1}, \ldots, \alpha^n$ that can be used to complete F. If we take any such completion

$$\{h^1, \ldots, h^k, \alpha^{k+1}, \ldots, \alpha^j, \ldots, \alpha^n\}$$

and replace any particular α^j, then if F is *interior* to Top, the resulting simplex will also be in Top. If F is not *interior*, there is some completion so that if we replace a particular α^j the resulting simplex will not be in Top.

We have the following useful conclusion:

Lemma 1. *Let* $F = \{h^1, \ldots, h^k\}$ *be a* $k-1$ *dimensional face of* Top. *F is* interior *if, and only if,*

$$\min_{j=1,\ldots,k} a_i h^j < 0 \ \textit{ for all } \ i = 1, \ldots, n.$$

Proof. Let us consider an arbitrary completion of F given by the simplex in $K_0(A)$:

$$S = \{0, h^1, \ldots, h^k, \alpha^{k+1}, \ldots, \alpha^n\}.$$

According to our convention the row minima of the matrix associated with S lie on the main diagonal, so that in particular

$$\min_{j=1,\ldots,k} a_i h^j = a_i h^i < 0 \quad \text{for} \quad i = 1, \ldots, k \quad \text{and}$$

$$\min_{j=k+1,\ldots,n} a_i \alpha^j = a_i \alpha^i < 0 \quad \text{for} \quad i = k+1, \ldots, n.$$

Our first task is to show that if F is an interior $k-1$ face then

$$\min_{j=1,\ldots,k} a_i h^j < 0 \quad \text{for} \quad i = k+1, \ldots, n \quad \text{as well.}$$

Assume to the contrary that F is an interior face and $\min_{j=1,\ldots,k} a_i h^j > 0$ for some particular row i, with $k+1 \leq i \leq n$. Let us remove column α^i from the simplex S and replace it with column $\widehat{\alpha}^i$, obtaining a new simplex $\widehat{S}$. According to the general properties of the replacement step, the smallest entry in row i of the matrix associated with $\widehat{S}$ is equal to the *second* smallest element in row i of the corresponding matrix for S. There are two possible cases:

1. The second smallest element in row i of the matrix associated with S does not lie in any of the columns $k+1, \ldots, n$. But since $\min_{j=1,\ldots,k} a_i h^j > 0$ the second smallest element in row i must therefore lie in column 0. It follows that the replacement for α^i leads to a simplex not in $K_0(A)$ and therefore F is not an interior face.

2. The second smallest element in row i of the matrix associated with S does lie in one of the columns $k+1, \ldots, n$. In this case we reach a new completion $\widehat{S}$ in which the minimum entry in row i of the associated matrix has strictly increased. This can only occur for a finite number of steps; we therefore ultimately return to the previous case and obtain a contradiction to the assumption that F is an interior face and $\min_{j=1,\ldots,k} a_i h^j > 0$.

On the other hand if

$$\min_{j=1,\ldots,k} a_i h^j < 0$$

for all $k+1 \leq i \leq n$, then the second smallest entry in row $i \geq k+1$ of any completion

$$S = \{0, h^1, \ldots, h^k, \alpha^{k+1}, \ldots, \alpha^n\}$$

of F will not lie in column 0 of the associated matrix. The replacement for α^i will lead to a new simplex in $K_0(A)$ for any completion and F is therefore an interior face. ■

The same argument can be used to produce the following property of Top.

Lemma 2. *Let $F = \{h^1, \ldots, h^k\}$ be a $k-1$ dimensional boundary face of* Top. *Then F has a lattice translate $F + h$, with $a_0 h > 0$, which is interior to* Top.

Proof. There may be several lattice translates of F on the boundary of Top. Let us assume that F has been selected from this set so that no lattice translate $F + h$ with $a_0 h > 0$ is on the boundary of Top. Since F is on the boundary there must be a completion $S = \{0, h^1, \ldots, h^k, \alpha^{k+1}, \ldots, \alpha^n\}$ with the replacement for a particular α^j leading to a simplex not in $K_0(A)$, say

$$S' = \{-h, h^1, \ldots, h^k, \alpha^{k+1}, \ldots, 0, \ldots, \alpha^n\}$$

with $a_0 h > 0$. But then $F + h \in$ Top, and it must therefore be interior. ■

5. Lattice Translates of Interior Simplicies

We have the following remarkably simple conclusion:

Lemma 3. *Let $F = \{h^1, \ldots, h^k\}$ with $a_0 h^j > 0$ be a $k-1$ dimensional* interior *face of* Top *and let $h \in Z^n$ with $a_0 h > 0$. Then $\{h^1 + h, \ldots, h^k + h\}$ is **not** a face of* Top.

Proof. The smallest body of the form $Ax \geq b$, containing $0, h^1+h, \ldots, h^k+h$ has

$$b_0 = 0 \quad \text{and}$$
$$b_i \leq \min_{j=1,\ldots,k} a_i(h^j + h) < a_i h \quad \text{for} \quad i = 1, \ldots, n,$$

because $\min_{j=1,\ldots,k} a_i h^j < 0$. But then h is contained in this body and $\{h^1 + h, \ldots, h^k + h\}$ is not a face in Top.

This result implies that if F and $F+h$, with $a_0 h > 0$ are both faces of Top, then $F \in \partial\,\text{Top}$. There cannot be a pair of lattice translates of a $k-1$ face both of which are interior simplicies of Top. ■

6. Intervals in Top $[g]$

Let g be a generator of the smallest cone containing the set of neighbors $\{h\}$ with $a_0 h > 0$.

Theorem 4. *Let $E+g$ be the first $n-2$ face of the form $E+lg$ contained in* Top $[g]$ *and $E+tg$, with $t \geq 1$, the last such face. Then the set of $n-2$ simplicies $E+lg \in$* Top $[g]$, *consists of an interval*

$$E+g, \ldots, E+tg.$$

Moreover the faces

$$E, E+g, \ldots, E+(t-1)g$$

are all in ∂ Top. *The last face in the interval, $E+tg$, may be interior to* Top *or on the boundary. If $E+tg$ is a boundary interval, the value of t is given by*

$$t = \left\lfloor \min_j \left(-a_j h^j / a_j g\right) : j > 0,\ a_j g > 0) \right\rfloor .$$

Proof. A Preliminary Observation. Let $\{g, E+g\}$ be the first $n-1$ simplex of the form $\{g, E+ug\}$ in Top $[g]$. We assume, as customary, that the row minima of

$$\{0, g, E+g\} = \{0, h^1+g, \ldots, g, \ldots, h^n+g\} \in K_0(A)$$

are on the main diagonal (with "g" in column i). Let us show that E is also in Top. To obtain this conclusion we first observe that a_0 can be varied – without changing the simplicial complex – so that $a_0 g$, while positive, is arbitrarily close to 0. This tells us that the second smallest entry in row 0 of the matrix associated with $\{0, g, E+g\}$ is in column i. If column 0 is removed from this simplex we obtain a new simplex

$$\{g, h, E+g\} = \{g, h^1+g, \ldots, h, \ldots, h^n+g\},$$

with row minima on the main diagonal and with g the smallest entry in row 0. It follows that

$$\{0, h - g, E\} = \{0, h^1, \ldots, h - g, \ldots, h^n\}$$

is a maximal lattice free body, again with row minima on the main diagonal. E is therefore in Top.

From the assumption that the row minima of $\{0, g, E + g\}$ lie on the main diagonal we also know that $a_i g < 0$.

Our next observation is that a completion of $E + tg$ can be found with row minima on the main diagonal. Let $T = \{0, h^1 + tg, \ldots, \beta, \ldots, h^n + tg\}$ be a completion of $E + tg$. There are two cases to consider:

1. $E + tg$ is *interior* to Top. In this case, according to Lemma 2, $\min_{j \neq i} a_k(h^j + tg) < 0$ for $k = 1, \ldots, n$. For $k \neq i$ the minimum is reached at $j = k$ and we have

$$\min_{j \neq i} a_k(h^j + tg) = a_k(h^k + tg) < 0.$$

   On the other hand, if the smallest element in row i of T is in column i then the row minima of T lie on the main diagonal. If the row minimum in row i of T is in column k then the smallest entry in row k of T must be in column i and the second smallest entry in row k is $a_k(h^k + tg)$. If we then replace β in T, we obtain a new $T^* = \{0, h^1 + tg, \ldots, \beta^*, \ldots, h^n + tg\} \in K_0(A)$ with row minima on the main diagonal. In this event let the completion be T^* rather than T.

2. $E + tg$ is on the *boundary* of Top. Then from Lemma 2 we must have $\min_{j \neq i} a_k(h^j + tg) > 0$ for some $k = 1, \ldots, n$. We shall show that this k is equal to i. If k is different from i, the smallest entry in row k of the matrix associated with T, being negative, must be in column i and therefore $a_k(h^k + tg) > 0 > a_k g$. But this is impossible since the row minima of $\{0, g, E + g\}$ lie on the main diagonal implying that $a_k(h^k + g) < 0$. It follows that $k = i$ and the row minima of T lie on the main diagonal.

Now let us turn to the main argument. Let $0 < s < t$ and define

$$T_s = \{0, h^1 + sg, \ldots, g, \ldots h^n + sg\}.$$

We want to show by induction, starting with $s = 1$ and continuing to $s = t$, that T_s is a simplex in $K_0(A)$, and that the row minima of T_s lie on the main diagonal.

Let us assume that these two features are correct for $T_1, \ldots, T_{s-1}$ and show that they are also correct for T_s. Since they hold for $s = 1$, by assumption, this demonstrates the theorem.

The row minima of the matrix associated with $\{0, h^1, \ldots, h - g, \ldots, h^n\}$ lie on the main diagonal. Therefore for any $k \neq 0, i$, we have

$$a_k h^k < a_k h^j \quad \text{for} \quad j \neq 0, k, i$$

It follows that

$$a_k(h^k + sg) < a_k(h^j + sg) \quad \text{for} \quad j \neq 0, k, i.$$

We also know that

$$\begin{aligned} a_k h^k &< 0 \quad \text{for} \quad k \neq 0, i \quad \text{and} \\ a_k(h^k + tg) &< 0 \quad \text{for} \quad k \neq 0, i. \end{aligned}$$

If we average these inequalities we see that $a_k(h^k + sg) < 0$ and also $a_k\big(h^k + (s-1)g\big) < 0$, (from which it follows that $a_k(h^k + sg) < a_k g$), for $0 < s < t$ and $k \neq 0, i$. Therefore the smallest element in row $k \neq 0, i$ of T_s is in column k.

By induction $E + (s-1)g$ is a face of Top so that it must be in $\partial\,$Top and its associated matrix must have one row which is entirely positive. But this must be row i, since, by induction,

$$a_k\big(h^k + (s-1)g\big) < 0 \quad \text{for} \quad k \neq 0, i.$$

It follows that row i is the row with positive entries and

$$\begin{aligned} a_i\big(h^j + (s-1)g\big) &> 0 \quad \text{for} \quad j \neq 0, i \text{ and therefore} \\ a_i(h^j + sg) &> a_i g \end{aligned}$$

Combining this observation with the previous remark that $a_k(h^k + sg) < a_k g$ for $k \neq 0, i$ we see that the row minima of T_s lie on the main diagonal.

In order to show that T_s is a simplex in $K_0(A)$, we need to show that there are no lattice points interior to the smallest body $Ax \geq b$ containing the vertices of T_s. The vector b for this body is given by

$$\begin{aligned} b_0 &= 0 \\ b_k &= a_k(h^k + sg) \quad \text{for} \quad k \neq 0, i \quad \text{and} \\ b_i &= a_i g. \end{aligned}$$

If the lattice point ξ, not a multiple of "g", is strictly contained in this body then

$$\begin{aligned} &a_0\xi > 0 \\ &a_k\xi > a_k(h^k + sg) \quad \text{for} \quad k \neq 0, i \quad \text{and} \\ &a_i\xi > a_i g. \end{aligned}$$

But then $\xi - sg$ is strictly contained in the body

$$\{0, h - g, E\} = \{0, h^1, \ldots, h - g, \ldots, h^n\},$$

since

$$\begin{aligned} &a_0(\xi - sg) > 0 \quad \text{for} \quad a_0 g \quad \text{close to} \quad 0 \\ &a_k(\xi - sg) > a_k h^k \quad \text{for} \quad k \neq 0, i, \quad \text{and} \\ &a_i(\xi - sg) > (1 - s)a_i g \geq 0 > a_i(h - g) \end{aligned}$$

This contradicts the fact that $\{0, h - g, E\} \in K_0(A)$ and demonstrates that T_s is a simplex in $K_0(A)$. The induction is complete.

In order to complete the proof of Theorem 4, we need to show that if $E + tg$ is a boundary interval then

$$(*) \qquad t = \Big\lfloor \min_j \big(-a_j h^j / a_j g\big) : j > 0, \; a_j g > 0 \Big\rfloor.$$

First of all, we know that $T_t = \{0, h^1 + tg, \ldots, g, \ldots, h^n + tg\} \in K_0(A)$, with row minima on the main diagonal. Therefore $a_j(h^j + tg) < 0$ for $j \neq 0, i$. Since $a_j h^j < 0$, for such j, this inequality is certainly correct if $a_j g < 0$. But if $a_j g > 0$, the inequality implies that $t < \big(-a_j h^j / a_j g\big)$, and therefore

$$t \leq \Big\lfloor \min_j \big(-a_j h^j / a_j g\big) : j > 0, \; a_j g > 0 \Big\rfloor.$$

We shall show that if the inequality is strict then the row minima of the matrix associated with

$$T_{(t+1)} = \{0, h^1 + (t+1)g, \ldots, g, \ldots, h^n + (t+1)g\}$$

lie on the main diagonal and that $T_{(t+1)}$ is a simplex in $K_0(A)$, contradicting the assumption of Theorem 4.

If the inequality is strict so that

$$t + 1 \leq \left\lfloor \min_j \left(-a_j h^j / a_j g \right) : j > 0,\ a_j g > 0 \right\rfloor,$$

then $a_j\left(h^j + (t+1)g\right) < 0$ for $j \neq 0, i$. Since $a_j(h^j + tg) < 0$ for $j \neq 0, i$, we see that $a_j\left(h^j + (t+1)g\right) < a_j g$ for $j \neq 0, i$, and therefore the row minima of the matrix associated with $T_{(t+1)}$ lie on the main diagonal for all rows other than possibly row i.

But since $E + tg$ is on the boundary of Top, it follows from Lemma 1 that $\min_{k \neq 0,i}(a_j h^k + tg) > 0$ for some row j and this can only be row i. Therefore the row minimum in row i in the matrix associated with $T_{(t+1)}$ is in column i, so that the row minima lie in different columns. The smallest body $Ax \geq b$ containing the vertices of $T_{(t+1)}$ is therefore given by

$$\begin{aligned} b_0 &= 0 \\ b_k &= a_k\left(h^k + (t+1)g\right) \quad \text{for} \quad k \neq 0, i \quad \text{and} \\ b_i &= a_i g. \end{aligned}$$

Arguments identical to those previously given show that there are no lattice points interior to this body. Therefore $T_{(t+1)}$ is a simplex in $K_0(A)$, again contradicting the assumption of Theorem 4. ■

Note: We cannot have

$$\min_j \left(-a_j h^j / a_j g \right) : j > 0,\ a_j g > 0 \text{ equal to an integer } t,$$

because then

$$a_j(h^j + tg) = 0 \text{ for some } j.$$

Since $h^j + tg$ is a neighbor, this violates the assumption of genericity $A2$.

7. Recovering Top from Top / Top [g]

It is very easy to recover all of the simplicies in Top from a knowledge of the simplicies in Top / Top [g]. It suffices, of course, to recover the $n-1$ simplicies. We assume that a list, say L, of the $n-1$ simplicies of Top / Top [g] is known. Figure 17 shows these simplicies in our standard example. Let us begin by recovering the boundary intervals in Top.

7.1. Boundary Intervals in Top [g]

Begin by making a list of those $n-2$ faces E of Top / Top [g], which are on the boundary of Top. (We remind the reader that Lemma 1 permits us to determine if a particular face E is on the boundary of Top, even if we do not yet have a full list of the simplicies in Top.) Each such E is contained in a simplex

$$\{0, h^1, \ldots, h, \ldots, h^n\} = \{0, h, E\} \in \mathrm{Top} / \mathrm{Top}\,[g],$$

with h in column i, arranged so that the row minima are on the main diagonal. Because each such face is on the boundary the second smallest entry in row i of the matrix associated with $\{0, h, E\}$ is in column 0. We are looking for those boundary faces E, of Top / Top [g] which initiate a boundary interval

$$\{h, E\}, \{g, g+E\}, \ldots, \{g, tg+E\}$$

with each $E+ug$ in $\partial\,$Top. In order to find the length of the interval we use $(*)$ to calculate the appropriate value of

$$t = \left\lfloor \min_j \left(-a_j h^j / a_j g\right) : j > 0,\ a_j g > 0 \right\rfloor.$$

If $t = 0$, E does not initiate a boundary interval.

7.2. Interior Intervals in Top $[g]$

Examine the $n-2$ faces of Top / Top $[g]$ to see if there is a pair E, $E+tg$ (with $t>0$) which are congruent mod (g). Such a pair will give rise to an interior interval

$$\{h, E\}, \{g, g+E\}, \ldots, \{g, tg+E\}$$

with $\{g, g+E\}, \ldots, \{g, tg+E\} \in$ Top $[g]$ and $tg+E$ interior to Top.

After the interior intervals have been added there will be no remaining pairs E, $E+ug \in$ Top / Top $[g]$.

8. The Behavior of Top under Continuous Changes in a_0

The plane $a_0x = 0$ supports the cone C generated by the set of neighbors with $a_0h > 0$. As we vary the normals to this plane, $K_0(A)$ will remain the same until the plane touches one of the generators of the cone, say, g. After the plane passes through g (and no other neighbor) the generator g is replaced by $-g$; the other generators have more dramatic replacements. In this section, I will describe, first without proofs, how to calculate the changes in the entire simplicial complex Top after this perturbation. Let us use the notation Top* for the simplicial complex after the perturbation, and Top*$[-g]$ for the collection of simplicies in Top* containing $-g$ as a vertex. We shall recover Top*$[-g]$ using the techniques of the last section.

- Every $n-1$ simplex in Top $[g]$ disappears without an image in Top*.

Let $S = \{0, h^1, \ldots, g, \ldots, h^n\}$ be a simplex in $K_0(A)$, with row minima on the main diagonal. After the plane passes through g (and no other neighbor), a_0g changes sign. The smallest body of the form $Ax \geq b$ containing $0, h^1, \ldots, g, \ldots, h^n$ now has $b_0 = a_0g$ so that 0 is properly contained in this body and $\{0, h^1, \ldots, g, \ldots, h^n\}$ is not a simplex.

- Each of the remaining $n-1$ simplicies in Top / Top $[g]$ (in this instance 4 simplicies) is translated by an integral amount. Specifically, if $F = \{h^1, \ldots, h^n\}$ is an $n-1$ simplex in Top / Top $[g]$ before this perturbation, it is replaced by $F + tg$ after the perturbation, where

$$t = \Big\lfloor \min_j \big(-a_jh^j/a_jg \big) : j > 0,\ a_jg > 0 \Big\rfloor .$$

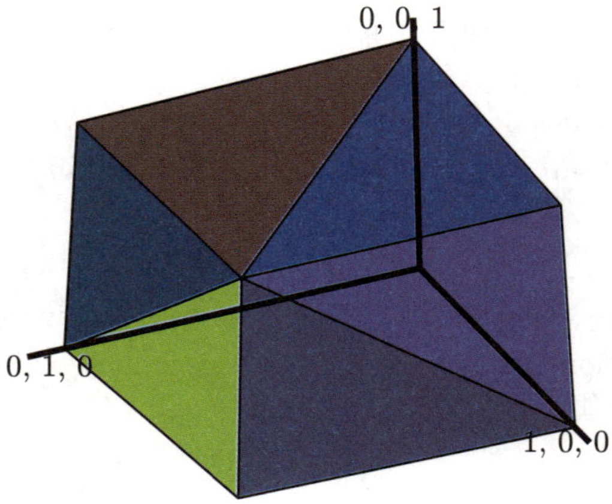

Fig. 15. Top

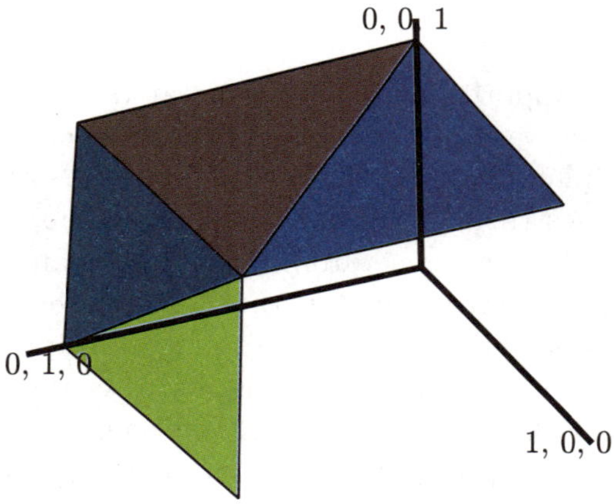

Fig. 16. Top / Top $[g]$

Let $F = \{h^1, \ldots, h^n\} \in$ Top / Top $[g]$, and let $Ax \geq b$ be the smallest body of this form containing $S = \{0, F\}$, with row minima of the matrix associated with S on the main diagonal. The right hand side b does not change after the plane passes though g, but $-a_0 g > 0$, and lattice points of the form $-ug$ (with $u > 0$) may now be contained in this body if

$$-ua_j g > a_j h^j \quad \text{for} \quad j = 1, \ldots, n, \text{ or}$$

$$u \leq t = \Big\lfloor \min_j \big(-a_j h^j / a_j g \big) : j > 0, \ a_j g > 0 \Big\rfloor .$$

Since there are no other lattice points in this body, it follows that $\{-tg, h^1, \ldots, h^n\} \in K^*(A)$ and therefore $F + tg \in \mathrm{Top}^*$.

As before, we cannot have

$$t = \min_j(-a_j h^j / a_j g) \text{ for some } j,$$

for this would violate the assumption that the matrix A is generic, after $a_0 g$ changes sign.

The above rule requires us to examine every simplex in $\mathrm{Top} / \mathrm{Top}\,[g]$. But it is very easy to work with and gives us a complete description of $\mathrm{Top}^* / \mathrm{Top}^*[-g]$. How can we find the simplicies in $\mathrm{Top}^*[-g]$?

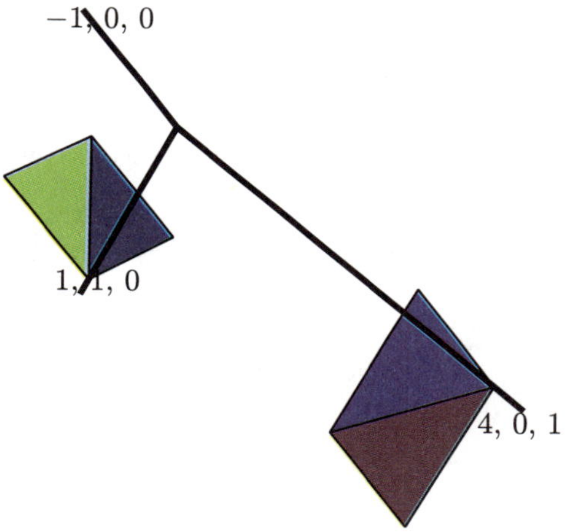

Fig. 17. $\mathrm{Top}^* / \mathrm{Top}^*[-g]$

Boundary Intervals We make a list of those $n-2$ faces E of $\mathrm{Top}^* / \mathrm{Top}^*[-g]$ which are on the boundary of Top^* and such $E + u(-g)$ is not in $\mathrm{Top}^* / \mathrm{Top}^*[-g]$ for $u > 0$. Each such E initiates a boundary interval

$$\{h, E\}, \{-g, -g + E\}, \ldots, \{\,-g, t(-g) + E\}$$

of $\mathrm{Top}^*[-g]$, with

$$t = \left\lfloor \min_j \left(a_j h^j / a_j g\right) : j > 0,\ a_j g < 0\right) \right\rfloor.$$

This is our earlier t with g replaced by $-g$.

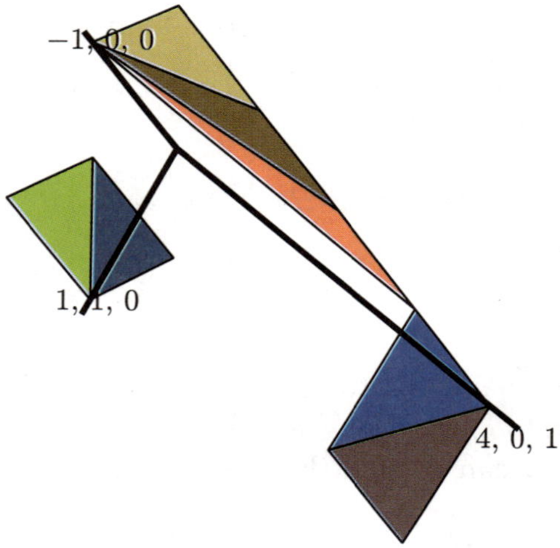

Fig. 18. Adding the Boundary Interval

Our example contains a unique boundary interval which we add to the previous figure.

Interior Intervals As previously described, an interior interval of $\text{Top}^*[-g]$ is determined by a sequence

$$E + tg, E + (t-1)g, \ldots, E$$

with

1. The $n-2$ simplicies $E+tg, E+(t-1)g, \ldots, E+g$ in the boundary of Top^*,
2. The $n-1$ simplicies $\{-g, E+(t-1)g\}, \ldots, \{-g, E\}$ in $\text{Top}^*[-g]$,
3. E interior to Top^*,
4. The first simplex $E + tg$ is not in $\text{Top}^*[-g]$.

We have the full list of simplicies in $\text{Top}^* / \text{Top}^*[-g]$. In order to find an interior interval, we examine the list to find pairs of $n-2$ simplicies

$$E + tg, E$$

in $\text{Top}^* / \text{Top}^*[-g]$. Any such pair will generate an interior interval. Our example has three interior intervals which are added one at a time in each of the following figures.

This is the essential step in a homotopy algorithm.

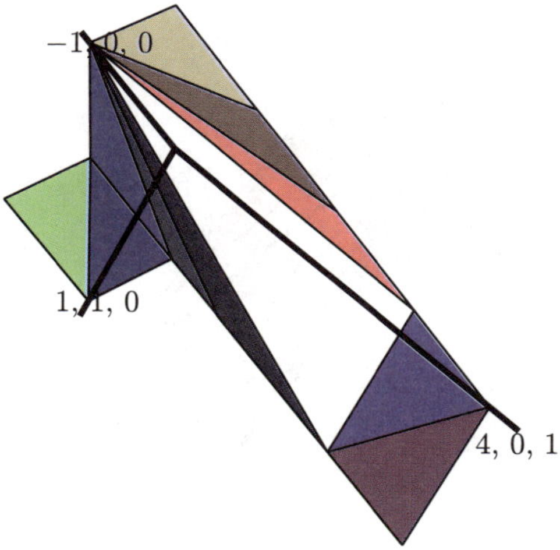

Fig. 19. Adding Interior Interval 1

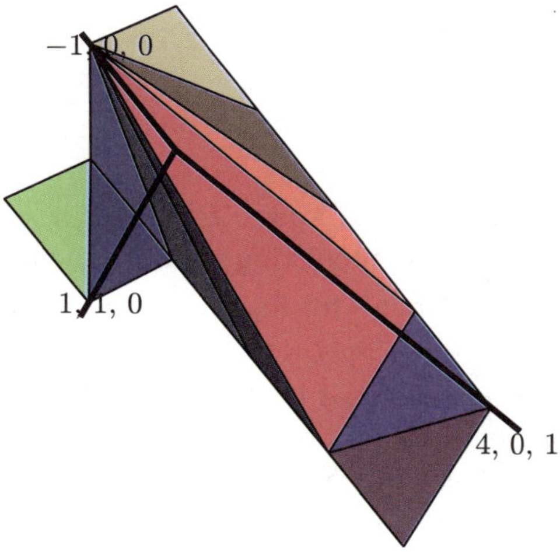

Fig. 20. Adding Interior Interval 2

9. What is Next?

Let us return to the Frobenius problem based on the positive integer vector $c = (c_1, c_2, \ldots, c_n)$. Let

$$G = \{b \colon b = ch, \text{ for some } h \in Z^n_+\},$$

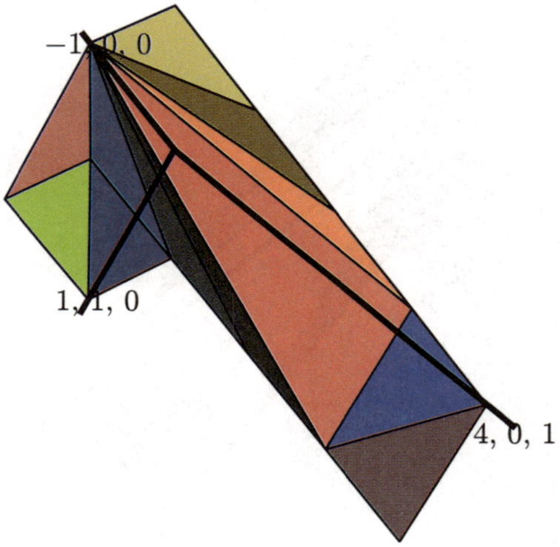

Fig. 21. Adding Interior Interval 3

and define the generating function

$$f(x) = \sum_{b \in G} x^b .$$

There are two remarkable results about $f(x)$, which seem to be at arm's length. It would be extremely interesting to unite them.

In [1] the authors show that – for fixed n – the generating function $f(x)$ can be written as the sum of a polynomial number of rational functions.

To state the second result, let A be an $n \times (n-1)$ integral matrix whose columns generate the lattice $L = \{ h = (h_1, \ldots, h_n) \in \mathbb{Z}^n : ch = 0 \}$ and let $K_0(A)$ be the associated set. Let S be a j dimensional simplex $S \in K_0(A)$ defined by $j+1$ lattice points $\{0, h^1, \ldots, h^j\}$ with $a_0 h^k > 0$ for $k = 1, \ldots, j$. The smallest body of the form $Ax \geq b$ containing $\{0, h^1, \ldots, h^j\}$ will have

$$b(S) = \text{Min}\,[0, Ah^1, \ldots, Ah^j].$$

[2] and [17] showed that if A is generic, then

$$f(x) = \frac{\sum_{S \in K_0(A)} (-1)^{\dim(S)} x^{-c \cdot b(S)}}{\prod_i (1 - x^{c_i})}.$$

What is surprising about these two results is that in general dimension, the number of simplicies in $K_0(A)$ is definitely not small in the size of the matrix A. This suggests that the complex $K(A)$ has sufficient structure so that we can combine the terms in the generating function into a small number of rational functions. What can this structure be? Can it be related to the collection of interior and boundary intervals that characterize Top $[g]$?

References

[1] Alexander Barvinok and Kevin Woods, Short Rational Generating Functions for Lattice Point Problems, *J. Amer. Math. Soc.,* **16** (2003), 957–979.

[2] Dave Bayer and Bernd Sturmfels, Cellular Resolutions of Monomial Modules, *Journal für die Reine und Angewandte Mathematik,* **502** (1998), 123–140.

[3] Dave Bayer, Irena Peeva and Bernd Sturmfels, Monomial resolutions, *Mathematical Research Letters,* **5** (1998), 36–41.

[4] Anders Björner, Face numbers of Scarf complexes, *Discrete and Computational Geometry,* **24** (2000), 185–196.

[5] Imre Bárány, Roger Howe and Herbert Scarf, The Complex of Maximal Lattice Free simplicies, *Mathematical Programming,* Vol. 66, No. 3 (1994), 273–281.

[6] Imre Bárány and Herbert Scarf, Matrices with Identical Sets of Neighbors, *Mathematics of Operations Research,* Vol. 23, No. 4 (1998), 863–873.

[7] Imre Bárány, Herbert Scarf and David Shallcross, The Topological Structure of Maximal Lattice Free Convex Bodies: The General Case, *Mathematical Programming,* Vol. 80, No. 1, January (1998), 1–15.

[8] David Einstein, Daniel Lichtblau, Adam Strzebonski and Stan Wagon, Frobenius numbers by lattice point enumeration, *Integers,* **7** (2007), available at http://www.integers-ejcnt.org/

[9] Anders Jensen, Niels Lauritzen and Bjarke Roune, *Maximal lattice free bodies, test sets and the Frobenius problem,* preprint, (2007).

[10] Ravi Kannan, Test Sets for Integer Programs with ∀, ∃ Sentences, *DIMACS Series,* Vol. 1, AMS (1990), 39–47.

[11] Bjarke Roune, Solving Thousand Digit Frobenius Problems Using Grobner Bases, *Journal of Symbolic Computation,* volume 43, issue 1, (2008), 1–7.

[12] Herbert Scarf, The Core of an N-Person Game, *Econometrica,* Vol. 35, No. 1, (1967), 50–69.

[13] Herbert Scarf, Production Sets with Indivisibilities, Part I: Generalities, *Econometrica,* Vol. 49, No. 1, (1981), 1–32.

[14] Herbert Scarf, Production Sets with Indivisibilities, Part II: The Case of Two Activities, *Econometrica,* Vol. 49, No. 2, (1981), 395–423.

[15] Herbert Scarf, Neighborhood Systems for Production Sets with Indivisibilities, *Econometrica,* Vol. 54, No. 3, (1986), 507–532.

[16] Herbert Scarf and David Shallcross, The Frobenius Problem and Maximal Lattice Free Bodies, *Mathematics of Operations Research,* Vol. 18, No. 3, (1993), 511–515.

[17] Herbert Scarf and Kevin Woods, Neighborhood Complexes and Generating Functions for Affine Semigroups, *Discrete and Computational Geometry,* Vol. 35, (2006), 385–403.

[18] Bernd Sturmfels and Rekha Thomas, Variation of cost functions in integer programming, *Mathematical Programming,* Vol. 77, (1997), 357–387.

[19] Rekha Thomas, A geometric Buchberger algorithm for integer programming, *Mathematics of Operations Research,* Vol. 20, (1995), 864–884.

Herbert E. Scarf
Cowles Foundation for Research in Economics
Yale University
New Haven
CT 06520-8281
U.S.A.

e-mail: herbert.scarf@yale.edu

BOLYAI SOCIETY
MATHEMATICAL STUDIES, 19

Building Bridges
pp. 487–498.

Graph Invariants in the Edge Model

ALEXANDER SCHRIJVER

To my always inspiring friend Laci

We sharpen the characterization of Szegedy of graph invariants

$$f_b(G) = \sum_{\phi:\, EG\to[n]} \prod_{v\in VG} b(\phi(\delta(v))),$$

where b is a real-valued function defined on the collection of all multisubsets of $[n] := \{1,\dots,n\}$.

1. Introduction

Laci Lovász is a main inspirator of the new area of graph limits and graph connection matrices and their relations to graph parameters, partition functions, mathematical physics, reflection positivity, and extremal combinatorics. Prompted by Lovász's questions, Balázs Szegedy [5] characterized graph invariants in the 'edge model'. His proof is based on a highly original combination of methods from invariant theory and real algebraic geometry. It answers a question formulated in [1], which considers the corresponding 'vertex model'.

In this paper we give a sharpening of Szegedy's theorem and give a slightly shorter proof, although parts of our proof follow the scheme of Szegedy's proof. New elements of the present paper are the connections between linking of graphs and differentiation of polynomials and the use of a deep theorem of Procesi and Schwarz [4] in real invariant theory.

Let $\mathcal{G}$ be the collection of all finite graphs, where two graphs are considered to be the same if they are isomorphic. Graphs may have loops and multiple edges. Moreover, 'pointless' edges are allowed, that is, loops without a vertex. We use the notation

$$[n] := \{1, \ldots, n\} \tag{1}$$

for any $n \in \mathbb{N}$, where $\mathbb{N} = \{0, 1, 2, \ldots\}$.

A *graph invariant* is any function $f\colon \mathcal{G} \to \mathbb{R}$. In this paper, as in Szegedy [5], we consider graph invariants obtained as follows. Let $n \in \mathbb{N}$ and let A_n be the collection of all multisubsets of $[n]$, that is, of all multisets with elements from $[n]$. (So each element of $[n]$ has a 'multiplicity' in any $\alpha \in A_n$. There is a one-to-one relation between A_n and $\mathbb{N}^n$, given by the multiplicities of $i \in [n]$ in a multiset $\alpha \in A_n$.)

For $b\colon A_n \to \mathbb{R}$ define $f_b\colon \mathcal{G} \to \mathbb{R}$ by

$$f_b(G) = \sum_{\phi\colon EG \to [n]} \prod_{v \in VG} b\big(\phi\big(\delta(v)\big)\big) \tag{2}$$

for $G \in \mathcal{G}$. Here VG and EG denote the sets of vertices and edges of G, respectively, $\delta(v)$ is the set of edges incident with v, and $\phi\big(\delta(v)\big)$ is the multiset of ϕ-values on $\delta(v)$, counting multiplicities. (Actually, also $\delta(v)$ is a multiset, as loops at v occur twice in $\delta(v)$.)

Several graph invariants are equal to f_b for some appropriate b. For instance, the number of proper n-edge-colourings of a graph G is equal to $f_b(G)$, where, for $\alpha \in A_n$, $b(\alpha) = 1$ if all elements of $[n]$ have multiplicity 0 or 1 in α, and $b(\alpha) = 0$ otherwise. The number of perfect matchings in G is equal to $f_b(G)$ for $n = 2$ and $b(\alpha) = 1$ if the multiplicity of 1 in α is equal to 1, and $b(\alpha) = 0$ otherwise. For more background, see de la Harpe and Jones [3] and Freedman, Lovász, and Schrijver [1].

We characterize which graph invariants f satisfy $f = f_b$ for some $n \in \mathbb{N}$ and $b\colon A_n \to \mathbb{R}$, extending the characterizing of Szegedy [5]. We also prove that $f_b = f_c$ if and only if c arises from b by an orthogonal transformation. (Szegedy proved sufficiency here.)

2. The Characterization

To describe the characterization, call a graph invariant f *multiplicative* if $f(K_0) = 1$ and $f(GH) = f(G)f(H)$ for any $G, H \in \mathcal{G}$. Here K_0 is the graph with no vertices and no edges, and GH denotes the disjoint union of G and H.

We also need the following operation. Let u and v be distinct vertices of a graph G, and let π be a bijection from $\delta(u)$ to $\delta(v)$. (This obviously requires that $\deg(u) = \deg(v)$.) Let $G_{u,v,\pi}$ be the graph obtained as follows (where we consider the graph as topological space). Delete u and v from G, and for each $e \in \delta(u)$, reconnect e to $\pi(e)$. So the open ends of e and $\pi(e)$ are glued together with a new topological point (which however will not be a vertex). It might be that $e = \pi(e)$ (so e connects u and v), in which case we create a pointless loop.

We need a repeated application of this operation, denoted as follows. Let $u_1, v_1, \ldots, u_k, v_k$ be distinct vertices of graph G and let $\pi_i\colon \delta(u_i) \to \delta(v_i)$ be a bijection, for each $i = 1, \ldots, k$. Then we set

$$G_{u_1,v_1,\pi_1,\ldots,u_k,v_k,\pi_k} := \big(\cdots(G_{u_1,v_1,\pi_1})\cdots\big)_{u_k,v_k,\pi_k}. \tag{3}$$

Now define the $\mathcal{G} \times \mathcal{G}$ matrix $M_{f,k}$ by

$$(M_{f,k})_{G,H} := \sum_{u_1,v_1,\pi_1,\ldots,u_k,v_k,\pi_k} f\big((GH)_{u_1,v_1,\pi_1,\ldots,u_k,v_k,\pi_k}\big) \tag{4}$$

for $G, H \in \mathcal{G}$, where the sum extends over all distinct $u_1, \ldots, u_k \in VG$, distinct $v_1, \ldots, v_k \in VH$, and bijections $\pi_i\colon \delta_G(u_i) \to \delta_H(v_i)$, for $i = 1, \ldots, k$.

Theorem 1. *Let $f\colon \mathcal{G} \to \mathbb{R}$. Then $f = f_b$ for some $n \in \mathbb{N}$ and some $b\colon A_n \to \mathbb{R}$ if and only if f is multiplicative and $M_{f,k}$ is positive semidefinite for each $k = 0, 1, \ldots$.*

The positive semidefiniteness of $M_{f,k}$ can be seen as a form of 'reflection positivity' of f. In Section 6, we derive Szegedy's characterization from Theorem 1. In Section 7 we prove that b is uniquely determined by f, up to certain orthogonal transformations.

3. Some Framework

Let $\mathcal{Q}$ denote the collection of all formal real linear combinations $\sum_G \gamma_G G$ of graphs (with at most finitely many γ_G nonzero). These are called *quantum graphs.* By taking the disjoint union GH as multiplication, $\mathcal{Q}$ becomes a commutative algebra. The function f can be extended linearly to $\mathcal{Q}$.

For $G, H \in \mathcal{G}$ and $k \in \mathbb{N}$, define the quantum graph $\lambda_k(G,H)$ by

$$(5) \qquad \lambda_k(G,H) := \sum_{u_1,v_1,\pi_1,\ldots,u_k,v_k,\pi_k} (GH)_{u_1,v_1,\pi_1,\ldots,u_k,v_k,\pi_k},$$

where the sum is taken over the same set as in (4). We can extend $\lambda_k(G,H)$ linearly to a bilinear function $\mathcal{Q} \times \mathcal{Q} \to \mathcal{Q}$. As

$$(6) \qquad M_{f,k}(G,H) = f\big(\lambda_k(G,H)\big),$$

the positive semidefiniteness of $M_{f,k}$ is equivalent to the fact that

$$f\big(\lambda_k(\gamma,\gamma)\big) \geq 0$$

for each $\gamma \in \mathcal{Q}$.

For each $\alpha \in A_n$, we introduce a variable x_α. For each $G \in \mathcal{G}$, define the following polynomial in $\mathbb{R}[x_\alpha \mid \alpha \in A_n]$:

$$(7) \qquad p_n(G) := \sum_{\phi\colon EG \to [n]} \prod_{v \in VG} x_{\phi(\delta(v))}.$$

So $f_b(G) = p_n(G)(b)$ for any $b\colon A_n \to \mathbb{R}$. We extend p_n linearly to $\mathcal{Q}$.

Let $\mathcal{O}(n)$ be the group of (real) orthogonal $n \times n$ matrices. The group $\mathcal{O}(n)$ acts (linearly) on $\mathbb{R}[y_1,\ldots,y_n]$, and via the bijection

$$(8) \qquad x_\alpha \leftrightarrow \prod_{i \in \alpha} y_i$$

between the variables x_α and monomials in $\mathbb{R}[y_1,\ldots,y_n]$, $\mathcal{O}(n)$ also acts on $\mathbb{R}[x_\alpha \mid \alpha \in A_n]$. Then, by the First Fundamental Theorem of invariant theory for the orthogonal group $\mathcal{O}(n)$ (cf. Goodman and Wallach [2]), we have

$$(9) \qquad p_n(\mathcal{Q}) = \mathbb{R}[x_\alpha \mid \alpha \in A_n]^{\mathcal{O}(n)}$$

(the latter denotes as usual the set of polynomials in $\mathbb{R}[x_\alpha \mid \alpha \in A_n]$ invariant under $\mathcal{O}(n)$), as can be seen using the connection (8).

4. Derivatives

For any polynomial $p \in \mathbb{R}[x_\alpha \mid \alpha \in A_n]$, let dp be its derivative, being an element of $\mathbb{R}[x_\alpha \mid \alpha \in A_n] \otimes_{\mathbb{R}} L_n$, where L_n is the space of linear functions in $\mathbb{R}[x_\alpha \mid \alpha \in A_n]$. Then $d^k p \in \mathbb{R}[x_\alpha \mid \alpha \in A_n] \otimes_{\mathbb{R}} L_n^{\otimes k}$.

Let $\langle .,. \rangle$ be the inner product on L_n given by

$$\langle x_\alpha, x_\beta \rangle := c_\alpha \delta_{\alpha,\beta} \tag{10}$$

for $\alpha, \beta \in A_n$, where

$$c_\alpha := \prod_{i=1}^{n} \mu_i(\alpha)!, \tag{11}$$

where $\mu_i(\alpha)$ denotes the multiplicity of i in α. This induces an inner product on $L_n^{\otimes k}$ for each k. With the usual product of polynomials in $\mathbb{R}[x_\alpha \mid \alpha \in A_n]$, this gives an inner product on $\mathbb{R}[x_\alpha \mid \alpha \in A_n] \otimes_{\mathbb{R}} L_n^{\otimes k}$ with values in $\mathbb{R}[x_\alpha \mid \alpha \in A_n]$.

The following lemma is basic to our proof, and is used several times in it.

Lemma 1. *For all graphs G, H and $k, n \in \mathbb{N}$:*

$$p_n\big(\lambda_k(G,H)\big) = \big\langle d^k p_n(G), d^k p_n(H) \big\rangle. \tag{12}$$

Proof. We expand $d^k p_n(G)$:

$$\begin{aligned} d^k p_n(G) &= \sum_{\alpha_1,\dots,\alpha_k \in A_n} \sum_{\phi\colon EG \to [n]} \frac{d}{dx_{\alpha_1}} \cdots \frac{d}{dx_{\alpha_k}} \\ &\quad \times \Big(\prod_{v \in VG} x_{\phi(\delta(v))} \Big) \otimes x_{\alpha_1} \otimes \cdots \otimes x_{\alpha_k} \\ &= \sum_{\alpha_1,\dots,\alpha_k \in A_n} \sum_{\phi\colon EG \to [n]} \sum_{\substack{u_1,\dots,u_k \in VG \\ \forall i\colon \phi(\delta(u_i)) = \alpha_i}} \\ &\quad \times \Big(\prod_{v \in VG \setminus \{u_1,\dots,u_k\}} x_{\phi(\delta(v))} \Big) \otimes x_{\alpha_1} \otimes \cdots \otimes x_{\alpha_k} \end{aligned} \tag{13}$$

$$= \sum_{u_1,\ldots,u_k \in VG} \sum_{\phi\colon EG \to [n]} \Big(\prod_{v \in VG \setminus \{u_1,\ldots,u_k\}} x_{\phi(\delta(v))} \Big)$$

$$\otimes x_{\phi(\delta(u_1))} \otimes \cdots \otimes x_{\phi(\delta(u_k))}.$$

Here $u_1, \ldots, u_k$ are distinct. Now for any $\phi\colon EG \to [n]$ and $\psi\colon EH \to [n]$ and any $u \in VG$ and $v \in VH$, $\langle x_{\phi(\delta(u))}, x_{\psi(\delta(v))} \rangle$ is equal to the number of bijections $\pi\colon \delta(u) \to \delta(v)$ such that $\psi \circ \pi = \phi \mid \delta(u)$. This implies (12). ■

5. Proof of Theorem 1

To see necessity, let $b\colon A_n \to \mathbb{R}$ and $f = f_b$. Then, trivially, f is multiplicative. Positive semidefiniteness of $M_{f,k}$ follows from

$$(14) \qquad f_b\big(\lambda_k(G,H)\big) = p_n\big(\lambda_k(G,H)\big)(b) = \big\langle d^k p_n(G)(b), d^k p_n(H)(b) \big\rangle,$$

using Lemma 1.

We next show sufficiency. First we have:

Claim 1. Let γ be a quantum graph consisting of k-vertex graphs. If $f\big(\lambda_k(\gamma,\gamma)\big) = 0$ then $f(\gamma) = 0$.

Proof. We prove the claim by induction on k. So assume that the claim holds for all quantum graphs made of graphs with less than k vertices.

We can assume that all graphs occurring in γ with nonzero coefficient have the same degree sequence $d_1, \ldots, d_k$, since if we would write $\gamma = \gamma_1 + \gamma_2$, where all graphs in γ_1 have degree sequence different from those in γ_2, then $\lambda_k(\gamma_1, \gamma_2) = 0$, whence $f\big(\lambda_k(\gamma_i, \gamma_i)\big) = 0$ for $i = 1, 2$.

Now $f\big(\lambda_k(\gamma,\gamma)\big) = 0$ implies, by the positive semidefiniteness of $M_{f,k}$:

$$(15) \qquad f\big(\lambda_k(\gamma, H)\big) = 0 \quad \text{for each graph } H.$$

Let P be the graph with $2k$ vertices $1, 1', \ldots, k, k'$, where for each $i = 1, \ldots, k$, there are d_i parallel edges connecting i and i'. If $d_1, \ldots, d_k$ are all distinct, we are done, since then γ is a multiple of $\lambda_k(\gamma, P)$, implying with (15) that $f(\gamma) = 0$ – but generally there can be vertices of equal degrees.

The sum in (5) for $\lambda_k(\gamma, P)$ can be decomposed according to the set I of those components of P with both vertices chosen among $v_1, \ldots, v_k$ and to the set J of those components of P with no vertices chosen among $v_1, \ldots, v_k$ (necessarily $|I| = |J|$). Let K denote the set of components of P, and for $J \subseteq K$, let P_J be the union of the components in J. Then

$$\lambda_k(\gamma, P) = \sum_{\substack{I,J \subseteq K \\ I \cap J = \emptyset,\ |I|=|J|}} \alpha_{I,J} \gamma_I P_J, \tag{16}$$

where $\alpha_{I,J} \in \mathbb{N}$ with $\alpha_{\emptyset,\emptyset} \neq 0$, and where

$$\gamma_I := \lambda_{2|I|}(\gamma, P_I). \tag{17}$$

Now for each $I \subseteq K$, we have $\lambda_{k-2|I|}(\gamma_I, \gamma_I) = \lambda_k(\gamma, \gamma_I P_I)$. Hence

$$f\big(\lambda_{k-2|I|}(\gamma_I, \gamma_I)\big) = f\big(\lambda_k(\gamma, \gamma_I P_I)\big) = 0, \tag{18}$$

by (15). So by induction, if $I \neq \emptyset$ then $f(\gamma_I) = 0$. Therefore, by (16), since $f\big(\lambda_k(\gamma, P)\big) = 0$ and $\alpha_{\emptyset,\emptyset} \neq 0$, $f(\gamma) = f(\gamma_\emptyset P_\emptyset) = 0$. ∎

Let O be the graph just consisting of the pointless loop.

Claim 2. $f(O) \in \mathbb{N}$.

Proof. Suppose not. Then we can choose a $k \in \mathbb{N}$ with $\binom{f(O)}{k} < 0$. For each $\pi \in S_k$, let G_π be the graph with vertex set $[k]$ and edges $\{i, \pi(i)\}$ for $i = 1, \ldots, k$. (So G_π is 2-regular.) Define

$$\gamma := \sum_{\pi \in S_k} \operatorname{sgn}(\pi) G_\pi. \tag{19}$$

Then for any n:

$$\begin{aligned} p_n(\gamma) &= \sum_{\pi \in S_k} \operatorname{sgn}(\pi) p_n(G_\pi) = \sum_{\pi \in S_k} \operatorname{sgn}(\pi) \sum_{\phi\colon EG_\pi \to [n]} \prod_{v \in VG_\pi} x_{\phi(\delta(v))} \\ &= \sum_{\pi \in S_k} \operatorname{sgn}(\pi) \sum_{\phi\colon [k] \to [n]} \prod_{v \in [k]} x_{\{\phi(v), \phi\circ\pi^{-1}(v)\}} \\ &= \sum_{\phi\colon [k] \to [n]} \sum_{\pi \in S_k} \operatorname{sgn}(\pi) \prod_{v \in [k]} x_{\{\phi(v), \phi\circ\pi(v)\}}. \end{aligned} \tag{20}$$

Now if ϕ is not injective, then $\phi = \phi \circ \pi$ for some $\pi \in S_k$ with $\text{sgn}(\pi) = -1$, and hence the last inner sum is 0. So if $n < k$ then $p_n(\gamma) = 0$. If $n = k$, then $p_n(\gamma)$ contains the term $x_{\{1,1\}} \cdots x_{\{k,k\}}$ with nonzero coefficient, so $p_k(\gamma) \neq 0$.

Since $\lambda_k(\gamma, \gamma)$ is a sum of graphs with no vertices, we know that $\lambda_k(\gamma, \gamma) = q(O)$ for some polynomial $q \in \mathbb{R}[y]$, of degree at most k. Then, if $n < k$,

$$\text{(21)} \qquad q(n) = q\big(p_n(O)\big) = p_n\big(q(O)\big) = p_n\big(\lambda_k(\gamma, \gamma)\big) = 0,$$

with Lemma 1, as $p_n(\gamma) = 0$. Moreover, $q(k) = p_k\big(\lambda_k(\gamma, \gamma)\big) > 0$, as $p_k(\gamma) \neq 0$ (again using Lemma 1). So $q(y) = c\binom{y}{k}$ for some $c > 0$. Therefore,

$$\text{(22)} \qquad f\big(\lambda_k(\gamma, \gamma)\big) = f\big(q(O)\big) = q\big(f(O)\big) = c\binom{f(O)}{k} < 0,$$

contradicting the positive semidefiniteness condition. ■

This gives us n:

$$\text{(23)} \qquad n := f(O).$$

Then Claim 1 implies:

(24) there is linear function $\hat{f} \colon p_n(\mathcal{Q}) \to \mathbb{R}$ such that $f = \hat{f} \circ p_n$.

Otherwise, there is a quantum graph γ with $p_n(\gamma) = 0$ and $f(\gamma) \neq 0$. We can assume that $p_n(\gamma)$ is homogeneous, that is, all graphs in γ have the same number of vertices, k say. Hence, since $\lambda_k(\gamma, \gamma)$ is a polynomial in O, and since $f(O) = n = p_n(O)$,

$$\text{(25)} \qquad f\big(\lambda_k(\gamma, \gamma)\big) = p_n\big(\lambda_k(\gamma, \gamma)\big) = 0,$$

by Lemma 1. So by Claim 1, $f(\gamma) = 0$. This proves (24).

In fact, $\hat{f}$ is an algebra homomorphism, since for all $G, H \in \mathcal{G}$:

$$\text{(26)} \qquad \begin{aligned} \hat{f}\big(p_n(G)p_n(H)\big) &= \hat{f}\big(p_n(GH)\big) = f(GH) = f(G)f(H) \\ &= \hat{f}\big(p_n(G)\big)\,\hat{f}\big(p_n(H)\big). \end{aligned}$$

Then for all $G, H \in \mathcal{G}$, using Lemma 1:

$$\hat{f}\big(\langle\, dp_n(G), dp_n(H)\rangle\,\big) = \hat{f}\big(p_n\big(\lambda_1(G,H)\big)\big) = f\big(\lambda_1(G,H)\big) = \left(M_{f,1}\right)_{G,H}. \tag{27}$$

Since $M_{f,1}$ is positive semidefinite, (27) implies that for each $q \in p_n(\mathcal{Q})$:

$$\hat{f}\big(\langle dq, dq\rangle\big) \geq 0. \tag{28}$$

Now choose $\Delta \in \mathbb{N}$. Let $\mathcal{G}_\Delta$ be the set of graphs of maximum degree at most Δ, and let $\mathcal{Q}_\Delta$ be the set of all formal linear combinations of graphs in $\mathcal{G}_\Delta$. Define

$$A_{n,\Delta} := \big\{\, \alpha \in A_n \mid |\alpha| \leq \Delta \big\}. \tag{29}$$

By (28) and since the inner product $\langle .,.\rangle$ on L_n is $\mathcal{O}(n)$-invariant, the theorem of Procesi and Schwarz [4] (which we can apply in view of (9)) implies the existence of a $b \in \mathbb{R}^{A_{n,\Delta}}$ such that $\hat{f}(p) = p(b)$ for each $p \in p_n(\mathcal{Q}_\Delta)$. So

$$f(G) = \hat{f}\big(p_n(G)\big) = p_n(G)(b) = f_b(G) \tag{30}$$

for each $G \in \mathcal{G}_\Delta$.

This can be extended to the collection $\mathcal{G}$ of all graphs. For each d, let $K_2^{(d)}$ be the graph with two vertices, connected by d parallel edges. Then any $b\colon A_n \to \mathbb{R}$ with $f = f_b$ satisfies

$$b(\alpha)^2 \leq f\Big(K_2^{(|\alpha|)}\Big) \tag{31}$$

for $\alpha \in A_n$.

For each $\Delta \in \mathbb{N}$, define

$$B_\Delta := \Big\{ b\colon A_n \to \mathbb{R} \mid f(G) = f_b(G) \text{ for each } G \in \mathcal{G}_\Delta, \; b(\alpha)^2 \leq f\Big(K_2^{(|\alpha|)}\Big) \text{ for each } \alpha \in A_n \Big\}. \tag{32}$$

By the above, $B_\Delta \neq \emptyset$ for each Δ. As each B_Δ is compact by Tychonoff's theorem, and as $B_\Delta \supseteq B_{\Delta'}$ if $\Delta \leq \Delta'$, we know $\bigcap_\Delta B_\Delta \neq \emptyset$. Any b in this intersection satisfies $f = f_b$. ∎

Note that the positive semidefiniteness of $M_{f,k}$ for $k \neq 1$ is only used to prove Claims 2 and 1. If we know that $f = \hat{f} \circ p_n$ for some linear function $\hat{f}\colon p_n(\mathcal{Q}) \to \mathbb{R}$ and some n, then it suffices to require that $M_{f,1}$ is positive semidefinite and f is multiplicative.

6. Derivation of Szegedy's Theorem

We now derive as a consequence the theorem of Szegedy [5]. Consider some $k \in \mathbb{N}$. A *k-exit graph* is a pair (G, u) of an undirected graph G and an element $u \in VG^k$ such that the u_i are distinct vertices, each of degree 1. Let $\mathcal{G}_k$ denote the collection of k-exit graphs.

If (G, u) and (H, v) are k-exit graphs, then $(G, u)\cdot(H, v)$ is the undirected graph obtained by taking the disjoint union of G and H, and, for each $i = 1, \ldots, k$, deleting u_i and v_i and adding a new point connecting the ends left by u_i and v_i.

Let $\mathcal{G}_k$ be the collection of k-exit graphs. For $f\colon \mathcal{G} \to \mathbb{R}$, define the $\mathcal{G}_k \times \mathcal{G}_k$ matrix $N_{f,k}$ by

$$(N_{f,k})_{(G,u),(H,v)} := f\big((G, u) \cdot (H, v)\big) \tag{33}$$

for $(G, u), (H, v) \in \mathcal{G}_k$. Then ([5]):

Corollary 1a. *Let $f\colon \mathcal{G} \to \mathbb{R}$. Then $f = f_b$ for some $n \in \mathbb{N}$ and some $b\colon A_n \to \mathbb{R}$ if and only if f is multiplicative and $N_{f,k}$ is positive semidefinite for each $k \in \mathbb{N}$.*

Proof. Necessity follows similarly as in Theorem 1. To see sufficiency, let for any graph G and any $d \in \mathbb{N}$, G_d be the quantum d-exit graph being the sum of all d-exit graphs obtained as follows. Choose a vertex v of degree d, delete v topologically from G, and add vertices of degree 1 to the loose ends. Let F be the graph obtained this way. Order these new vertices in a vector in $u \in VF^d$. The sum of these (F, u) makes G_d. So G_d is a sum of precisely $d!m$ d-exit graphs, where m is the number of vertices of degree d in G.

We can repeat this to define the quantum $d_1+\cdots+d_k$-exit graph $G_{d_1,\ldots,d_k}$ for any $d_1, \ldots, d_k \in \mathbb{N}$, where we concatenate the exit vectors. Then

$$\lambda_k(G, H) = \sum_{d_1,\ldots,d_k} c_{d_1,\ldots,d_k} G_{d_1,\ldots,d_k} \cdot H_{d_1,\ldots,d_k} \tag{34}$$

for some $c_{d_1,\ldots,d_k} > 0$ (namely, the inverse of the number of permutations $\pi \in S_k$ with $d_{\pi(i)} = d_i$ for each $i \in [k]$). Hence for any quantum graph γ:

$$f\big(\lambda_k(\gamma,\gamma)\big) = \sum_{d_1,\ldots,d_k} c_{d_1,\ldots,d_k} f(\gamma_{d_1,\ldots,d_k} \cdot \gamma_{d_1,\ldots,d_k}) \geq 0, \tag{35}$$

by the positive semidefiniteness of the $N_{f,l}$. ∎

7. Uniqueness of b

We finally consider the uniqueness of b, and extend a theorem of Szegedy [5] (who showed sufficiency). As usual, b^U denotes the result of the action of U on b.

Theorem 2. *Let $b\colon A_n \to \mathbb{R}$ and $c\colon A_m \to \mathbb{R}$. Then $f_b = f_c$ if and only if $n = m$ and $c = b^U$ for some $U \in \mathcal{O}(n)$.*

Proof. Sufficiency can be seen as follows. Let $n = m$ and $c = b^U$ for some $U \in \mathcal{O}(n)$. Then for any graph G, using (9),

$$f_b(G) = p_n(G)(b) = p_n(G)^{U^{-1}}(b) = p_n(G)(b^U) = f_{b^U}(G) = f_c(G). \tag{36}$$

Conversely, let $f_b(G) = f_c(G)$ for each graph G. Then $n = f_b(O) = f_c(O) = m$. We show that for each $\Delta \in \mathbb{N}$, there exists $U \in \mathcal{O}(n)$ such that $c \mid A_{n,\Delta} = b^U \mid A_{n,\Delta}$, where $A_{n,\Delta}$ is as in (29). As $\mathcal{O}(n)$ is compact, this implies that there exists $U \in \mathcal{O}(n)$ with $c = b^U$.

Suppose that $c \mid A_{n,\Delta} \neq b^U \mid A_{n,\Delta}$ for each $U \in \mathcal{O}(n)$. Then the sets

$$S := \big\{ b^U \mid A_{n,\Delta} \mid U \in \mathcal{O}(n) \big\} \quad \text{and} \quad T := \big\{ c^U \mid A_{n,\Delta} \mid U \in \mathcal{O}(n) \big\} \tag{37}$$

are compact and disjoint subsets of $\mathbb{R}^{A_{n,\Delta}}$. So, by the Stone-Weierstrass theorem, there exists a polynomial $q \in \mathbb{R}[x_\alpha \mid \alpha \in A_{n,\Delta}]$ such that $q(s) \leq 0$ for each $s \in S$ and $q(t) \geq 1$ for each $t \in T$. Replacing q by

$$\int_{\mathcal{O}(n)} q^U \, d\mu(U) \tag{38}$$

(where μ is the normalized Haar measure on $\mathcal{O}(n)$), we can assume that $q^U = q$ for each $U \in \mathcal{O}(n)$. Hence by (9), $q \in p_n(\mathcal{Q})$, say $q = p_n(\gamma)$ with $\gamma \in \mathcal{Q}$. Then $f_b(\gamma) = p_n(\gamma)(b) = q(b) \leq 0$ and $f_c(\gamma) = p_n(\gamma)(c) = q(c) \geq 1$. This contradicts $f_b = f_c$. ■

Acknowledgement. I am indebted to Jan Draisma for pointing out reference [4] to me.

References

[1] M. H. Freedman, L. Lovász and A. Schrijver, Reflection positivity, rank connectivity, and homomorphisms of graphs, *Journal of the American Mathematical Society,* **20** (2007), 37–51.

[2] R. Goodman and N. R. Wallach, *Representations and Invariants of the Classical Groups,* Cambridge University Press, Cambridge (1998).

[3] P. de la Harpe and V. F. R. Jones, Graph invariants related to statistical mechanical models: examples and problems, *Journal of Combinatorial Theory, Series B,* **57** (1993), 207–227.

[4] C. Procesi and G. Schwarz, Inequalities defining orbit spaces, *Inventiones Mathematicae,* **81** (1985), 539–554.

[5] B. Szegedy, Edge coloring models and reflection positivity, *Journal of the American Mathematical Society,* **20** (2007), 969–988.

Alexander Schrijver
CWI and University of Amsterdam
Mailing address:
CWI, Kruislaan 413
1098 SJ Amsterdam
The Netherlands

e-mail: lex@cwi.nl

BOLYAI SOCIETY
MATHEMATICAL STUDIES, 19

Building Bridges
pp. 499–513.

Incidences and the Spectra of Graphs*

JÓZSEF SOLYMOSI†

In this paper we give incidence bounds for arrangements of curves in $\mathbb{F}_q^2$. As an application, we prove a new result that if $(x, f(x))$ is a Sidon set then either $A+A$ or $f(A)+f(A)$ should be large. The main goal of the paper is to illustrate the use of graph spectral techniques in additive combinatorics.

1. Introduction

The main goal of the paper is to illustrate the use of graph spectral techniques in additive combinatorics. The problem of finding non-trivial incidence bounds on lines and curves in $\mathbb{F}_q^2$ is closely related to sum-product estimates. In the first section we will prove Garaev's sum-product bound [14] using combinatorial arguments. Such techniques were used in similar context by Vu [27] and by Vinh [26]. Vu gave incidence bounds on polynomial curves and Vinh reproved Garaev's result, an improvement on the Bourgain–Katz–Tao incidence bound for large (larger than q) sets of points and lines in $\mathbb{F}_q^2$.

In Section 3 we sketch a spectral proof for Roth's theorem, that every dense set of integers contains three-term arithmetic progressions. There are several examples where one can choose between the Fourier method or a proof based on eigenvalues. A classical example is a discrepancy theorem for arithmetic progressions by Roth [23], who used the Fourier transform.

*The research was conducted while the author was member of the Institute for Advanced Study. Funding provided by The Charles Simonyi Endowment.

†The research was supported by NSERC and OTKA grants and by Sloan Research Fellowship.

Later, Lovász and Sós proved the theorem using eigenvalues. (see in [3] or in [8] on page 20.)

In the third part of the paper we present new results. We partially answer a question of Bourgain [6], giving incidence bounds similar to Garaev's, but for a more general family of curves. It is a finite field extension of a theorem of Elekes, Nathanson, and Ruzsa. Applying Elekes' incidence method [11], Elekes, Nathanson, and Ruzsa proved in [13] the following. Let $f\colon \mathbb{R} \to \mathbb{R}$ be a convex function. Then for any finite set $A \subset \mathbb{R}$,

$$\max\left\{|A+A|, \left|f(A)+f(A)\right|\right\} \geq c|A|^{5/4} \tag{1}$$

In the inequality $A+A$ denotes the set of pairwise sums, $A+A=\{a+b\colon a,b\in A\}$ and $f(A)=\left\{f(a)\colon a\in A\right\}$. We don't have the notion of a convex function in $\mathbb{F}_q$, so we will use a weaker condition on f to get results in $\mathbb{F}_q$ similar to (1).

2. The Sum-Product Problem

An old conjecture of Erdős and Szemerédi states that if A is a finite set of integers then the sumset or the productset should be large. The sumset of A was defined earlier and the productset is defined in a similar way,

$$A\cdot A=\{ab \mid a,b\in A\}.$$

Erdős and Szemerédi conjectured that the sumset or the productset is almost quadratic in the size of A, i.e.

$$\max\left(|A+A|, |A\cdot A|\right) \geq c|A|^{2-\delta}$$

for any positive δ.

Bourgain, Katz, and Tao proved a nontrivial, $|A|^{1+\varepsilon}$, lower bound for the finite field case [5]. Let $A\subset \mathbb{F}_p$ and $p^{\alpha}\leq |A|\leq p^{1-\alpha}$. Then there is an $\varepsilon>0$ depending on α only, such that

$$\max\left(|A+A|, |A\cdot A|\right) \geq c|A|^{1+\varepsilon}.$$

It is important that p is prime, otherwise one could select A being a subring in which case both the product set and the sum set are small, equal to $|A|$. For the case, $\mathbb{F}_q$, where q is a power of an odd prime, the best known bound is due to Garaev [14]. It follows from a construction of Ruzsa, that his bound is asymptotically the best possible in the range $|A| \geq q^{2/3}$. Garaev's proof uses bounds on exponential sums. We are going to derive similar sum-product estimates using spectral bounds for graphs.

Sum-product bounds have important applications, not only to number theory, but to computer science, Ramsey theory, and cryptography.

2.1. The Sum-Product graph

The vertex set of the sum-product graph G_{SP} is the Cartesian product of the multiplicative subgroup and the field, $V(G_{SP}) = \mathbb{F}_q^* \times \mathbb{F}_q$ (as before, q is a power of an odd prime). Two vertices, $u = (a, b)$ and $v = (c, d) \in V(G_{SP})$, are connected by and edge, $(u, v) \in E(G_{SP})$, iff $ac = b + d$. This multigraph (there are a few loops) has a very special structure which makes it easy to compute the second largest eigenvalue of the graph. The set of eigenvalues are given by the eigenvalues of the adjacency matrix of the graph. The matrix is symmetric, so all $q(q-1)$ eigenvalues are real, we can order them, $\mu_0 \geq \mu_1 \geq \dots \geq \mu_{q^2-q-1}$. The second largest eigenvalue, λ, is defined as $\lambda = \max\left(\mu_1, |\mu_{q^2-q-1}|\right)$. Using λ, one can write isoperimetric inequalities on the graph. In order to do so, we give a bound on λ. First, observe that for any two vertices, $u = (a, b)$ and $v = (c, d) \in V(G_{SP})$, if $a \neq c$ and $b \neq d$, then the vertices have exactly one common neighbor, $N(u, v) = (x, y) \in V(G_{SP})$.

The unique solution of the system

$$\text{(2)} \qquad \left.\begin{aligned} ax &= b + y \\ cx &= d + y \end{aligned}\right\} \quad \text{is given by} \quad \begin{aligned} x &= (b-d)(a-c)^{-1} \\ 2y &= x(a+c) - b - d. \end{aligned}$$

If $a = c$ or $b = d$, then the vertices, u, v, have no common neighbors. Let M denote the adjacency matrix of G_{SP}, that is $a_{ij} = 1$ if $(v_i, v_j) \in E(G_{SP})$, and $a_{ij} = 0$ otherwise. M is a symmetric matrix, moreover

$$M^2 = J + (q-2)I - E,$$

where J is the all-one matrix, I is the identity matrix, and E is the "error matrix", the adjacency matrix of the graph, G_E, where for any two vertices, $v_i = (a, b)$ and $v_j = (c, d) \in V(G_{SP})$, $(v_i, v_j) \in E(G_E)$ iff $a = c$ or $b = d$. As G_{SP} is a $(q-1)$-regular graph, $q-1$ is an eigenvalue of M with the all-one eigenvector, $\vec{1}$. The matrix M is symmetric, so that eigenvectors of other eigenvalues are orthogonal to $\vec{1}$. It is a corollary of the Spectral Theorem, that there is an orthonormal basis, V, consisting of eigenvectors of M. Let θ denote the second largest eigenvalue of M. The graph, G_{SP}, is connected so the eigenvalue $q-1$ has multiplicity one, and the graph is not bipartite, so for any other eigenvalue, θ, $|\theta| < q-1$. A corresponding eigenvector is denoted by $\vec{v_\theta}$. Let us multiply both sides of the matrix equation above by $\vec{v_\theta}$. The "trick" is that $J\vec{v_\theta} = 0$, as the eigenvectors are orthogonal to the all-one vector, so we get:

$$(\theta^2 - q + 2)\vec{v_\theta} = E\vec{v_\theta}.$$

Note that E has the same set of eigenvectors as M has. G_E is a $2q-1$-regular graph, so any eigenvalue of E is at most $2q-1$ in absolute value.

$$\theta^2 - q + 2 \leq 2q,$$

$$|\theta| < \sqrt{3q}.$$

2.2. The spectral bound

The small value of the second largest eigenvalue shows us that G_{SP} is a quasirandom graph and we can bound the number of edges between large vertex sets efficiently. We are going to use following Cheeger-type discrepancy bound; For any two sets of vertices, $S, T \subset V(G_{SP})$,

$$\left| e(S,T) - \frac{|S|\,|T|}{q} \right| \leq \lambda\sqrt{|S|\,|T|}, \tag{3}$$

where $e(S,T)$ is the number of edges between S and T. (see e.g. in [10] or [1].) Inequality 3 and the bound on λ imply that

$$e(S,T) \leq \frac{|S|\,|T|}{q} + \sqrt{3q|S|\,|T|}. \tag{4}$$

From (4) we can deduct Garaev's sum-product bound [14]. We can suppose that $0 \notin A$, WLOG. Set $S = (AA) \times (-A)$ and $T = (A^{-1}) \times (A + A)$. There is an edge between any two vertices $(ab, -c) \in S$ and $(b^{-1}, a+c) \in T$, therefore the number of edges between S and T is at least $|A|^3$. On the other hand

$$|A|^3 \le e(S,T) \le \frac{|S|\,|T|}{q} + \sqrt{3q|S|\,|T|}$$
$$= \frac{|AA|\,|A+A|\,|A|^2}{q} + \sqrt{3q|AA|\,|A+A|\,|A|^2}.$$

After rearranging the inequality we get the desired sum-product bound.

$$|A+A|\,|AA| \gg \min\left\{q|A|, \frac{|A|^4}{q}\right\}.$$

In particular, if $|A| \approx q^{2/3}$, then $\max\{\,|AA|, |A+A|\} \gg |A|^{5/4}$.

3. 3-term Arithmetic Progressions

In the previous example it was enough to show that the second largest eigenvalue is small. There are cases where we can not guarantee that the second eigenvalue is small; however when it is large then we might find some structure in the graph. To illustrate this, we will sketch one of the several possible proofs of Roth's theorem [22].

Theorem 3.1 (Roth's Theorem). *For any $N \ge 3$ if $S \subset [1,\dots,N]$ and $|S| \gg N/\log\log N$ then S contains a 3-term arithmetic progression.*

Note, that it is enough to prove Roth's theorem modulo a prime p. For any $p \ge 3$ if $S \subset \mathbb{F}_p$ and $|S| \gg p/\log\log p$ then S contains a 3-term arithmetic progression. Indeed, choose p that $p \ge 3N$ and translate S that it is in the middle third of the interval $[1,\dots,p]$. In this way any arithmetic progression modulo q is also a "regular" arithmetic progression.

3.2. The 3-AP graph

To prove the "mod p" variant, we define a graph, G_{3AP}, on $2p-1$ vertices. We label the vertices by $v_0, v_1, \dots v_{p-1}$, and $v_{-1}, v_{-2}, \dots, v_{-p+1}$., A way to think of the vertices if they were the $(2p-1)$-th roots of unity, assigning v_j to $\exp\left(\frac{2\pi\imath}{2p-1}j\right)$. The neighbours of v_0 are defined by the set S in the following way; v_i is connected to v_0 by an edge iff $|i| \in S$. (Suppose that $0 \notin S$.) Extend the graph by adding the edges necessary that the mapping, $i \mapsto i+1 \pmod{2p-1}$, is an automorphism of G_{3AP}. Using the roots of unity notation, it means that multiplying the vertices by $\exp(\frac{2\pi\imath}{2p-1}j)$ is an automorphism of the graph for any integer j. (It is the Cayley graph of $\mathbb{Z}/(2p-1)\mathbb{Z}$ with respect to S.) For graphs with a "nice" automorphism group, finding the eigenvectors and eigenvalues is not a hard task. (see Exercise 8. in [18], Chapter 16 in [4], or [19] for a more detailed description) In our case it is easy to check that for this circulant graph, $2p-1$ linearly independent eigenvectors are given by the vectors

$$\left[\exp\left(\frac{2\pi\imath k}{2p-1}\right), \exp\left(\frac{4\pi\imath k}{2p-1}\right), \exp\left(\frac{6\pi\imath k}{2p-1}\right), \ldots, \exp\left(\frac{2(2p-1)\pi\imath k}{2p-1}\right)\right]^T,$$

where $0 \le k \le 2p-2$. Then the eigenvalues of G_{3AP} are given by the sums

$$\theta_k = \sum_{s\in S} \exp\left(\frac{2\pi\imath sk}{n}\right) + \sum_{s\in S} \exp\left(\frac{-2\pi\imath sk}{n}\right).$$

There are two possibilities. Either the second largest eigenvalue is large or all eigenvalues but the largest, $\mu_0 = 2|S|$, are small. In the former case, most of the summands have large positive real part. It implies that there is a long arithmetic progression having a very large intersection with S. We won't explore this case here, instead we show that if all eigenvalues are small then there is a 3-term arithmetic progression in S. The interested reader will find the details for the density increment case in Roth's original paper [22], or in one of the many books discussing Roth's theorem, like [15],[25], or [16]. Our moderate plan here is to show that if $|S|^2/(2p-1) > \lambda$ then S contains a 3AP.

We can find a relation between the assumption that S has no 3-term arithmetic progressions (it is *3AP-free*) and the structure of the graph G_{3AP}. We show that if S is 3AP-free then there are large vertex sets spanning less than expected edges. For every edge we can define its halving point. Consider the edges as arcs between points on the unit circle. The points are the vertices, represented by the roots of unity and the edges are the shorter circular arcs. The halving point of the edge is the geometric halving point of the circular arc. The number of possible halving points is $4p-2$. The number of edges is $|S|(2p-1)$, so there is a point which is the halving point of at least $\lceil |S|/2 \rceil$ edges. Note that if we had two edges sharing the same halving point, such that there is another edge between the two-two endvertices separated by the halving point, that would imply that there is a 3AP in S. Indeed, if there are two edges sharing the same halving point, h, then the endpoints of the edges can be written as $h + d_1, h - d_1$ and $h + d_2, h - d_2$. If $h + d_2$ and $h - d_1$ are connected by an edge, it means that $d_1 + d_2$ is in S with $2d_1$ and $2d_2$, forming a 3AP.

If S is 3AP-free then between the two $\lceil |S|/2 \rceil$-size sets of endvertices, A and B, there are exactly $\lceil |S|/2 \rceil$ edges. Inequality 3 implies that

$$\left| e(A,B) - \frac{2|S| \lceil |S|/2 \rceil^2}{2p-1} \right| \leq \lambda l \lceil |S|/2 \rceil,$$

from where we get that

$$\frac{|S|^2}{2p-1} \leq \lambda,$$

as we wanted to show.

4. Sidon functions

In this section we extend a result of Elekes, Nathanson, and Ruzsa [13] to the finite field case.

Theorem 4.1 (Elekes, Nathanson, and Ruzsa). *Let $f: \mathbb{R} \to \mathbb{R}$ be a convex function. Then for any finite set $A \subset \mathbb{R}$,*

$$\max \left\{ |A+A|, \left| f(A) + f(A) \right| \right\} \geq c|A|^{5/4}$$

4.2. Sidon functions

We need a notation which substitutes convexity in finite fields. The graph of a convex function is a Sidon set in $\mathbb{R}^2$, this is the property we are going to use for finite fields. A set $H \subset \mathbb{F}_q \times \mathbb{F}_q$ is a *Sidon set* if for any $h_i, h_j, h_k, h_l \in H$ the equation

$$h_i - h_j \equiv h_k - h_l \pmod{q}$$

implies $i = k$ and $j = l$. A function, $f: S \to \mathbb{F}_q$ for some $S \subset \mathbb{F}_q$, is said to be a *Sidon function* if its graph $H = \{(x, f(x)) : x \in S\}$ is a Sidon set. Note that the graph of any convex function in $\mathbb{R}^2$ forms a Sidon set.

Theorem 4.3. *For any integer, k, and for any $S \subset \mathbb{F}_q$, $|S| \geq q - k$, if $f: S \to \mathbb{F}_q$ is a Sidon function, then for any set $A \subset S$, and sets $B, C \subset \mathbb{F}_q$,*

$$|A + B||f(A) + C| \geq \min\left\{\frac{q|A|}{2}, \frac{|A|^2|B||C|}{8(k+1)q}\right\}.$$

Using the right substitution for C and D, Theorem 4.3 gives the following corollaries.

Corollary 4.4. *For any integer, k, there is a constant, $c = c(k)$, such that for any $S \subset \mathbb{F}_q$, $|S| \geq q - k$, if $f: S \to \mathbb{F}_q$ is a Sidon function, then for any set $A \subset S$,*

$$|A + A||f(A) + f(A)| \geq c \min\left\{q|A|, \frac{|A|^4}{q}\right\}.$$

It is remarkable that the inequality above matches to the Elekes–Nathanson–Ruzsa bound for sets A such that $|A| \approx q^{2/3}$. It has a single term variant, which we state in a separate statement.

Corollary 4.5. *For any integer, k, there is a constant, $c = c(k)$, such that for any $S \subset \mathbb{F}_q$, $|S| \geq q - k$, if $f: S \to \mathbb{F}_q$ is a Sidon function, then for any set $A \subset S$,*

$$|A + f(A)| \geq c \min\left\{\sqrt{q|A|}, \frac{|A|^2}{\sqrt{q}}\right\}.$$

4.6. A bipartite incidence graph

The proof of Theorem 4.3 is based on the following incidence bound. Let f be a function, $S \to \mathbb{F}_q$, for some $S \subset \mathbb{F}_q$. The *graph of f* is the set of points $\{(x, f(x)) \in \mathbb{F}_q \times \mathbb{F}_q : x \in S\}$. A translate of f by a vector $u = (u', u'') \in \mathbb{F}_q \times \mathbb{F}_q$, is the set $T_u(f) = \{(x + u', f(x) + u'') : x \in S\}$. The translate of the *mirror graph of f* is defined as $T_u(f)^\top = \{(u' - x, u'' - f(x)) : x \in S\}$.

Lemma 4.7. *For any integer, k, and for any $S \subset \mathbb{F}_q$, $|S| = q-k$ if $f\colon S \to \mathbb{F}_q$ is a Sidon function, then for any set $P \subset \mathbb{F}_q \times \mathbb{F}_q$, the number of incidences between P and s translates of f, the set $\left\{T_{u_i}(f)\right\}_{i=1}^{s}, u_i = (u_i', u_i'')$, is bounded as follows;*

$$\sum_{i=1}^{s} \left|\left\{x \in S\colon \left(x+u_i', f(x)+u_i''\right) \in P\right\}\right| \le \frac{|P|s}{q} + \sqrt{2(k+1)q|P|s}.$$

Proof. Define a bipartite graph, $G(A,B)$, as follows. The vertex set of G consists of two copies of $\mathbb{F}_q \times \mathbb{F}_q$.

The edges of $G(A,B)$ are given by the graph of f. Two vertices, $u = (u', u'') \in A = \mathbb{F}_q \times \mathbb{F}_q$, and $v = (v', v'') \in B = \mathbb{F}_q \times \mathbb{F}_q$, are connected by an edge in G if

$$f(v' - u') = v'' - u''. \tag{5}$$

The neighborhood of a vertex $u \in A$ is given by $N(u) = T_u(f) \subset B$, and neighborhood of a vertex $u \in B$ is described by $N(v) = T_v(f)^{\tau} \subset A$. The graph, $G(A,B)$, is a $(q-k)$-regular bipartite graph. The spectra of $G(A,B)$ is symmetric. For this graph the second largest eigenvalue is defined as $\lambda = \mu_1$. As the graph is $(q-k)$-regular, the largest and the smallest eigenvalues are $q-k$ and $k-q$. Similarly as we did in the sum-product example, we can bound λ by examining the $q^2 \times q^2$ adjacency matrix of $G(A,B)$, denoted by M. The function f is a Sidon function, therefore the neighborhoods of two vertices in A or in B intersect in at most one vertex. A translate, $T_u(f)$, covers $\binom{q-k}{2}$ vertex pairs. All translates (the neighborhoods of vertices) cover $2\binom{q-k}{2}q^2$ pairs out of the $2\binom{q^2}{2}$ vertex pairs in A and in B. Let us define an error graph, H, which has two components, one in A and one in B, and two vertices, u and v are connected by an edge iff there is no vertex connected to both in $G(A,B)$. The error graph, H, has $2(\binom{q^2}{2} - q^2\binom{q-k}{2})$ edges and it is regular of degree $q^2 - 1 - (q-k)(q-k-1)$. Its adjacency matrix is denoted by E.

$$M^2 = \begin{bmatrix} J & 0 \\ 0 & J \end{bmatrix} + (q-k-1)I - E.$$

As in the first example, we can multiply the equation by an eigenvector of M, which belongs to the second largest eigenvalue.

$$E\overrightarrow{v_\lambda} = (q-k-1-\lambda^2)\overrightarrow{v_\lambda}.$$

We know that H is $(2kq+q-k^2-k-1)$-regular, therefore any eigenvalue of E is less or equal to $2kq+q-k^2-k-1$.

$$|q-k-1-\lambda^2| \le 2kq+q-k^2-k-1,$$

so

$$\lambda < \sqrt{2(k+1)q}.$$

4.8. The spectral bound

For bipartite graphs, like $G(A,B)$, inequality (3) is slightly different. If $G(A,B)$ is a r-regular bipartite graph on n vertices, then for any subsets $A' \subset A$ and $B' \subset B$,

$$\left| e(A',B') - \frac{2r|A'|\,|B'|}{n} \right| \le \lambda\sqrt{|A'|\,|B'|}.$$

Now we are ready to complete the proof of Lemma 4.7, to state a bound on incidences between a set of points, P, and s translates, $\left\{T_{u_i}(f)\right\}_{i=1}^{s}$, $u_i = (u_i', u_i'')$. An edge in the graph $G(A,B)$ represents an incidence between a point and a translate.

$$\sum_{i=1}^{s} \left|\left\{x \in S\colon \left(x+u_i', f(x)+u_i''\right) \in P\right\}\right| \le \frac{|P|s}{q} + \sqrt{2(k+1)q|P|s}. \qquad \blacksquare$$

Proof of Theorem 4.3. Let us consider the Cartesian product $(A+B)\times(f(A)+C)$. It has $|B|\,|C|$ translates of the smaller product $A\times f(A)$, which contains an $|A|$-element subset of the graph of f. The $|B|\,|C|$ translates determine $|A|\,|B|\,|C|$ incidences in $(A+B)\times(f(A)+C)$. Now we apply Lemma 4.7 with substitutions $s=|B|\,|C|$, $|P| = |A+B|\big|f(A)+C\big|$, and with $|A|\,|B|\,|C|$ incidences. $\blacksquare$

Note that Theorem 4.3 generalizes Garaev's point-line incidence bound, since the mapping $(x,y)\mapsto(x,y+x^2)$ maps any line, $ax+by=c, a\neq 0$, to a translate of the parabola, $y=x^2$, which is a Sidon function.

For any set $P\subset \mathbb{F}_q\times\mathbb{F}_q$, the number of incidences between P and s lines is bounded by

$$O\left(\frac{|P|s}{q} + \sqrt{q|P|s}\right). \tag{6}$$

5. Incidence Bounds on Pseudolines

Incidence bounds in geometries have various applications. The celebrated theorem of Szemerédi and Trotter [24] gives sharp incidence bound for the number of point-line incidences in the Euclidean plane. The Szemerédi–Trotter Theorem was extended to pseudolines. For the details about variants of the planar Szemerédi–Trotter theorem we refer to [21].

Our goal here is finding non-trivial incidence bounds for pseudolines in $\mathbb{F}_q^2$. First we give a definition of pseudolines which form a partial geometry in $\mathbb{F}_q^2$. The incidence graph will be a strongly regular graph, therefore we can use standard spectral bounds to estimate incidences.

5.1. The incidence bound

The following is a standard, (however not the only) definition of pseudolines in the Eucledean plane, see in e.g. [2].

A collection $\mathcal{L}$ of x-monotone unbounded Jordan curves in the plane is called a family of pseudolines if every pair of curves intersects in at most one point.

To find a proper definition of pseudolines in finite fields isn't so straightforward. We are going to use one possible definition which has interesting applications. Instead of x-monotone unbounded Jordan curves we consider "lines", $l_i = \{(x, f(x)) : x \in \mathbb{F}_q\}$, where $f : \mathbb{F}_q \to \mathbb{F}_q$.

Definition 5.2. A collection $\mathcal{L}$ of subsets of $\mathbb{F}_q^2$, $\mathcal{L} = \{l_1, l_2, \ldots, l_k\}$ is called a family of pseudolines if the following conditions hold

a, For every $a \in \mathbb{F}_q$, any set, l_i, has exactly one point with x-coordinate a.

b, Every pair of sets, l_i and l_j, intersects in at most one point.

c, If $l_i \in \mathcal{L}$, then its y-translates are also in the arrangement, $l_i + (0, a) \in \mathcal{L}$ for any $a \in \mathbb{F}_q$.

The last condition implies that the size of a family of pseudolines is divisible by q.

Theorem 5.3. *Let a family of pseudolines, $\mathcal{L}$, and a family of points, P, in $\mathbb{F}_q^2$ be given. Suppose that $|\mathcal{L}| = kq$, and $|P| = n$. Then the number of* *incidences between m pseudolines and n points is bounded by*

$$I(m,n) \leq n\sqrt{km/q} + \sqrt{qnm}.$$

The incidence bound for pseudolines implies a new bound on point line incidences. It is better than inequality 6 in line arrangements with a few distinct slopes only.

Corollary 5.4. *Let a family of pseudolines, $\mathcal{L}$, and a family of points, P, in $\mathbb{F}_q^2$ be given. Suppose that the lines have no more than k different slopes. Then the number of incidences between the s lines and points of P is bounded by*

$$I\big(s,|P|\big) \leq |P|\sqrt{ks/q} + \sqrt{q|P|s}.$$

The bound is better than inequality (6) if $k < s/q$. To see how Theorem 5.3 implies Corollary 5.4, observe that the set all lines with slopes from a given set, forms a family of pseudolines.

The incidence bound in Theorem 5.3 is a corollary of the following statement which is proved in the next subsection.

Theorem 5.5. *Given a family of pseudolines, $\mathcal{L}$, and two sets of points, P_1 and P_2 in $\mathbb{F}_q^2$. Suppose that $|\mathcal{L}| = kq$, $|P_1| = n_1$, and $|P_2| = n_2$. Then the number of collinear pairs in $P_1 \times P_2$ is bounded as*

$$\big|\big\{\,(p_i,p_j)\colon p_i \in P_1, p_j \in P_2, \exists\,\ell \in \mathcal{L}\colon p_i,p_j \in \ell\big\}\,\big| \leq \frac{kn_1n_2}{q} + q\sqrt{n_1n_2}.$$

Proof of Theorem 5.3. Suppose that the m pseudolines are incident to $t_1,t_2,\ldots,t_m$ points in P. Then, the number of copseudolinear pairs in P is at least $\sum_{i=1}^m \binom{t_i}{2}$. On the other hand, $I(m,n) = \sum_{i=1}^m t_i$, so the number of copseudolinear pairs is at least $m\binom{I(m,n)/m}{2} \sim I(m,n)^2/m$. Using the inequality from Theorem 5.5 we have

$$\frac{I(m,n)^2}{m} \leq \frac{kn^2}{q} + qn,$$

concluding the proof of Theorem 5.3. ∎

5.6. Strongly regular graphs

In [7] Bose introduced the notation of partial geometries. A set of points and lines is a finite partial geometry if there are integers such that:

i For each two different points p and q, there is at most one line incident with both of them.

ii Each line is incident with $r+1$ points.

iii Each point is incident with $s+1$ lines.

iv If a point p and a line L are not incident, then there are exactly t points on L collinear to p.

Lemma 5.7. *Any family of pseudolines is a partial geometry.*

The easy proof is left to the reader.

Proof of Theorem 5.5. The incidence graph, $G(\mathcal{L})$, of a family of pseudolines is defined as follows. G has q^2 vertices, the elements of $\mathbb{F}_q^2$. Two vertices v and u are connected iff the points are collinear, i.e. there is a line, $l \in \mathcal{L}$, such that $v, u \in l$. As we observed earlier, the number of lines is divisible by q. There is an integer, k, $1 \leq k \leq q$, such that $|\mathcal{L}| = kq$.

$G(\mathcal{L})$ is a strongly regular graph, where each vertex has degree $k(q-1)$. Two collinear (adjacent) vertices have $q-2+(k-1)(k-2) = q+k^2-3k$ common neighbors and non-adjacent vertices have k^2-k common neighbors. The adjacency matrix of the graph is denoted by A.

$$A^2 = (q+k^2-3k)A + (k^2-k)(J-A-I) + k(q-1)I.$$

The usual trick – multiplying both sides by an eigenvector – helps us to find the eigenvalues. The adjacency matrix of this graph has only three distinct eigenvalues. The largest is $k(q-1)$ and the other two are $q-k$ and $-k$. (For more details about such graphs we refer to [17].) In our applications $q \gg k$, so the second largest eigenvalue is $q-k$. From this, Theorem 5.5 follows immediately by applying inequality (3). ∎

Acknowledgement. I am indebted to the anonymous referee for his valuable comments on my previous draft.

REFERENCES

[1] N. Alon and J. Spencer, *The Probabilistic Method,* Wiley, 3rd. ed. 2008.

[2] P. K. Agarwal and M. Sharir, *Pseudoline Arrangements: Duality, Algorithms, and Applications,* 2001.

[3] J. Beck and V. T. Sós, Discrepancy theory, in: *Handbook of Combinatorics,* Chapter 26, eds., Graham, Grötschel, Lovász, North-Holland (1995), 1405–1446.

[4] N. Biggs, *Algebraic Graph Theory,* Cambridge University Press, 1974, 1993.

[5] J. Bourgain, N. Katz, and T. Tao, A sumproduct estimate in finite fields and applications, *GAFA,* **14** (2004), 27-57.

[6] J. Bourgain, (personal communication, October 10, 2007)

[7] R.C. Bose, Strongly regular graphs, partial geometries and partially balanced designs, *Pacific J. Math.,* Volume **13**, Number **2** (1963), 389–419.

[8] B. Chazelle, *The Discrepancy Method: Randomness and Complexity,* Cambridge University Press, 2000.

[9] J. Cheeger, A lower bound for the smallest eigenvalue of the Laplacian, *Problems in Analysis,* Papers dedicated to Salomon Bochner, 1969, Princeton University Press, Princeton, 195–199.

[10] F. R. K. Chung, *Spectral Graph Theory,* Providence, RI: American Mathematical Society. no 92. (1997).

[11] Gy. Elekes, On the Number of Sums and Products, *Acta Arithmetica,* **LXXXI.4** (1997), 365–367.

[12] P. Erdős and E. Szemerédi, On sums and products of integers, *Studies in Pure Mathematics,* 213–218, Birkhäuser, Basel, 1983.

[13] Gy. Elekes, M. B. Nathanson, and I. Z. Ruzsa, Convexity and Sumsets, *Journal of Number Theory,* **83** (1999), 194–201.

[14] M. Z. Garaev, An Explicit Sum-Product Estimate in $\mathbb{F}_q$, *International Mathematics Research Notices* (2007), Vol. 2007, 1–11.

[15] R. Graham, B. Rothschild, and J. Spencer, *Ramsey Theory,* Wiley-Interscience Series in Discrete Mathematics, second ed. (1990),

[16] A. Granville, M. B. Nathanson, and J. Solymosi, Additive Combinatorics, *CRM Proceedings & Lecture Notes, AMS,* vol. **43** (2007).

[17] C. Godsil and G. Royle, Algebraic Graph Theory, §10 Strongly Regular Graphs, Springer, *Graduate Text in Mathematics,* **207** (2001).

[18] L. Lovász, Combinatorial problems and exercises, *AMS Chelsea Publishing* (1979), 639 pp.

[19] L. Lovász, Spectra of graphs with transitive groups, *Periodica Mathematica Hungarica,* Vol. **6 (2)** (1975), 191–195.

[20] L. Lovász, J. Spencer and K. Vesztergombi, Discrepancy of set systems and matrices, *European J. Combinatorics,* **7** (1986), 151–160.

[21] J. Pach and P. Agarwal, *Combinatorial Geometry,* Wiley, New York, 1995.

[22] K. F. Roth, On certain sets of integers, I, *Journal of the London Mathematical Society,* **28** (1953), 104-109.

[23] K. F. Roth, Remark concerning integer sequences, *Acta Arithmetica,* **9** (1964), 257–260.

[24] E Szemerédi and W. T. Trotter, Extremal problems in discrete geometry, *Combinatorica,* **3** (1983), 381–392.

[25] T. Tao and V. Vu, *Additive Combinatorics,* Series: Cambridge Studies in Advanced Mathematics (No. 105) Cambridge University Press (2006).

[26] L. A. Vinh, *Szemeredi–Trotter type theorem and sum-product estimate in finite fields,* preprint. arXiv:0711.4427v1 [math.CO]

[27] V. H. Vu, Sum-product estimates via directed expanders, *Math. Res. Lett.,* **15** (2008), no. 2, 375-388.

József Solymosi

Department of Mathematics
UBC, 1984 Mathematics Road
Vancourver, BC
Canada V6T 1Z2

e-mail: solymosi@@math.ubc.ca

BOLYAI SOCIETY
MATHEMATICAL STUDIES, 19

Building Bridges
pp. 515–524.

The Maturation of the Probabilistic Method

JOEL SPENCER

In this historical review we discuss probability results of László Lovász and Svante Janson. These results have, we feel, played a central role in the development of the Probabilistic Method.

1. Introduction

As we celebrate the sixtieth birthday of László Lovász it is natural to examine the many places where his theorems and insights have had a vital influence on our field. In this note I examine the Local Lemma, discovered by Lovász in the early 1970s. I also discuss the Janson Inequality, discovered by Svante Janson July 2, 1989. Both results have short arguments which are given (albeit, not in the most general forms) in this note. The similarities of the arguments makes one wonder if they aren't simply examples of some deeper phenomenon which we have yet to understand. In this personal account I shall include my own small role in these events.

2. Common Elements

Let A_i, $i \in I$ be a finite set of events in some probability space. We say that a graph G with vertex set I is a *Dependency Graph* for the events if for every i the event A_i is mutually independent of the set of events A_j with $j \neq i$ and j not adjacent to i.

The Dependency Graph is not uniquely defined. Indeed, the complete graph on vertex set I is always trivially a dependency graph. However, in many situations there will be a natural Dependency Graph. Often we are examining a random coloring of some finite set S. The colors of the $s \in S$ need not have the same distribution (though often they do) but they do need to be mutually independent. Let A_i be an event that refers to the coloring on a set $B_i \subset C$. For example, A_i could be that B_i is entirely Red. We can then create a dependency graph by letting i, j be adjacent if and only if $B_i \cap B_j \neq \emptyset$.

For the Janson Inequality we have the above situation and in addition we assume the following *Correlation Inequalities.*

1. For every $i \in I$ and every $J \subseteq I - \{i\}$

$$\Pr\left[A_i \,\Big|\, \bigwedge_{j\in J} \overline{A_j}\right] \leq \Pr\left[A_i\right] \tag{1}$$

2. For every distinct $i, k \in I$ and every $J \subseteq I - \{i, k\}$

$$\Pr\left[A_i \,\Big|\, A_k \wedge \bigwedge_{j\in J} \overline{A_j}\right] \leq \Pr\left[A_i \mid A_k\right] \tag{2}$$

Here the frequent example is a random subset R of some finite set S. Each $s \in S$ can be in the subset with different probability (though often these probabilities are the same) but the events $s \in R$ need to be mutually independent. The event A_i is that a certain $B_i \subseteq S$ has all of its elements in R.

In both cases our object of concern is the event we shall call `NEEDLE`. We define

$$\texttt{NEEDLE} = \wedge_{i\in I}\overline{A_i} \tag{3}$$

Our notion is that the A_i are our "bad" events and we want to "sieve out" an object that has no bad events. In many interesting cases there are many bad events and the probability of `NEEDLE` is exponentially small. We have the notion of finding a needle in an exponential haystack. The Lovász Local Lemma gives a *lower bound* on $\Pr[\texttt{NEEDLE}]$. In application the critical point is that this bound is not zero, so that $\texttt{NEEDLE} \neq \emptyset$. Thus, by Erdős Magic, there is an object with no bad events. In the Janson framework, the Correlation Inequalities give an immediate lower bound. $\Pr[\texttt{NEEDLE}]$ is at least what it would be if the events were mutually independent. The importance of the Janson Inequality is that it gives an *upper bound* on Pr [NEEDLE]. In many (not all!) cases the upper and lower bounds are quite close and one gets an asymptotic evaluation of Pr [NEEDLE].

The Lovász and Janson have another similarity. Label $I = \{1, \ldots, n\}$ for convenience. Then

$$\Pr[\texttt{NEEDLE}] = \prod_{i=1}^{n} \Pr\left[\overline{A_i} \;\middle|\; \bigwedge_{j=1}^{i-1} \overline{A_j}\right] \tag{4}$$

(For $i = 1$ the term is simply $\Pr[\overline{A_1}]$.) The difficulty, and the beauty, lies in finding a way to approximate the conditional probabilities in (4). Our book [1] gives these results and generalities in more detail.

3. Lovász

Perhaps it is just human nature to believe that a subject really got started at the time one began working in it. Probabilistic Methods had a clear beginning, with the two page 1947 paper [4] of Paul Erdős giving the lower bound on the Ramsey function $R(k, k)$. For two decades Erdős gave a series of beautiful results. The methods were ingenious, even brilliant, but in retrospect we can see that the use of probability was fairly elementary. Linearity of Expectation, Chernoff Bounds and Chebyschev's Inequality were the main tools. In the early 1970s this changed. A key event was a sieve method, the Lovász Local Lemma.

Theorem 3.1. *Let A_i, $i \in I$ be events with a Dependency Graph G as defined earlier. Let $p \in [0, 0.25)$ and integer $d \geq 1$ be such that:*

1. $\Pr[A_i] \leq p$ *for all* $i \in I$
2. *Every vertex i of G has degree at most d*
3. $4dp < 1$

Then

$$\Pr[\texttt{NEEDLE}] > (1 - 2p)^{|I|} \tag{5}$$

In particular

$$\texttt{NEEDLE} \neq \emptyset \tag{6}$$

The idea is to show that for every i and every $J \subseteq I - \{i\}$

$$\Pr\left[A_i \,\middle|\, \bigwedge_{j\in J} \overline{A_j}\right] \le 2p \tag{7}$$

Suppose this holds. For convenience, write $I = \{1,\ldots,n\}$. All the conditional probabilities in (4) would then have probability at least $1-2p$ so that (5) follows. The result (7) is shown by induction on $|J|$, the number of events in the conditioning. For $J = \emptyset$ we have simply $\Pr[A_i] \le p \le 2p$. Otherwise, for convenience, let us renumber so that $J = \{1,\ldots,i-1\}$. Let us further renumber so that $D = \{1,\ldots,d'\}$ is the set of $j \in J$ adjacent to i. Thus $d' \le d$. If $D = \emptyset$ then from the mutual independence

$$\Pr\left[A_i \,\middle|\, \bigwedge_{j=1}^{i-1} \overline{A_j}\right] = \Pr[A_i] \le p \le 2p \tag{8}$$

The heart of the argument is when $D \ne \emptyset$. Given any three events A, B, C we have $\Pr[A \mid B \wedge C] = \Pr[A \wedge B \mid C]/\Pr[B \mid C]$. Here we let $B = \wedge_{j\in D}\overline{A_j}$, peeling off the "dependent part."

$$\Pr\left[A_i \,\middle|\, \bigwedge_{j=1}^{i-1} \overline{A_j}\right] = \frac{\Pr\left[A_i \wedge \bigwedge_{j=1}^{d'} \overline{A_j} \mid \bigwedge_{k=d'+1}^{i-1} \overline{A_k}\right]}{\Pr\left[\bigwedge_{j=1}^{d'} \overline{A_j} \mid \bigwedge_{k=d'+1}^{i-1} \overline{A_k}\right]} \tag{9}$$

Let NUM, DEN denote the numerator and denominator respectively of (9). We bound the numerator from above. We first strip off the $\overline{A_j}$, $1 \le j \le d'$ and then use that A_i is mutually independent of the $\overline{A_k}$:

$$\texttt{NUM} \le \Pr\left[A_i \,\middle|\, \bigwedge_{k=d'+1}^{i-1} \overline{A_k}\right] \le \Pr[A_i] \le p \tag{10}$$

We first switch the denominator to its complement:

$$\texttt{DEN} = 1 - \Pr\left[\bigvee_{j=1}^{d'} A_j \,\middle|\, \bigwedge_{k=d'+1}^{i-1} \overline{A_k}\right] \tag{11}$$

As the probability of a disjunction is at most the sum of the probabilities:

$$\texttt{DEN} \ge 1 - \sum_{j=1}^{d'} \Pr\left[A_j \,\middle|\, \bigwedge_{k=d'+1}^{i-1} \overline{A_k}\right] \tag{12}$$

Now we use induction on $|J|$. Each of the addends in (12) has a conditioning on fewer than $|J|$ events. Thus each is at most $2p$. Thus

$$\texttt{DEN} \geq 1 - \sum_{j=1}^{d'} 2p \geq 1 - 2pd' \geq 1 - 2pd \geq \frac{1}{2} \tag{13}$$

Thus

$$\frac{\texttt{NUM}}{\texttt{DEN}} \leq \frac{p}{1/2} \leq 2p \tag{14}$$

completing the induction hypothesis for (7).

I first heard about this result before I had met László Lovász. The early 1970s were part of the Cold War era and communication between West and East was generally difficult. I had already met Paul Erdős and had begun work on Probabilistic Methods. Thus it was with some joy that I received a letter from János Komlós. He described the breakthrough of Lovász that I have given above. The Local Lemma first appeared in joint work [3] with Paul Erdős. One could see immediately that it would have wide application. I recalled the already classic Erdős 1947 result [4]:

Theorem 3.2. *Assume*

$$\binom{n}{k} 2^{1-\binom{k}{2}} < 1 \tag{15}$$

There exists a two-coloring of the edges of the complete graph K_n with no monochromatic K_k.

Using now standard notation, the conclusion is that the Ramsey Function $R(k,k) > n$. The proof is today just two words: Color Randomly! For each set S of k vertices let A_S be the event that all of the edges of S are the same color. Then

$$\Pr\left[\bigvee_S A_S\right] \leq \sum_S \Pr\left[A_S\right] = \binom{n}{k} 2^{1-\binom{k}{2}} < 1 \tag{16}$$

Some calculation gives

$$R(k,k) > \big(1 + o(1)\big) \frac{e}{k\sqrt{2}} \sqrt{2}^k \tag{17}$$

But now there was an improvement!

Theorem 3.3. *Assume*

$$4\binom{k}{2}\binom{n}{k-2}2^{1-\binom{k}{2}} < 1 \tag{18}$$

There exists a two-coloring of the edges of the complete graph K_n with no monochromatic K_k.

There is a natural dependency graph on the events A_S. S, T are adjacent if $|S \cap T| \geq 2$. Otherwise, the events involve coloring of disjoint sets of edges. For each S, $\binom{k}{2}\binom{n}{k-2}$ is an upper bound on the number of adjacent T. I rushed to make the calculation which gave:

$$R(k,k) > \big(1 + o(1)\big)\frac{e\sqrt{2}}{k}\sqrt{2}^k \tag{19}$$

This was disappointing. The upper bound, known to Erdős and Szekeres from the earliest days, is roughly $R(k,k) < 4^{n(1+o(1))}$. The true asymptotics of $R(k,k)$ remain a mystery. The lower bound (19) has never been improved. My paper [7] was, of course, a trivial consequence of the deep result of Lovász. It had the very positive effect for me of strengthening my relation with Hungary and Hungarian Mathematics, a relationship that has been a cornerstone of my professional and personal life.

4. Janson

The main importance of the Lovász result, Theorem 3.1, is in showing that `NEEDLE` is nonempty. In the Janson set-up this follows immediate from the Correlation Inequalities. The new result is the *upper* bound on Pr [`NEEDLE`].

We set

$$M = \prod_{i \in I} \Pr\left[\overline{A_i}\right] \tag{20}$$

We think of M as what Pr [`NEEDLE`] would be *if* the A_i were mutually independent. We set

$$\mu = \sum_{i \in I} \Pr\left[A_i\right] \tag{21}$$

Thus μ is the expected number of the events A_i to hold when a random sample is taken from the distribution. We set

$$\Delta = \sum \Pr[A_i \wedge A_j] \tag{22}$$

where the sum is over those ordered pairs $i \neq j$ with $\{i,j\} \in G$. (The expression $\Delta/2$ is then the sum over unordered pairs.) The value Δ gives a quantitative measure of the dependency of the A_i.

Theorem 4.1. *Let A_i, $i \in I$ with dependency graph G, satisfying the Correlation Inequalities* $(1,2)$ *above, and with M, μ, Δ as above. Then*

$$M \leq \Pr[\texttt{NEEDLE}] \leq e^{-\mu+(\Delta/2)} \tag{23}$$

For both upper and lower bounds we set $I = \{1, \ldots, n\}$ for convenience and consider the product (4). Correlation Inequality (1) gives that the conditioning can only increase $\Pr[\overline{A_i}]$ so that

$$\Pr[\texttt{NEEDLE}] \geq \prod_{i=1}^{n} \Pr[\overline{A_i}] = M \tag{24}$$

In the other direction we need an upper bound on the probability of $\overline{A_i}$ conditional on $\wedge_{j<i}\overline{A_j}$. Equivalently, we seek a lower bound on the probability of A_i conditional on $\wedge_{j<i}\overline{A_j}$. We use the inequality $\Pr[A \mid B \wedge C] \geq \Pr[A \mid C] \Pr[B \mid A \wedge C]$. Let i be fixed. Let D be the set of $j < i$ which are adjacent to i in G. Let E be the set of $k < i$ which are not adjacent to i in G. We pull the dependent events across the conditioning:

$$\Pr\left[A_i \,\middle|\, \bigwedge_{j<i} \overline{A_j}\right] \geq \Pr\left[A_i \,\middle|\, \bigwedge_{j\in E} \overline{A_j}\right] \Pr\left[\bigwedge_{D} \overline{A_j} \,\middle|\, A_i \wedge \bigwedge_{E} \overline{A_k}\right] \tag{25}$$

From the mutual independence $\Pr\left[A_i \mid \bigwedge_{j\in E}\overline{A_j}\right] = \Pr[A_i]$. The second term in the product (25) is (similarly to Lovász!) bounded from below by

$$\Pr\left[\bigwedge_{D} \overline{A_j} \,\middle|\, A_i \wedge \bigwedge_{E} \overline{A_k}\right] \geq 1 - \sum_{j\in D} \Pr\left[A_j \,\middle|\, A_i \wedge \bigwedge_{E} \overline{A_k}\right] \tag{26}$$

Correlation Inequality (2) gives that each addend is at most $\Pr[A_j \mid A_i]$ so that

$$\Pr\left[A_i \,\Big|\, \bigwedge_{j<i} \overline{A_j}\right] \geq \Pr[A_i]\Big(1 - \sum_{j\in D} \Pr[A_j \mid A_i]\Big) \tag{27}$$

Reversing

$$\Pr\left[\overline{A_i} \,\Big|\, \bigwedge_{j<i} \overline{A_j}\right] \leq 1 - \Pr[A_i] + \sum_{j\in D} \Pr[A_j \wedge A_i] \tag{28}$$

We switch to exponentials, using the inequality $1 + z \leq e^z$, valid for all z. Then

$$\Pr\left[\overline{A_i} \,\Big|\, \bigwedge_{j<i} \overline{A_j}\right] \leq \exp\left[-\Pr[A_i] + \sum_{j\in D} \Pr[A_j \wedge A_i]\right] \tag{29}$$

The product (4) then gives

$$\Pr\left[\bigwedge_{i} \overline{A_i}\right] \leq \exp\left[\sum_{i} \Big[-\Pr[A_i] + \sum_{j\in D} \Pr[A_j \wedge A_i]\Big]\right] \tag{30}$$

The terms $-\Pr[A_i]$ sum to $-\mu$. There is one term $\Pr[A_j \wedge A_i]$ for each unordered pair i, j which are adjacent in G. Thus these terms sum to $\Delta/2$. Thus

$$\Pr\left[\bigwedge_{i} \overline{A_i}\right] \leq e^{-\mu+(\Delta/2)} \tag{31}$$

as claimed.

My connection with the Janson Inequality was quite strong. Let $G \sim G(n,p)$ be the usual Random Graph and let NOTRI be the event that G does not contain a triangle. Let $\mu = \binom{n}{3}p^3$ denote the expected number of triangles. Suppose, first, that $p \sim \frac{c}{n}$ with c a constant. Then $\mu \sim c^3/6$. Paul Erdős and Alfred Rényi, in their 1960 masterwork [5] showed that $\Pr[\texttt{NOTRI}] \sim e^{-\mu}$. They used the method of moments. Letting X denote the number of triangles they showed that the factorial moments satisfied $E\big[X^{(r)} \sim \mu^r$ for each fixed r. This implied that X had a limiting Poisson distribution so that, in particular, $\Pr[\texttt{NOTRI}] = \Pr[X = 0] \sim e^{-\mu}$. But what happens when $p \gg n^{-1}$ so that μ is not a constant but rather goes to infinity with n. While the argument for each moment (as long as p is not too large) goes through, the conclusion that X is asymptotically Poisson can no longer be made. I had pondered this question for quite some time when I came to the Random Structures and Algorithms meeting in Poznań, Poland in 1989. On the evening of July 2 (from Janson's personal recollection) I spoke with Janson about the question. How many times do we have mathematical conversations that are fascinating at the time but ultimately lead nowhere. Not this time! Janson came down the next morning with what we now call Janson's Inequality. His original proof used the Stein–Chen method, the more elementary argument above is due to Ravi Boppana and myself [2] and was derived from Janson's approach.

Let $G(n,p)$ be the underlying probability space. We think of $G(n,p)$ as a random subset of the set of $\binom{n}{2}$ potential edges, thus fitting the Janson scheme. Let the index set I be the set of triples of vertices and for each such triple $S = \{x,y,z\}$ let A_S be the event that x,y,z forms a triangle. Now `NEEDLE` is the event that $G(n,p)$ has no triangles. The critical calculation is of Δ, as defined in (22). Here the sum is over pairs of triangles $S = \{x,y,z\}$, $T = \{x,y,w\}$ that overlap on an edge. There are $\binom{n}{3} \cdot 3 \cdot (n-2) = \Theta(n^4)$ such pairs and for each $\Pr[A_S \cap A_T] = p^5$, as five edges must lie in $G(n,p)$. Thus $\Delta = \Theta(n^4 p^5)$. (Note: The calculation of Δ looked only at *pairs* of events, the more complex interactions which the Erdős-Rényi approach examined did not concern us!) When $p \sim cn^{-1}$, $M \sim e^{-\mu} \sim e^{-c^3/6}$, $\Delta \sim 0$, so Theorem 4.1 gives the result of Erdős and Rényi that $\Pr[\texttt{NEEDLE}] \sim e^{-\mu}$. But now allow p to be much bigger, suppose only that $p = o(n^{-4/5})$. Calculation still gives $M \sim e^{-\mu}$. Most importantly, we still have $\Delta \sim 0$. Therefore Theorem 4.1 *still* gives the result $\Pr[\texttt{NEEDLE}] \sim e^{-\mu}$. Note that if, say, $p = n^{-0.9}$, this probability is extremely small, yet Janson's Inequality gives its asymptotic values. This was greatly extended in [6] where bounds on the probability that $G(n,p)$ did not contain a copy of a fixed graph H were given for all H and all p.

References

[1] N. Alon and J. Spencer, *The Probabilistic Method,* 3rd ed., John Wiley, 2008.

[2] R. Boppana and J. Spencer, A Useful Elementary Correlation Inequality, *J. Comb. Th. - Ser. A,* **50** (1989), 305–307.

[3] P. Erdős and L. Lovász, Problems and results on 3-chromatic hypergraphs and some related questions, in: *Infinite and Finite Sets* (A. Hajnal et al., eds.), North-Holland (1975), pp. 609–628.

[4] P. Erdős, Some remarks on the theory of graphs, *Bull. Amer. Math. Soc.*, **53** (1947), 292–294.

[5] P. Erdős and A. Rényi, On the evolution of random graphs, *Magyar Tud. Akad. Mat. Kutató Int. Közl.*, **5** (1960), 17–61.

[6] S. Janson, T. Łuczak and A. Ruciński, An exponential bound for the probability of nonexistence of a specified subgraph in a random graph, in: *Random graphs '87* (Poznań, 1987), Wiley (1990), pp. 73–87.

[7] J. Spencer, Ramsey's Theorem – A new lower bound, *J. Comb. Th. – Ser. A,* **18** (1975), 108–115.

Joel Spencer

Courant Institute of Mathematical Sciences
New York

e-mail: spencer@cims.nyu.edu

BOLYAI SOCIETY
MATHEMATICAL STUDIES, 19

Building Bridges
pp. 525–545.

A Structural Approach to Subset-Sum Problems

VAN VU*

To Kati and Laci

We discuss a structural approach to subset-sum problems in additive combinatorics. The core of this approach are Freiman-type structural theorems, many of which will be presented through the paper. These results have applications in various areas, such as number theory, combinatorics and mathematical physics.

1. Introduction

Let $A = \{a_1, a_2, \dots\}$ be a subset of an additive group G (all groups discussed in this paper will be abelian). Let S_A be the collection of subset sums of A

$$S_A := \Bigl\{ \sum_{x \in B} x \mid B \subset A,\ |B| < \infty \Bigr\}.$$

Two related notions that are frequently considered are

$$lA := \{a_1 + \cdots + a_l \mid a_i \in A\}$$

$$l^*A := \{a_1 + \cdots + a_l \mid a_i \in A,\ i \neq j\}.$$

We have the trivial relations

$$l^*A \subset lA \quad \text{and} \quad \cup_l\, l^*A = S_A.$$

*V. Vu is supported by NSF Career Grant 0635606.

One can have similar definitions for A being a sequence (repetitions allowed).

Example.

$$A = \{0,1,4\}, \quad G = \mathbb{Z}, \quad 2A = \{0,1,2,4,5,8\},$$
$$2^*A = \{1,4,5\}, \quad S_A = \{0,1,4,5\}.$$

$$A = \{0,1,4\}, \quad G = \mathbb{Z}_5, \quad 2A = G, \quad 2^*A = \{0,1,4\} = S_A.$$

Now let A be a sequence:

$$A = \{1,1,9\}, \quad G = \mathbb{Z}, \quad 3A = \{3,11,19,27\},$$
$$3^*A = \{11\}, \quad S_A = \{1,1,2,9,10,11\}.$$

Notice that for a large l, lA can be significantly different from S_A and l^*A. In general, it is easier to handle than the later two.

Many basic problems in additive combinatorics have the following form: *If A is sufficiently dense in G, then S_A (or l^*A or lA) contains a special element (such as 0 or a square), or a large structure (such as a long arithmetic progression G itself).*

The main question is to find the threshold for "dense". As examples, we present below a few well-known results/problems in the area. In the whole paper, we are going to focus mostly on two special cases: (1) $G = \mathbb{Z}_p$, where $\mathbb{Z}_p$ denotes the cyclic group of residues modulo a large prime p; (2) $G = \mathbb{Z}$, the set of integers.

Following the literature, we say that A is *zero-sum-free* if $0 \notin S_A$. Furthermore, A is *complete* if $S_A = G$ and *incomplete* otherwise. The asymptotic notation is used under the assumption that $|A| \to \infty$.

A basic result concerning zero-sum-free sets is the following theorem of Olson [53] and Szemerédi [60] from the late 1960s, addressing a problem of Erdős and Heilbronn [23].

Theorem 1.1 (Olson–Szemerédi). *Let A be a subset of $\mathbb{Z}_p$ with cardinality $C\sqrt{p}$, for a sufficiently large constant C. Then S_A contains zero.*

To see that order $\sqrt{p}$ is necessary, consider $A := \{1,2,\dots,n\}$, where $n \approx \sqrt{2p}$ is the largest integer such that $1 + \cdots + n < p$.

Concerning completeness, Olson [52], proved the following result

Theorem 1.2 (Olson). *Any subset A of $\mathbb{Z}_p$ with cardinality at least $\sqrt{4p-3}+1$ is complete.*

To see that the bound is close to optimal, take $A := \{-m, \dots, -1, 0, 1, \dots, m\}$ where m is the largest integer such that $1 + \cdots + m < \lfloor p/2 \rfloor$.

Another classical result concerning zero sums is that of Erdős–Ginburg–Ziv [42], again from the 1960s.

Theorem 1.3 (Erdős–Ginburg–Ziv). *If A is a sequence of $2p-1$ elements in $\mathbb{Z}_p$, then p^*A contains zero.*

This theorem is sharp by the following example: $A = \{0^{[p-1]}, 1^{[p-1]}\}$ Furthermore, instead of 0 and 1, one can use any two different elements of $\mathbb{Z}_p$. (Here and later $x^{[k]}$ means x appears with multiplicity k in A.)

Now we discuss two problems involving integers. Set $[n] := \{1, 2, \dots, n\}$. An old and popular conjecture concerning subset sums of integers is Folkman's conjecture, made in 1966 [25]. Folkman's conjecture is a strengthening of a conjecture by Erdős [20] about finding a necessary and sufficient condition for a sequence A such that S_A contains all but finite exception of the positive integers.

Conjecture 1.4 (Folkman's conjecture). The following holds for any sufficiently large constant C. Let A be an strictly increasing sequence of positive integers with (asymptotic) density at least $C\sqrt{n}$ (namely, $|A \cap [n]| \geq C\sqrt{n}$ for all sufficiently large n). Then S_A contains an infinite arithmetic progression.

Cassels [10] and Erdős [20] showed that density $\sqrt{n}$ is indeed needed; thus, Folkman's conjecture is sharp up to the value of C. For more discussion about Folkman's conjecture and its relation with Erdős' conjecture, we refer to [25] and the monograph [21] by Erdős and Graham.

Finally, a problem involving a non-linear relation, posed by Erdős in 1986 [19].

Problem 1.5 (Erdős' square-sum-free problem). A set A of integers is square-sum-free if S_A does not contain a square. Find the largest size of a square-sum-free subset of $[n]$.

Erdős observed that one can construct such a square-sum-free subset of $[n]$ with at least $\Omega(n^{1/3})$ elements. To see this, consider $A := \{q, 2q, \dots, kq\}$ with q prime, $(k+1)k < 2q$, $kq \leq n$. Since the sum of all elements of A is less than q^2, S_A does not contain a square. Erdős [19] conjectured that the truth is close to this lower bound.

Problems involving subset sums such as the above (and many others) have been attacked, with considerable success, using various techniques: combinatorial, harmonic analysis, algebraic etc. The reader who is interested in these techniques may want to look at [3, 57, 64, 48] and the references therein.

The goal of this paper is to introduce the so-called "structural approach", which has been developed systematically in recent years. This approach is based on the following simple plan

Step I: Force a structure on A. In this step, one tries to show the following: If A is relatively dense (close to the conjectured threshold but not yet there) and S_A does not contain the desired object, then A has a very special structure.

Alternately, one can try to

Step I': Find a structure in S_A. If A is relatively dense (again close to the conjectured threshold but not yet there) then S_A contains a special structure.

Step II: Completion. Since $|A|$ is still below the threshold, we can add (usually many) new elements to A. Using these elements together with the existing structure, one can, in most cases, obtain the desired object in a relatively simple manner.

The success of the method depends on the quality of the information we can obtain on the structure of A (or S_A) in Step I (or I'). In several recent studies, it has turned out that one can frequently obtain something close to a *complete characterization* of these sets. Thanks to these results, one is able to make considerable progresses on many old problems and also reprove and strengthen several existing ones (with a better understanding and a complete classification of the extremal constructions).

The rest of this paper is devoted to the presentation of these structural theorems and their representative applications.

2. Freiman's Structural Theorem

A corner stone in additive combinatorics is the structural theorem of Freiman (sometime referred to as Freiman's inverse theorem), which writes down the structure of sets with small doubling.

A *generalized arithmetic progression* (GAP) of rank d in a group G is a set of the form

$$\{a_0 + a_1x_1 + \cdots + a_dx_d \mid M_i \leq x_i \leq N_i\},$$

where a_i are elements of G and $M_i \leq N_i$ are integers. It is intuitive to view a GAP Q as the image of the d-dimensional integral box $B := \{(x_1, \ldots, x_d) \mid M_i \leq x_i \leq N_i\}$ under the linear map

$$\Phi(x_1, \ldots, x_d) = a_0 + a_1x_1 + \cdots + a_dx_d.$$

We say that Q is *proper* if Φ is one-to-one. It is easy to see that if Q is a proper GAP of rank d and A is a subset of density δ of Q, then $|2A| \leq C(d,\delta)|A|$. Indeed,

$$|2A| \leq |2Q| \leq |2B| = 2^d|B| = 2^d|Q| \leq \frac{2^d}{\delta}|A|$$

since the volume of a box increases by a factor 2^d if its sizes are doubled.

Freiman's theorem shows that this is the *only* construction of sets with constant doubling.

Theorem 2.1 (Freiman's theorem) [27]. *For any positive constant C, there are positive constants $d = d(C)$ ad $\delta = \delta(C)$ such that the following holds. Let A be a finite subset of a torsion-free group G such that $|2A| \leq C|A|$. Then there is a proper GAP Q of dimension d such that $A \subset Q$ and $|A| \geq \delta|Q|$.*

Freiman theorem has been extended recently to the torsion case by Green and Ruzsa [35]. [64, Chapter 5] contains a detailed discussion of both theorems and related results.

One can use Freiman's theorem iteratively to treat the sumset lA, for $l > 2$. For simplicity, assume that $l = 2^s$ is a power of 2. Thus, the set $A_s := lA = 2^sA$ can be viewed as $2A_{s-1}$ where $A_{s-1} := 2^{s-1}A$. Using a multi-scale analysis combined with Fremain's theorem, one can obtain useful structural information about lA or A itself. For an example of this technique, we refer to [61] or [64, Chapter 12].

The treatment of l^*A and S_A is more difficult. However, one can still develop structural theorems in these cases. While the content of most theorems in this direction are quite different from that of Freiman's, they do bear a similar spirit that somehow the most natural construction happens to be (essentially) the only one.

3. Structure of Zero-Sum-Free Sets

Let A be a zero-sum-free subset of $\mathbb{Z}_p$. We recall the example following Theorem 1.1. Let $A := \{1, 2, \ldots, n\}$. If $1 + \cdots + n < p$, then obviously S_A does not contain 0. This shows that a zero-sum-free set can have close to $\sqrt{2p}$ elements. In [61], Szemerédi and Vu showed that *having elements with small sum is essentially the only reason for a set to be zero-sum-free.* More quantitative versions of this statement were worked out in [49] and [50]. For example, we have [50, Theorem 2.2]

Theorem 3.1. *After a proper dilation (by some non-zero element), any zero-sum-free subset A of $\mathbb{Z}_p$ has the form*

$$A = A' \cup A''$$

where the elements of A' (viewed as integers between 0 and $p-1$) are small, $\sum_{x \in A'} x < p$, and A'' is negligible, $|A''| \leq p^{6/13+o(1)}$.

One can perhaps improve the constant 6/13 by tightening the analysis in [50]. It is not clear, however, what would be the best constant here. In most applications, it suffices to have any constant strictly less than 1/2.

The dilation is necessary. Notice that if A is zero-sum-free (incomplete), then the set $A_x := \{xa \mid a \in A\}$ is also zero-sum-free (incomplete) for any $0 \neq x \in \mathbb{Z}_p$.

We can also prove similar results for lA and l^*A, and for A being a sequence (see [50] for details). In the rest of this section, we present few applications of these results.

3.1. The size of the largest zero-sum-free set in $\mathbb{Z}_p$

Let m_p denote the size of the largest zero-sum-free set in $\mathbb{Z}_p$. The problem of determining m_p was posed by Erdős and Heilbronn [23] and has a long history. Szemerédi proved that $m_p \leq C\sqrt{p}$, for some sufficiently large C independent of p [60]. Olson showed that $C = 2$ suffices [53]. Much later, Hamidoune and Zémor [37] showed that $m_p \leq \sqrt{2p} + 5\log p$, which is asymptotically sharp. Using an earlier version of Theorem 3.1, Szemerédi, Nguyen and Vu [49] recently obtained the exact value of m_p.

Theorem 3.2. *Let n_p be the largest integer so that $1 + \cdots + (n_p - 1) < p$.*

- *If $p \neq \frac{n_p(n_p+1)}{2} - 1$, then $m_p = n_p - 1$.*
- *If $p = \frac{n_p(n_p+1)}{2} - 1$, then $m_p = n_p$. Furthermore, up to a dilation, the only zero-sum-free set with n_p elements is $\{-2, 1, 3, 4, \ldots, n_p\}$.*

The same result was obtained by Deshouillers and Prakash (personal communication by Deshouillers) at about the same time.

3.2. The structure of relatively large zero-sum-free sets

Let us now consider the structure of zero-sum-free sets of size close to $\sqrt{2p}$. Let $\|x\|$ denote the integer norm of x. In [15], Deshouillers proved

Theorem 3.3. *Let A be a zero-sum-free subset of Z_p of size at least $\sqrt{p}$. Then (after a proper dilation)*

$$\sum_{x\in A,\ x<p/2} \|x/p\| \leq 1 + O(p^{-1/4}\log p)$$

$$\sum_{x\in A,\ x>p/2} \|x/p\| \leq O(p^{-1/4}\log p).$$

Deshouillers showed (by a construction) that the error term $p^{-1/4}$ cannot be replace by $o(p^{-1/2})$. Using an earlier version of Theorem 3.1, Nguyen, Szemerédi and Vu [49] improved Theorem 3.3 to obtain the best possible error term $O(p^{-1/2})$, under a stronger assumption on the size of $|A|$.

Theorem 3.4. *Let A be a zero-sum-free subset of Z_p of size at least $.99\sqrt{2p}$. Then (after a proper dilation)*

$$\sum_{x\in A,\ x<p/2} \|x/p\| \leq 1 + O(p^{-1/2})$$

$$\sum_{x\in A,\ x>p/2} \|x/p\| \leq O(p^{-1/2}).$$

The constant .99 is, of course, ad-hoc and can be improved by redoing the analysis carefully. On the other hand, it is not clear what the best assumption on $|A|$ should be.

3.3. Erdős–Ginburg–Ziv revisited

Using a version of Theorem 3.1 for sequences, Nguyen and Vu [50] obtained the following characterization for a sequence of size slightly more than p and does not contain a subsequence of p elements summing up to 0.

Corollary 3.5 [50, Theorem 6.2]. *Let ε be an arbitrary positive constant. Assume that A is a p-zero-sum-free sequence and $p+p^{12/13+\varepsilon} \leq |A| \leq 2p-2$. Then A contains two elements a and b with multiplicities m_a, m_b satisfying $m_a + m_b \geq 2\big(|A| - p - p^{12/13+\varepsilon}\big)$.*

The interesting point here is that the structure kicks in very soon, when A has just slightly more than p elements. Few years ago, Gao, Panigrahi, and Thangdurai [43] proved a similar statement under the stronger assumption that $|A| \geq 3p/2$.

It is easy to deduce Erdős–Ginburg–Ziv theorem from Corollary 3.5, together with a complete characterization of the extremal sets. The reader may want to consider as an exercise or check [50] for details.

4. Incomplete Sets

Now we turn our attention to incomplete sets, namely sets A where $S_A \neq \mathbb{Z}_p$. The situation here is very similar to that with zero-sum-free sets. Szemerédi and Vu [61] showed that *having elements with small sum is essentially the only reason for a set to be incomplete.* More quantitative versions of this statement were worked out in [49] and [50]. For example, in [50], the following analogue of Theorem 3.1 was proved

Theorem 4.1. *After a proper dilation (by some non-zero element), any incomplete subset A of Z_p has the form*

$$A = A' \cup A''$$

where the elements of A' are small (in the integer norm), $\sum_{x \in A'} \|x/p\| < 1$ and A'' is negligible, $|A''| \le p^{6/13+o(1)}$.

The reader can find similar results for lA and l^*A and for A being a sequence in [50]. We next discuss some applications of these results.

4.1. The structure of relatively large incomplete sets

Theorem 4.1 enables us to prove results similar to those in the last section for incomplete sets. The problem of determining the largest size of an incomplete set in $\mathbb{Z}_p$ was first considered by Erdős and Heilbronn [23] and essentially solved by Olson (Theorem 1.2). da Silva and Hamidoune [12] tightened the bound to $\sqrt{4p-7}+1$, which is best possible. We are not going to go into these results here, but note that one can perhaps obtain a new proof (with a characterization of the extremal sets) using Theorem 4.1.

Concerning the structure of relatively large incomplete sets, Deshouillers and Freiman [17] proved

Theorem 4.2. *Let A be an incomplete subset of $\mathbb{Z}_p$ of size at least $\sqrt{2p}$. Then (after a proper dilation)*

$$\sum_{x \in A} \|x/p\| \le 1 + O(p^{-1/4} \log p).$$

They conjectured that the error term may be replaced by $O\left(\sqrt{p}\right)$, which would be best possible due to a later construction of Deshouillers [16].

Using Theorem 4.1, Nguyen and Vu [50] confirmed this conjecture, provided that A is sufficiently close to $2\sqrt{p}$.

Theorem 4.3. *Let A be an incomplete subset of Z_p of size at least $1.99\sqrt{p}$. Then (after a proper dilation)*

$$\sum_{x \in A} \|x/p\| \le 1 + O(p^{-1/2}).$$

Similar to the constant .99 (in the previous section), the constant 1.99 is ad-hoc and can be improved by redoing the analysis carefully. On the other hand, it is not clear what the best assumption on $|A|$ is.

4.2. The structure of incomplete sequences

Let us now discuss (rather briefly) the situation with sequences. The main difference between sets and sequences is that a sequence can have elements with high multiplicities. It has turned out that when the maximum multiplicity of incomplete sequence A is determined, one can obtain strong structural information about A.

Let $1 \le m \le p$ be a positive integer and A be an incomplete sequence of $\mathbb{Z}_p$ with maximum multiplicity m. Trying to make A as large as possible, we come up with the following example,

$$B_m = \{ -n^{[k]}, (n-1)^{[m]}, \dots, -1^{[m]}, 0^{[m]}, 1^{[m]}, \dots, (n-1)^{[m]}, n^{[k]} \}$$

where $1 \le k \le m$ and n are the unique integers satisfying

$$2m(1+2+\cdots+n-1) + 2kn < p \le 2m(1+2+\cdots+n-1) + 2(k+1)n.$$

It is clear that any subsequence of B_m is incomplete and has multiplicity at most m. In [50], we proved that any incomplete sequence A with maximum multiplicity m and cardinality close to $|B_m|$ is essentially a subset of B_m.

Theorem 4.4. *Let $6/13 < \alpha < 1/2$ be a constant. Assume that A is an incomplete sequence of $\mathbb{Z}_p$ with maximum multiplicity m and cardinality $|A| = |B_m| - O\big((pm)^\alpha\big)$. Then after a proper dilation, we can have*

$$A = A' \cup A''$$

where $A' \subset B_m$ and $|A''| = O\big((pm)^{(\alpha+1/2)/2}\big)$.

4.3. Counting problems

Sometime one would like to count the number of sets which forbid certain additive configurations. A well-known example of problems of this type is the Cameron–Erdős problem [11], which asked for the number of subsets of $[n] = \{1, 2, \ldots, n\}$ which does not contain three different elements x, y, z such that $x + y - z = 0$. Cameron an Erdős noticed that any set of odd numbers has this property. Thus, in $[n]$ there are at least $\Omega(2^{n/2})$ subsets with the required property. They conjectured that $2^{n/2}$ is the right order of magnitude. There were several partial results [2, 9, 22] before Green settled the conjecture [34].

Using structural theorems such as Theorem 3.1, one can obtain results of similar spirit for the number of zero-sum-free or incomplete sets and sequences. For example, using an earlier version of Theorem 3.1 and standard facts from the theory of partitions [1], Szemerédi and Vu [61] proved

Corollary 4.5. *The number of incomplete subsets of* $\mathbb{Z}_p$ *is*

$$\exp\left(\left(\sqrt{\frac{2}{3}}\pi + o(1)\right)\sqrt{p}\right).$$

Using Theorem 4.4, one obtains the following generalizations [50].

Corollary 4.6. *The number of incomplete sequences A with highest multiplicity m in $\mathbb{Z}_p$ is* $\exp\left(\left(\sqrt{\left(1 - \frac{1}{m+1}\right)\frac{4}{3}}\pi + o(1)\right)\sqrt{p}\right)$.

It is an interesting question to determine the error term $o(1)$.

5. Incomplete Sets in a General Abelian Group

Let us now consider the problem of finding the largest size of an incomplete set in a general abelian group G, which we denote by $\mathrm{In}\,(G)$ in the rest of this section. The situation with a general group G is quite different from that with $\mathbb{Z}_p$, due to the existence of non-trivial subgroups. It is clear that any such subgroup is incomplete. Thus, $\mathrm{In}\,(G) \geq h$, where h is the largest non-trivial divisor of $|G|$. The intuition behind the discussion in this section is that *a large incomplete set should be essentially contained in a proper subgroup.*

In 1975, Diderrich [13] conjectured that if $|G| = ph$, where $p \geq 3$ is the smallest prime divisor of $|G|$ and h is composite, then $c(G) = h + p - 2$. (The cases where $p = 2$ or h is a prime is simpler and were treated earlier, some by Diderrich himself [13, 47, 14].) Didderich's conjecture was settled by Gao and Hamidoune in 1999 [29].

The following simple fact explains the appearance of the term $p - 2$.

Fact. If $S_{A \cap H} = H$ for some maximal subgroup H of (prime) index q, then $|A| \leq |H| + q - 2$.

To verify this fact, notice that A/H is a sequence in the group $\mathbb{Z}_q$. It is easy to show (exercise) that if B is a sequence of $q - 1$ non-zero elements in $\mathbb{Z}_q$, then $S_B \cup \{0\} = \mathbb{Z}_q$.

We say that subset A of G is *sub-complete* if there is a subgroup H of prime index such that $S_{A \cap H} = H$.

Once we know that an incomplete set A is sub-complete, we can write down its structure completely. There is a subgroup H with prime index q such that $|A \backslash H| \leq q - 2$, and the sequence A/H is incomplete in $\mathbb{Z}_q$. (The structure of such a sequence was discussed in the previous section.) It is natural to pose the following

Problem 5.1. Find the threshold for sub-completeness.

Recently, Gao, Hamidoune, Lladó and Serra [30] showed (under some weak assumption) that any subset of at least $\frac{p}{p+2}h + p$ elements is sub-complete. Furthermore, one can choose H to have index p, where p is the smallest prime divisor of $|G|$. Vu [68] showed (again under some weak assumption)that $\frac{5}{6}h$ is sufficient to guarantee sub-completeness. It is not clear, however, that what the sharp bound is.

The situation is much better if we assume that $|G|$ is sufficiently composite. In particular, if the product of the two smallest prime divisors of $|G|$ is significantly smaller than $\sqrt{|G|}$, then one can determine the sharp threshold for sub-completeness.

Theorem 5.2 [68]. *For any positive constant δ there is a positive constant $D(\delta)$ such that the following holds. Assume that $|G| = p_1 \dots p_t$, where $t \geq 3$ and $p_1 \leq p_2 \cdots \leq p_t$ are primes such that $p_1 p_2 \leq \frac{1}{D(\delta)} \sqrt{|G| / \log |G|}$. Then any incomplete subset A of G with cardinality at least $(1+\delta)\frac{n}{p_1 p_2}$ is subcomplete. Furthermore, the lower bound $(1+\delta)\frac{n}{p_1 p_2}$ cannot be replaced by $\frac{n}{p_1 p_2} + n^{1/4-\alpha}$, for any constant α.*

6. Structures in S_A

As mentioned in the introduction, an alternative way to implement our plan is to find a structure in S_A rather than in A (Step I'). A well-known result concerning the structure of S_A is the following theorem, proved by Freiman [28] and Sárközy [55] independently.

Theorem 6.1. *There are positive constants C and c such that the following holds for all sufficiently large n. Let A be a subset of $[n] := \{1, \dots, n\}$ with at least $C\sqrt{n \log n}$ elements. Then S_A contains an arithmetic progression of length $c|A|^2$.*

It is clear that the bound on the length of the arithmetic progression (AP) is optimal, as one can take A to be an interval. The lower bound on $|A|$, however, can be improved to $C\sqrt{n}$, as showed by Szemerédi and Vu [62].

Theorem 6.2. *There are positive constants C and c such that the following holds for all sufficiently large n. Let A be a subset of $[n] := \{1, \dots, n\}$ with at least $C\sqrt{n}$ elements. Then S_A contains an arithmetic progression of length $c|A|^2$.*

The assumption $|A| \geq C\sqrt{n}$ is optimal, up to the value of C, as one can construct a set $A \subset [n]$ of $\varepsilon\sqrt{n}$ elements, for some small constant ε, such that S_A does not contain any arithmetic progression of length larger than $n^{3/4}$ (see [62] or [63, Section 3.4]).

Theorem 6.2 can be extended considerably. Szemerédi and Vu [63] showed that for any set $A \subset [n]$ and any integer l such that $l^d|A| \geq n$ for some constant d, the sumset l^*A contains a large proper generalized arithmetic progression (GAP). The parameters of this GAP is optimal, up to a constant factor (see [63, Section 3] for more details).

Theorem 6.3 [63, Theorem 7.1]. *For any fixed positive integer d there are positive constants C and c depending on d such that the following holds. For any positive integers n and l and any set $A \subset [n]$ satisfying $l^d|A| \geq Cn$, l^*A contains a proper GAP of rank d' and volume at least $cl^{d'}|A|$, for some integer $1 \leq d' \leq d$.*

There are variants of Theorem 6.3 for finite fields, and also for sums of different sets (see [63, Section 5] and [63, Section 10]). In the following subsections, we discuss few applications of Theorems 6.2 and 6.3.

6.1. Folkman conjectures on infinite arithmetic progressions

Let us recall to the conjecture of Folkman, mentioned in the introduction.

Conjecture 6.4 (Folkman's conjecture). The following holds for any sufficiently large constant C. Let A be an strictly increasing sequence of positive integers with (asymptotic) density $A(n)$ at least $C\sqrt{n}$ (namely $A(n) := \big|A \cap [n]\big| \geq C\sqrt{n}$ for all sufficiently large n). Then S_A contains an infinite arithmetic progression.

Folkman [25] showed that the conjecture holds under a stronger assumption that $A(n) \geq n^{1/2+\varepsilon}$, where ε is an arbitrarily small positive constant. (An earlier result of Erdős [20] on a closely related problem can perhaps be adapted to give a weaker bound $n^{(\sqrt{5}-1)/2}$.) Hegyvári [45] and Łuczak and Schoen [46], independently, reduced the density $n^{1+\varepsilon}$ to $C\sqrt{n \log n}$, using Theorem 6.1.

Using the stronger Theorem 6.2, together with some additional arguments, Szemerédi and Vu [62] proved the full conjecture.

Theorem 6.5. *Conjecture* 6.4 *holds.*

In the same paper [25], Folkman also made a related conjecture for increasing, but not strictly increasing sequences. Let $A(n)$ now be the number of elements of A (counting multiplicities) at most n.

Conjecture 6.6 (Folkman's second conjecture). The following holds for any sufficiently large constant C. Let A be an increasing sequence of positive integers with such that $A(n) \geq Cn$ for all sufficiently large n. Then S_A contains an infinite arithmetic progression.

Despite the huge change from $\sqrt{n}$ to n in the density bound, this conjecture is also sharp [25], and (for some time) appeared more subtle than the first one (see a discussion in [21, Chapter 6]). Folkman [25] proved the conjecture under the stronger assumption that $A(n) \geq n^{1+\varepsilon}$. It does not seem that one can obtain the analogue of Hegyvári and Łuczak–Schoen results due to the lack of a "sequence" variant of Theorem 6.1. However, the method in [63] is sufficiently robust to enable one to obtain such a variant for the stronger Theorem 6.2. With the help of this result, one can settle Conjecture 6.6

Theorem 6.7 [63, Section 6]. *Conjecture* 6.6 *holds.*

The strategy for the proofs of Theorems 6.5 and 6.7 is the following. We first find a sufficient condition for a sequence A such that S_A contains an infinite AP.

We say that an infinite sequence A admits a *good* partition if it can be partitioned into two subsequences A' and A'' with the following two properties

- There is a number d such that $S_{A'}$ contains an arbitrary long arithmetic progression with difference d.
- Let $A'' = b_1 \le b_2 \le b_3 \le \dots$. For any number K, there is an index $i(K)$ such that $\sum_{j=1}^{i-1} b_j \ge b_i + K$ for all $i \ge i(K)$.

Lemma 6.8. *If a sequence A admits a good partition then S_A contains an infinite AP.*

The second assumption is easy to satisfied given that A has proper density. Thus, the key is the first assumption. The main feature here is that in this assumption, we only need to guarantee the existence of long (but finite) APs. So, Theorem 6.2 and its variants can be used with full power to achieve this goal.

6.2. Erdős conjecture on square-sum-free sets

In this section, we return to Erdős conjecture on square-sum-free sets, mentioned in the introduction. Let $SF(n)$ denote the size of the largest subset A of $[n]$ such that S_A does not contain a square (or A is square-sum-free). Erdős [19] observed that $SF(n) = \Omega(n^{1/3})$ and conjectured that the truth is close to this lower bound. Since then, there have been several attempts on his conjecture. Alon [4] proved that $SF(n) = O\left(\frac{n}{\log n}\right)$. In [40] Lipkin improved the bound to $SF(n) = O(n^{3/4+\varepsilon})$. Later, Alon and Freiman [5] obtained another improvement $SF(n) = O(n^{2/3+\varepsilon})$. About fifteen years ago, Sárközy [56] showed $SF(n) = O\left(\sqrt{n}\log n\right)$.

Let us now address the problem from our structural approach point of view. Theorem 6.2 is no longer useful, as we are dealing with sets of size around $n^{1/3}$, way below the lower bound $\sqrt{n}$ required in this theorem. Fortunately, we have a more general result, Theorem 6.3, which enables us to find structures in S_A for any set of size n^{δ}, for any constant δ. In particular, we can deduce from this theorem the following corollary.

Corollary 6.9. *There are positive constants C and c such that the following holds for all sufficiently large n. Let A be a subset of $[n]$ with cardinality at least $Cn^{1/3}$. Then S_A contains either an AP of length $c|A|^2$ or a proper GAP of rank 2 and volume $c|A|^3$.*

Combining this corollary with some number theoretic arguments, Nguyen and the author [51] can get close to the conjectured bound.

Theorem 6.10. *There is a constant C such that $SF(n) \leq n^{1/3} \log^C n$.*

We strongly believe that the log term can be removed. Details will appear elsewhere.

7. Inverse Littlewood–Offord Theorems and Random Matrices

In this final section, we discuss a problem with a slightly different nature. Let A be a sequence of non-zero integers. Now we are going to view S_A as a multi-set of 2^n elements. We denote by M_A be the largest multiplicity in S_A. For example, if $A = \{1, \dots, 1\}$, then $M_A = \binom{n}{\lfloor n/2 \rfloor} = \Theta\big(2^n/\sqrt{n}\big)$.

The problem of bounding M_A originated from Littlewood and Offord's work on random polynomials [41]. In particular, they proved that $M_A = O\big(2^n \log n/\sqrt{n}\big)$. The log term was removed by Erdős [18], who obtained a sharp bound for M_A. Many extensions of this result were obtained by various researcher: Erdős–Moser [24], Sárközy–Szemerédi [58], Katona [38] Kleitman [39], Halász [44], Griggs et. al. [36], Frankl–Füredi [26], Stanley [59] etc. Among others, it was showed that the bound on M_A keeps improving, if one forbids more and more additive structures in A. For example, Erdős and Moser [24] showed that if the elements of A are different (i.e., A is a set), then $M_A = O\big(2^n \log n/n^{3/2}\big)$. In general, the following can be deduced from results of [44] (see also [64, Problem 7.2.8])

Theorem 7.1. *For any fixed integer k there is a constant C such that the following holds. Let $A = \{a_1, \dots, a_n\}$ and R_k be the number of roots of the equation*

$$\varepsilon_1 a_{i_1} + \cdots + \varepsilon_{2k} a_{i_{2k}} = 0$$

with $\varepsilon_i = \pm 1$ and $i_1, \dots, i_{2k} \in [n]$. Then $M_A \leq Cn^{-2k-1/2} R_k 2^n$.

In [66], Tao and Vu introduced the notion of Inverse Littlewood–Offord theorems. The intuition here is that if M_A is large (of order $2^n/n^C$ for any constant C, say), then A should have a very strong structure.

The most general example we found with large M_A is the following. Let Q be a proper GAP of constant rank d and volume V. If A is a subset of Q, it is easy to show that $M_A = \Omega(\frac{2^n}{n^{d/2}V})$ (in order to see this, view the elements of S_A as random sums $\sum_{i=1}^{n} \xi_i a_i$ where a_i are elements of A and ξ_i are iid random variables taking values 0 and 1 with probability 1/2). Thus, if the volume of Q is small, then M_A is large.

In [66], Tao and Vu proved the inverse statement, asserting that *having A as a subset of a small GAP is essentially the only way to guarantee make M_A large.*

Theorem 7.2 [66]. *For any constant C and ε there are constants B and d such that the following holds. Let A be a sequence of n elements in a torsion-free group G. If $M_A \geq 2^n/n^C$ for some constant C, then all but at most $n^{1-\varepsilon}$ elements of A is contained in a proper GAP Q of rank d and cardinality n^B.*

In a more recent paper [67], the same authors obtained a (near) optimal relationship between the parameters C, ε, d and B. As a corollary, one can deduce (asymptotic versions of) many earlier results, such as Theorem 7.1. (In spirit, this process is somewhat similar to the process of using Theorem 3.1 to reprove, say, Erdős–Ginburg–Ziv theorem.)

We would like to conclude this survey with a rather unexpected application. Let us leave combinatorial number theory and jump to the (fairy remote) area of mathematical physics. In the 1950s, Wigner observed and proved his famous semi-circle law concerning the limiting distribution of eigenvalues in a *symmetric* random matrix [69]. A brother of this law, the so-called circular law for *non-symmetric* random matrices, has been conjectured, but remains open since that time.

Conjecture 7.3 (Circular Law Conjecture). Let ξ be a random variable with mean 0 and variance 1 and M_n be the random matrix whose entries are iid copies of ξ. Then the limiting distribution of the eigenvalues of $\frac{1}{\sqrt{n}}M_n$ converges to the uniform distribution on the unit disk.

Girko [31] and Bai [6] obtained important partial results concerning this conjecture. These results and many related results are carefully discussed in the book [7]. There has been a series of rapid developments recently by

Götze–Tikhomirov [32, 33], Pan–Zhou [54], and Tao–Vu [65]. In particular, Tao and Vu [65] confirmed the conjecture under the slightly stronger assumption that the $(2+\eta)$-moment of ξ is bounded, for any $\eta > 0$.

Theorem 7.4. *The Circular Law holds (with strong convergence) under an additional assumption that*

$$\mathbf{E}\big(|\xi|^{2+\eta}\big) < \infty$$

for some fixed $\eta > 0$.

The key element of this proof is a variant of Theorem 7.2, which enables us to count the number of sequences A with bounded elements such that M_A (more precisely a continuous version of it) is large. For details, we refer to [65].

References

[1] G. E. Andrews, *The theory of partitions,* Cambridge University Press (1998).

[2] N. Alon, Independent sets in regular graphs and sum-free subsets of abelian groups, *Israel Journal Math.,* **73** (1991), 247–256.

[3] N. Alon, Combinatorial Nullstellensatz, Recent trend in combinatorics (Mátraháza, 1995), *Combin. Probab. Comput.,* **8** (1999), 7–29.

[4] N. Alon, Subset sums, *Journal of Number Theory,* **27** (1987), 196–205.

[5] N. Alon and G. Freiman, On sums of subsets of a set of integers, *Combinatorica,* **8** (1988), 297–306.

[6] Z. D. Bai, Circular law, *Ann. Probab.,* **25** (1997), no. 1, 494–529.

[7] Z. D. Bai and J. Silverstein, Spectral analysis of large dimensional random matrices, *Mathematics Monograph Series,* **2**, Science Press, Beijing (2006).

[8] A. Bialostocki and P. Dierker, On Erdős–GinzburgiZiv theorem and the Ramsey number for stars and matchings, *Discrete Mathematics,* **110** (1992), 1–8.

[9] N. Calkin, On the number of sum-free sets, *Bull. London Math. Soc.,* **22** (1990), 141–144.

[10] J. W. S Cassels, On the representation of integers as the sums of distinct summands taken from a fixed set, *Acta Sci. Math. Szeged,* **21** (1960), 111–124.

[11] P. Cameron and P. Erdős, On the number of sets of integers with various properties, *Number Theory* (Banff, AB 1988), 61–79, de Gruyter, Berlin (1990).

[12] D. da Silva and Y. O. Hamidoune, Cyclic spaces for Grassmann derivatives and additive theory, *Bull. London Math. Soc.,* **26** (1994), no. 2, 140–146.

[13] G. T. Diderrich, An addition theorem for abelian groups of order pq, *J. Number Theory,* **7** (1975), 33–48.

[14] G. T. Diderrich and H. B. Mann, Combinatorial problems in finite Abelian groups, *Survey of combinatorial theory* (Proc. Internat. Sympos., Colorado State Univ., Fort Collins, Colo., 1971), pp. 95–100. North-Holland, Amsterdam, 1973.

[15] J-M. Deshouilers, Quand seule la sous-somme vide est nulle modulo p. (French) [When only the empty subsum is zero modulo p] *J. Thor. Nombres Bordeaux,* **19** (2007), no. 1, 71–79.

[16] J.-M. Deshouillers, Lower bound concerning subset sum wich do not cover all the residues modulo p, *Hardy- Ramanujan Journal,* Vol. 28 (2005), 30–34.

[17] J.-M. Deshouillers and G. Freiman, When subset-sums do not cover all the residues modulo p, *Journal of Number Theory,* **104** (2004), 255–262.

[18] P. Erdős, On a lemma of Littlewood and Offord, *Bull. Amer. Math. Soc.* **51** (1945), 898–902.

[19] P. Erdős, Some problems and results on combinatorial number theory, *Proceeding of the first China conference in Combinatorics* (1986).

[20] P. Erdős, On the representation of large interges as sums of distinct summands taken from a fixed set, *Acta. Arith.,* **7** (1962), 345–354.

[21] P. Erdős and R. Graham, *Old and new problems and results in combinatorial number theory.* Monographies de L'Enseignement Mathmatique [Monographs of L'Enseignement Mathmatique], 28. Universit de Genve, L'Enseignement Mathmatique, Geneva, 1980.

[22] P. Erdős and A. Granville, *Unpublished.*

[23] P. Erdős and H. Heilbronn, On the addition of residue classes mod p, *Acta Arith.,* **9** (1964), 149–159.

[24] P. Erdős and L. Moser, P. Erdős, Extremal problems in number theory. 1965 Proc. Sympos. Pure Math., Vol. VIII pp. 181–189 Amer. Math. Soc., Providence, R.I.

[25] J. Folkman, On the representation of integers as sums of distinct terms from a fixed sequence, *Canad. J. Math.,* **18** (1966), 643–655.

[26] P. Frankl and Z. Füredi, Solution of the Littlewood–Offord problem in high dimensions. *Ann. of Math.,* (2) **128** (1988), no. 2, 259–270.

[27] G. Freiman, Foundations of a structural theory of set addition. Translated from the Russian. *Translations of Mathematical Monographs,* Vol 37. American Mathematical Society, Providence, R. I., 1973. vii+108 pp.

[28] G. Freiman, New analytical results in subset sum problem, *Discrete mathematics,* **114** (1993), 205–218.

[29] W. Gao and Y. O. Hamidoune, On additive bases, *Acta Arith.,* **88** (1999), no. 3, 233–237.

[30] W. Gao, Y. O. Hamidoune, A. Llad and O. Serra, Covering a finite abelian group by subset sums, *Combinatorica,* **23** (2003), no. 4, 599–611.

[31] V. Girko, Circle law, (Russian) *Teor. Veroyatnost. i Primenen.*, **29** (1984), no. 4, 669–679.

[32] F. Götze and A. N. Tikhomirov, On the circular law, *preprint.*

[33] F. Götze and A. N. Tikhomirov, The Circular Law for Random Matrices, *preprint.*

[34] B. Green, The Cameron–Erdős Conjecture *Bull. London Math. Soc.*, **36** (2004), no. 6, 769–778.

[35] B. Green and I. Ruzsa, Freiman's theorem in an arbitrary abelian group, *Jour. London Math. Soc.*, **75** (2007), no. 1, 163–175.

[36] J. Griggs, J. Lagarias, A. Odlyzko and J. Shearer, On the tightest packing of sums of vectors, *European J. Combin.*, **4** (1983), no. 3, 231–236.

[37] Y. Hamidoune and G. Zémor, On zero-free subset sums, *Acta Arithmetica,* **78** (1996), no. 2, 143–152.

[38] G. Katona, On a conjecture of Erdős and a stronger form of Sperner's theorem, *Studia Sci. Math. Hungar.*, **1** (1966), 59–63.

[39] D. Kleitman, On a lemma of Littlewood and Offord on the distributions of linear combinations of vectors, *Advances in Math.*, **5** (1970), 155–157 (1970).

[40] E. Lipkin, On representation of r-powers by subset sums, *Acta Arithmetica,* **52** (1989), 114–130.

[41] J. E. Littlewood and A. C. Offord, On the number of real roots of a random algebraic equation. III. *Rec. Math. [Mat. Sbornik] N.S.*, **12** (1943), 277–286.

[42] P. Erdős, A. Ginzburg and A. Ziv, Theorem in the additive number theory, *Bull. Res. Council Israel,* **10F** (1961), 41–43.

[43] W. D. Gao, A. Panigrahi and R. Thangadurai, *On the structure of p-zero-sum-free sequences and its application to a variant of Erdős–Ginzburg–Ziv theorem,* Proc. Indian Acad. Sci. Vol. 115, No. 1 (2005), 67–77.

[44] G. Halász, Estimates for the concentration function of combinatorial number theory and probability, *Period. Math. Hungar.*, **8** (1977), no. 3–4, 197–211.

[45] N. Hegyvári, On the representation of integers as sums of distinct terms from a fixed set, *Acta Arith.*, **92** (2000), no. 2, 99–104.

[46] T. Łuczak and T. Schoen, On the maximal density of sum-free sets, *Acta Arith.*, **95** (2000), no. 3, 225–229.

[47] H. B. Mann and Y. F. Wou, An addition theorem for the elementary abelian group of type (p, p), *Monatsh. Math.*, **102** (1986), no. 4, 273–308.

[48] M. Nathanson, *Elementary methods in number theory,* Springer 2000.

[49] H. H. Nguyen, E. Szemerédi and V. Vu, Subset sums in $\mathbf{Z}_p$, to appear in *Acta Arithmetica.*

[50] H. H. Nguyen and V. Vu, Classification theorems for sumsets modulo a prime, *submitted.*

[51] H. Nguyen and V. Vu, On square-sum-free sets, *in preparation.*

[52] J. E. Olson, Sums of sets of group elements, *Acta Arith.,* **28** (1975/76), no. 2, 147–156.

[53] J. E. Olson, An addition theorem modulo p, *J. Combinatorial Theory,* **5** (1968), 45–52.

[54] G. Pan and W. Zhou, Circular law, Extreme singular values and potential theory, *preprint.*

[55] A. Sárközi, Finite addition theorems I, *J. Number Theory,* **32** (1989), 114–130.

[56] A. Sárközy, Finite Addition Theorems, II, *Journal of Number Theory,* **48** (1994), 197–218.

[57] A. Sárközy and C. Pomerance, Combinatorial number theory, Chapter 20, *Handbook of Combinatorics* (eds. R. Graham, M. Grötschel and L. Lovász), North-Holland 1995.

[58] A. Sárközy and E. Szemerédi, Über ein Problem von Erdős und Moser, *Acta Arithmetica,* **11** (1965), 205–208.

[59] R. Stanley, Weyl groups, the hard Lefschetz theorem, and the Sperner property, *SIAM J. Algebraic Discrete Methods,* **1** (1980), no. 2, 168–184.

[60] E. Szemerédi, On a conjecture of Erdős and Heilbronn, *Acta Arith.,* **17** (1970), 227–229.

[61] Endre Szemerédi and V. Vu, Long arithmetic progression in sumsets and the number of x-free sets, *Proceeding of London Math Society,* **90** (2005), 273–296.

[62] E. Szemerédi and V. Vu, Finite and infinite arithmetic progression in sumsets, *Annals of Math,* **163** (2006), 1–35.

[63] E. Szemerédi and V. Vu, Long arithmetic progressions in sumsets: Thresholds and Bounds, *Journal of the A.M.S,* **19** (2006), no. 1, 119–169.

[64] T. Tao and V. Vu, *Additive Combinatorics,* Cambridge Univ. Press (2006).

[65] T. Tao and V. Vu, Random matrices: The Circular Law, *Communication in Contemporary Mathematics,* **10** (2008), 261–307.

[66] T. Tao and V. Vu, Inverse Littlewood–Offord theorems and the condition number of random matrices, *to appear in Annals of Mathematics.*

[67] T. Tao and V. Vu, *paper in preparation.*

[68] V. Vu, Structure of large incomplete sets in abelian groups, to appear in *Combinatorica.*

[69] Wigner, On the distribution of the roots of certain symmetric matrices, *Annals of Mathematics,* (2) **67** (1958), 325–327.

Van Vu

Department of Mathematics
Rutgers, Piscataway
NJ 08854

e-mail: vanvu@math.rutgers.edu

Nyomás: Regiszter Kiadó és Nyomda Kft.